Camel Clinical Biochemistry and Hematology

AF839276

Bernard Faye • Mohammed Bengoumi

Camel Clinical Biochemistry and Hematology

Springer

Bernard Faye
Centre de Coopération International en Recherche Agronomique pour le Développement Research Unit SELMET (Livestock system in Mediterranean and tropical milieu)
Montpellier, France

Mohammed Bengoumi
FAO Regional Office
Tunis, Tunisia

IAV Hassan II
Rabat, Morocco

ISBN 978-3-030-07058-8 ISBN 978-3-319-95562-9 (eBook)
https://doi.org/10.1007/978-3-319-95562-9

© Springer International Publishing AG, part of Springer Nature 2018
Softcover re-print of the Hardcover 1st edition 2018
This work is subject to copyright. All rights are reserved by the Publisher, whether the whole or part of the material is concerned, specifically the rights of translation, reprinting, reuse of illustrations, recitation, broadcasting, reproduction on microfilms or in any other physical way, and transmission or information storage and retrieval, electronic adaptation, computer software, or by similar or dissimilar methodology now known or hereafter developed.
The use of general descriptive names, registered names, trademarks, service marks, etc. in this publication does not imply, even in the absence of a specific statement, that such names are exempt from the relevant protective laws and regulations and therefore free for general use.
The publisher, the authors and the editors are safe to assume that the advice and information in this book are believed to be true and accurate at the date of publication. Neither the publisher nor the authors or the editors give a warranty, express or implied, with respect to the material contained herein or for any errors or omissions that may have been made. The publisher remains neutral with regard to jurisdictional claims in published maps and institutional affiliations.

Printed on acid-free paper

This Springer imprint is published by the registered company Springer Nature Switzerland AG
The registered company address is: Gewerbestrasse 11, 6330 Cham, Switzerland

Acknowledgement

The authors are grateful to Dr. Gaukhar Konuspayeva, Professor of Biochemistry at Al-Farabi Kazakh National University, Almaty (Kazakhstan), for her encouragement, her constant support in bibliographical research and her contribution to the layout of the final document.

They also thank all their colleagues sharing their passion for camels all along their respective careers.

Disclaimer

The contents of this book are the sole responsibility of the authors and can under no circumstances be regarded as reflecting the position of their originated institutions.

Contents

Chapter 1
Introduction

The idea to use blood biochemical parameters to evaluate the health status of one individual is an old practice in human medicine (cf., e.g., Carmalt et al. 1970). In veterinary medicine, blood analyses have been proposed to help the diagnosis of cattle diseases since nearly one century (Abderhalden 1898). But they were initially developed only in the 1960s in specialized research laboratories (Michel 1980), in Veterinary Services, and later on with the development of dry chemistry methods and miniaturization of the analyzers in the veterinary cabinets. Beyond the diagnosis needs (Kerr 2001), veterinary surgeons and zootechnicians are also wondering about the relevance of blood analyses to evaluate the nutritional status of animals (Rowlands 1980), particularly during investigations in farms in which it is difficult to have reliable information on feeding (Barnouin et al. 1994). Currently, a whole set of analyses that are critical for vets' diagnostics and farmers' decision-making is proposed to the livestock and animal health actors to define metabolic, biochemical, or blood profiles.

Moreover, in addition to blood and urine, these analyses may include other biological substrates. However, considering blood as a main drain for tissue metabolites, blood analyses constitute an excellent indicator of body metabolism and functions. From this point of view, cattle undoubtedly represents the best known species among farm animals. Data on sheep and goat are numerous, but it is already more disparate on tropical breeds. For domestic species confined in the extensive or marginal zones of the tropical and arid areas, data on clinical biochemistry is not very frequent. The dromedary is among those species.

The present book precisely aims at filling this gap by proposing a complete state of art as well as the available data in the bibliography using a synthesis of the work completed by the authors of the book.

The objective is to provide a rather complete view of the available field knowledge on the clinical and nutritional biochemistry of the dromedary camel. Also, it is hoped that this book would convince experts or scientists that several gaps are still not filled. This quasi-exhaustive publication of the available data will contribute to identify gaps and enhance research on this topic. The stake appears to be the most

© Springer International Publishing AG, part of Springer Nature 2018
B. Faye, M. Bengoumi, *Camel Clinical Biochemistry and Hematology*,
https://doi.org/10.1007/978-3-319-95562-9_1

important to us as the dromedary often presents a rather frustrate, ubiquist, and non-characteristic symptomatology on the pathological level. Also, the biochemical survey can represent an effective tool for diagnosis, allowing to use the dromedary with efficiency. In addition, in spite of its existence in some relatively marginal intensive sectors in the countries of the Gulf and in the Middle East, the dromedary is still linked to desert spaces and to an extensive mode of breeding. The knowledge of its feeding system and its nutrition in rangelands thus proves difficult to be acquired. This makes us interested to consolidate our experience in surveying the nutritional status.

This book is meant to be didactic and pragmatic at the same time. We wish to satisfy the intellectual curiosity of the researcher, which is explained by the large call to the existing literature and the professional concern of the stockbreeders, veterinary surgeons, and livestock technicians, which is explained by the insistence on a certain number of technical points. To facilitate the reading of this book, we will classically proceed by an inventory of the analyses according to the usual nomenclature and parameter by parameter while insisting on their clinical or nutritional interest, the principles of dosage (if necessary), the usual values, and the factors of physiological and the pathological variations. We do not claim to define the term ad vitam aeternam as the biological standards of the camel. One uncertainty knows that it is related to the concept of normality in clinical biochemistry (Farver 1997). At most, we hope to provide some supports to the decision-making and the interpretation of the biological results for a species worthy of interest.

1.1 Restraining and Samplings

1.1.1 Restraint Techniques

The dromedary is an animal which is not always easy to control, particularly entire males. It can be necessary, especially for blood sampling or biopsies, to ensure a severe restraining of the animal. If the operator is fast and skillful, and the animal is naturally calm or accustomed to be handled by a man, such as animals of experimental stations or animals compelled with daily activities like milking, the blood sampling can be carried out by a very light application (animal standing up, members tied) and even without restraint. However, this case of figure is far to be the most frequent. In almost all the cases, the application of large camelids constitutes, in time and energy, the most considerable investment in the implementation of the protocols of sampling (Photo. 1.1).

The natural steady position of large camelids is sitting with legs folded under the body, the animal being placed in sternal *decubitus*. In case of traditional constraining, it is important to take care and impose such a position by force or persuasion. Generally, the know-how of the breeder is enough to encourage the animal to sit down, whether by the voice or by a simple pressure on the halter. In addition to the halter, it can be necessary to maintain the forelimb folded up by a

Photo. 1.1 Restraining a Bactrian Camel with one rope and four men running around the animal

second man. Generally, the sitting is spontaneously done under these conditions. It is then enough to block the members when this position is maintained to prevent the rising up of the animal at the moment of the blood sampling. However, the animal can be recalcitrant or anxious and refuse the injunctions of its master in this context. Regarding rutting males or females just after delivery, the exercise can become utterly difficult, without a minimum of know-how.

It is consequently advisable to intervene more firmly. One of the ways that are largely used by the camel farmers to force the sitting is to pass a rope behind the rear limbs by two men positioned on each side of the animal, while a third man folds one of the forelimbs. The first two people draw the rope in order to push the rear limbs forwards the animal, thus obliging it to fold the whole of its members and go down on its sternal pad (Drawing 1). When the animal is seated down, it is maintained in this position by the rope tied around the two rear limbs in addition to the forelimbs. Indeed, to be raised, camelids proceed in two times, the first being the extension of the rear limbs. Any restraint thus considerably limits the ability of the animal to be raised. If necessary, the upper lip can be held firmly at the time of the sampling to ensure total immobilization of the animal. Sometimes, it is necessary to tranquillize the animal by means of a small rope provided with a slipknot and passed around the neck. This rope could cause a constriction of the blood flow on the level of the jugular vein, which leads to a discomfort calming the animal. Also, it has the advantage of causing a swelling of the jugular vein which is favorable for blood collection. Restraining of folded up forelimb can also be enough without being brought to force the sitting down.

1.1.2 Blood Sampling

The blood on the standing animal is preferably situated on a tense neck that is pulled down to facilitate venous stasis. The forelimbs are constrained because some animals have the ability to "kick" to the front in this position.

Comparatively on the sitting animal, blood sampling is easily made on the neck when folded against the body of the animal. Such a position makes it difficult during any untimely movement and makes the standing up impossible, but large camelids use their neck as a powerful balance to take up their standing position. The sampling on the jugular vein zone is easily identifiable especially after even light pressure exerted at the base of the neck or, preferably, at a mid-distance between the thorax and the head (Photo. 1.2).

The easiest sampling point is located close to the head. However, this anatomical region in the male has abundant and long hair which can make the tactile perception of the jugular vein difficult. The same problem is encountered with the Bactrian camel especially during winter, as their fur is quite long at that place. Moreover, a confusion of the jugular vein with the distended part of the throat in an angry or

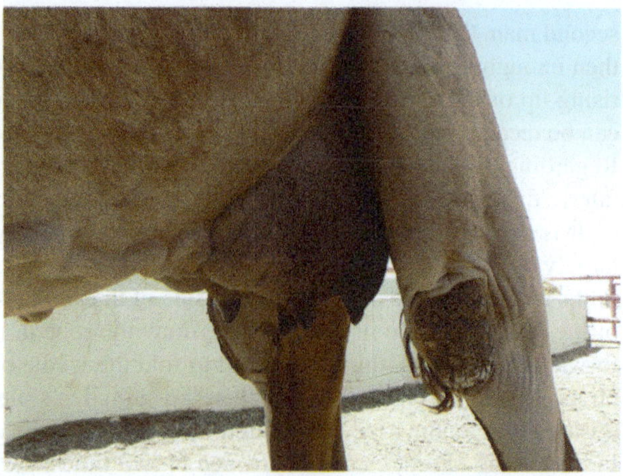

Photo. 1.2 In a lactating camel, the mammary vein is easily visible, and blood sampling can be done without restraining the animal

stressed-out animal is possible. One must be well to feel the vein under the fingers before proceeding with the sampling. Blood can also be taken in other places. In particular, Higgins and Kock (1984) suggested the medial metacarpal vein (visible on the medial place of the carpus) and the dorsal metatarsal vein (visible on the edge between the extensor tendons metatarsal cranio-lateral). In intensive dairy farms where animals are accustomed to enter the milking corridor, it could be easier to sample blood at the mammary vein.

The use of vacutainer tubes allows the use of thinner needles, which are less traumatic for the animal. Moreover, the resistance of red blood cells in the dromedary is much higher than in other species on average, which considerably limits the risk of hemolysis. As a result, it is not necessarily essential to use sophisticated systems of blood sampling. In particular, the use of Sarstedt© tubes, which is not obligatory and generally advocated for the determination of trace elements, is used to avoid latex caps and considered as a source of well-known zinc contamination (Lamand 1987).

The good reputation of the rusticity of animals belonging to the genus *Camelus* should not avoid the need for respecting the conventional rules of hygiene during sampling, namely, skin disinfection with alcohol possibly after cutting the excess of hair in the intervention area, the use of one needle per animal, or well disinfection with alcohol between each use. In the latter case, one should know that needles quickly become blunt and the camels' skin is very thick and sometimes indurated by common skin diseases in species like mange.

Assays are carried out either on the whole blood or on serum or plasma. According to the studied parameters, we will or will not use an anticoagulant (heparin, liquemin, fluoride/oxalate, EDTA). If analyses are to be delayed, it is imperative to preserve samples in the best possible conditions. In particular, the enzyme and hormonal parameters badly support a break in the cold chain. From a general point of view, there is no specific rule for blood collection and storage of samples in camelids. The rules in question remain identical to those implemented in other species of zootechnical interest. However, the ecology of the dromedary (arid and semiarid areas of difficult access) and its mode of management (nomadism, transhumance) require additional precautions to ensure the storage of samples in the best possible conditions.

In the traditional farming system, the main difficulty is not in dealing with the technical aspects of contention and sampling, nor is in the fact of farmers being often efficient for restraining their animals. The main problem lies rather in persuading farmers to carry out a blood sampling on animals for which they have deep emotional links, generally much stronger than for other species (especially small ruminants). In some pastoral populations in the Horn of Africa belonging to the omotic group, the practice of bleeding used to consume fresh blood or blood mixed with milk is very common, usually in cattle but also sometimes in camels. In this case, having access to the animals for a blood sampling is much easier. However, this practice is never observed in Muslim populations.

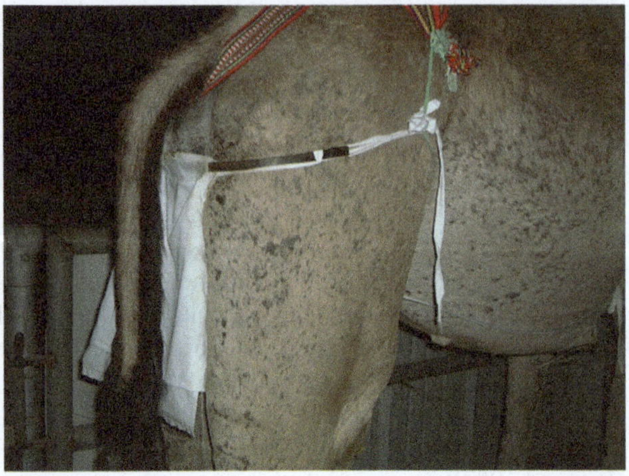

Photo. 1.3 Device for urine collection in dromedary camel

It will therefore be important to convince farmers about the safety of sampling by preliminary negotiations through the chiefs, if necessary. In any case, the sampling has to be achieved in the rules of art.

1.1.3 Urine Sampling

Unlike human medicine, the collection of urine in animals is less easy than blood sampling. The determination of the biochemical parameters of clinical interest is thus much less used, particularly in the dromedary, as the relevance of the results needs a urine collection over a period of 24 hours. This collection technique was proposed by Bengoumi (1992) and consists of setting up a plastic bag with a form that is adapted to the urogenital anatomy and hence different depending on the sex of the animal. As shown in Photo. 1.3, the plastic bag is fixed using glue and a piece of string. A similar system consisting of a bag equipped with a device of urine discharge was also proposed in India (Rai and Khanna 1988; Khanna 1993). It is of course obvious that such a technique is operational only in experimental conditions. It seems difficult and ultimately of little interest as a technique adapted to the field investigation (Photo. 1.3).

The capacity of nutrients' recycling in the dromedary makes the kidney an organ of considerable interest. The determination of urinary parameters consequently informs about the health status of the kidney but can also appreciate the level of excretion of some nutrients through a liquid way. However, according to the technical difficulties mentioned above, urine samples are virtually not performed routinely.

1.1.4 Milk Sampling

Milk represents a specific biological substrate of female mammals, which is easily taken, as opposed, reverse to urine collection. It is however of a low clinical relevance and feeding partially influences its chemical composition. However, it may constitute an emunctory for abiotic elements or some parameters showing a deviation of the metabolism. Accordingly a few studies are available in the literature. The collection of milk in the camel can be sometimes difficult without the presence of the camel calf, if the animal is not accustomed to be milked. An injection of oxytocin may be necessary to facilitate the milk release.

1.1.5 Fecal Sampling

As urine, feces reflect the excretion of the elements provided by the food supply or related to internal metabolism. Their analysis is therefore of certain interest in case of gastrointestinal contamination or assessment of various nutrient excretions. However, fecal sampling is however primarily used for parasitic diagnosis, which is not the topic of this book. Fecal collection is easy and can be directly achieved per rectum. The humidity is particularly low in the excrement of the dromedary, thus the easiness of its conservation.

1.1.6 Liver Biopsy

Most of the organs may be submitted to biopsy, but only liver biopsy is occasionally practiced in veterinary clinic. Sampling a liver tissue requires precautions and a certain amount of know-how. Several methods for the adult dromedary (Bucci et al. 1982) and the camel calf (Cherrier et al. 1991) have been described in the literature.

In this work, we describe the methodology used in Morocco (Bengoumi et al. 1998; Faye and Bengoumi 1997). According to this method, the animal is blocked to receive intravenous injection of a general sedative (1 ml Xylazine, Rompun N.D.) and then placed in sternal decumbency. The skin is shaved, degreased with alcohol, and disinfected with iodine tincture. After local anesthesia with an injection of 5 ml xylocaine solution 2%, a small skin incision of approximately 1 cm is practiced. The puncture point is precisely located on the right side at the level of the ninth intercostal space, which is either the third from the last side placed above the xiphoid apex with approximately 15 cm or slightly below the horizontal line joining the point of shoulder to point of the hip.

After crossing the intercostal muscles and the peritoneum, the biopsy probe is pressed slightly forward. The operator recognizes the liver by its soft consistency compared to the neighboring organs (diaphragm pillars, gastric compartments).

After pressing the biopsy probe, the operator performs a sharp back movement to perform the biopsy. This method allows for collecting about 100–500 mg of fresh tissue stored in 0.5 ml of sulfuric acid. A point of suture on the skin is sufficient for a quick skin wound healing, but this is not necessarily required. The spray of an antiseptic product (Aluspray N.D.) on the surgical wound and the intramuscular injection of a long-acting antibiotic (oxytetracycline) at 5 mg/kg help to avoid possible postoperative infections.

The risk of bleeding is minimal. At the end of the sedative effect, the animal gets up and begins to eat. No complication, loss of appetite, and long-term aftereffect are generally observed. The biopsy procedure may be repeated 2 weeks later without apparent effect on the animal.

1.1.7 Hump Biopsy

Also, the hump of the camel may sometimes be subjected to biopsy, but the nature of the target tissue (adipose tissue) authorizes the sampling of adipocytes only, which is of an interest in highly specific research protocols. However, hump biopsy could be useful for determining lipophilic pollutants (Faye et al. 2013). Moreover, it does not pose any particular difficulty, thanks to the lack of nerve or blood irrigation in this anatomical area (Photo 1.4).

The procedure consists of the following stages. First, animals are tranquilized with an IM sedative injection (Seton 2% ©, 20 mg Xylazine in solution) to facilitate the contention. The intervention is done on the standing animal if a corridor is available or on the sitting animal if not. After 10 minutes, the sedative shows the effect on the animal.

Then, the place of biopsy on the hump is widely shaved, and the skin is disinfected with an iodine solution. Around the place of biopsy, a local anesthesia is achieved by the subcutaneous injection of 10 cc Xylocaine solution in 5–6 different places "in crown" around the place where incision is projected to be done.

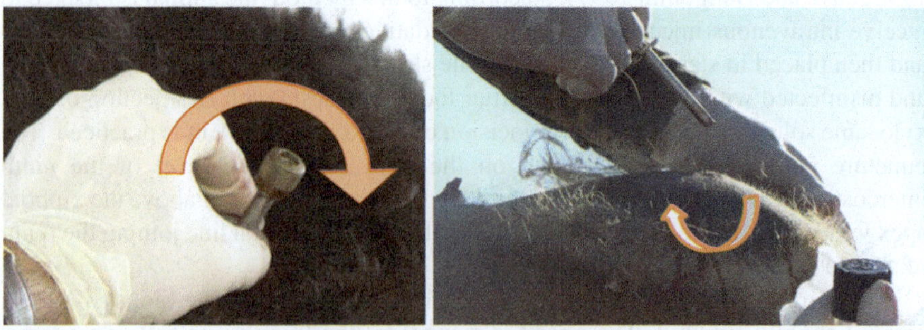

Photo. 1.4 Introduction of cannula with twisting movement

A small skin incision (no more than 1 cm large) is achieved approximately at the middle of the side of the hump (left or right is without importance). Then, the trocar is introduced through the wound straight in the fat of the hump (only the cannula without the trocar). The cannula is turned in the hump fat during the progressive introduction in order to cut the fat and to get a cylindrical piece of hump tissue (see photos).

More fat tissues can be extracted by the introduction of the cannula in different directions through the same wound according to the same twisting procedure. The cannula was withdrawn and the fat is collected with a Luer spoon. For each coring, approximately 0.5 to 1 g of fat could be collected.

Then one suture is done using a ½ circle surgical needle (big size for large animals). Two or three points of suture are sufficient, but more points are needed if the incision is longer than 1 cm. After suture, the wound is disinfected with blue spray. The camel remains quiet for 3 to 4 hours but can stand up, return to his box, or go to the field as soon as the biopsy was finished.

1.1.8 Other Biological Substrates

Usually, the peritoneal, pleural, pericardial, or cerebrospinal fluid can be requested for certain analyses of clinical interest. With the exception of peritoneal fluid, these samples are rarely implemented with the dromedary and require specialized techniques performed under anesthesia. The collection of the seminal fluid may be sometimes considered (Singh et al. 1994). Therefore, it is sufficient to use an artificial vagina for the collection of the semen and seminal fluid by simple centrifugation separation.

1.2 The Rules for Convenient Interpretation

The usefulness of biochemical patterns may be disparaged, and critical arguments are not without merit. However, these critics were often fed by the failure to comply with certain rules. If the later are not taken into consideration, the quality of the interpretation of results will be limited. Without being exhaustive, a certain number of points are necessary to consolidate the interest of biochemical investigations. These are:

- **To clarify the purpose of the analyses**: (sampling procedure plan) the sampling type is not the same for the diagnosis on noninfectious health disorder diagnosis, for determining the nutritional status of an aggregate of animals, or for assessing the physiological standards in a species.
- **To take into account the physiological variation factors** for convenient interpretation of the results. It includes effects related to gender, age, physiological

status, and level of milk production (Fayet et al. 1986). The establishment of the sampling protocol has to take into account these factors of variation, either to assign to each factor the potential explanatory part of observed variations or to remove the effect: for example, sample only multiparous females for a given breed at an early lactation and with a determined milk potential.

- **To ensure the respect of sampling protocol**: The feeding time may influence some parameters (e.g., glucose). Then, it is important to implement the sampling according to the feeding times. Generally, camel herds are found in grazing areas, and the best time for sampling is at the morning before leaving for pastures where animals are still fasting. Some parameters are very sensitive to contamination (e.g., trace elements). It is therefore imperative to respect the rules designed to limit this kind of risk. Probably, some results in the literature regarding dromedary can be seriously questioned. Moreover, some parameters can be changed by the stress status of sampled animals. Considering the fact that camelids are particularly difficult to handle, it is essential to limit all untimely handling and to ensure an optimal comfort for sampling.
- **To interpret the results on a set of animals**: The between-individual variability is often important, and the analytical results are meaningful in many cases. The results are interpreted on a set of individuals subjected to the same food or health constraints (typically the herd or any other relevant aggregate).
- **To insist on the concept of "pattern"**: In many cases, the interpretation cannot be satisfied by the reading of a single parameter. High blood sugar is not the same biological sense if accompanied by low or high uremia or by high or not circulating rate of free fatty acids. Therefore, the combinations of parameters give generally better information on the nutritional or clinical status of the animal or herd.
- **To avoid confusion between statistical difference and biological significance**: In many publications, the authors insist on the significance level of differences observed between two or more animal statuses (e.g., gender). But if the values are within the normal physiological range, the biological significance is of low interest.

Taking these precautions into consideration, the clinical and nutritional biochemistry can be an excellent tool to better understand and manage the nutrition and health status of an animal as large camelids.

In the present book, the following chapters are developed: (1) hematology, (2) energetic parameters, (3) nitrogen and proteins, (4) enzymes, (5) electrolytes and macro-minerals, (6) trace elements, (7) vitamins, and (8) hormones. For each group of parameters, the usual values, the physiological variation effects (gender, age, physiological status, season), and pathological effect (diseases) are reported. If most of the parameters are measured in the blood, the other substrates and organs are evoked to be exhaustive, even if their clinical interest is not central. To facilitate the comparisons between the numerous references, the same units were used.

References

Abderhalden E (1898) Zur quantitativen vergleichenden analyse des blutes. Z Physiol Chem 25:65–115

Barnouin J, Chacornac JP, Aissaoui C, El Idilbi N, Mazur A (1994) Comment dépister les déséquilibres biologiques et les troubles de santé chez la vache laitière dans le cadre d'études écopathologiques. Vet Res 25:104–109

Bengoumi M (1992) Biochimie clinique du dromadaire et mécanismes de son adaptation à la deshydratation. Thèse de doctorat ès Sciences Agronomiques, I.A.V. Hassan II, rabat, Maroc, 184 p

Bengoumi M, Essamadi K, Tressol JC, Faye B (1998) Comparative study of copper and zinc metabolism in cattle and camel. Biol Trace Element Res 63:81–94

Bucci JT, Botros BAM, Gaines JF, Atrash S (1982) Technique for liver biopsy in the dromedary camel. Vet Rec 110:200–201

Carmalt MHB, Freeman P, Stephens AJH, Whitehead TP (1970) Value of routine multiple blood tests in patients attending the general practitioner. Br Med J:620–623

Cherrier R, Sumboro M, Faye B (1991) Note sur une méthode de biopsie chez le jeune dromadaire. Rev. Elev. Méd. Vét. Pays Trop., 44, 131–133

Farver TB (1997) Concepts of normality in clinical biochemistry (Chapter 1). In: Kaneko J, Harvey J, Bruss M (eds) Clinical biochemistry of domestic animals, 5th edn. Academic Press, New York, p 932

Faye B, Bengoumi M (1997) Données nouvelles sur le métabolisme des principaux éléments-traces chez le dromadaire. Rev Elev Méd Vét Pays Trop 50:47–53

Faye B, Nurseitova M, Konuspayeva G, Alsharary BS, Jurjanz S (2013) Hump biopsy on large camelids (*Camelus dromedarius* and *Camelus bactrianus*). J Camel Pract Res 20(2):1–4

Fayet JC, Genest M, Chacornac JP, Chonion N, Faye B, Barnouin J, Paccard P, Chassagne M, Brochart M (1986) Enquête écopathologique continue: 6. Influence de facteurs zootechniques et spatio-temporels sur quelques variables biochimiques chez les vaches laitières. Ann Rech Vet 17:215–223

Higgins AJ, Kock RA (1984) The camel in health and disease. I. A guide to the clinical examination, chemical restraint and medication of the camel. Br Vet J 140:485–504

Kerr MG (2001) Veterinary laboratory medicine, 2nd edn. Wiley-Blackwell, New Jersey, 392 pp

Khanna D (1993) Adaptative responses of camel to dehydration and rehydration following water restriction. PhD Thesis, Ajmer University, India, 307 p

Lamand M (1987) Place du laboratoire dans le diagnostic des carences en oligo-éléments. Rec Med Vet 163:1071–1082

Michel MC (1980) Utilisation des profils métaboliques dans l'élevage bovin. Bull Techn CRZV Theix-INRA 41:23–31

Rai AK, Khanna ND (1988) National Research Camel Center, Bikaner (India). Annual report 1988, 76 p

Rowlands GJ (1980) A review of variations in the concentrations of metabolites in the blood of beef and dairy cattle associated with physiology, nutrition and disease, with particular reference to the interpretation of metabolic profiles. World Rev Nutr Diet 35:172–235

Singh AP, Sharma SN, Taneja M (1994) Status of zinc (*Camelus dromedarius*) with reference to blood serum, seminal plasma and hair. Indian J Anim Sci 64:750–751

Chapter 2
Hematology

In all domestic species, blood carries a large number of very useful information for the clinician or researcher in physiological or pathological investigation, due to its transport functions of nutrients and oxygen to the organs, tissues, and cells in the body. Three types of chemical or structural parameters are usually explored in the blood tissue, namely, biochemical, serological and hematological parameters. This chapter focuses on the hematology of the camel, i.e., the study of different blood cells that play a key role in the transport of oxygen (red blood cells) and in the immune defense of the organism (white blood cells).

The conditions of the blood tissue sampling have been previously described. The blood submitted to hematological examination must be collected in tubes containing an anticoagulant product, usually EDTA (ethylenediamine-tetra-acetic acid). But, heparin, sodium citrate, or fluoride/oxalate could also be used according to the literature. Obviously, the sampling and hematological analysis techniques are not different from those of other animal species of economical interest.

Hematological Examination

Classically, the hematological examinations include the count of red blood cells, hemoglobin, hematocrit, the white blood cells, and the blood formula. The dosage of total proteins may be added, although they may be admitted to the group of biochemical parameters.

The **count of red blood cells** is generally carried out by automatic counting using a "Coulter counter" device equipped with a photoelectric cell. However, because of the special morphology of the camel's red cells and the variability of their size as we will see below, their count may be faulty (Chartier et al. 1986). Also, the classic technique with the help of the Malassez cell after dilution appears to be more appropriate. Some authors also use the Neubauer hemocytometer. Nowadays, automated blood cell counters are available and can be used in big laboratories analyzing a large amount of samplings. The results are displayed in "number of red blood cells per mm^3."

Hemoglobin is determined by the cyanide method which allows for the transformation of free hemoglobin (Hb symbolized) into cyanmethemoglobin after

© Springer International Publishing AG, part of Springer Nature 2018

B. Faye, M. Bengoumi, *Camel Clinical Biochemistry and Hematology*,
https://doi.org/10.1007/978-3-319-95562-9_2

hemolysis. Also, it can be easily measured by photometry after dilution and lysis of red blood cells. The scanning method has been also proposed, and it has been demonstrated that the linearity, precision, and sensitivity of this method are suitable for clinical use (Mezzou et al. 2006). The results are expressed in g/100 ml.

The **hematocrit**, which expresses the ratio between the volume of red blood cells and plasma, is the simplest hematologic measure. This can be achieved by a simple passive sedimentation, but more generally the hematocrit is quantified to microhematocrit on a capillary tube. The microhematocrit is frequently of the Hawksley type and is subjected to a centrifugation of 10,000 rpm for about 3 min. The results are read directly through a reading disc on the centrifuge rotor and are given as a percentage. This method provides accurate values with less than 0.2% error. Moreover, the development of a small centrifuge resistant to shocks and vibrations allows to easily achieve hematocrit measurements in the field conditions.

The **white blood cell count** relies on an apparatus similar to that of red blood cells, namely, a "Coulter counter," which allows for automatic counting. The count can also be achieved with a hemocytometer of Neubauer, after proper dilution and lysis of red blood cells. The results are expressed in number of cells per mm^3.

If count reflects the overall share of the cells of the white line in the blood, the blood formula gives its composition, thus expressing the number and the proportion of different types of cells. The smears are colored with MGG (May Grünwald Giemsa) and then read under a microscope. The results are expressed in numbers and as a percentage.

The modern automatic blood cell counter can provide a largest number of hematological parameters as follows:

WBC	Number of white blood cells
W SCC	Number of small white blood cells, lymphocyte
W MCC	Number of middle white blood cells, monocyte
W LCC	Number of large white blood cells, gametocyte
W SCR	Percent ratio of small white blood cell
W MCR	Percent ratio of middle white blood cell
W LCR	Percent ratio of large white blood cell
RBC	Number of red blood cells
HGB	Hemoglobin concentration
HCT	Hematocrit
WBC	Number of white blood cells
W SCC	Number of small white blood cells, lymphocyte
W MCC	Number of middle white blood cells, monocyte
W LCC	Number of large white blood cells, gametocyte
W SCR	Percent ratio of small white blood cell
W MCR	Percent ratio of middle white blood cell
W LCR	Percent ratio of large white blood cell
RBC	Number of red blood cells
HGB	Hemoglobin concentration
HCT	Hematocrit volume

MCV	Mean corpuscular volume
MCH	Mean corpuscular hemoglobin
MCHC	Mean corpuscular hemoglobin concentration
RDW	Red blood cell dimension width
PLT	Number of platelets
PCT	Platelet crit value
MPV	Mean platelet volume
PDW	Platelet dimension width

2.1 Red Blood Cells (Erythrocytes)

The shape and number of red blood cells provide information on the type and severity of the anemia, namely, hemorrhagic anemia, deficiency in iron or vitamins, and infectious or parasitic anemia. These are highly variable and their etiology is far from unequivocal.

2.1.1 Properties of Red Blood Cells of the Dromedary

2.1.1.1 Shape, Structure, and Dimensions

Unlike all other species of mammals, the red blood cells in the camelid family are elliptical or oval shaped and not circular. This observation was reported very long time ago (Mandl 1838) and strongly signals the hematologic peculiarity of the camel (Singh et al. 1997).

Although usually not nucleated, very rare red blood cells, which have a larger size, a round nucleus as well as a compact and very dark color, have been observed (Sergent and Poncet 1942). Also, there are punctuated red blood cells and round-shaped globulins that contain small irregular purple granules in their center. Marginal bands and their constituent microtubules representing at least 3% of the cell were also described (Cohen and Terwilliger 1979).

Generally, the red cells of the camel do not contain intracellular organelles, except for a few mitochondria which are sometimes observed but still in small number (no more than three in the same red cell). One could also notice the presence of marginal bands composed of microtubules, which is unique to mammals. This microstructure could be related to the capacity of the erythrocytes' resistance to strong plasma osmotic changes observed during a massive drinking and following a period of dehydration (Abdo et al. 1990). It is well known, indeed, that the camel is able to drink an impressive water quantity in record time. Yagil et al. (1974a) described the case of a dromedary camel of 600 kg having lost 200 kg of its body weight after 14 days of deprivation of water. To recover the losses, the animal drank

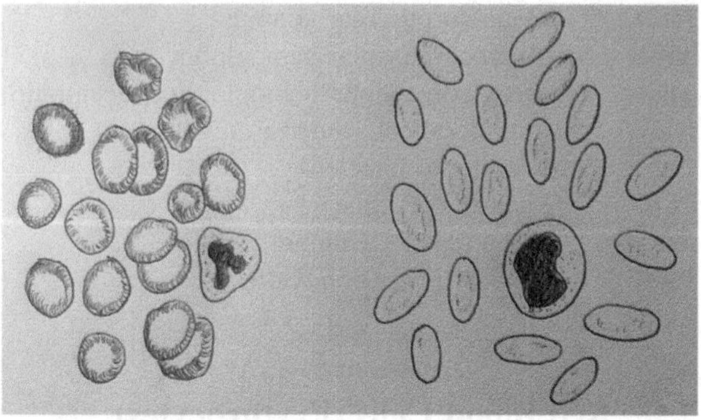

Fig. 2.1 Erythrocytes shape in human and camel (Drawing: B. Faye)

200 l in 3 min. Such a physiological feat involves a particular resistance of red blood cells.

This resistance could be linked to the composition of the red blood external membrane. Indeed, there is a significant difference in the lipid composition of the erythrocyte membranes of the camel compared to the desert goat and sheep which have no differences among themselves. In the erythrocyte membranes of the camel, the two classes of phospholipids (the phosphatidylcholine and the sphingomyelin), cholesterol, and proteins are proportionally much higher than in the other two species of small ruminants (Al-Qarawi and Mousa 2004). For example, the rate of proteins in these membranes is 5.67×10^{-10} mg/cell in the dromedary, while it is 4.08 and 1.95 in the goat and sheep, respectively. In addition, neither fasting nor dehydration can alter these constituents. However, the classes of phospholipids are directly related to the fragility of red blood cells. Protein/fat has the highest ratio among the camel than in small ruminants and also plays a prominent role in the stability of erythrocytes.

The red cells of the dromedary are rather flat, and their overall shape is not changed by dehydration. On the other hand, a quick rehydration causes a dramatic swelling of these cells that become circular (Yagil et al. 1974b) (Fig. 2.1).

Different sizes of red blood cells, which are probably related to different physiological contexts, were identified in the literature: 5.3×1.4 µm for smaller dimensions according to Abdo et al. (1990), 8.3×3.45 µm as stated by Nassar et al. (1977), and 7.55×4.15 µm as reported by Sergent and Poncet (1942) who measured 2000 red cells from 65 camels in the Sahara. Undoubtedly, the hydration status has a significant impact on the observed size of red blood cells. Yagil et al. (1974b) observed an average size of 10.1×6.4 µm or an area of 50.6 µm^2 for a camel watered ad libitum. After 7 days of dehydration, those dimensions decreased to 8.8×5.4 µm and 37.3 µm^2, respectively, without change of form. Four hours after the rehydration, the swelling of the red blood cell was clearly visible, and its area reached 41 µm^2.

Finally, according to Banerjee et al. (1962), the red blood cells of the dromedary (7.7 × 4.2 μm) would be substantially larger than that of the Bactrian camel (7.2 × 3.5 μm).

2.1.1.2 Osmotic Fragility

Like all cells, red blood cells resist more or less to the extreme concentrations in the mineral salts of the environment in which they are immersed (Long 2007). In a little concentrated salt, which is a hypotonic environment, red blood cells tend to increase their volume and be hemolyzed. Hemolyzed cells are observed from hypotonic solutions of 0.2%, and total hemolysis is obtained with a 0.1% solution (Yagil et al. 1974b). However, these values can vary from one author to another. Indeed, total hemolysis is obtained with solutions at 0.37–0.40% according to Boué (1948), while it is obtained with 0.25% solutions according to Soliman and Shaker (1967).

Compared to other mammals, the red blood cells of the camel well resist to the hypotonicity, which allows for a large and brutal blood tissue dilution during a massive watering. A water supply of 50 l within 1 min does not lead to a hemolysis of red blood cells. Generally, the critical mass that the red blood cells of the camel can reach before hemolysis in hypotonic solutions is 2.3 times the original volume, which is significantly higher than what is observed in humans (Yagil et al. 1974c). In young camels, two types of erythrocytes were described: one having the same shape as in adults and another which is more resistant to hypotonicity, but never observed in adult's blood (Perk 1966).

The remarkable resistance of the camel's red blood cells is also observed in case of hypertonicity. The cells keep a normal form in 10% hypertonic solutions, although their volume decreases slightly; and they become predominantly abnormal from 20% solutions (Yagil et al. 1974b). By comparison, the human red blood cells cannot resist to a tonicity that is higher than 1.5%. So, the camel has hematological characteristics enabling him to survive over severe dehydration conditions without affecting his blood cells' functions. Comparing camels from drought areas to camels that are normally watered, no difference was observed in osmotic fragility (Kataria and Kataria 2004).

Knowing that seawater contains 3.5% of salt, it appears that the camel can survive by drinking such water without major disruption. However, the hemoparasites' infestations (first at all trypanosomosis) increase the osmotic fragility of erythrocytes (Jatkar and Purohit 1971). At reverse, there is no effect of pollution (Ghoke et al. 2013) nor of anesthesia (Peshin et al. 2010) on the osmotic fragility. However, erythrocytes' osmotic fragility increases in blood samples collected with EDTA. An effect of storage time, cold, heat, acidity, and hydrogen peroxide was also reported (Lektib et al. 2016). In addition, this osmotic fragility is influenced by age, sex, season, and the vitamins' concentrations in blood. So, the quality of the osmotic fragility test may easily be influenced by several biological, environmental, and technical factors. These factors induce damage to the RBC cell membrane and may disturb the results of other blood analyses (Oyewale et al. 2011; Lektib et al. 2016).

2.1.1.3 Life Span of the Camel Erythrocytes

The life span of red blood cells depends on climatic and hydration conditions (Yagil et al. 1974a), which is 90 days on average for camels that are normally watered in a cold season, 120 days in similar watering conditions in a hot season, and 150 days in case of chronic dehydration at summer time. This change in the erythrocytes' life span according to hydration status underlines an important mechanism of adaptation to heat and aridity in camels. Indeed, in extreme conditions, the camel has to minimize its water loss and its metabolic heat production. Yet, the destruction of red blood cells necessitates water, first for excreting the products' degradation and then for producing new erythrocytes. So, the camel could have the ability to preserve its physiological water by increasing the life span of its red blood cells (Yagil 1985).

2.1.2 Red Blood Cell Count (RBC)

According to different published sources, the RBC in the camel varies between 6 and $10 \times 10^6/\text{mm}^3$ (Table 2.1) with extreme values between 5 (Chartier et al. 1986, Mauritania) and $12.5 \times 10^6/\text{mm}^3$ (Sharma et al. 1973—India). Beyond the uncertainty about the counting methods, the pathological and geoclimatic factors could explain the observed variations. It is known that the altitude plays a primordial role when it comes to RBC, provoking a "polyglobuly of altitude." Therefore, small camelids that are living in Andean Altiplano with 4000 m above sea level have very high RBC (Hajduk 1992). Compared to the dromedary camel, the Bactrian camel living at higher altitudes on average has also higher values of RBC: 10 to $19 \times 10^6/\text{mm}^3$ vs 6 to $9 \times 10^6/\text{mm}^3$ (Banerjee et al. 1962).

The variability of the values observed by various authors from different regions far away from each other may also suggest the idea of geographic variations, related to climatic differences, even if the normal distribution place of the camel remains attached to the arid and semiarid areas from Mauritania to India. For example, Durand and Kchouk (1959) observed that the animals of the desert have a higher red blood cell count of two million per mm^3 of blood compared to those from less arid regions of the Sahel. The dehydration status seems to have no significant impact on the RBC. After 25 days of water dehydration, the difference observed before and at the end of the dehydration experiment is not significant (Mohamed et al. 1984), but this observation was limited to only three animals.

From a general point of view, the high variability of the RBC in camels is a phenomenon that is very similar to what is observed in the majority of ruminants (Soliman and Shaker 1967). It should be noted that, despite the existence of many camel's breeds often associated with a given geographical area, hematological changes have been studied in the light of the racial diversity. In the rare available references, no breed difference was observed for RBC as well as for other hematological parameters (Aichouni et al. 2010). However, a significant phenotypic

Table 2.1 Red blood cell count, hematocrit, and hemoglobin rate in camel according to different authors

References	RBC (nb/mm^3)	Hematocrit (%)	Hemoglobin (g/100 ml)	*n*
Soni and Aggarwal (1958)—Inde	8.2 ± 2.2	–	15.5 ± 2.4	95
Durand and Kchouk (1959)—Tunisia	7.3–9.4	–	10.4–14.2	26
Lakhotia et al. (1964)—India	5.4–6.5	30.1–31.5	11.5–11.8	60
Soliman and Shaker (1967)—Egypt	7.2 ± 0 0.1	43 ± 1.1	13.2 ± 0.8	8
Hassan (1968)—Sudan	8.8	–	–	45
Barakat and Abdel-Fattah (1970)—Egypt	7.7 ± 0.18	–	13.05 ± 0.12	260
Jatkar and Purohit (1971)—India	9.8	28.9	12.5	25
Ghodsian et al. (1978)—Iran	7–7.2	28–29	11–11.5	99
Gupta et al. (1979)—India	5.1 ± 0.3	21.5 ± 1.1	9.2 ± 0.5	13
Majeed et al. (1980)—Pakistan	6.7 ± 0.17	–	11.1 ± 0.3	20
Raisinghani et al. (1981b)—India	–	30.6 ± 0.4	10.2 ± 0.2	9
Musa and Mukhtar (1982)—Sudan	6.1 ± 1.5	25.9 ± 4.5	11.6 ± 2.5	174
Faye et al. (1986)—Ethiopia	–	22–28	–	52
Chartier et al. (1986)—Mauritania	5	29.2–36.5	11.9–14	130
Yagoub (1988)—Sudan	9 ± 1.6	26.4 ± 3.4	12.5 ± 1.5	97
Abdalla et al. (1988)—UAE	8.8 ± 0.7	31.3 ± 2.8	15.2 ± 1.3	33
Abdel Samee (1987)—Egypt	10.5 ± 2.2	28.4 ± 0.4	12.2 ± 0.15	174
Ibrahim et al. (1992)—Bahrein	8.3 ± 1.6	28.5 ± 4	11.2 ± 1.5	301
Al-Ani et al. (1992)—Iraq	9.4 ± 1.8	29.7 ± 3.1	13.1 ± 1.2	15
Khanna (1993)—India	7–11	28.5–30	11–15.5	–
Alhadrami (1997)—UAE	9.04 ± 0.2	–	12.6 ± 0.3	20
Nyang'ao et al. (1997)—Kenya	8.5 ± 1.6	27.1 ± 3	11.2 ± 1.7	21
Sarwar and Majeed (1997)—Pakistan	7.96 ± 0.2	26.1 ± 0.4	13.3 ± 0.17	56
Rezakhani et al. (1997)—Iran	10.7 ± 0.9	33.2 ± 4.9	14.2 ± 1.7	83
Ayoub and Saleh (1998)—UAE	9.7 ± 0.93	31 ± 2.9	14.5 ± 1.56	3
Naeini and Nafizi (2001)—Iran	9.5 ± 1.5	29 ± 4	12.7 ± 14.6	50
Liu (2003)—China[a]	12.5 ± 3.6	31.2 ± 3.1	10.5 ± 2.4	50
Mohammed et al. (2008)—Nigeria	9.4 ± 1.1	29.9 ± 3.4	11.6 ± 1.1	11
Al-Sultan (2008)—Saudi Arabia	7.4 ± 0.3	33 ± 3	9.3 ± 0.3	50
Faye et al. (2009)—UAE	–	27.3 ± 3.6	12.8 ± 5.5	44
Nazifi et al. (2009)—Iran	10.7 ± 0.9	33.5 ± 3.0	14.3 ± 7.5	20
Al-Mujalli et al. (2011)—Saudi Arabia	9.1 ± 0.45	–	12.5 ± 0.6	20
Farooq et al. (2011)—Pakistan	7.31 ± 0.6	32.8 ± 3.7	11.3 ± 0.9	30
Hussein et al. (2012)—Saudi Arabia	9.44 ± 1.8	32 ± 3.1	13.1 ± 1.2	209
Auer et al. (2015)—South Africa	10.3 ± 0.8	43 ± 3.5	–	11

[a]Bactrian camel

difference in RBC, which is higher in the black-coat camel (Al-Majaheem breed) than in the white-coat (Waddha breed) and brown-coat (Homor breed) camel, was reported in Saudi Arabia. Such a difference was also observed for hemoglobin concentration and hematocrit (Hussein et al. 2012).

2.1.2.1 Seasonal, Age, and Sex Effect

The season could play a role, according to some authors. For example, Mehrotra and Gupta (1989) reported the variations from simple to double RBC for the same sample of 30 animals. Such a difference is less marked according to others. More precisely, a very slight decrease was reported in a dry season (Mutugi et al. 1993) or during winter (Abdoun et al. 2012). In reverse, Knight et al. (1994) found lower values of RBC during summer in UAE. Comparing the four seasons (July = rainy; September = hot rainy; October = dry wet; April = dry hot), Babeker et al. (2013) has found significant lower values of RBC in September ($2.25 \times 10^6/mm^3$) compared to around $3 \times 10^6/mm^3$ in the other seasons. However, the RBC values in this publication are quite lower than those reported in the literature, as shown in Table 2.1. However, other authors did not report any significant difference between winter and summer (Salman and Afzal 2004).

It seems that there is no sex effect on the number of red blood cells in the dromedary (Petrelli et al. 1982; Abdalla et al. 1988; Mutugi et al. 1993) and in Bactrian (Liu 2003), although a few authors attributed a lower count to the male (Lakhotia et al. 1964; Farooq et al. 2011). This count can be associated with the sexual activity (Khan and Kohli 1978). There is no RBC difference in males affected by abnormal spermatozoids (azoospermia or oligospermia) (Ali et al. 2015).

The red blood cell gradually increases from $4.5 \times 10^6 /mm^3$ at birth to $9.5 \times 10^6 /mm^3$ up to the age of 1 year, remains stable up to 4 years, and then decreases with age to reach $6.6 \times 10^6 /mm^3$ up to 16 years (Petrelli et al. 1982). Such a changing pattern was disputed by Yagoub (1988) who observed lower values for animals between 1 and 5 years compared to camels belonging to upper and lower age groups. Finally, other authors noted no influence of age (Durand and Kchouk 1959; Chartier et al. 1986). In a 1-week neonate camel, an RBC of $8.5 \times 10^6/mm^3$ was reported, while a significant higher value was reached in a second camel on his seventh day of life (Elkhair and Elmgboul 2013). For Omer et al. (2016), the RBC value is higher in weaned camel calves ($7.45 \pm 1.4 \times 10^6/mm^3$) compared to the suckling ones ($6.22 \pm 0.9 \times 10^6/mm^3$) and lactating dams ($5.56 \pm 1.2 \times 10^6/mm^3$). In racing camels, Alhadrami (1997) noted lower values in adults (9.04 ± 0.2) compared to camel calves ($11.7 \pm 0.2 \times 10^6/mm^3$).

2.1.2.2 Effect of Effort

In the camel race, Snow et al. (1988) did not reveal any effect of physical exercise on the erythrocyte count. Yet, for horse racing, it was observed that intensive training is associated with an increase in RBC (Persson 1967). In the Emirates, in resting camel, Knight et al. (1994) reported that a significant RBC increase is observed in January (9.6 vs 8.7 $\times$ 10^6/mm^3 in summer). However, the month of January, which is the coolest month of the year, coincides with the high-season for racing.

2.1.2.3 Seasonal Effect

Blood volume would decrease during the winter (Ghosal et al. 1973; Aichouni et al. 2011; Abdoun et al. 2012), which leads to an increase in the concentration of erythrocytes. Therefore, it is likely that, there is a combination of seasonal effects and physical activity. The notable absence of changes in RBC during physical exertion in the camel indicates a failure in the storage of erythrocytes in the spleen, unlike other racing animals as horses or dogs who release red cells stored in the spleen during exercise. In camel, as in human, the ability to fix oxygen in the blood is not therefore increased by the influx of erythrocytes.

2.1.2.4 Parasitism Effect

Gastrointestinal parasitism also affects the hematological parameters but would be more marked in females. Thus, Ibrahim et al. (1992) observed an RBC decrease in 114 infested females compared to 72 noninfested ones. This difference was not observed in 115 male samples including 80% with parasites.

Trypanosomosis, a common camel disease caused by *Trypanosoma evansi*, leads to significant changes of hematological parameters. In particular, a very significant effect on RBC was witnessed with 9.8 $\times$ 10^6/mm^3 vs 4.5 $\times$ 10^6/mm^3, respectively, in non-infected and infected camels (Jatkar and Purohit 1971). Similar observation (7.44 $\pm$ 1.04 in healthy camels vs 4.25 $\pm$ 0.07 $\times$ 10^6/mm^3 in affected animals) was reported by Hussain et al. (2016a, b). The lower erythrocyte count has been reported in trypanosomiasis-positive camels by many other authors (Chaudhary and Iqbal 2000; Fatihu et al. 2000; Al-Mujalli 2007; Abd El-Baky and Salem 2011; Saleh et al. 2011; Padmaja 2012; Eyob and Matios 2013; Ahmadi-Hamedani et al. 2014a; Varia et al. 2015).

Trypanosomes, which are extraglobular parasites, cause hemolysis of red blood cells and therefore hemolytic anemia in infested camel. The most direct consequence is a reduction in the number of red blood cells. However, the hemolytic action would be more related to the toxins released by the trypanosome than to the parasite itself, which would explain the lack of correlation between parasitemia and hematological changes (Raisinghani et al. 1981a).

Reticulocytes, which are immature red blood cells, may provide useful information. They are present in the case of regenerative anemia. Raisinghani et al. (1981a) observed no reticulocytes in healthy animals. However, in animals inoculated with *Trypanosoma evansi*, the reticulocyte rate may reach 6% after 88 days of infestation (Raisinghani et al. 1981b) and even more than 11% in some cases (Jatkar and Purohit

1971). The trypanosomiasis is accompanied by hyperplasia of the bone marrow (Raisinghani et al. 1981b), in consequence of the above-mentioned hemolytic anemia. It is a regenerative anemia, which explains the appearance of immature red blood cells in the peripheral circulation (Jatkar and Purohit 1971). The presence of normoblasts and erythroblasts, which are cells at the origin of the red blood cells, confirms the hypothesis of erythropoietic hyperplasia. Trypanosomosis provokes oxidative stress, causing an increase of peroxides. Malondialdehyde (MDA) is the most abundant lipid peroxidation product used as an indicator of oxidative stress linked to trypanosomosis infestation in cattle (Igbokwe 1994). In camel, a significant negative correlation was observed with all blood parameters (RBC, PCV, Hb), those parameters decrease linearly when MDA increases (Saleh et al. 2009).

Other hemoparasites could have similar effects. For example, theileriosis contributed to significantly decrease PCV in camels from 9.05 ± 0.27 in healthy camels to $7.23 \pm 0.12 \times 10^6/\text{mm}^3$ in affected ones (Youssef et al. 2015). RBC is slightly affected by copper deficiency induced by sulfur excess (Li and Hai 2014). At reverse, camels supplemented with trace elements (Cu, Zn, Co, and Se) presented higher RBC (Kosanovic et al. 2014). In case of selenium toxicosis, signs of anemia appeared from 8 mg/day of supplementation leading to a significant decrease in Hb and PCV (Faye et al. 2009).

The use of tranquilizers, such as promazine, xylazine, acepromazine, and chlorpromazine, seems to affect the number of red blood cells (Ali et al. 1989), but the observed 10% decline was not significant.

2.1.3 Hematocrit: Pack Cell Volume (PCV)

As RBC, hematocrit may be a good indicator of hemorrhagia, anemia, or polycythemia. But more generally, it provides information on the volume of circulating liquids (hemodilution, hemoconcentration) during the deprivation of water supplies or even of therapeutic solutes. Because of the camel biotope peculiarities (hot and arid climate ecosystem), the assessment of hematocrit can provide useful information on the dehydration status of the animal. Normal values vary between 25 and 30% with extremes ranging from 22 to 43% (Table 2.1). These values are comparable to those of other domestic herbivorous but lower than those of most other mammals. If the opinions could diverge about the effect of dehydration on RBC, it is not the case for hematocrit as discussed in the chapter on the effect of dehydration.

Like RBC, hematocrit does not obviously change with age. According to some observations, hematocrit is lower in young camels: 22.3% vs 27.2% in adults (Mutugi et al. 1993). These results are inconsistent with those of Petrelli et al. (1982) which reported that hematocrit was 16.5% at birth, 22% at 1 year old, and 20.6% at 16 years old. For Omer et al. (2016), suckling camels have higher PCV (26.7%) than weaned ones (24.8%) and their dams (23.7% on average). Hematocrit is stable up to 1 month in the postpartum period (Elias and Yagil 1984).

However, in the same conditions of breeding, no sex difference was observed. According to Chartier et al. (1986), females of more than 7 years old have lower values than youngest ones and those of males of the same age group. This effect of sex was also confirmed by other authors (Nassar et al. 1977) who reported higher values for PCV in males (37.2%) compared to females (32.8%) (Farooq et al. 2011). However, no difference was observed between castrated and whole males (Ali et al. 2015).

Seasonal effects mentioned by some authors (Yagil et al. 1974b; Mehrotra and Gupta 1989; Salman and Afzal 2004) are attributed to temperature (reduced number of RBC during the hottest summer and the coldest winter), but these differences are subtle (Knight et al. 1994; Aichouni et al. 2011) or nonsignificant (Babeker et al. 2013). Winter reduction of hematocrit was also attested by Ghosal et al. (1973), but not in the summer, which would be the opposite of what was observed in other tropical species (Pandey and Roy 1969). The increase of hematocrit in the summer would ensure better blood oxygen fixation and thus a better tissue oxygenation in the dromedary. This increase would be essentially linked to that of the mean corpuscular volume (Butt and Afzal 1992).

Meanwhile, physical effort leads to a significant increase of hematocrit (Snow et al. 1988); the effect was positively correlated to the duration of the effort (Rose et al. 1994). Tissue anoxia associated with effort, as well as acceleration of metabolism, stimulates hematopoiesis and so increases hematocrit (Ghosal et al. 1973). However, it is more likely that the physical effort causes a decrease in plasma volume (Snow 1992), which alone would explain the change in hematocrit. It is noted, however, that the reduction in plasma volume during intense effort is much less important in camel than in the other species which are supposed to lose a lot of water by sweating or breathing. This physiological characteristic contributes to strengthen the adaptability of the dromedary to desert environment even during intense physical activity, by promoting the storage of heat instead of its elimination using evaporation mechanisms. Hematocrit would decrease significantly by around 13% when the camel is working as a cart animal during the hot season compared to other seasons of the year (Padheriya and Wadhwani 2013).

However, hematocrit would decline in animals infested by gastrointestinal parasites (Ibrahim et al. 1992) or coccidiosis (Athar Khan et al. 1992). In animals affected by trypanosomosis, hematocrit fall may exceed 65%; for example, Raisinghani et al. (1981b) reported a hematocrit of 10% in infected camels vs 29.7% in healthy control group. Similar decreases, although less spectacular, have been usually cited in the literature (Jatkar and Purohit 1971; Adam et al. 1974; Raisinghani et al. 1981a; Ahmadi-Hamedani et al. 2014a, b; Hussain et al. 2016a, b). A trypanosomosis survey in UAE showed a decrease in PCV from 31.5% in negative control camels to 22.6% in Suratex-positive animals and 18.2% in smear-positive camels (Chaudhary and Iqbal 2000). This fall is obviously associated with important hemolysis (Faye et al. 1986). Decrease in hematocrit was also observed in case of theileriosis, with PCV passing from 30.44 ± 0.53 in healthy animals *to* 26.83 ± 0.40% in infected camels (Youssef et al. 2015). Furthermore, it was witnessed in case of secondary copper deficiency, passing from 39.6 ± 3.1 to 31.6 ± 4.2% in non-deficient and deficient Bactrian camels, respectively (Li and Hai 2014). Trace element supplementation contributes significantly to the improvement of PCV passing from 27.8 to 30% in non-supplemented and supplemented camels, respectively (Kosanovic et al. 2014). In case of selenium toxicosis, signs of anemia appeared since 8 mg/day of supplementation leading to a significant decrease in Hb and PCV (Faye et al. 2009).

2.1.4 Hemoglobin

The dosage of hemoglobin (Hb) is interesting to detect the different forms of anemia. Hb concentration varies in the majority of the references between 9.3 and 15.5 g/dl (Table 2.1). As for hematocrit, Hb rate in camel is comparable to that of other domestic ruminants and therefore lower than in most other mammals (including domestic carnivores). It appears to be significantly highest in summer than in winter (Majeed et al. 1980). However, the seasonal variation was not confirmed by Butt and Afzal (1992), Salman and Afzal (2004), Aichouni et al. (2011), and Babeker et al. (2013).

According to Petrelli et al. (1982), Hb is higher in 1-year-old camel (10.6 g/dl) than in the newborn calf (8.5 g/dl) and in adult dromedary (8.9 g/dl). During the first week of life, Hb concentration is regularly decreasing from 13 to 9.1 g/dl (Elkhair and Elmgboul 2013). Suckling camels have higher value of Hb (11.42 g/dl) than weaned camels and their dams (10.6 g/dl on average) (Omer et al. 2016). In lactating camels, Hb concentration is stable during the first month of the postpartum period (Elias and Yagil 1984).

A discreet sex effect is also noted, females showing significantly lower rates than males (Nassar et al. 1977; Chartier et al. 1986). During a rut, however, hemoglobin drops significantly from 12.6 to 10.8 g/dl in males (Khan and Kohli 1978). There is no impact of male infertility (Ali et al. 2015).

The increase of hemoglobin observed in racing camels after a physical effort is slight but significant (Snow et al. 1988). Similarly to hematocrit, it is related to the low decline of plasma volume after race. There is also a significant effect of the load in carting animals, where the hemoglobin concentration decreases when the cart load is passing from 1.5 tons to 2 tons (Padheriya and Wadhwani 2013).

Moreover, like hematocrit and red blood cell, hemoglobin decreases in animals infested with gastrointestinal parasites (Ibrahim et al. 1992; Ahmadi-Hamedani et al. 2014a). Camels affected by theileriosis (11.25 g/dl) have less Hb than those non-infected (14.14 g/dl) (Youssef et al. 2015). With values in the order of 7.6 g/dl, the average hemoglobin also decreased in camels strongly affected by coccidiosis (Athar Khan et al. 1992). Such a decline, although less pronounced, is observed in camels infested with larvae of *Cephalopina titillator* (Hussein et al. 1983). In case of trypanosomosis, the fall of hemoglobin in the camel blood is usually around 50% (Jatkar and Purohit 1971; Adam et al. 1974; Raisinghani et al. 1981a, b; Zia-ur-Rahman 1992; Varia et al. 2015). Similar decreases of all hematological parameters including RBC, hematocrit, hemoglobin and others, occurred in case of infection with Mycoplasma spp. (Nazifi et al. 2009).

Secondary copper deficiency (sulfur induced) is linked to a significant decrease of Hb passing from 12.7 in non-affected camels to 9.67 g/dl in deficient camels (Li and Hai 2014). In trace element-supplemented camel, Hb concentration was 13.9 g/dl on average compared to 12.5 g/dl in non-supplemented camel (Kosanovic et al. 2014). In camels affected by perivascular dermatitis, an increase of Hb concentration was reported (Mathur et al. 2012).

2.1.5 Erythrocyte Indices of Wintrobe

These indices are calculated from previous data. They are used for typing anemia.

- MCV (mean corpuscular volume) is measured in fentoliters (fl).
- MCH (mean corpuscular hemoglobin), which is expressed in picograms (pg), corresponds to the mean weight of hemoglobin within a red cell.
- MCHC (mean corpuscular hemoglobin concentration) is expressed in g/dl or in % Hb in red blood cell.

In camel, the mean MCV values in camel vary according to the authors between 28.5 fl (Yagil et al. 1974b) and 60 fl (Soliman and Shaker 1967). However, most of the citations are between 30 and 45 fl (Al-Ani et al. 1992; Ibrahim et al. 1992; Musa and Mukhtar 1982; Gupta et al. 1979; Al-Sultan 2008; Aichouni et al. 2011). A very important seasonal effect was reported by Babeker et al. (2013) with a quite lower value (39.9 fl) in dry and hot summer compared to other seasons (72.3–103.9 fl).

If the hemoparasites tend to act on the variability of the observed values (Jatkar and Purohit 1971), age and sex appear to play only a minor role (Lakhotia et al. 1964; Yagoub 1988; Ibrahim et al. 1992; Ghodsian et al. 1978; Omer et al. 2016). In contrast, trypanosomosis leads to a marked increase of MCV, from 35.3 fl in healthy camels to 53.5 fl in infested animals (Jatkar and Purohit 1971). It is noted that the highest number of immature cells (normoblasts, erythroblasts, reticulocytes) of larger size explains the increase in the mean corpuscular volume.

The mean values of the MCH in camel are between 12 and 18 pg (Al-Sultan 2008), but extreme values between 9 (Sharma et al. 1973), 21.5 pg (Lakhotia et al. 1964), and even 38.5 pg (Babeker et al. 2013) were reported in the literature, with a significant seasonal effect, where this maximum value was observed at the hot rainy summer.

Regarding MCHC in camels, common values range between 40 and 50 g/dl (Al-Sultan 2008) with extreme values reported between 27.1 (Sharma et al. 1973) and 54.4 g/dl (Yagil et al. 1974b). The highest values were reported during summer, but still in the normal range (Babeker et al. 2013). An increase of MCV and MCHC linked to the inflammatory process has been mentioned in camels affected by perivascular dermatitis (Mathur et al. 2012).

Table 2.2 Range values of hematological parameters in different species [from after Schalm et al. 1975]

Species	RBC ($10^6/mm^3$)	Hb (g/dl)	PCV (%)	MCV (fl)	MCH (pg)	MCHC (%Hb)
Camel	5–11	9.3–15.5	21.5–43	28.3–60	9–21.5	27.1–54.4
Cattle	5–10	8.0–15.0	24.0–46	40.0–60	11–17	26. 0–36.0
Sheep	8–16	8.0–16.0	24.0–50	23. 0–48	9–12	29.0–38.0
Goat	8–18	8.0–14.0	19.5–48	15.5–37	7–8	30.0–42.0
Dog	5.5–8.5	12–18.0	37.0–55	60.0–77	19.5–24.5	31.0–36.0

Overall, erythrocyte indices, including the MCV, are characterized in camel, by a big interindividual variability (Gupta et al. 1979) but low interracial variability (Hussein et al. 2012). These indices are very sensitive to watering status of the animal, dehydration and rehydration leading to significant variations (Yagil et al. 1974b). The average values of the erythrocyte indices are rather lower than those of the cattle but higher than those of small ruminants (Bono et al. 1983; Schalm et al. 1975), except for the MCHC, and higher in camels (Table 2.2). Higgins and Kock (1984) proposed almost identical values for the dromedary and the Bactrian camel but higher MCHC in small camelids living in altitude, allowing them to increase their capacity for oxygen fixation.

Overall, erythrocyte indices decrease from birth to 1-year age, remain stable up to 4 years, and then increase in adults (Petrelli et al. 1982). Except MCH, these indices are significantly influenced by the season with higher values in winter (Aichouni et al. 2011).

As for RBC, PCV, and Hb, these parameters are also sensitive to copper deficiency (Li and Hai 2014) and to trace element supplementation (Kosanovic et al. 2014).

2.2 White Blood Cells

2.2.1 Numeration and Leukocyte Formula

White blood cells of the camel do not show any apparent functional adaptation to life in the desert. They have the same functions as in other mammals (Yagil 1985). They are also involved in the inflammatory process and mechanisms of defense against infections.

2.2.1.1 Leukocyte Numeration (White Blood Cells: WBC)

The values reported in the literature are part of a fairly wide range between 9.7 and 20.1 × $10^3/mm^3$ (Table 2.3). Normal values are usually between 10.5 and 15.5 × $10^3/mm^3$, which are on average higher than those of ruminants (Higgins and Kock 1984; Yagil 1985). Indeed, in cattle and sheep, blood leukocyte number falls within a range of 4 and

Table 2.3 Leukocyte numeration reported by different authors in camel

References	WBC (in $10^3/mm^3$)	*n*
Soni and Aggarwala (1958)—India	20.7 ± 3.6	95
Durand and Kchouk (1959)—Tunisia	15.5	26
Lakhotia et al. (1964)—India	11.3 to 12.8[a]	60
Soliman and Shaker (1967)—Egypt	12.5 ± 0.9	8
Hassan (1968)—Sudan	15.5	45
Jatkar and Purohit (1971)—India	17.2	25
Sharma et al. (1973)—India	15.2 ± 0.7	6
Ghodsian et al. (1978)—Iran	15.2 to 16.7[b]	99
Gupta et al. (1979)—India	10.4 ± 0.6	9F and 4 M
Majeed et al. (1980)—Pakistan	10.5 ± 0.4	10F and 10 M
Musa and Mukhtar (1982)—Sudan	12.6 ± 5.2	174
Chartier et al. (1986)—Mauritania	13.8 to 16.8[a]	130
Yagoub (1988)—Sudan	12.9 ± 2	97
Abdalla et al. (1988)—UAE	14.6 ± 2	33
Abdel Samee (1987)—Egypt	14.2 ± 0.2	174
Ibrahim et al. (1992)—Bahrein	12.9 ± 3 (F)	186F and 115 M
	9.7 ± 2 (M)	
Al-Ani et al. (1992)—Iraq	10.0 ± 1.8	15
Knight et al. (1994)—UAE	11.9 ± 2.8	–
Sarwar and Majeed (1997)—Pakistan	19.04 ± 0.56	56
Ayoub and Saleh (1998)	14.13 ± 1.19	3
Tabatabaei Naeini and Nafizi (2001)—Iran	10.9 ± 2.7	50
Liu (2003)—China[c]	6.86 ± 2.26	50
Al-Busadah (2007)—Saudi Arabia	19.6 ± 0.51	60
Al-Sultan (2008)—Saudi Arabia	17.9 ± 0.2	50
Mohammed et al. (2008)—Nigeria	13.4 ± 2.8	11
Faye et al. (2009)—UAE	16.4 ± 4.5	44
Nazifi et al. (2009)—Iran	8.67 ± 1.42	20
Aichouni et al. (2011)—Algeria	15.3 ± 1.2	40
Al-Mujalli et al. (2011)—Saudi Arabia	8.37 ± 0.99	20
Farooq et al. (2011)—Pakistan	12.9 ± 1 (F)	30F and 30 M
	12.4 ± 0.9 (M)	
Al-Sultan and El-Bahr (2012)—Saudi Arabia	10.0 ± 0.6	6
Omer et al. (2016)—Sudan	12.6 ± 3.4	48

[a]According to age classes and sex
[b]According to age classes
[c]Bactrian camel

$12 \times 10^3/mm^3$, with an average of $8 \times 10^3/mm^3$, and that of goats with an interval of 4 and $13 \times 10^3/mm^3$ and an average of $9 \times 10_3/mm^3$ (Schalm et al. 1975). In Bactrian camel, reported values are on average higher and range between 8.6 and 16.5 (Higgins and Kock 1984).

WBC increases from birth up to 2 years from $12.4 \times 10^3/mm^3$ to $19.1 \times 10^3/mm^3$, then decreases to $16.3 \times 10^3/mm^3$ up to 4 years, and stabilizes in adults (Petrelli et al. 1982). According to other authors, the number of white blood cells would be higher in young camels of less than 1 year (Ghodsian et al. 1978). Except for Lakhotia et al. (1964) and Nassar et al. (1977) for whom the number of leukocytes is $17.5 \times 10^3/mm^3$ in males vs $22 \times 10^3/mm^3$ in females, most authors did not report greater WBC in males (Majeed et al. 1980; Ibrahim et al. 1992). However, the number of white blood cells increased significantly in rutting males from 10.5 to $14.2 \times 10^3/mm^3$ (Khan and Kohli 1978) without change of leukocyte formula. A similar change was reported by Zeidan and Abbas (2003): 9.5 ± 1.05 vs $14.9 \pm 0.58 \times 10^3/mm^3$ in non-rutting and rutting males, respectively. This increase is regarded as a response to the stress of the rutting time helping the animal to resist against exhaustion and infection (El-Mougy et al. 1984). At reverse, there is no significant difference between pregnant and nonpregnant camels (Muhammad et al. 2011). Also there is no breed effect (Al-Busadah 2007) as well as seasonal effect (Salman and Afzal 2004).

A marked leukocytosis is described in the dromedary camel infested by *Trypanosoma evansi*, linked in large part to the rise of polynuclear eosinophils (Jatkar and Purohit 1971) and lymphocytes (Zia-ur-Rahman 1992). After treatment with Cymelarsan®, the WBC values return to normal levels after 3–4 weeks (Njiru et al. 2000).

At reverse, there is no difference in WBC between camels affected by mange ($9.3–9.9 \times 10^3/mm^3$) and healthy ones ($8.8–9 \times 10^3/mm^3$), whatever the season is (Mal et al. 2006). Also there is no effect of high zinc dose in the diet on the WBC count (Al-Sultan and El-Bahr 2012).

Kataria and Kataria (2004) observed a significant decrease in WBC in camels leaving in area affected by drought, the mean values passing from 18.99 ± 1.01 to $15.09 \pm 1.01 \times 10^3/mm^3$.

2.2.1.2 Leukocyte Formula

Proportions of the different types of white blood cells (named "leukocyte formula") are highly variable from one author to another, as can be seen in Table 2.4. However, regardless of the observed values, there is an important characteristic of the camel in comparison to other domestic herbivores: the predominance of polynuclear neutrophils representing 37–60% of leukocytes in the camel while lymphocytes predominate in cattle and small ruminants (Schalm et al. 1975). Higgins and Kock (1984) attributed this difference to the stress related to the blood sampling in this species as restraint is often necessary, but this assumption is not confirmed. Moreover, this characteristic seems to be more pronounced among the Bactrian camel in which the polynuclear neutrophil rate is between 55 and 79% (Higgins and Kock 1984).

The **lymphocyte** proportion is usually between 29 and 63% with an average below 50%, in contrary to the other domestic herbivores. This value is even lower in the Bactrian camel, as the average values are between 18 and 33% only (Higgins and Kock 1984). The proportion of lymphocytes would tend to increase in adult animals

Table 2.4 Leukocyte formula in camel according to different authors

References	Neutrophils	Eosinophils	Basophils	Lymphocytes	Monocytes
Sergent and Poncet (1942)	54.5	3.7	0	30.2	11.6
Soni and Aggarwal (1958)	38.7 ± 8.8	9.5 ± 4.7	<1	46 ± 9.7	5.7 ± 3.3
Soliman and Shaker (1967)	31.7 ± 1.1	2.2 ± 0.04	0.7 ± 0.01	63 ± 2.2	2.4 ± 0.3
Hassan (1968)	37 ± 5	7 ± 3	<1	52 ± 8	4 ± 1
Barakat and Abdel-Fattah (1970)	26.8 ± 0.8	3.4 ± 0.23	<1	66.5 ± 0.82	2.7 ± 0.13
Sharma et al. (1973)	43.8 ± 7.2	13.8 ± 6.6	0.2 ± 0.14	39.6 ± 8.9	2.6 ± 0.3
Pegram and Scott (1976)	23–66	2–13	–	30–72	0–6
Ghodsian et al. (1978)	51–58	5	<1	29–38	3–3.5
Gupta et al. (1979)	55 ± 4.6	2.6 ± 0.6	0	28.8 ± 4.8	3.7 ± 0.6
Majeed et al. (1980)	44.6 ± 1.4	7.2 ± 0.4	0.05 ± 0.0	47.5 ± 1.4	1.2 ± 0.1
Musa and Mukhtar (1982)	55.1 ± 11.5	1.5 ± 0.8	0.16 ± 0.5	33.9 ± 11.4	4.5 ± 1.6
Yagil (1985)	33–70	0–4	0–3	21–62	1–7
Chartier et al. (1986)	50–60	4–6	<1	30–40	1–2
Abdel Samee (1987)	41.5 ± 0.9	4.9 ± 0.9	0.03 ± 0	52.6 ± 0.9	1 ± 0.1
Yagoub (1988)	54.2 ± 9.5	5.4 ± 4.4	0.5 ± 0.1	37.7 ± 9.2	2.2 ± 1.6
Ibrahim et al. (1992)	47.0 ± 11	5.9 ± 4.5	0	44.4 ± 11.8	2.5 ± 1.4
Al-Ani et al. (1992)	43.3 ± 7.1	4.6 ± 1.1	0	43.2 ± 8.5	8.8 ± 1.4
Nyang'ao et al. (1997)	38.4 ± 4.0	1.0 ± 0.2	0	56 ± 5	0.5 ± 0.2
Sarwar and Majeed (1997)	36.6 ± 1.7	5.63 ± 0.4	0.61 ± 0.1	51.8 ± 1.8	4.7 ± 7.6
Rezakhani et al. (1997)	46.1 ± 16.5	6.55 ± 5.4	0	44.6 ± 16.1	2.06 ± 1.7
Tabatabaei Naeini and Nafizi (2001)	46.0 ± 5.9	9.3 ± 3.1	0.18 ± 0.2	42.1 ± 7.9	2.42 ± 2.1
Liu (2003)[a]	58.8 ± 9.3	6.76 ± 2.56	0.49 ± 0.6	31.2 ± 5.2	0.62 ± 0.7
Al-Busadah (2007)	37.45 ± 0.7	4.86 ± 1.22	0.52 ± 0.2	50.13 ± 1.7	7.0 ± 0.62
Mohammed et al. (2008)	42.6 ± 16.9	2.8 ± 2.7	0	54.0 ± 17.2	0.63 ± 0.6
Al-Sultan (2008)	49.1 ± 0.1	2.8 ± 0.6	0.69 ± 1.7	35.8 ± 0.07	7.7 ± 0.04
Faye et al. (2009)	61.4 ± 10.5	6.0 ± 4.5	0	32.1 ± 12.3	1.8 ± 0.7
Nazifi et al. (2009)	46.3 ± 11.6	6.73 ± 2.5	0	44.9 ± 12.47	2.05 ± 1.6
Aichouni et al. (2010)	41.7 ± 0.71	3.9 ± 1.2	0.03 ± 0.2	46.6 ± 5.7	7.7 ± 0.65
Farooq et al. (2011)	43.6 ± 1.3	7 ± 0.4	<0.1	48.6 ± 1.5	1 ± 0.1
Hussein et al. (2012)	43.6 ± 3.6	4.6 ± 1.2	–	43.0 ± 3.5	8.8 ± 1.4

[a]Bactrian camel

(Ghodsian et al. 1978), but other authors are not in agreement with this statement (Chartier et al. 1986; Farooq et al. 2011). It is also higher among the male according to some authors (Majeed et al. 1980; Mohammed et al. 2008).

The number of **polynuclear neutrophils** is lower, and according to the authors, their proportion varies from 0 to 1%, which is similar to that in the other species. Contrary data were reported on the effect of age on the rate of neutrophils. According to Ghodsian et al. (1978), this rate decreases in adults, while Chartier et al. (1986) did not find any effect of age. Finally, the lymphocytes/neutrophils ratio would not be the same in females compared to males, with the neutrophils being higher in females (Nassar et al. 1977; Mohammed et al. 2008).

The very high variability in the rate of **polynuclear eosinophils** (from 1.5 to 13.8%) is obviously in relation with parasitic infestations that are frequent in this species, in particular the gastrointestinal ones (Ibrahim et al. 1992), the nasal myiasis due to *Cephalopina titillator* (Hussein et al. 1983), and the trypanosomosis due to *Trypanosoma evansi* (Raisinghani et al. 1981a, b; Njiru et al. 2000). However, some other authors did not find such a change. Neutrophils, lymphocytes, and eosinophils could increase in the same proportion in camels affected by trypanosomosis (Njiru et al. 2000). At reverse, smear-positive camels presented higher neutrophils (82.2% compared to 56.2% in negative animals) and lower lymphocytes (14% compared to 36.7% in healthy animals) (Chaudhary and Iqbal 2000). Similar trend of the ratio lymphocyte/neutrophil was reported in camels having theileriosis (Youssef et al. 2015) or mange (Mal et al. 2006), but eosinophilia also increased in infected animals. Parasitism significantly changes leukocyte formula, especially a spectacular rise in the rate of eosinophils (Chartier et al. 1986; Jain 1993). The return to normal is effective in a fortnight after an anti-parasite treatment (Raisinghani and Lhoda 1980).

Also significant positive correlation was observed between eosinophilia and selenium supplementation in camels (Faye et al. 2009).

Some references did not report any effect of sex in leukocyte formula (Liu 2003; Farooq et al. 2011). For others, eosinophils could be higher in females compared to males (Majeed et al. 1980) and castrated males compared to whole ones (Gupta et al. 1979).

The rate of **monocytes** generally varies between 1 and 11.6% but seems to increase until the age of 2 years (Petrelli et al. 1982).

Seasonal variations have been reported in leukocyte formula (Majeed et al. 1980; Mohammed et al. 2008; Aichouni et al. 2011; Babeker et al. 2011). However, these seasonal variations were not always observed. For example, Ghosal et al. (1973) did not observe the summer eosinophilia, and Butt and Afzal (1992) recorded no seasonal variation in racing camel. At reverse, Mehrotra and Gupta (1989) reported a decline in the number of leukocytes during the hottest summer month. In fact, it is difficult to assess the right way in those variations, especially as a high individual variability is observed. Most of the results reported in the literature deal with small number of animals, which limits considerably the convenience of the results and their variations. Moreover, the seasons are not always delimited alike (Table 2.5).

Table 2.5 Seasonal variation of leukocyte formula according to seasons

	Neutrophils	Eosinophils	Basophils	Lymphocytes	Monocytes
Majeed et al. (1980)—India					
Summer	41.3 ± 2.8	8.5 ± 1.1	–	50.3 ± 3.1	1.7 ± 0.3
Autumn	38.9 ± 1.2	6.9 ± 0.5	–	53.3 ± 1.2	1.0 ± 0.2
Winter	38.8 ± 1.8	6.4 ± 0.7	–	53.9 ± 1.7	1.0 ± 0.2
Mohammed et al. (2008)					
Dry season	39.3 ± 15.1	3.3 ± 3.0	–	56.2 ± 16.1	0.73 ± 0.62
Wet season	47.6 ± 18.6	1.97 ± 1.8	–	50.6 ± 17.8	0.44 ± 0.61
Aichouni et al. (2011)					
Summer	36.6 ± 1.5	4.2 ± 0.4	0.04 ± 0.02	51.05 ± 1.6	8.1 ± 0.6
Winter	50.1 ± 1.6	3.9 ± 0.4	0.03 ± 0.02	38.44 ± 1.5	7.5 ± 0.6
Babeker et al. (2011)					
Rainy season (Jul)	40.4 ± 0.9	6.27 ± 0.3	1.13 ± 0.2	40.07 ± 1.0	12.2 ± 0.5
Rainy hot (Sept)	32.9 ± 1.25	8.6 ± 0.5	2.3 ± 0.15	47.10 ± 2.06	10.30 ± 0.6
Dry wet wint (Oct)	36.67 ± 0.6	8.40 ± 0.3	0.93 ± 0.15	41.6 ± 0.92	10.53 ± 0.4
Dry hot sum (Apr)	43.82 ± 0.8	4.65 ± 0.4	1.12 ± 0.15	40.24 ± 1.03	10.35 ± 0.7

In trace element-supplemented camels, the leukocyte formula was significantly changed with an increase of neutrophils (from 42.4 to 56.8%) and a decrease of lymphocytes (from 46.1 to 39.9%) without any change in the WBC (Kosanovic et al. 2014).

In an area affected with drought, a marked lymphocytopenia, monocytopenia, and eosinopenia were reported associated with an increase of neutrophil percentage (Kataria and Kataria 2004).

It seems that there is no significant change in the formula during the first month postpartum both in lactating camel and suckling baby (Elias and Yagil 1984) as well as according to the pregnancy status of the female camel (Muhammad et al. 2011). In newborn camels, the part of lymphocytes (14–32%) is lower than that in adults, while neutrophils (60–78%) are quite higher (Elkhair and Elmgboul 2013).

However, lymphocytes tend to increase with age passing from 29% in 0–1-year-old camels to 38% in adults, while the values were 58 and 51% for neutrophils in young and adult camels, respectively (Ghodsian et al. 1978). A slight age effect was also reported by Rezakhani et al. (1997), with highest neutrophils in young animals but the change for lymphocytes percentage is not linear with age.

There is few data regarding a potential breed effect; Al-Busadah (2007), Aichouni et al. (2010), and Hussein et al. (2012) did not observe any significant variation. In addition there is no clear effect of the water deprivation (Mohamed et al. 1984).

2.3 Other Hematological Parameters

Several other parameters can be analyzed for their clinical or physiological information. They are, however, less frequently requested than the blood formula in routine clinical examinations.

2.3.1 *Blood Mass and Volume*

The determination of the blood mass and the volume has no clinical interest because it requires the sacrifice of the animal.

The weight of blood collected by bleeding and reported to body weight varies from 1/28 to 1/30 kg in adult dromedaries (Boué 1948), or about 15.5 kg of blood for an animal of 370 kg live weight. This ratio is higher in young camel and obviously lower in fatty animals. These values are however different from the true blood mass, all of the blood being not collected at the time of bleeding.

Blood volume in camels is 93 ml per kg of body weight (Djegham and Belhadj 1986), which is a higher value than that observed in most other domestic species.

2.3.2 *Erythrocyte Sedimentation Rate (ESR)*

The erythrocyte sedimentation rate represents the stability of erythrocyte suspension in plasma, and its variability depends mainly on plasma proteins. The most important factor of change is the fibrinogen concentration with which the sedimentation rate is proportional. The role of the globulins and albumin is secondary. However, factors other than protein fractions also influence ESR, especially red blood cell sizes (Soliman and Shaker 1967; Gupta et al. 1979) which may strongly vary in the dromedary camel as it has been underlined previously. In sick animals, ESR increases during all inflammatory processes especially chronic ones.

Usual values proposed in the literature are variable, but the measurement conditions are often different from one author to another. Therefore, it is difficult to propose a single normative value (Table 2.6). But, as for cattle, there is a strong individual variability. For example, Gupta et al. (1979) reported an ESR from 10 to 67 mm in 24 h according to the animals. There is no breed effect on ESR (Al-Busadah 2007).

Bono et al. (1983) reported average values ranging from 0–0.7 after 1 h, 2–10.2 after 2 h, and 8–70 after 24 h. These values are on average higher than those of cattle and other ruminants in similar breeding conditions (Gupta et al. 1979). However, they are lower than those in other species (domestic carnivores and monogastrics) where red blood cells do not aggregate in rolls as in ruminants and camelids (Soliman and Shaker 1967).

Table 2.6 Erythrocyte sedimentation rate (ESR) in camel according to different authors

References	*n*	1 h	2 h	4 h	6 h	8 h	24 h
Lakhotia et al. (1964)—India	60	1.0	1.3				
Barakat and Abdel-Fattah (1970)—India	260	0.64	1.53				
Jatkar and Purohit (1971)—India	25	0.9					
Soliman and Shaker (1967)—Egypt	8	1	2.5				
Sharma et al. (1973)—India	6	0					
Gupta et al. (1979)—India	5	2.2	4.4	9.6		19	33.8
Majeed et al. (1980)—Pakistan	20	1.4				8.9	
Elias and Yagil (1984)—Israel	7	<1			6.3		18.5
Khanna (1993)—India	–					22	22.5
Al-Busadah (2007)—Saudi Arabia	60					8.1	

Age and sex appear to play a minor role. Lakhotia et al. (1964) reported a sedimentation rate of 1.05 mm in 1 h on average in camel calves vs 1.16 in adult females and 1.31 in adult males. Therefore, there is no breed effect on ESR (Al-Busadah 2007). According to Majeed et al. (1980), the sedimentation rate would be faster in females. However, the season seems to be the most important factor of change (Majeed et al. 1980; Mehrotra and Gupta 1989; Khanna 1993; Babeker et al. 2011), with the sedimentation rate being higher in the summer. The trypanosomosis is accompanied by an increase of ESR, passing from 0.9 in healthy animals to 3.5 mm/h in sick animals (Jatkar and Purohit 1971).

2.3.3 *Blood Density*

Whole blood (1.051–1.056), serum (1.024–1.026), and plasma densities (1.027–1.029) are not affected by age and sex (Lakhotia et al. 1964). During dehydration, blood density increases from 1.054 to 1.064 after 6 days of deprivation of water (Ghosal et al. 1975). This development is explained by the increase in the concentration of electrolytes and serum proteins. The values observed in the summer are usually higher than winter values (Ghosal 1971). Blood density is higher in the dromedary than that of other species (Soliman and Shaker 1967).

2.3.4 *Platelets and Clotting Time*

The number of platelets in the dromedary is on average around 300,000–400,000/ml, and their size is 1–2 μm (Yagil 1985). The number of platelets is lower in dromedary camels (230,000–360,000) than in Bactrian camels (220,000–526,000/ml) (Higgins and Kock 1984). They are smaller and also flatter than those of other mammals. Their number seems to vary depending on the season (Fig. 2.2) with high differences over

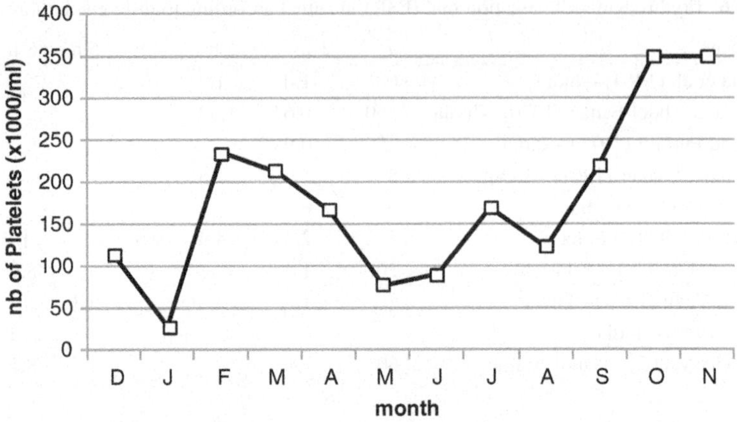

Fig. 2.2 Monthly changes in platelet number [according to Mehrotra and Gupta (1989)]

the year from 33,500 in January to 350,000 in October (Mehrotra and Gupta 1989). The platelet number seems to be higher in suckling camels (316,000–388,000/ml) compared to their dams (162,000 to 174,000/ml) (Elias and Yagil 1984).

The average clotting time, which is inversely proportional to the sedimentation rate, is 244 ± 1.7 s (4 min and 4 s). But it varies strongly according to the season, with the winter season being marked by a significant increase (Majeed et al. 1980). Compared to other species, the clotting time is longer than that for cow and sheep (Soliman and Shaker 1967).

2.3.5 Osmolality

The osmotic concentration of the plasma is around 300 mosmol/l in the dromedary. This concentration does not always change as expected in case of dehydration followed by rehydration (see below the chapter on effect of dehydration).

The dromedary and Bactrian camel have hematological characteristics that distinguish them from domestic ruminants in particular and from many mammals in general. Thus, the elliptical shape of the red blood cells, their ability to deform during sudden changes of the plasma tonicity (consequence of dehydration or massive rehydration, for example), and their resistance to hemolysis are the main features testifying to the adaptability of this species to desert life conditions.

The normal hematological profiles of the camel are as follows:

- The number of red blood cells varies between 6 and 10 × 10^6/mm^3.
- Hematocrit is between 25 and 30%.
- The hemoglobin rate is from 9.3 to 15.5 g/100 ml.
- The blood shows a predominance of polynuclear neutrophils.
- The number of leukocytes varies from 10.5 to 15.5 × 10^3/mm^3.

Table 2.7 Normal hematological parameters of dromedary and Bactrian camel from after Higgins and Kock (1984)

Parameters	*C. dromedarius*	*C. bactrianus*
Total erythrocytes ($\times 10^6$/mm^3)	7.6–11.0	8.5–13.4
Reticulocytes (%)	0–0.7	0–0.5
Hematocrit (%)	24–42	25–39
Hemoglobin concentration (g/100 ml)	11.4–14.2	11.1–17.4
Mean corpuscular volume (fl)	27.5–29.4	25.3–31.6
Mean corpuscular Nb (pg)	12.1–13.7	10.6–14.3
Mean corpuscular Hb concentration (g/dl)	42.1–49.6	37.0–47.0
Total leukocytes ($\times 10^3$/mm^3)	2.9–9.7	8.6–16.5
Neutrophils (%)	33.0–70.0	55.0–79.0
Lymphocytes (%)	21.0–62.0	18.0–33.0
Eosinophils (%)	0–4.0	0–9.0
Basophils (%)	0–3.0	0–1.0
Monocytes (%)	0–7.0	0–4.0
Erythrocyte sedimentation rate (mm/h)	0–1.0	0
Platelets ($\times 10^3$/mm^3)	230–360	220–526
Fibrinogen (mg/100 ml)	200–400	210–270

However, these parameters may strongly vary from one camel to another, both for physiological and pathological reasons already reported (Table 2.7). According to the ecology of the camel and his mode of life, his hydration status can play a considerable role in the variability of the hematological parameters. This aspect has been widely studied by many authors for a long time (cf. in particular the studies of Yagil et al. (1974a, b, c). We have already referred above the effect of dehydration/rehydration on the structure of red blood cells, but hematocrit, RBC and WBC, as well as erythrocyte Wintrobe indices can also be affected by the water imbalance. That's why a particular chapter will be devoted to the effect of dehydration and fast rehydration on hematological parameters of the camel.

2.4 The Effect of Dehydration on Hematological Parameters

The resistance of the camel to water deprivation is proverbial and is the most popular physiological feature of this species. In addition to this ability, it is also as noted the remarkable ability to ingest massive amounts of water after a long period of deprivation. Physiologically, this adaptation translates into (Richard et al. 1985):

- The ability to vary the peripheral body temperature up to 6.2 °C of gap (from 34.5 to 40.7 °C) between the hottest hours of the day and the coldest ones at night, which limits energy expenditures and water in consequence.

- A limitation on the water evaporation by the respiratory tract by maintaining a low respiratory rate in case of high temperature.
- A particular behavior toward the sun to minimize body exposure.
- A limitation of urinary, fecal, and sedative losses.
- The ability to raise its water reserves by transferring water from one fluid compartment to another (Djegham and Belhadj 1986).
- To compensate for water loss by a massive watering (representing up to 30% of body weight) at intervals which are sometimes very long.

Dehydration does not affect the health of the animals as long as it does not exceed the (very large) physiological limits of the species. Only weight loss after prolonged water fasting (25 days) can be observed (Mohamed et al. 1984). Hassan (1971) described the case of an animal having lost 32% of its body weight after 51 days of water deprivation with only an apparent clinical sign, namely, the occurrence of a capricious appetite the last week before watering.

2.4.1 Effect on RBC and Hemoglobin Rate

The number of red blood cells appears to increase during dehydration (Hassan 1971), with an increase from 8.2 to 10.9 $\times$ $10^6/mm^3$ after 50 days of water deprivation and an increase three times larger in winter (10% increase) than in summer (Khanna 1993). After rehydration, the number of red blood cells goes back to its previous level. However, these developments are not observed by all authors. Yagil et al. (1974b) did not reveal any variation after 7 days of water deprivation. On the other hand, the number of red blood cells would suddenly fall (by 31%) a few hours after a quick rehydration. Mohamed et al. (1984) did not observe, however, any changes in hematological parameters. Without comparable protocols, it is difficult to have a final opinion on the effect of the dehydration on RBC. In any case, it is important to know the dehydration status and the last drinking day of the camel to interpret the observed values of RBC.

Hemoglobin concentration should be affected by the dehydration/rehydration cycle since Hb decreases during water deprivation phase, and this decrease is accentuated during the drinking period (Yagil et al. 1974b). At reverse, Bengoumi (1992) observed a significant increase of Hb from 13.0 to 25.6 g/100 ml after 14 days of water deprivation during the hottest period of summer and then a decrease to 20.8 and 17.2 g/100 ml, respectively, after 2 and 12 h following the watering.

2.4.2 Effect on Hematocrit

As for RBC, the authors also did not agree also on the effect of dehydration/rehydration on hematocrit. Surprisingly, some authors observed a decrease in

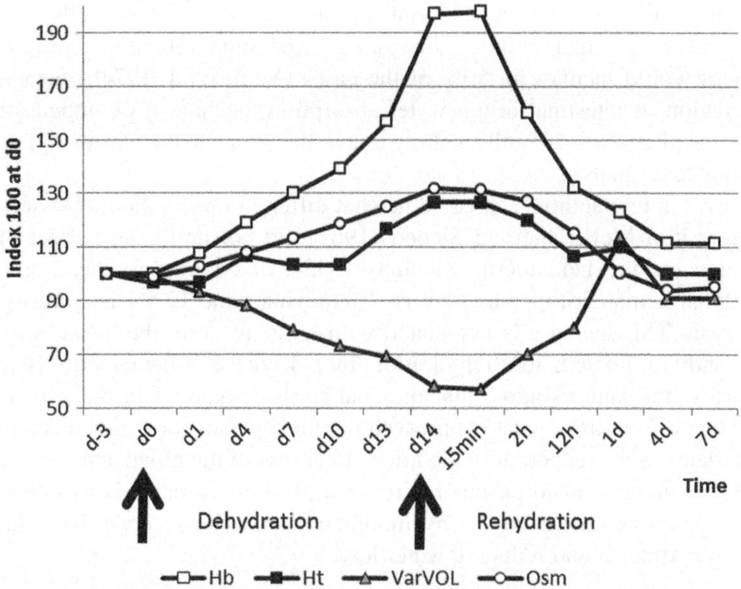

Fig. 2.3 Relative effects of 2 weeks of dehydration and then 1 week of rehydration on hematocrit (ht), hemoglobin concentration (Hb), change in plasma volume (VarVOL), and osmolality (Osm) [from after Bengoumi (1992)]

hematocrit by 4% after 6 days of water deprivation and 7% after 16 days (Ghosal et al. 1975). This decrease was also observed for hemoglobin (Yagil et al. 1974b). For Mohamed et al. (1984), the hematocrit is the only hematological parameter to be clearly affected by the status of hydration with a decrease of 3 to 4% after 1 to 2 weeks without water deprivation. Similar observations were reported by Yagil et al. (1974b) who noted a drop in hematocrit by 8.8% after dehydration. At reverse, Abdoun et al. (2010) reported an increase of PCV by 70% (passing from 20 to 34%) after 1 week of water deprivation. Similar trends were reported in young camel (Al-Haidary 2005). Bengoumi (1992) affirmed that hematocrit does not change during the first week of dehydration and then increases significantly passing from 30 to 38 after 14 days of water deprivation. After 4 days of rehydration, it comes back to normal value (30). For Bengoumi (1992), the low increase at the beginning of dehydration stage is linked to the decrease of RB size due to the cellular water loss (Yagil et al. 1976). Indeed, the decrease of plasmatic volume and the increase of osmolality (Abdoun et al. 2010) during dehydration favor transfer of water from cells to the plasma (Fig. 2.3).

The dehydration status causes a transient hypertonicity of blood (in particular by the increase of plasma sodium ion concentration) resulting in a narrowing of red blood cells and therefore reducing in their volume. Yagil (1985) also formulated the hypothesis of an increased absorption of water in case of dehydration from the digestive tract (including the intestines) to the blood, regulated by ADH and aldosterone hormones. This would cause a dilution of red blood cells and thus a relative decline in their volume. Just after rehydration (30 min approximately), hematocrit would increase in 50% of the cases (Yagil et al. 1974b) meaning that the regulation of intestinal origin water absorption mechanisms compensates only partially the phenomena of cell swelling due to the return to the plasma hypotonicity in rehydrated animals.

However, a few authors shared somewhat different observations. Hassan (1971) and before him McFerlane and Siebert (1967) did not notice any effect of water deprivation on the hematocrit. Khanna (1993), in contrast to these references, reported that hematocrit increases by 16.7% in winter and 12.5% in summer during dehydration. This increase is associated with an increase in the mean corpuscular volume and is followed, after rehydration, by a decrease of hematocrit (14.3%).

Globally, the authors agree that the change in hematocrit in the camel during dehydration differs from what is observed in other species for which water deprivation translates into a reduction of the noncellular part of the blood and, correlatively, by a relative increase in corpuscular part and thus of the hematocrit. Conversely, the camel compensates these effects by mobilizing its water reserves from the other water compartments and reducing water losses.

2.4.3 Effect on Erythrocyte Index of Wintrobe

Narrowing of the RBC in dehydration period logically leads to a reduction of the MCV (mean corpuscular volume), and conversely, their swelling during massive watering leads to a considerable increase, by more than 25% (from 26 to 33.5 fl) (Yagil et al. 1974b). Mean corpuscular hemoglobin (MCH) content decreases from 15.5 to 12.1 pg during water deprivation then returns to its normal value during rehydration, reflecting the decline in hemoglobin in dehydrated animals. In the same way, mean corpuscular hemoglobin concentration (MCHC) declines during dehydration, but unlike the previous index, it continues to decline after watering, probably as a result of the passage of water in red blood cells, contributing thus to the dilution of hemoglobin (Yagil et al. 1974b).

2.4.4 Effect on the Leukocyte Count

The effects appear even less clear on the WBC where interaction with the season is observed. Thus, for Khanna (1993), the number of white blood cells decreases by around 10% during winter dehydration and increase by 20% during summer dehydration. Moreover, whatever the season, the WBC decreases after rehydration. These variations appear in part due to transient eosinopenia during dehydration. This important decrease (by more than 60%) is attributed to "water stress" (Ghosal et al. 1975). However, the mechanisms of these contradictory variations are not clarified.

2.5 Conclusion

Globally, the dromedary camel is therefore able to lose 25% of its total body water without manifesting major symptoms of dehydration with maintaining volume at about 93 ml/kg of body weight by transferring of intracellular water to blood. Unlike the other species, the camel is able to withstand to hemoconcentration during water deprivation, observable phenomenon through the weak change of serum protein concentration and the reduction, or even maintaining the hematocrit. Thus, the interpretation of the hematological results must take into account the hydration status of the camel.

References

Abdalla OM, Wasfi IA, Gadir FA (1988) The Arabian race camel normal parameters. I. Haemogram, enzymes and minerals. Comp Biochem Physiol 90A:237–239

Abdel Samee AA (1987) Blood parameters in camel in health and disease, MVSc thesis. Faculty of Veterinary Medicine, Cairo University, Cairo

Abd El-Baky AA, Salem SI (2011) Clinicopathological and cytological studies on naturally infected camels and experimentally infected rats with *Trypanosoma evansi*. World appl Sci J 14:42–50

Abdo MS, Ali A, Prentis PF (1990) Fine structure of camel erythrocytes in relation to its functions. Z Mikrosk Anat Forsch 104:440–448

Abdoun KA, Alameen A, Elmagbol W, Makkawi T, Al-Haidary A (2010) Effects of hydration status on osmolality and minerals profile of serum and forestomach liquor in dromedary camel (*Camelus dromedarius*). J Camel Pract Res 17(2):235–240

Abdoun KA, Samara EM, Okab AB, Al-Haidary AI (2012) A comparative study on seasonal variation in body temperature and blood composition of camels and sheep. J Anim Vet Adv 11 (6):769–773

Adam SEI, Obeid HM, Ashour N, Tartour G (1974) Serum enzyme activities and haematology of normal and diseased ruminants in the Sudan. Acta Vet Brno 43:225–231

Ahmadi-Hamedani A, Ghazvinian K, Darvishi MM (2014a) Hematological and serum biochemical aspects associated with a camel (*Camelus dromedarius*) naturally infected by *Trypanosoma evansi* with severe parasitemia in Semnan, Iran. Asian Pac J Trop Biomed 4(9):743–745

Ahmadi-Hamedani M, Ghazvinian K, Kokhaei P, Barati M, Mahdavi A (2014b) Comparison of effects of age and sex on serum protein electrophoretic pattern in one-humped camels (*Camelus dromedarius*) in Semnan, Iran. Open Vet J 4(1):4–8

Aichouni A, Jeblawi R, Dellal A, Hammou H, Aggad H (2010) Breed variation in blood constituents of the one-humped camel (*Camelus dromedarius*) in Algeria. J Camelid Sci 3:19–25

Aichouni A, Dellal A, Jebmawi R (2011) Influence de la saison sur les paramètres hématologiques du dromedaire (*Camelus dromedarius*) algérien. Rev Med Vet 162(7):327–332

Al-Ani FK, Al-Azzawi WARA, Jermukly MS, Razzaq KK (1992) Studies on some haematological parameters of camel and llama in Iraq. Bull Anim Prod Afr 40:103–106

Al-Busadah KA (2007) Some biochemical and haematological indices in different breeds of camel in Saudi Arabia. Sci J King Faisal Univ 8(1):131–142

Alhadrami GA (1997) Comparative haematology in the camel calf and adult racing camel (*Camelus dromedarius*). J Camel Pract Res 4(1):13

Al-Haidary A (2005) Effect of dehydration on core body temperature of young Arabian camel (*Camelus dromedarius*). J King Saud Univ Agric Sci 18(1):1–7

Ali BH, El Sanhouri AA, Musa BE (1989) Some clinical, haematological and biochemical effects of four tranquilizers in camels (*Camelus dromedarius*). Rev Elev Méd Vét Pays Trop 42:13–17

Ali A, Refaat D, Tharwat M, Al-Sobayil F (2015) Impotentia generandi in male dromedary camels: breeding soundness, haematology, biochemistry and testosterone level. Comp Clin Pathol 24:727–731

Al-Mujalli MA (2007) Hematological and biochemical changes in camels affected by trypanosomiasis in Saudi Arabia. Vet Med J Giza 55(1):155–159

Al-Mujalli AM, Al-Naeem AA, Al-Ghamdi G, Al-Swailem A, Al-Yamani E, Shehata AM, Al-Dubaib MA, Hashad M, El-Lithy DA, Mahmoud OM, Alfayez M (2011) Cellular and biochemical blood profile in camels suffering from dubduba syndrome. Sci J King Faisal Univ (Basic Appl Sci) 12(2):165–172

Al-Qarawi AA, Mousa HM (2004) Lipid concentrations in erythrocyte membranes in normal, starved, dehydrated and rehydrated camels (*Camelus dromedarius*), and in normal sheep (*Ovis aries*) and goats (*Capra hircus*). J Arid Environ 59:675–683

Al-Sultan SI (2008) Studies on hematological and certain serum biochemical values in young Magaheim dromedary camel at Al-Ahsa province. Vet Res 2(3–4):34–37

Al-Sultan SI, El-Bahr SM (2012) Effect of high zinc dose on haematological, lipids and protein patterns in dromedary camel. J Camel Pract Res 19(2):207–212

Athar Khan M, Sarwer Khan M, Ashfaq-Ul-Rehman (1992) The incidence and effects of coccidiosis on certain blood parameters in camels of Pakistan. In: International camel conference, Dubaï (UAE), 2–6 Feb 1992. R&W Publications, England, p 406

Auer R, Gleiß A, Windberger U (2015) Towards a basic understanding of the properties of camel blood in response to exercise. Emir J Food Agric 27(3):302–311

Ayoub MA, Saleh AA (1998) A comparative physiological study between camels and goats during water deprivation. In: Alhadrami G, Ayoub MA (eds) Proceedings of the 3rd annual meeting for animal production under arid conditions, 2–3 May 1998, vol 1. Al-Ain, UAE, pp 71–87

Babeker EA, Elmansoury YHA, Suleem AE (2011) The influence of seasons on blood constituents of dromedary camel (*Camelus dromedarius*). Online J Anim Feed Res 3(1):1–8

Babeker EA, Elmansoury YHA, Suleem AE (2013) The influence of seasons on blood constituents of dromedary camel (*Camelus dromedarius*). Online J Anim Feed Res 3(1):1–8

Banerjee S, Bhattarchajee RC, Singh TI (1962) Haematological studies in the normal Indian camel (*Camelus dromedarius*). Am J Phys 203:1185

Barakat MZ, Abdel-Fattah M (1970) Biochemical analysis of normal camel blood. Zbl Vet Med A 17:550–557

Bengoumi M (1992) Biochimie clinique du dromadaire et mécanismes de son adaptation à la deshydratation, Thèse Doct. Sciences agronomiques, IAV Hassan II, Rabat, Maroc

Bono G, Mohamed AS, Dahir AM (1983) Raccolta comparativa dei principali parametri fisiologico-clinici del dromedario (*Camelus dromedarius*). Boll Sci Fac Zoot Vet Univ Naz Somalia 4:51–85

Boué A (1948) La résistance des hématies à l'hémolyse chez le chameau. Rev Elev Méd Vét Pays Trop 2:118

Butt RS, Afzal M (1992) Seasonal variations in haematological and serum biochemical parameters in racing camels. In: International camel conference, Dubai, UAE, 2–6 Feb 1992. R&W Publications, England, 413 p

Chartier C, Chartier F, Lepers JP, Pesce JL (1986) Etude préliminaire de quelques paramètres sanguins usuels du dromadaire Mauritanien (*Camelus dromedarius*). Rev Elev Méd Vét Pays Trop 39:395–401

Chaudhary ZI, Iqbal J (2000) Incidence, biochemical and haematological alterations induced by natural trypanosomosis in racing dromedary camels. Acta Trop 77(2):209–213

Cohen WD, Terwilliger NB (1979) Marginal bands in camel erythrocytes. J Cell Sci 36:97–107

Djegham M, Belhadj O (1986) Comportement de thermorégulation et résistance à la privation d'eau chez le dromadaire. Variations saisonnières des profils biochimiques et hématologiques chez le dromadaire. Maghreb Vétérinaire 10:11–16

Durand M, Kchouk M (1959) Quelques constantes hématologiques chez le dromadaire tunisien. Arch Inst Pasteur Tunis 36:183–194

Elias E, Yagil R (1984) Haematological and serum biochemical values in lactating camels (*Camelus dromedarius*) and their newborn. Refuah Vet 41(1):7–13

Elkhair NM, Elmgboul WE (2013) Haematological parameters of neonate camels (*Camelus dromedarius*) during the first week of life. In: Al-Tariki Y, Al-Abdullrahman A, Al-Naeem A (eds) Proceedings of the "conference of sustainability of camel population and production (CSCPP)", College of Agricultural and Food Sciences, 17–20 Feb 2013. Al-Ahsa University, Saudi Arabia, pp 290–295

El-Mougy SA, El-Megaury S, Ismail AA, Radwan VM, El-Nagar EL (1984) Effect of breeding season on LH, testosterone and sole blood parameters in male one humped camels. In: Proceedings of 1st international conference on physiology science, Cairo, Egypt, pp 1–18

Eyob E, Matios L (2013) Review on camel trypanosomosis (Surra) due to *Trypanosoma evansi*: epidemiology and host response. J Vet Med Anim Health 5:334–343

Farooq U, Samad HA, Khurshid A, Sajjad S (2011) Normal reference hematological values of one-humped camels (*Camelus dromedarius*) kept in Cholistan desert. J Anim Plant Sci 21 (2):157–160

Fatihu MY, James AS, Abubakar MB (2000) Hematological data of healthy and natural *Trypanosoma evansi* infected camels (*Camelus dromedarius*) in Sokoto. Nigeria J Prot Res 10(1):1–5

Faye B, Grillet C, Tessema A (1986) Teneur en oligo-éléments dans les fourrages et le plasma des ruminants domestiques en Ethiopie. Rev Elev Méd Vét Pays Trop 39:227–237

Faye B, Seboussi R, Alhadrami G (2009) Meta-analysis of the interactions between selenium status and haematological and mineral parameters in camel blood. Trends Comp Biochem Physiol 14:25–34

Ghodsian I, Nowrouzian I, Schels HF (1978) A study of some haematological parameters in the Iranian camel. Trop Anim Health Prod 10:109–110

Ghoke SS, Jadhav KM, Thorat KS (2013) Assessing the osmotic fragility of erythrocytes of rural and semiurban camels (*Camelus dromedarius*). Camel Int J Vet Sci 1(1):75–78

Ghosal AK (1971) Water metabolism in camel (*Camelus dromedarius*). PhD thesis, V-daipm University, India

Ghosal AK, Appana TC, Dwaraknath PK (1973) Seasonal variations in the blood constituents of the Indian camel (*Camelus dromedarius*). Indian J Anim Sci 43:642–644

Ghosal AK, Appana TC, Dwaraknath PK (1975) A note on the effect of short-term water deprivation on certain blood characteristics in the camel (*Camelus dromedarius*). Indian J Anim Sci 45:105–108

Gupta GC, Joshi BP, Rai P (1979) Observations on haematology of camel (*Camelus dromedarius*). Indian Vet J 56:269–272

Hajduk P (1992) Haematological reference values for alpacas. Aust Vet J 69:89–90

Hassan YM (1968) Some of the physiological factors affecting milk production in the Sudan. Sudan J Vet Sci & Anim Husb 9:29–34

Hassan YM (1971) A note on the effect of dehydration on a camel. Sudan J Vet Sci Anim Husb 12:111–112

Higgins AJ, Kock RA (1984) The camel in health and disease. I. A guide to the clinical examination, chemical restraint and medication of the camel. Br Vet J 140:485–504

Hussain M, Saeed Z, Gulsher M, Shaikh RS, Ali M, Ahmad AN, Hussain I, Akhtar M, Iqbal F (2016a) A report on the molecular detection and seasonal prevalence of Trypanosoma brucei in dromedary camels from Dera ghazi Khan District in southern Punjab (Pakistan). Trop Biomed 33(2):268–275

Hussain R, Khan A, Abbas RZ, Ghaffar A, Abbas G, Rahman TU, Ali F (2016b) Clinico-Hematological and biochemical studies on naturally infected camels with trypanosomiasis. Pak J Zool 48(2):311–316

Hussein MF, Hassan H, Bilal HK, Basmae'il SM, Younis TM, Al Motlaq AR, Al-Sheikh MA (1983) *Cephalopina titillator* (Clark 1797) infection in Saudi Arabian camels. Zbl Vet Med B 30:553–558

Hussein YA, Al-Eknah MM, Al-Shami SA, Mandour MA, Fouda TA (2012) Coat color breed variation in blood constituents among indigenous Saudi Arabia camel strains. Mansoura Vet Med J 14(1):191–204

Ibrahim AB, Abdel Gaffar AA, Gameel AA, Nayel NM, Abdel Gaffar A, El Galaini M (1992) A note on the haemogram of the dromedary camel in Bahrain. Rev Elev Méd Vét Pays Trop 45:318–320

Igbokwe IO (1994) Mechanisms of cellular injury of African trypanomiasis. Vet Bull 64:611–620

Jain NC (1993) Essentials of veterinary hematology. Lea and Febiger, Philadelphia, PA, p 417

Jatkar PR, Purohit MS (1971) Pathogenesis of anaemia in *Trypanosomia evansi* infection. Part I Haematology. Indian Vet J 48:239–244

Kataria N, Kataria AK (2004) Use of blood analytes in assessment of stress due to drought in camel. J Camel Pract Res 11(2):129–133

Khan AA, Kohli S (1978) A note on some haematological studies on male camel (*Camelus dromedarius*) before and during rut. Indian J Anim Sci 48:325–326

Khanna D (1993) Adaptative responses of camel to dehydration and rehydration following water restriction. PhD Thesis, Ajmer University, India, 307 p

Knight PK, Cluer D, Manefield GW, Gorde AK (1994) Haematology in the racing camel at rest: seasonal and training variations. In the racing camel. Physiology, metabolic functions and adaptations. Acta Physiol Scand 150(suppl):19–23

Kosanovic M, Rao MV, Al-Shamisi NS, Tinson A (2014) Effect of supplementation with trace elements (copper, cobalt, zinc and selenium) on blood cell count in camels (*Camelus dromedarius*). J Trace Elem Anal 3(1):23–31

Lakhotia R, Bhargava AK, Mehrothra PN (1964) Normal ranges for some blood constituents of the Indian camel. Vet Rec 76:121–122

Lektib I, Bargaâ R, Chakir Y, Belhouari A, Hammoumi A, El Khasmi M (2016) Study of incubation conditions for erythrocytes osmotic fragility testing in dromedary camel (*Camelus dromedarius*). Int J Res Environ Sci 2(2):22–32

Li WX, Hai WJ (2014) Studies of sulphur-induced copper deficiency in the Bactrian camel. J Anim Vet Adv 13(14):886–890

Liu ZP (2003) Studies of the haematology and trace element status of adult Bactrian camels (*Camelus bactrianus*) in China. Vet Res Commun 27:397–405

Long CA (2007) Evolution of function and form in camelid erythrocytes. In: Proceedings of the 2007 WSEAS international conference on cellular and molecular biology—biophysics and bioengineering, Athens, Greece, 26–28 Aug 2007, pp 18–24

MacFerlane WV, Siebert DD (1967) Hydration and dehydration of desert animals. Aust J Exp Biol Med Sci 45:29

Majeed MA, Hur G, Rahman Z, Ahmad A (1980) Effect of sex and season on 10 haematological values of normal adult one-humped camel. Rev Elev Méd Vét Pays Trop 37:313–317

Mal G, Sena S, Sahani MS (2006) Haemato-biochemical changes in camels infested with mange during winter and summer season. J Camel Pract Res 13(2):173–174

Mandl E (1838) Globules du sang de forme elliptique observés chez deux espèces de mammifères. CR Acad Sci 7:1060

Mathur M, Dadhich H, Khare S, Singh AP, Meena SS (2012) Pathological and hemato-biochemical observations of perivascular dermatitis in camel (*Camelus dromedarius*). Vet Pract 13(1):69–71

Mehrotra V, Gupta ML (1989) Seasonal variations in certain blood constituents in camel. Indian J Anim Sci 59:1559–1561

Mezzou H, Khelifa AB, Neffati F, Douki W, Amor AB, Najjar MF (2006) Détermination de l'hémoglobine plasmatique et évaluation spectrophotométrique de l'hemolyse en biochimie clinique. Rev Francoph Lab 386:59–64

Mohamed MH, Mohamed AH, Locatelli A (1984) Water deprivation effects on the hematological and hematochemical pictures of *Camelus dromedarius*. Rev Elev Méd Vét Pays Trop 37 (3):313–317

Mohammed AK, Sackey AKB, Tekdek LB, Gefu JO (2008) Mean haematological characteristics of healthy adult one humped camel (*Camelus dromedarius*) introduced into a sub-humid climate in Nigeria. J Camel Pract Res 15(2):187–190

Muhammad BF, Aliyu D, Njidda AA, Madigawa IL (2011) Some haematological, biochemical and hormonal profile of pregnant and non-pregnant she-camels (*Camelus dromedarius*) raised in a Sudan savanna zone of Nigeria. J Camel Pract Res 18(1):73–77

Musa BE, Mukhtar AM (1982) Studies on the normal haemogram and some blood electrolytes in camels (*Camelus dromedarius*). Sudan J Vet Sci Anim Husb 23:38–43

Mutugi MW, Olaho-Mukani W, Kuto B, Alushula H, Njogu AR (1993) Haematological characteristics of the one-humped camel, *Camelus dromedarius*, in Kenya. Bull Anim Prod Afr 41:181–184

Naeini AT, Nafizi S (2001) Biochemical and cytological properties of blood and peritoneal fluid in clinically healthy adult camels (*Camelus dromedarius*). J Camel Pract Res 8(2):123–126

Nassar SM, Mansour SA, Lofti LA (1977) Influence of sex on the normal blood picture of adult Egyptian camel (*Camelus dromedarius*). Assiut Vet Med J 4:43–50

Nazifi S, Oryan A, Bahrami S, Razavi SM (2009) Evaluation of haematological and serum biochemical parameters in Iranian camels (*Camelus dromedarius*) infected with haemotrophic mycoplasma (Eperythrozoon) spp. Comp Clin Pathol 18:329–332

Njiru ZK, Olaho-Mukani W, Khaemba BM, Ochieng RS, Ndung'u JM (2000) Haematological and serological changes during acute *Trypanosoma evansi* infection in dromedary camel (*Camelus dromedarius*). J Camel Pract Res 7(1):113–116

Nyang'ao JMN, Olaho-Mukani W, Maribei JM, Omuse JK (1997) A study of some haematological and biochemical parameters of the normal dromedary camel in Kenya. J Camel Pract Res 4 (1):31–33

Omer SA, Khougali SME, Agab H, Samad GHA (2016) Studies on some biochemical and haematological indices of Sudanese camels (*Camelus dromedarius*). http://www.sustech.edu/staff_publications/20090615152717900.pdf

Oyewale J, Dzenda T, Yaqub L, Akanbi D, Ayo J, Owoyele O, Minka N, Dare T (2011) Iterations in the osmotic fragility of camel and donkey erythrocytes caused by temperature, pH and blood storage. Vet Archiv 81(4):459–470

Padheriya YD, Wadhwani KN (2013) Blood profile of Kachchi camel during exercise. J Camel Pract Res 20(2):265–269

Padmaja K (2012) Haemato-biochemical studies and therapy of trypanosomiasis in camels. Vet World 5:356–358

Pandey MD, Roy A (1969) Studies on the adaptability of buffalo to tropical climate. Indian Vet J 43:795–800

Pegram RG, Scott JM (1976) The prevalence and diagnosis of *Trypanosoma evansi* infection in camels in southern Ethiopia. Trop Anim Health Prod 8:20–27

Perk K (1966) Osmotic hemolysis of the camel's erythrocytes. I. A microcinematographic study. J Exp Zool 163(3):241–246

Persson SGB (1967) On blood volume and working capacity of horses. Acta Vet Scand Suppl 19:461–476

Peshin PK, Kumar A, Singh S (2010) Effect of thiopentone on osmotic fragility of erythrocytes in camels. J Camelid Sci 3:62–65

Petrelli F, Dahir AM, Mohamed AS, Moretti P (1982) Blood values in clinically normal African camels (*Camelus dromedarius*) of various age. Boll Scient Fac Zootech Vet Univ Naz Somalia 3:133–137

Raisinghani PM, Lhoda KR (1980) Prognostic values of some haematological and biochemical parameters of camels affected with Surra following the treatment with antrypol, antrycide methyl sulphate, naganol and berenil. Indian Vet J 57:479–484

Raisinghani PM, Lhodha KR, Bathia JS, Dwaraknath PK, Vyas U (1981a) Note on some haematological and biochemical studies in experimental Surra in camel. Indian J Anim Sci 51:1108–1112

Raisinghani PM, Lhodha KR, Bathia JS, Dwaraknath PK (1981b) Variation in haematological and serum electrolyte levels during first 20 bouts of experimental Surra in camels. Indian J Anim Sci 51:724–729

Rezakhani A, Nazifi Habibabadi S, Maghrebi Ghojogh M (1997) Studies on normal haematological and biochemical parameters of Turkmen camel in Iran. J Camel Pract Res 4(1):41–44

Richard D, Hoste CH, Peyre de Fabregues B (1985) Le dromadaire et son élevage. Coll. Etudes et synthèses. IEMVT, Maisons-Alfort, 161 p

Rose RJ, Evans DL, Henckel P, Knight PK, Cluer D, Saltin B (1994) Metabolic responses to prolonged exercise in the racing camel. In: Saltin B, Rose RJ (eds) The racing camel. Physiology, metabolic functions and adaptations. Acta Physiol Scand Suppl 150(s.617):49–60

Saleh MA, Al-Salahy MB, Sanousi SA (2009) Oxidative stress in blood of camels (*Camelus dromedaries*) naturally infected with *Trypanosoma evansi*. Vet Parasitol 162:192–199

Saleh MA, Mahran OM, Al-Salahy MB (2011) Circulating oxidative stress status in dromedary camels infested with sarcoptic mange. Vet Res Commun 35(1):35–45

Salman R, Afzal M (2004) Seasonal variations in hematological and serum biochemical parameters in racing camels. J Camel Sci 1:63–65

Sarwar A, Majeed MA (1997) Interrelationships between 30 parameters of blood in normal one-humped camel in summer. J Camel Pract Res 4(1):35–39

Schalm OW, Jain NC, Carroll EJ (1975) Veterinary hematology, 3rd edn. Lee and Febiger, Philadelphia, 807 p

Sergent E, Poncet A (1942) Etude morphologique du sang des dromadaires sahariens. Arch Inst Pasteur Algérie 20:204–208

Sharma DP, Malik PD, Sapra KL (1973) Age-wise and species-wise haematological studies in farm animals. Indian J Anim Sci 43:289–295

Singh G, Singh Y, Nagpal SK (1997) Ultrastructure of formed elements of blood of camel (*Camelus dromedarius*). J Camel Pract Res 4(1):1–12

Snow DH (1992) An introduction to the racing camel. In: International camel conference, Dubai, UAE, 2–6 Feb 1992. R&W Publications, England, pp 215–217

Snow DH, Billah A, Ridah A (1988) Effects of maximal exercise on the blood composition of the racing camel. Vet Rec 123:311–312

Soliman MK, Shaker M (1967) Cytological and biochemical studies on the blood of adult she camel. Vet Rec 123:311–312

Soni BK, Aggarwal AC (1958) Studies in the physiology of the camel (*Camelus dromedarius*). Part I. Cellular blood constituents. Indian Vet J 35:209–214

Tabatabaei Naeini A, Nafizi S (2001) Biochemical and cytological properties of blood and peritoneal fluid in clinically healthy adult camels (*Camelus dromedarius*). J Camel Pract Res 8(2):123–126

Varia T, Prajapati A, Raval S, Patel S, Chauhan V, Pandey S (2015) Haematological alterations in trypanosoma infection in camels of Gujarat. Life Sci Leaflets 69:40–43

Yagil R (1985) The desert camel: comparative physiological adaptation. Karger, Basel, 163 p

Yagil R, Sod-Moriah VA, Meyerstein N (1974a) Dehydration and camel blood. I. Red blood cell survival in the one-humped camel (*Camelus dromedarius*). Am J Phys 226:298–300

Yagil R, Sod-Moriah VA, Meyerstein N (1974b) Dehydration and camel blood. II. Shape, size and concentration of blood cells. Am J Phys 226:301–304

Yagil R, Sod-Moriah VA, Meyerstein N (1974c) Dehydration and camel blood. III. Osmotic fragility, specific gravity and osmolality. Am J Phys 226:305–308

Yagil R, Etzion Z, Berlyne GM (1976) Effect of dehydration on the urea nitrogen metabolism in the camel. Isr J Med Sci 12:1514

Yagoub IA (1988) Studies on haematological parameters and some biochemical blood constituents of Sudanese camels (*Camelus dromedarius*). Acta Vet 38:99–106

Youssef SY, Yasien S, Ali Mousa WH, Nasr SM, El-Kelesh EAM, Mahran KM, Abd-El-Rahman AH (2015) Vector identification and clinical, hematological, biochemical, and parasitological characteristics of camel (*Camelus dromedarius*) theileriosis in Egypt. Trop Anim Health Prod 47:649–656

Zeidan AEB, Abbas HE (2003) Physio-biochemical changes in the male dromedary camels during breeding (rutting) and non-breeding season. J Camel Pract Res 10(2):183–190

Zia-ur-Rahman M (1992) Serum biochemical, enzymes and haematological changes in one-humped camels infected with Surra. In: International camel conference, Dubai, UAE, 2–6 Feb 1992. R&W Publications, 405 p

Chapter 3
Energetic Parameters

Camels (*Camelus dromedarius* and *C. bactrianus*) as the other domestic ruminants take the main part of their energy from volatile fatty acids (VFA) produced in the rumen from the degradation of carbohydrates. Dry matter, fibers, and digestion in the dromedary camel are similar to that observed in the other ruminants with better digestion of crude fibers (Rutagwenda et al. 1990; Heller et al. 1986). Digestion of crude fiber increases during dehydration.

The main volatile fatty acids found in the rumen of camels are acetic acid (77%), propionic acid (16%), and butyric acid (7%). Ketogenesis is very low in dromedary camels and is mainly located in the kidney and secondary in the liver (Emmanuel 1980, 1981). The absorption rate of VFA is two to three times higher in camels than in sheep and goat (Maloiy and Clemens 1980).

Despite the higher degradation of glucose into VFA in the rumen (Valleras and Stevens 1971), the plasma concentration of glucose is higher in camels than in the other ruminants (Chandrasena et al. 1979a). The major part of the plasma and tissue glucose is biosynthesized via gluconeogenesis, which is very active in the camel. Biosynthesis of glucose from pyruvic acid, glutamic acid, and propionic acid is, respectively, two times and three times more active in the kidney and liver of camels than in those of sheep (Emmanuel 1981).

The dromedary stores its energy reserves mainly in the hump (Mirgani 1977a; Kamili et al. 2006; Faye et al. 2012) and abdominal fat (Emmanuel and Nahapetian 1980). Lipid composition of the mesentery of camels, determined by the technique of crystallization at low temperatures, showed a predominance of palmitic (29%), stearic (27%), and oleic (26%) acids (Mirgani 1977a; Kadim et al. 2002). That of the hump, via thin-layer chromatography (TLC) and gas-liquid chromatography (GLC), revealed that triglycerides (TG) are the major component with only traces of phospholipids. Fatty acid composition of the hump includes palmitic acid (35.3%), followed by stearic acid (25.6%) and oleic acid (23.6%). Saturated fatty acids account for 74% of total fatty acids (Mirgani 1977a). Other studies have shown that saturated fatty acid contents in the hump are 60.2% (Orlov et al. 1985), 60.5% (Rawdah et al. 1994), and 63.4% (Kadim et al. 2002). The same rate was reported for

© Springer International Publishing AG, part of Springer Nature 2018

B. Faye, M. Bengoumi, *Camel Clinical Biochemistry and Hematology*,
https://doi.org/10.1007/978-3-319-95562-9_3

abdominal fat (63.6%) (Emmanuel and Nahapetian 1980). The hump's fat has lower polyunsaturated fatty acid (PUFA) content (Emmanuel 1981; Orlov et al. 1985; Rawdah et al. 1994).

During growth, there are changes in the proportions of fat, water, and connective tissue (Wood 1984). The difference between the saturated fatty acid and unsaturated fatty acid percentages varies with age and site of adipose deposition. According to Kadim et al. (2002), there is no significant difference between fatty acid composition of the hump and abdominal fat, and the age has no effect on fatty acid composition.

The weight of the hump is not correlated with the size of its adipocytes. No significant correlation was established between the weight of the hump and the number of adipocytes (Faye et al. 2001a).

3.1 Regulation of the Energetic Metabolism

Carbohydrate metabolism is controlled by several hyperglycemic hormones (glucagon, adrenaline, growth hormone, glucocorticoids, etc.) and insulin, the only hypoglycemic hormone.

Body fat distribution changes are not well known in dromedaries. Physical exercises and nutritional and physiological status (age, sex, pregnancy, and castration) have a considerable effect (Congiu 1953; Chilliard 1989). When the animal restores its reserves, the hump size increases, and the percentage of body water decreases. The distribution of fat in different anatomical parts seems rather to be due to differences in the number of adipocytes, than to variations of the size of the cells (Faye et al. 2001b).

The ability of ruminants to mobilize and replenish their lipid reserves is widely used in intensive livestock systems. When animals are underfed, they mobilize their body reserves, which are restricted during unfavorable seasons, allowing them to start a new breeding cycle. Lipid metabolism is reactive and well regulated, to provide the energy needed for survival and production. This regulation is homeostatic in the short term, linked to nutritional status and changes in the environment. In the long term, it is linked to the physiological state. The establishment of long-term homeostasis, provided by feedback control mechanisms, promotes the return to normal state before any change in lipid reserves (Chilliard et al. 2000). For this purpose the body involves a certain shadow of metabolic and endocrine factors, the regulation of which depends on the nutritional state, the physiological stage, the species, and the anatomical site.

Dromedaries are known for their ability to withstand fasting for long periods, and to use energy efficiently. Their basic energy requirements are low (314 kJ/kg$^{0.75}$) (Guerouali and Filali 1992). This value accounts for only two-thirds of cattle requirements (NRC 2001). This ability to fast and use energy as efficiently as possible may be due to behavioral and physiological features (Wilson 1984), but biochemical mechanisms may also contribute to this.

The daily intake of the fat, contained in fodder and seeds, varies depending on the species, the vegetative stage, and the amounts ingested (Wilson 1984; Faye 1997).

Polyunsaturated fatty acids (PUFA) are fully hydrogenated in the digestive reservoirs and processed into C18:0 that are absorbed and incorporated into the serum triglycerides (TG) with the other food fatty acids (especially C16:0). Camel lipoproteins are mainly light ones; the high-density lipoproteins are practically nil. Hepatic lipids are rich in TG and poor in phospholipids (Mirgani 1977a) as for a cow with a hepatic steatosis (Mazur et al. 1986). Fatty acids (C14 and C16) are the major constituents of hepatic TG, whereas C18:1 has only a minor portion (Mirgani 1977a, 1981). The liver of the other ruminants has an important lipogenic activity (Chilliard 1987); however, that of the dromedary has comparable lipogenic activities to that of the hump (Mirgani et al. 1987).

Insulin, glucagon, thyroxine, growth hormone, insulin-like growth factor I, prolactin, and leptin are the main hormones controlling lipid metabolism in the ruminants. In the dromedary camel, few data are available on the role of these hormones in the regulation of lipid metabolism (see Chap. 9). Leptin, mainly secreted by adipose tissue, plays a key role in the regulation of energy metabolism and body reserves (Chilliard et al. 2005). Leptin and its receptors were characterized in the adipose tissue, mammary gland, and liver of one-humped camel, and the sequenced cDNA was closely related to those of cow and water buffalo (Sayed-Ahmed et al. 2003; Bartha et al. 2005). Leptin was positively correlated in young camel to back-fat thickness (Al-Azraqi 2007) and differentially affected by season depending on the sex without apparent relation to insulin or glucose regulation (Al-Suhaimi et al. 2009). In experiments with different energy intakes and dehydration in camels, plasma leptin concentration was positively related to both energy intake level and hump adipocyte size, but to a lower extent than for other ruminant species (Bengoumi et al. 2005; Delavaud et al. 2013).

3.2 Energetic Blood Parameters

Several energetic blood parameters are regularly analyzed including carbohydrates as glucose, ketone bodies like β-hydroxybutyrate and lipids as triglycerides, non-esterified fatty acids (NEFA), and even cholesterol, which has no energetic role. Due to the effect of blood cell metabolism and some serum enzymes, it is recommended to use special anticoagulant tubes containing fluoride for inhibiting glycolysis. For lipids, it is recommended to use serum because of the effect of heparin on lipoproteins. In addition, serum or plasma should be stored at +4 °C or frozen at −20 °C if the analysis will not be performed on the same day.

3.2.1 *Glucose*

Plasma glucose concentration varies according to feeding status with a decrease during underfeeding. It increases during stress and dehydration as it was described for long time (Emmanuel 1979; Chandrasena et al. 1979a; Mohamed and Hussein 1999). The units used by different authors are mmol/l or mg/100 ml or g/l. For coherent comparison, all the values of literature are converted below in mg/100 ml. Plasma glucose concentration is measured using glucose oxidase method.

3.2.1.1 Normal Values

Plasma concentration in camels varies between 3.3 and 7.7 mmol/l (60–140 mg/100 ml) (Table 3.1). It is higher than that in the domestic ruminants which ranges from 2.5 to 4.5 mmol/l (45–80 mg/100 ml) and similar to that in monogastric domestic animals (4–7 mmol/l, i.e., 70–126 mg/100 ml) and llamas (4.7–8.3 mmol/l, i.e., 85–150 mg/100 ml) (Bengoumi 1992; Kaneko et al. 1997). The interpretation of its variability must be based on the feeding status of the animal, its stressing conditions, and other energetic parameters. Blood sampling itself, as stress factor, could influence the results. The values may change within a day with higher values in the morning (116 ± 8.7) than in the afternoon (105 ± 6.7 mg/100 ml) (Saeb et al. 2010).

3.2.1.2 Effect of Physiological Factors

Age Effect

Several authors confirmed that plasma glucose concentration is higher in young animals than adults (Elias and Yagil 1984; Bengoumi 1992). This is linked to the high consumption of milk's lactose, high concentration of growth hormone, and mainly stress during sampling. For example, Faye and Mulato (1991) reported 73.4 ± 23.7 in young camel and 58.5 ± 25.4 mg/100 ml in adult camels. The same results were reported in young other domestic ruminants before starting rumination (6 weeks) which starts in camel after 2 weeks. Ghodsian et al. (1978) do not report any age effect on plasma glucose concentration in camels. Al-Ali et al. (1988) also reported very high glycemia (138 ± 17.7 mg/100 ml) in 2–3-year-old camels. At reverse, Rezakhani et al. (1997) found higher but not significant glycemia in adult camels (more than 6 years old) than those of less than 3 years, respectively, 79.9 ± 25.3 and 68.7 ± 20.9 mg/100 ml. Souilem et al. (1999) reported similar results with lower values in young camels less than 2 years (71 ± 4 mg/100 ml) than in adults (111 ± 3 mg/100 ml).

Sex Effect

The effect of sex on plasma glucose concentration in camels is contradictory. Some authors (Faye and Mulato 1991) reported lower values in females (60 ± 25 mg/100 ml) than in males (78 ± 7 mg/100 ml) when others reported lower values in males: 106 ± 16 vs 112 ± 10 mg/100 ml (Barakat and Abdel Fattah 1971).

Table 3.1 Normal glycemia (in mg/100 ml) in camel according to different authors

References	Values	n	Country
Chavanne and Boué (1950)	91–104	–	North Africa
Durand and Kchouk (1958)	81–84	–	Tunisia
Kumar and Banerjee (1962)	110 ± 11.4	10	India
Soliman and Shaker (1967)	80	80	Egypt
Barakat and Abdel-Fattah (1970)	260	104 ± 0.9	Egypt
Maloiy (1972)	74–102	6	Kenya
Ghodsian et al. (1978)	107–109	99	Iran
Chandrasena et al. (1979a)	129	20	Iran
Abdelgadir et al. (1979)	49.8 ± 8.16	96	Sudan
Orliac (1980)	57–62	102	Algeria
Kouider and Kolb (1982a)	99.2 ± 2.9	6	Syria
Kouider and Kolb (1982b)	110.5 ± 4	6	Syria
Abdelgadir et al. (1984)	50 ± 8.2	–	Sudan
Chiericato et al. (1986)	94–96	24	Somalia
Bizzetti et al. (1988)	70 ± 24.5	44	Somalia
Snow et al. (1988)	122.5–133.3	9	UAE
Azwai et al. (1990)	31.2–128.7	142	Libya
Faye and Mulato (1991)	63.7 ± 25.6	52	Djibouti
Faye et al. (1992)	79.7 ± 5.4	32	Djibouti
Bengoumi (1992)	74–160	240	Morocco
Faye et al. (1995)	111 ± 10.2	65	France
Elmahdi et al. (1997)	129 ± 5.4	4	Sudan
Nyang'ao et al. (1997)	122.9 ± 22.7	–	Kenya
Rezakhani et al. (1997)	79.9 ± 25.5	31	Iran
Sarwar and Majeed (1997)	45.03 ± 2.34	56	Pakistan
Ayoub and Saleh (1998)	129.3 ± 10.3	3	UAE
Nazifi et al. (1998)	131.1 ± 2.8	43	Iran
Al-Qarawi (1999)	51.4–103.1	150	Saudi Arabia
Nazifi et al. (1999)	92.2–129.1	40	Iran
Shaheen (2001)	105.6 ± 2.3	–	UAE
Naeini and Nazifi (2001)	126.9 ± 1.5	50	Iran
Wensvoort et al. (2001)	110.9–149	3	UAE
Ayoub et al. (2003)	100.5 ± 1.03	8	UAE
Al-Sultan (2003)	48.1 ± 7.9	61	Saudi Arabia
Ismail et al. (2003)	120.1 ± 4.9	9	North America
Osman and Al-Busadah (2003)	134 ± 11	20	Saudi Arabia
Badryyah et al. (2005)	132.2 ± 4.45	31	Saudi Arabia
Asadi et al. (2009)	57.2–78.2	93	Iran
Mohammed et al. (2007)	41.2 ± 3.2	11	Nigeria
Ali et al. (2008)	186.9 ± 14.4	92	Pakistan
Ahmed et al. (2009)	162.2 ± 18.8	30	Egypt
El-Boshy et al. (2009)	95.6 ± 8.45	340	Egypt

(continued)

Table 3.1 (continued)

References	Values	*n*	Country
Aichouni et al. (2010)	92.7 ± 0.4	16	Algeria
Albomohsen et al. (2011)	120.4 ± 8.2	50	Iran
Hekmatimoghaddam et al. (2011)	61.8 ± 2.5	92	Iran
Sazmand et al. (2011)	65.4 ± 23.6	93	Iran
Al-Rukibat and Ismail (2014)	174.1 ± 46.3	100	Jordan
Faye et al. (2015a)	106.4 ± 28.9	56	Saudi Arabia
Badakhshan and Mirmahmoudi (2016)	60.11 ± 1.36	18	Iran
Moolchandani and Sareen (2016)	90.7 ± 2.0	6	India
Sahraoui et al. (2016)	90.1 ± 38.3	22	Algeria
Yousif et al. (2016)	86.4 ± 3.87	15	Sudan
Hamad et al. (2017)	117.3 ± 1.8	30	Algeria

Chiericato et al. (1986), Bengoumi (1992), Mohammed et al. (2007), and Souilem et al. (1999) did not report any significant effect of sex on plasma glucose concentration. However, according to Gupta et al. (1979a) and Al-Harbi (2012), glycemia was higher in camels during rutting season (108 ± 2.5 vs 118.7 ± 1.25 mg/100 ml).

Pregnancy-Lactation Effect

Plasma glucose concentration in camel increases with pregnancy and decreases during the first 2 weeks of lactation (Elias and Yagil 1984; Bengoumi 1992). For example, Souilem et al. (1999) found higher plasma glucose level in pregnant camels (111 ± 3 mg/100 ml) than in non-pregnant lactating ones (96 ± 4 mg/100 ml). The increase during pregnancy is due to the increase of neoglucogenesis. The decline of glucose in lactating animals is linked to its role as the only precursor for the synthesis of lactose in the mammary gland. However, at reverse, Saeed et al. (2009) found lower level in pregnant camel at term compared to non-pregnant, 84.5 ± 15.6 and 94.1 ± 6.0 mg/100 ml, respectively. Similar figure was revealed by Omidi et al. (2014a, b) on Bactrian camels from Iran. These authors attributed the lower level of glucose during pregnancy to developing fetus and mobilization of maternal glucose with fetal circulation. Moreover, some authors did not find significant difference between pregnant and non-pregnant camels (Ayoub et al. 2003).

Season Effect

The main effect of season on plasma glucose concentration in camels is linked to the seasonal change of feed intake with higher values during rainy season and lower values during dry ones (Barakat and Abdel Fattah 1971; Mehrotra and Gupta 1989; Bengoumi 1992; Mohammed et al. 2007; Nazifi et al. 1999). For example, in Pakistan, Amin et al. (2007) have found an average glycemia of 60.1 ± 0.2 in dry season vs 87.4 ± 0.2 mg/100 ml during green season. In Algeria also, Aichouni et al. (2013) reported a lower value in summer (dry season) than in winter (rainy season), 61.4 ± 0.5 and 89.4 ± 0.6, respectively. Similar results were reported in Morocco: 117.5 ± 6 in summer vs 151.3 ± 4.5 in winter (Bargaâ et al. 2016). In Egypt, Badawy et al. (2008) revealed higher glycemia in winter (119.2 on average) than in summer (74.1 mg/100 ml).

In fact, plasma glucose concentration increases slightly with energy intake (Faye and Mulato 1991; Bengoumi 1992; Bengoumi et al. 2004; Gupta et al. 2012). In case of feed deprivation, glycemia tends to decline but in a lower proportion than in goat and sheep (Shaheen 2001).

The external temperature could also explain the seasonal variation, especially when the conditions are extreme (much cold or hot), stressing camels. Thus, Nazifi et al. (1999) have found a higher glycemia in camels during summer. Al-Qarawi (1999) in Saudi Arabia observed higher glycemia in summer compared to winter whatever the breed: 91.8 ± 3.5 vs 76.5 ± 2.9 for Majaheem breed (black-coat camel), 103.1 ± 4.5 vs 51.4 ± 4.2 for Homor breed (brown-coat camel), and 105.7 ± 5.1 vs 80.6 ± 3.9 mg/100 ml for Waddah breed (white-coat camel). At reverse no seasonal variation was revealed by Yousif et al. (2016).

According to Delavaud et al. (2013), glycemia decreased in underfed camels when compared to control or overfed ones and increased in overfed when compared to control camels. Thus, the geographical variability observed by Faye and Mulato 1991) at Djibouti (from 49.8 to 98.7 mg/100 ml) could be explained by the variability of energy intake in the diet, the highest value being observed in peri-urban farms where the camels receive concentrates and the lowest in the most arid regions.

Dehydration and Fasting

Dehydration induces an important increase of plasma glucose concentration from 6.4 mmol/ to 16.1 mmol/l (1.6 to 2.9 g/l), much higher than hemoconcentration due to the plasma volume reduction with a maximum of 40%. This is explained by the decrease in plasma insulin concentration during water deprivation (Bengoumi 1992; Bengoumi et al. 1998). After 10 days of dehydration, Yagil and Berlyne (1977) reported glycemia increasing from 124 ± 65 to 150 ± 14 mg/100 ml.

Camel is more adapted to fasting than other domestic species. The impact of food deprivation on glycemia is less important in camel than in sheep (Chandrasena et al. 1979a). After 96 h of fasting, glucose concentration did not change significantly (Dahlborn et al. 1992), with the values remaining at around 100 mg/100 ml. Those results were confirmed by Wensvoort et al. (2001).

Obviously, glycemia is modulated by the energetic value of the diet. For example, in the comparison of the three types of diet with different levels of energy brought by a mixture of exogenous enzymes from anaerobic bacteria, Adel and El-Metwaly (2012) observed a significant increase of plasma glucose from 38.11 in control camel to 57.97 mg/100 ml in camel receiving additives.

Physical Exercise

The camel is known as a working animal for transportation, traction and agricultural works, and also as a race animal. Plasma concentration decreases during races (Bengoumi 1992).

Castration

Whole males have higher plasma glucose concentration than castrated ones, which could be explained by the stress and excitation of whole males (Gupta et al. 1979b).

3.2.1.3 Effect of Pathological Factors

The serum glucose increases normally in case of stress as it has been shown in transported camels (El Khasmi et al. 2013) where glycemia increased from 116 ± 2 to 171 ± 2.1 mg/100 ml after 2 h of transportation. This increase is proportional to the distance traveled: after 72–80 km, glycemia measured on 18 camels was 92.2 ± 0.5, then 141.8 ± 0.4 after 160–170 km, and 163.6 ± 0.6 mg/100 ml in long distance more than 350 km (El Khasmi et al. 2015). Thus, many stressing diseases are linked to hyperglycemia. For example, in a provoked jejunal obstruction in adult camels, glycemia increased regularly from approximately 70 (base value) to 140 mg/100 ml (72 h after obstruction) and returned to normal base, 72 h after release of the obstruction (Rana et al. 1998). Glycemia increased significantly in case of induced hypothyroidism passing from 86.6 ± 8.6 in the first month to 101.4 ± 9.7 mg/100 ml after the third month (Barsham et al. 2005). This increase could be linked to the reduction of insulin secretion induced by the decrease in thyroid activity (Yagil 1985). Glycemia increased also in case of mange, with the value passing from 90.6 ± 6.9 in healthy animal to 108.7 ± 9.3 mg/100 ml in affected camel during winter and from 95.8 ± 7.7 and 104.5 ± 4.9 mg/100 ml during summer (Mal et al. 2006). Significant higher glycemia was also observed in camel affected by wryneck, the level of glucose being twice in sick camels compared to healthy ones, 97.9 ± 0.9 vs 45.3 ± 0.1 mg/100 ml, respectively (Al-Sobayil and Mousa 2009). In racing camel affected by bone fracture, glycemia reached 181.8 ± 74 mg/100 ml, while it was 112.4 ± 25.4 in healthy camels (Alshamsi et al. 2015).

At reverse, in camels affected by trypanosomosis, glycemia declined by 6% (Moolchandani and Sareen 2016) and decreased from 65.4 ± 2.3 to 40.2 ± 7.1 mg/100 ml (Sazmand et al. 2011). Such decline was confirmed in an outbreak of abortion linked to *T. evansi* in Canary Islands (Gutierrez et al. 2005) where glycemia decreased from 89.4 ± 7.6 to 68.7 ± 5.6 mg/100 ml. A fall of glycemia occurred also in case of severe parasitic infection, the value being 64.9 ± 4.5 mg/100 ml when control animals showed value of 136.3 ± 6.0 mg/100 ml (Momenah 2014). Such decline could be linked to a lower efficient in energy absorption from the diet.

In camels affected by reproductive disorders, a decrease of glycemia was also described: from 88.7 ± 1.7 in healthy animals to 53.4 ± 3.5 mg/100 ml sick ones

(Zaher et al. 2017). A decline of glycemia is also related in camels affected by *Mycoplasma* sp., the values declining from 126.0 ± 2.3 in healthy camels to 56.0 ± 1.9 mg/100 ml in sick camels (Nazifi et al. 2009a). In camels affected by contagious ecthyma, glycemia declined significantly from 94.4 ± 2.1 to 84.9 ± 1.5 mg/100 ml (Narnawane et al. 2015).

Anti-inflammatory treatment, for example, dexamethasone injection, can provoke a hyperglycemia in camel (Wasfi et al. 1989) as well the administration of sedative drugs (Abdin-Bey 2001).

3.2.1.4 Glucose in Other Substrates

The determination of glucose in other substrates is generally having few clinical interests. In **peritoneal fluid**, the glucose level appeared lower than in blood: 90.2 ± 3.9 mg/100 ml (Naeini and Nazifi 2001).

In **cerebrospinal fluid**, glucose concentration was estimated 62.3 ± 2.0 mg/100 ml (Kataria et al. 1995). Highest concentrations (167.6 ± 2.0 mg/100 ml) were reported by Ahmed et al. (2009).

In **synovial fluid**, the glucose concentration was on average 85.5 ± 28.0 mg/100 ml with some differences between joints (Al-Rukibat and Ismail 2014). Glucose concentration in synovial fluid was determined at quite lower values: 0.7 ± 0.5 mg/100 ml (Nazifi et al. 1998). In follicular fluid, glucose concentration was significantly higher in a small-sized follicle (136.8 ± 4.0) compared to large ones (77.1 ± 4.3 mg/100 ml), but no effect of breeding season or age was revealed (Ali et al. 2008). A similar figure was published later (Albomohsen et al. 2011) with higher value in a small follicle (70.7 ± 7.3) than in large one (88.0 ± 13.4 mg/100 ml).

3.2.2 Ketone Bodies

Ketone bodies include three components: β-hydroxybutyric acid, acetoacetic acid, and acetone. They are synthetized in different tissues during ketosis due to prolonged energy deficiency with high lipolysis and limited concentration of glucose. In ruminant, when feeding resources are limited, tissue degradation is induced and generally causes production of ketone bodies as a result of lack of input of oxalate precursors. Acetone is volatile, and only concentration of β-hydroxybutyric and acetoacetic acids that are excreted in urine and milk can be measured. Plasma concentration of β-hydroxybutyric acid is a good indicator of ketosis. In camels, plasma concentration of β-hydroxybutyric acid is often lower than the detection limit because ketogenesis is very low due the lower activity of ketogenesis enzymes in tissues and limited excessive degradation of free fatty acids during lipolysis (Bengoumi 1992; Delavaud et al. 2013). Plasma β-hydroxybutyric acid is measured using β-hydroxybutyric acid kinase.

In the dromedary camel, fasting causes a small increase in serum non-esterified fatty acids (NEFA), while serum glucose level remains higher. Compared with sheep and cattle, it is assumed that these two factors are responsible elements for preventing increase in ketone body levels. It appears therefore that the dromedary camel has the capacity to control its lipogenic and glucogenic metabolism, in order to prevent ketosis.

As plasma concentrations of β-hydroxybutyric and acetoacetic acids in camel were lower than the detection limit of the analytical method (0.01 mmol/l), there are few references. Chandrasena et al. (1979b) reported values between 0.0085 and 0.012 mmol/l. Faye and Mulato (1991) gave a mean of 0.025 ± 0.026 mmol/l with neither age effect nor sex effect. However a significant regional difference linked to the quality of the diet was described.

After 96 h of fasting, ketone body concentration increased slightly in camel plasma but in a quite lower proportion than in sheep, from 0.089 to 0.129 mg/100 ml vs 2.97 to 5.03, respectively. At the same time, acetoacetic acid increased from 0.148 to 0.264 mg/100 ml in camel vs 0.63 to 0.91 mg/100 ml in sheep (Emmanuel 1984).

In case of trypanosomosis, the characteristic odor of the urine noticed in affected camel may be due to more elevated ketone bodies (Röttcher et al. 1987). Indeed, the conversion of butyrate to ketone bodies in camel is higher in the kidney than in the liver, contrary to the other ruminants (Emmanuel 1980).

3.2.3 Plasma Lipids

3.2.3.1 Cholesterol

Cholesterol is a sterol biosynthesized by all animal cells where it plays an essential role as structural component of cell membranes. Thus, cholesterol is essential to maintain both membrane structural integrity and fluidity. Cholesterol plays an important metabolic role as precursor of all steroid hormones, bile acids, and some vitamins. If in humans a high plasma cholesterol level is suspected to be linked with cardiovascular disease, its clinical significance in camel is not sufficiently investigated. Moreover, the authors did not mention, most of the time, if their results regarded total cholesterol or lipoprotein ones (VLDL/LDL/HDL). Plasma cholesterol concentration in dromedaries is lower than that in other animal species; however, hepatic cholesterol is higher in dromedary than in sheep and cattle (Mirgani 1981). Serum cholesterol concentration is mainly measured using cholesterol oxidase method.

Plasma cholesterol concentration reported in the literature varies from 0.46 to 3.88 mmol/l (18–150 mg/100 ml) (Table 3.2) and seems neither affected by age or sex (Abdelgadir et al. 1984; Al-Ani et al. 1992; Bengoumi 1992; Faye and Mulato (1991); Sarwar et al. 1991; Khadjeh et al. 1997; Saeed et al. 2004a; Mohamed 2008; Omidi et al. 2014a, b). For example, Perk and Lobl (1961) reported 54.9 ± 13.6 in

Table 3.2 Values of plasma cholesterol in camel according to different references (in mg/100 ml)

References	Values	*n*	Country
Mills and Taylaur (1971)[a]	11 (HDL)	–	ND
Barakat and Abdel Fattah (1971)	63–250	–	India
Leat and Northrop (1975)[a]	61.4	–	ND
Ateeq et al. (1984)	39.1–55.7	88	Syria
Mohamed et al. (1984)	32.1 ± 6.4	3	Somalia
Wasfi et al. (1987)	39.1 ± 13.3	19	Saudi Arabia
Bizzetti et al. (1988)	33.6 ± 10.9	44	Somalia
Al-Ali et al. (1988)	30.5 ± 8.7	20	Saudi Arabia
Faye and Mulato (1991)	19.5 ± 8.7	52	Djibouti
Azwai et al. (1990)	34.6–132.8	142	Libya
Hussein et al. (1992)	72.7 ± 11.5	19	Saudi Arabia
Abu Damir et al. (1993)	30 ± 8.8	7	Sudan
Osman and Al-Busadah (2000)	46.1 ± 11.5	22	Saudi Arabia
Naeini and Nazifi (2001)	31.5 ± 10.7	50	Iran
Al-Sultan (2003)	52.03 ± 2.78	62 F	Saudi Arabia
	55.74 ± 7.0	18 M	
Osman and Al-Busadah (2003)	58.4 ± 8.6	5	Saudi Arabia
Saeed et al. (2004a)	40.2 ± 8.8	82 F	UAE
	34.4 ± 8.4	85 M	
Abd El-Hag et al. (2005)	138.1 ± 39.3	10	Sudan
Badryyah et al. (2005)	100 ± 0.0	36	Saudi Arabia
Mal et al. (2006)	29.16 ± 3.05	8	India
Ali et al. (2008)	39.06 ± 5.82	92	Pakistan
Al-Sobayil and Mousa (2009)	34.6 ± 0.7	5	Saudi Arabia
Asadi et al. (2009)	9.23–61.15	93	Iran
Nazifi et al. (2009b)	28.8 ± 0.4	30	Iran
Shukla et al. (2009)	44.33 ± 13.86	16	India
Albomohsen et al. (2011)	38.3 ± 4.5	50	Iran
Hekmatimoghaddam et al. (2011)	33.4 ± 1.5	92	Iran
Nagpal et al. (2011)	26.54 ± 4.97	14	India
Sazmand et al. (2011)	34.2 ± 1.5	93	Iran
Tajik et al. (2013)	33.1 ± 1.15	180	Iran
Konuspayeva et al. (2014)	130 ± 18	8	Saudi Arabia
Omidi et al. (2014a)[a]	33.12 ± 3.1	20	Iran
Omidi et al. (2014b)	63.8–77.1	52	Iran
Faye et al. (2015a)	106.4 ± 28.9	56	Saudi Arabia
Narnawane et al. (2015)	36.8 ± 3.4	6	India
Ali et al. (2016)	76.03 ± 9.76	6	Saudi Arabia
Badakhshan and Mirmahmoudi (2016)	42.17 ± 1.68	18	Iran
Sahraoui et al. (2016)	38.4 ± 10.8	22	Algeria
Yousif et al. (2016)	36.36 ± 5.9	15	Sudan
Hamad et al. (2017)	33.2 ± 9.2	30	Algeria

[a]Bactrian camel

males vs 53.5 ± 7.7 mg/100 ml in females, and Ali et al. (2008) reported 36.9 ± 1.8 in young and 42.0 ± 2.6 mg/100 ml in adult camel. However, Mohamed (2008) reported higher cholesterol in plasma of adult camel (61.1 ± 14.6) compared to yearling (45.4 ± 8.1) and neonates (40.0 ± 1.4 mg/100 ml). Nazifi et al. (2000) also found higher plasma cholesterol in adult more than 6 years old (51.1 ± 13.5) than the youngest groups (30.8 to 35.4 mg/100 ml).

According to Osman and Al-Busadah (2000), plasma cholesterol was 46.1 ± 11.5 in dry she-camels and 42.3 ± 11.4 in lactating ones. However, in lactating camels, serum cholesterol increases relatively (101.5 in average the first 3 months) and then decreases gradually to 64.6 between 3 and 6 months, 39.6 between 6 and 9 months, and 57.7 mg/100 ml at the end of lactation (Hussein et al. 1992).

For Kamal and Salama (2009), cholesterol concentration in camel serum did not change significantly at the first month of lactation, except on day 30, the concentration being 46.1 ± 4.2 mg/100 ml, while it was 34.6 ± 1.9 mg/100 ml the first day of lactation. The highest value at d30 postpartum involved total cholesterol and all the fractions HDL, LDL, and VLDL (Kamal and Salama 2009). On average, there was no significant difference between lactating (72.7 ± 11.5) and non-lactating animals: (62.7 ± 11.4 mg/100 ml) (Hussein et al. 1992).

Regarding pregnancy, contrary to Bengoumi (1992) and Omidi et al. (2014a, b), some authors found significant lower level in plasma cholesterol in pregnant she-camels compared to non-pregnant ones (31.04 ± 5.2 vs 44.94 ± 10.96 mg/100 ml) (Saeed et al. 2009). However, if the lack of observed difference concerned total cholesterol, a significant lower HDL cholesterol value was reported in pregnant camels compared to non-pregnant or to male (Omidi et al. 2014a). However, whatever the type of cholesterol, this effect was not observed when comparing she-camels at the end of gestation (last 3 months) with non-pregnant camels (Omidi et al. 2014b).

It seems that an important difference occurs between total and free cholesterol. According to Abu-Sinna and Habaka (1974), total cholesterol in camel was 81.0 ± 4.0, while free cholesterol was 42.2 ± 1.9 mg/100 ml. Some references mentioned the type of cholesterol, HDL, LDL, and VLDL (Kamal and Salama 2009; Omidi et al. 2014a, b). The values were 11.9 ± 6.5, 4.6 ± 3.1, and 5.8 ± 0.3, respectively, in summer season and 28.4 ± 9.7, 10.8 ± 3.8, and 14.6 ± 3.3 mg/100 ml, respectively, in winter (Nazifi and Gheisari 1999). The distribution of HDL, LDL, and VLDL is changing according to the age of the camel (Fig. 3.1), with higher value of HDL in adult animals (Nazifi et al. 2000).

Seasonal Variation: The Effect of Rutting Period

In male camel, the effect of breeding season and castration on plasma cholesterol concentration is contradictory. During the mating season, Khan and Kohli (1973, 1981) reported a significant increase of cholesterol in males (from 65.6 ± 24.3 before the rut to 92.2 ± 6.2 mg/100 ml), while Bengoumi (1992) observed a decrease in whole males compared to castrated ones during rutting season (Fig. 3.2).

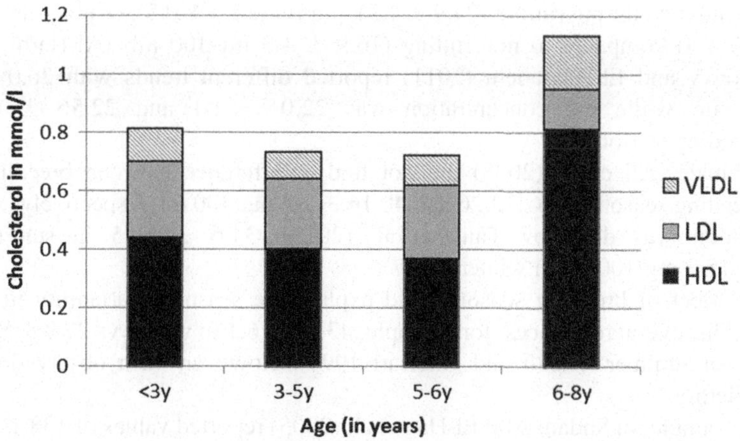

Fig. 3.1 Mean of different types of serum cholesterol in 87 Iranian male camels in different age groups [calculated from Nazifi et al. (2000)]

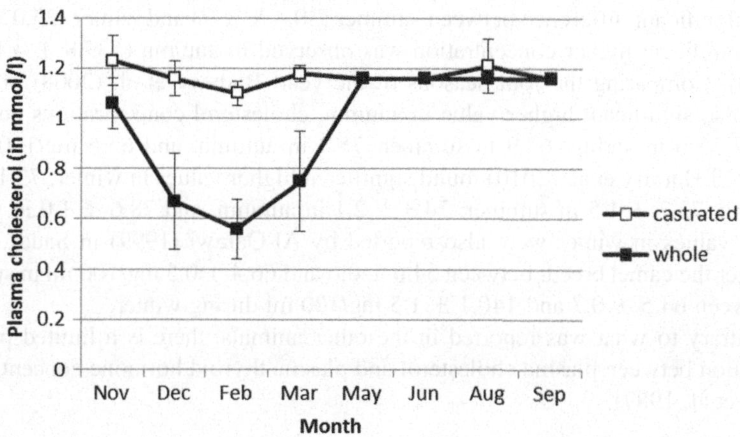

Fig. 3.2 Seasonal change of plasma cholesterol in castrated (open square) and whole (black square) camels [calculated from Bengoumi (1992)]

On average, cholesterol concentrations were 45 in castrated camel vs 38 mg/100 ml in whole camel all over the year (Bengoumi 1992). Yet, Kouider et al. (1988) reported also a higher cholesterolemia at breeding season (71.9 ± 7.3) than outside the breeding season (57.3 ± 5.8 mg/100 ml). Reporting quite higher values, Abd El-Salaam et al. (2011) found also an increase of cholesterol in camel plasma at breeding season (261.2 ± 24.2) than in non-breeding ones but with a significant difference between dry (228.7 ± 20.2) and humid season (123.7 ± 28.1 mg/100 ml). A similar figure was published by Zeidan et al. (2008), 209.4 ± 7.5, 175.5 ± 7.1, and 124.6 ± 5.6 mg/100 ml, respectively. Slightly higher values of plasma cholesterol were mentioned during rutting (24.9 ± 1.9), pre-rut (22.0 ± 0.5), or post-rut months (21.8 ± 1.3) compared to non-rutting (16.9 ± 1.3 mg/100 ml) (Al-Harbi 2012). El-Bahrawy and El-Hassanein (2011) reported different trends with 26.0 ± 2.0 during rut, while the concentration was 22.0 ± 1.1 and 22.56 ± 0.6 in non-breeding season.

In female, Ali et al. (2008) did not find a difference between breeding and non-breeding season, 38.6 ± 2.26 and 40.16 ± 2.2 mg/100 ml, respectively. Similar observation was done by Tajik et al. (2013): 34.6 ± 10.5 in summer vs 31.9 ± 15.3 mg/100 ml in winter.

The effect of breeding season could explain the seasonal variability in camel reported in several references, for example, 43.8 ± 16.1 in winter vs 27.7 ± 8.8 mg/100 ml in summer (Nazifi and Gheisari 1999). However, such observations are contradictory.

For example, in Sudan, Abd El-Hag et al. (2005) reported values of 138.1 ± 39.3 in dry season (mainly winter) and 140.9 ± 5.7 mg/100 ml in rainy season (Summer). Mal et al. (2006) did not observe also seasonal changes. From their side, Aichouni et al. (2013) found higher values in summer than in winter, 24.6 ± 11.5 vs 19.6 ± 7.7 mg/100 ml, respectively. In Sudan again, Yousif et al. (2016) did not report significant difference between summer (30.4 ± 3. 7) and winter (25.0 ± 3.1), but a significant higher concentration was observed in autumn (53.38 ± 4.16 mg/100 ml). Comparing the four seasons of the year, Badawy et al. (2008) in Egypt found also significant higher value in autumn, cholesterol concentrations being on average 56.6 in spring, 62.9 in summer, 75.3 in autumn, and 65.6 mg/100 ml in winter. El-Harairy et al. (2010) found significant higher values in winter: 72.1 ± 2.1 in spring, 72.7 ± 1.5 in summer, 74.3 ± 2.1 in autumn, and 78.6 ± 1.0 in winter. Higher values in winter were also reported by Al-Qarawi (1999) in Saudi Arabia whatever the camel breed: between 54.6 ± 0.6 and 66.4 ± 0.5 mg/100 ml in summer vs between 64.5 ± 0.7 and 140.1 ± 1.5 mg/100 ml during winter.

Contrary to what was reported in the other animals, there is a limited positive correlation between plasma cholesterol and plasma thyroid hormone concentrations (Wasfi et al. 1987).

Feeding Effect

The type of diet brought to the camel could influence the cholesterol status of the animal. In India, diets based on groundnut haulms and cluster bean straw were fed in one of three ratios, 75/25, 50/50, and 25/75 in three treatments, and concentrate mixture as per requirement of the camel: plasma cholesterol concentrations were, respectively, 48.7 ± 2.6, 40.0 ± 1.1, and 35.3 ± 2.2 mg/100 ml with significant differences (Gupta et al. 2012). At reverse, in young camels receiving three different isocaloric feed blocks containing 9.5, 12.0, and 14.5% crude proteins, no significant difference in blood cholesterol was observed: 26.54 ± 4.97, 24.83 ± 2.65, and 30.34 ± 3.68 mg/100 ml, respectively (Nagpal et al. 2011).

A slight but nonsignificant regional variability was described in Djibouti (from 17.7 to 21.9 mg/100 ml) probably in relationship with the variability of the diet (Faye and Mulato 1991).

After 14 days of dehydration, the cholesterol in camel serum increased by 250% (from 44.6 ± 0.1 to 160.4 ± 28.4 mg/100 ml) and returned progressively to normal after a week (Bengoumi 1992). The increase of cholesterol during dehydration was formerly observed by Mohamed et al. (1984) in Somalia, the values passing from 32.1 ± 6.4 at the beginning of the water privation trial to 53.1 ± 11.4 mg/100 ml after 25 days of dehydration. This increase is higher than the reduction of the plasma volume (40%).

Breed Effect

A slight but significant difference was revealed in Saudi camel breed characterized by their coat color. Cholesterol values were 142.3 ± 7.1 mg/100 ml in black-coat camel, 134.5 ± 1.6 mg/100 ml in brown-coat camel, and 132.3 ± 2.3 mg/100 ml in white-coat camel (Hussein et al. 2012). From his side, Al-Qarawi (1999) found higher cholesterol content in white camel than in black and brown ones, but the difference was more pronounced in winter than in summer. In Sudan, Mohamed (2008) reported a breed effect on serum cholesterol with 57.3 ± 8.1 mg/100 ml in Arabi breed vs 35.0 ± 11.5 mg/100 ml in Anafi breed.

Effect of Diseases

There is no change in plasma cholesterol of camels affected by mange (Mal et al. 2006). At reverse, in camels affected by wryneck syndrome, the plasma cholesterol is significantly higher: 59.2 ± 5.4 vs 34.6 ± 0.7 mg/100 ml in healthy animals (Al-Sobayil and Mousa 2009). There is no significant effect also of dystocia (Ali et al. 2016) or trypanosomosis (Sazmand et al. 2011).

Cholesterol in Milk

Regarding camel milk, many workers argue that camel milk contains less cholesterol than cow's milk (Kamal and Salama 2009; Raziq et al. 2008; Faye et al. 2015a), while others reported the reverse (Gorban and Izzeldin 1999; Konuspayeva et al. 2008). For example, on average, the total cholesterol content of camel milk was higher (31.32 mg/100 g) when compared to the total cholesterol content of cow's milk (25.63 mg/100 g). Moreover, the average free cholesterol content was 21.34 mg/100 g in camel milk ($n = 54$) vs 17.25 mg/100 g/l in 24 cow milk samples (Gorban and Izzeldin 1999). In Bactrian camel, the mean reported value was 37.15 ± 7.73 mg/100 g (Konuspayeva et al. 2008). In another context, lower values were reported: 5.64 ± 3.18 mg/100 g (Faye et al. 2015a). However, these authors have demonstrated

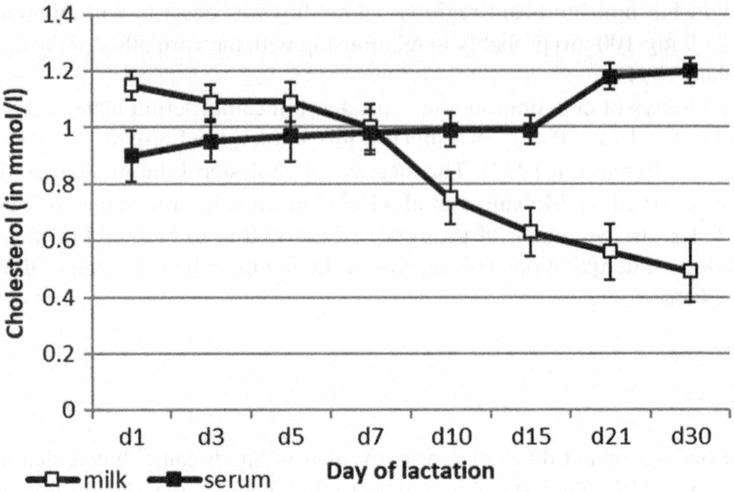

Fig. 3.3 Changes in camel serum and milk cholesterol (in mmol/l) during the first month of lactation [calculated from Kamal and Salama (2009)]

that the cholesterol content is obviously correlated to the fat content with a similar ratio cholesterol/fat matter in camel and cow milk (Faye et al. 2015b). A concentration of 11.85 ± 1.3 mg/100 ml was also reported recently (Konuspayeva et al. 2014).

There was no significant correlation between cholesterol in camel milk and in serum (Faye et al. 2015a).

The cholesterol content was higher in colostrum (Fig. 3.3) and then declined progressively from 44.2 ± 34.6 at birth to 18.8 ± 16.5 mg/100 ml after one month (Kamal and Salama 2009).

Cholesterol in Other Substrates

Peritoneal fluid contains a very low quantity of cholesterol: 2.7 ± 1.9 mg/100 ml (Naeini and Nazifi 2001).

The size of **follicle** has no influence on the cholesterol content in follicular fluid: 7.0 ± 0.88 in small follicle and 6.3 ± 0.7 in large ones (Albomohsen et al. 2011). These values are similar to that of Ali et al. (2008), 10.62 ± 1.08 and 16.72 ± 1.15 mg/100 ml in small and large follicle, respectively. However, the same authors observed significant higher cholesterol concentrations out of the breeding season (21.08 ± 1.11) than in the breeding season (6.25 ± 1.14 mg/100 ml), while no difference was observed in plasma concentration.

The cholesterol content in camel hump and meat is varying according to the authors but is generally lower than 50 mg/100 g (Elsanhoty et al. 2011; Kadim et al. 2008), although the observed range appeared higher in some references: 135–150 mg/100 g fresh weight (Abu-Tarboush and Dawood 1993).

In the comparative study regarding Bactrian and dromedary camel meat in Kazakhstan, Raiymbek (2013) found 53.7 ± 13.1 and 49.2 ± 12.7 mg/100 g, respectively.

The camel **liver** contains on average 6.3 mg/g wet weight cholesterol (Mirgani 1977b).

3.2.3.2 Triglycerides (TG)

Triglycerides (TG) are esters derived from glycerol and three fatty acids. They are the main constituents of body fat. In animals, TGs are used as an indicator of fat mobilization when feeding cannot cover energy requirements of the animal. However, as for cholesterol, the interpretation of TG concentration variations in camel is not really investigated. Serum TG concentration is mainly measured using lipase and glycerol kinase.

Serum triglyceride concentrations in camels vary globally from 0.12 to 0.88 mmol/l (i.e., 10–80 mg/100 ml). Such range of variation is mainly due to the difference in analytical methods and instability of triglycerides in blood due to several factors (Table 3.3).

As for glucose and cholesterol, to make easier the comparisons between published references, all the results in mmol/l are converted into mg/100 ml.

Physiological Variations

TG concentration in serum is influenced by age with higher values in young animals: 154.5 ± 5.6 in young calves, compared to 72 ± 9.1 in dried females and 36.3 ± 27.2 mg/100 ml in lactating camel (Osman and Al-Busadah 2000). Similar difference was reported recently in Algeria (Sahraoui et al. 2016) with double value in young camels 1–4 years old (181.8 ± 116.3) than in adults (96.3 ± 75.4 mg/100 ml).

However, comparing three age groups of camels (1–2 years, 2–4 years, and > 4 years), Saeed et al. (2004b) did not find significant differences. Asadi et al. (2009) also did not find significant effect of age in both sexes. At reverse, Faye and Mulato (1991) revealed a lowest value in young camel (21.5 ± 7.6) than in adults (36.2 ± 15.4 mg/100 ml). For Ali et al. (2008), the highest value observed in adults (47.5 ± 1.5) was not significantly different from that in young camels (40.6 ± 1.0 mg/100 ml). Mohamed (2008) found highest value in adults (92.7 ± 9.1) than in yearling (75.5 ± 7.3) and neonates (64.5 ± 8.2 mg/100 ml).

Castrated males have also lower plasma triglyceride concentration than whole males (Bengoumi 1992), but no sex difference was reported (Fig. 3.4): 23.4 ± 8.4 in male vs 21.8 ± 7.6 mg/100 ml in female (Saeed et al. 2004a). Similar results were observed by Khadjeh et al. (1997), Al-Sultan (2003), and Sahraoui et al. (2016).

For lactation or pregnancy effects, Chiericato et al. (1986), Bengoumi (1992), as well as Omidi et al. (2014b) did not observe any significant effect. However, the

Table 3.3 Values of serum triglycerides in camel according to different references (in mg/100 ml)

References	Values	*n*	Country
Orliac (1980)	18–41	102	Algeria
Mohamed et al. (1984)	36.6 ± 6.9	3	Somalia
Chiericato et al. (1986)	30.6	24	Somalia
Wasfi et al. (1987)	11.5 ± 2.1	20	Saudi Arabia
Snow et al. (1988)	22.6–24.4	9	UAE
Faye and Mulato (1991)	26.6 ± 12.9	52	Djibouti
Bengoumi (1992)	13.2–77.8	240	Morocco
Naeini and Nazifi (2001)	64.5 ± 15.4	50	Iran
Al-Sultan (2003)	32.8 ± 4.4	62 F	Saudi Arabia
	40.2 ± 6.9	18 M	
Osman and Al-Busadah (2003)	31.4 ± 3.0	5	Saudi Arabia
Saeed et al. (2004a)	21.8 ± 7.6	82 F	UAE
	23.4 ± 8.4	85 M	
Badryyah et al. (2005)	70 ± 14	36	Saudi Arabia
Mal et al. (2006)	9.7 ± 2.2	32	India
Elnahas (2008)	10.9–76.4	27	Egypt
Nazifi et al. (2009b)	43.6 ± 9.1	6	Iran
Asadi et al. (2009)	9.1–86.4	93	Iran
Aichouni et al. (2010)	58.4 ± 36.3	48	Algeria
Albomohsen et al. (2011)	34.3 ± 5.25	50	Iran
Sazmand et al. (2011)	40 ± 2.7	93	Iran
Konuspayeva et al. (2014)	50 ± 20	8	Saudi Arabia
Narnawane et al. (2015)	28.8 ± 2.3	6	India
Ali et al. (2016)	40.1 ± 2.8	6	Saudi Arabia
Badakhshan and Mirmahmoudi (2016)	39.9 ± 4.7	18	Iran
Sahraoui et al. (2016)	90.9 ± 35.5	22	Algeria
Yousif et al. (2016)	33.0 ± 3.1	15	Sudan
Hamad et al. (2017)	16.8 ± 8.9	30	Algeria

concentration of triglyceride (36.3 ± 7.3 mmol/l) in serum of pregnant camels at term (Saeed and Khan 2012) was significantly higher than that of non-pregnant camel (27.3 ± 7.3 mg/100 ml). The increase in the level of serum triglyceride prior to parturition might be due to overproduction of triglyceride as observed in other species.

Water deprivation increased plasma triglyceride concentrations from 20 ± 3.6 to 144.5 ± 61.8 mg/100 ml after 14 days of dehydration which indicates lipolysis. The values returned to normal level after a week of rehydration (Bengoumi 1992). After 25 days, Mohamed et al. (1984) observed also an increase of TG in camel blood serum (from 36.6 ± 6.9 to 78.7 ± 35.8 mg/100 ml). Lipomobilization increases at the final stage of water deprivation because of the important decrease of feed intake.

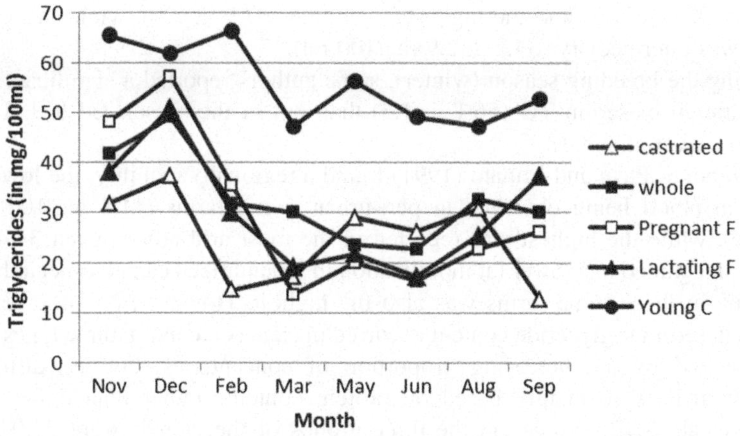

Fig. 3.4 Seasonal changes in triglycerides for different types of camels according to their age, sex, and physiological status [calculated from Bengoumi (1992)]

Feeding Effect

TGs being associated to fat balance, the energetic level of feeding could have an effect on their variability in serum. Thus, a comparison of the three types of diet including control animals fed with concentrate feed mixture, clover hay, and rice straw without additives and two tested animals receiving the same diet, plus 20 g additives (mixture of exogenous enzymes from anaerobic bacteria) or plus 40 g additives (Adel and El-Metwaly 2012): the mean TG concentrations were, respectively, 79.27, 104.84, and 111.51 mg/100 ml.

Seasonal Variation

Seasonal variation expresses food availability and seasonal variability of the nutritive values of pastures for camel. In consequence, interpretation of results of TG varies according to authors. Thus, lower values were reported in autumn and the highest in winter (Bengoumi 1992) (Fig. 3.1). Such significant seasonal variation is observed by other authors: in Iran, 40 ± 13.6 in summer vs 71.8 mg/100 ml in winter (Nazifi and Gheisari 1999). Decrease of serum TG concentration during summer or autumn corresponds to the season where camels increase their fat mobilization (i.e., mainly from the hump) because their requirements in energy cannot be covered by the lowest nutritive quality of the natural pasture (Chehma et al. 2010). At reverse, Aichouni et al. (2013), in Algeria, observed higher TG in summer (54.5 ± 18.1) than in winter (45.8 ± 18.0 mg/100 ml) probably because the rainfall conditions are not the same than in Iran. Obviously, higher TG was observed at the rainy season compared to dry season (Amin et al. 2007), 34.2 ± 1.5 vs 26.7 ± 1.5 mg/100 ml, respectively. In Sudan, Yousif et al. (2016) observed also the lowest value in summer (27.0 ± 1.9) and the highest in autumn (37.9 ± 2.6), while the value in winter was intermediary (34.2 ± 2.9 mg/100 ml).

During the breeding season (winter), some authors reported a significant higher concentration of serum TG (56.1 ± 1.3) than out of the season (31.9 ± 1.3 mg/100 ml) (Ali et al. 2008).

In Djibouti, Faye and Mulato (1991) found a regional variability, the lowest TG values in blood being observed in peri-urban camel farms (11.8 mg/100 ml on average), while the highest was reported in the most arid areas (mean 34.4). The authors concluded to a lower fat mobilization in sedentarized camel, especially since glycemia in these same farms was also the highest. However, on short term, no change in serum triglyceride content occurred in camels receiving three types of diet characterized by an increasing proportion of concentrates, but the differences between diets were mainly based on protein contents rather than crude energy (Gupta et al. 2012). Values in the three groups of the camels were 27.9 ± 6.2, 24.9 ± 4.9, and 22.8 ± 2.6 mg/100 ml, respectively.

Breed Effect

In Sudan, Arabi breed presented higher TG in their serum than Anafi breed, respectively, 99.1 ± 16.4 and 73.6 ± 18.1 mg/100 ml (Mohamed 2008).

Effect of Health Disorders

The camel affected by acute **mange** during winter season presented higher TG values (25.9 ± 4.7 mg/100 ml) than healthy animals (10.65 ± 1.5 mg/100 ml), but such difference did not occur in summer:, 9.2 ± 1.4 vs 8.7 ± 2.9 mg/100 ml, respectively (Mal et al. 2006)

In camel affected by **wryneck** syndrome, TG concentration increased up to 47.2 ± 9.1 compared to 28.2 ± 10.9 in healthy control animals (Al-Sobayil and Mousa 2009).

Reverse to cholesterol, TG increased significantly in camel affected by trypanosomosis (92.7 ± 30.9 mg/100 ml), i.e., a double value than in healthy camel (40 ± 2.7 mg/100 ml), probably because of the important lipomobilization in sick animals (Sazmand et al. 2011). Similar trend was observed in camel affected by theileriosis: from 44.5 ± 3.6 in healthy camel to 68.2 ± 21.6 mg/100 ml in affected one (Hekmatimoghaddam et al. 2011).

Triglycerides in Other Substrates

TGs are the main component of the hump. The total quantity of TGs varies from 157 to 388 mg/g wet weight (Mirgani 1977a).

In peritoneal fluid, the TG concentration was three times less than in serum: 20 ± 1.2 mg/100 ml compared to 64.5 ± 15.4 mg/100 ml in serum of the same camels (Naeini and Nazifi 2001).

In follicular fluid, the concentration of triglyceride is higher in small-sized follicle than in large ones, 31.3 ± 3.97 and 17.8 ± 4.23 mg/100 ml, respectively (Albomohsen et al. 2011). Reverse observation was done by Ali et al. (2008), 33.77 ± 1.01 and 38.87 ± 1.08 mg/100 ml, in small and large follicle, respectively.

In seminal plasma of male Bactrian camel, the TG concentration was determined at 101.6 ± 5.5 mg/100 ml (Mosaferi et al. 2005).

The triglyceride in camel liver was on average 8 mg/g wet weight (Mirgani 1977b).

3.2.3.3 Free Fatty Acids or Non-esterified Fatty Acids (NEFA)

NEFA are defined as free fatty acids, because they are not esterified with glycerol to form a glyceride. Few data are available on plasma concentration of NEFA in camels which varies on average between 0.01 and 0.30 mmol/l according to feeding energy intake. Contrary to other lipid parameters, only units in mmol/l are used in the literature. Usually, NEFA are increasing substantially in case of lipomobilization linked to underfeeding. This lipomobilization could have both hepatic and adipose origins. Serum NEFA concentration is determined using acyl coenzyme A synthase and acyl coenzyme A oxidase method.

References in the Literature

In 52 camels sampled in different regions of Djibouti, NEFA levels were reported to be between 0 and 0.57 mmol/l with an average of 0.17 ± 0.12 mmol/l (Faye and Mulato 1991). No age or sex or even regional effect was observed in this sample. Similar mean value was mentioned in a survey achieved in French camel farms: 0.15 ± 0.15 with extreme values between 0 and 0.9 mmol/l (Faye et al. 1995). In young camels receiving diet with different protein and mineral levels, plasma NEFA varied between 0.21 and 0.29 mmol/l (Faye et al. 1992). In Iran, values were determined on 93 camels with a range of 0.3–0.45 mmol/l (Asadi et al. 2009). No sex and age effect was also revealed. In camel submitted to sedation with detomidine-HCl, a significant increase of NEFA was observed 30 min after administration and then returned to pretreatment value (Abdin-Bey 2001), but the concentrations reported in this trial were quite higher (22 to 90 mmol/l).

Fasting Effect

In a comparative study between camelids (dromedary camel and llama) and bovine or ovine species, Wensvoort et al. (2001) showed that 5 consecutive days of fasting increased NEFA in all species with two times lower response in dromedary camels when compared to steers or sheep. During the fasting period, NEFA concentration reached no more than 3.5 mmol/l in camel vs more than 8 in cattle and 7.2 mmol/l in sheep.

According to Delavaud et al. (2013), plasma NEFA increased in the beginning of underfeeding up to a peak of 0.3 mmol/l and therefore decreases rapidly after 2–3 weeks with return to baseline (approx. 0.05 mmol/l) after 7–8 weeks.

Fasting in the dromedary causes a small increase in serum non-esterified fatty acids (NEFA), while serum glucose level remains higher. Compared to sheep and cattle, it is assumed that these two factors are responsible for preventing the increase in ketone body levels. It appears therefore that camels have the capacity to control lipogenic and glucogenic metabolism, so as to prevent the appearance of a state of camel ketosis.

In one comparative experiment on glucose tolerance, basal levels of NEFA did not differ between camel, sheep and ponies (approximately 0.35 mmol/l) (Elmahdi et al. 1997). NEFA concentrations dropped in all species 20–30 min after glucose infusion, and this decrease was less pronounced in camel (approximately 0.15 mmol/l). While NEFA concentrations returned to basal values after 180 and 210 min in sheep and ponies, plasma NEFA levels remained lower until the end of the experiment in camels. The authors concluded that high plasma glucose concentrations together with low insulin levels in camel compared to other herbivorous may be of benefit for animals with extremely poor diets where NEFA release from adipose tissue is increased due to inhibition of glucose uptake and reduced reesterification of NEFA.

Water Deprivation Effect

Prolonged water deprivation induces an important increase of plasma NEFA concentrations indicating lipolysis but with limited degradation of NEFA. After 14 days of water deprivation, NEFA concentration increased from 0.03 to 0.86 mmol/l and returned to normal (0.04) after a week of watering (Bengoumi 1992) (Fig. 3.5). Similar pattern was described by Delavaud et al. (2013) with a peak at 0.3 mmol/l approximately after 3 weeks of dehydration.

Exercise Effect

During physical exercise, there is an increase in the circulating glucose level, which favors the entry of this substrate to the muscles. On the other hand, the level of NEFA decreases to very low levels from 1.09 before race to 0.29 mmol/l after race (Snow et al. 1988). This finding suggests that during the initial stages of exercise, circulating NEFA is used as substrates. However, when the exercise continues, lipolysis in adipose tissue is unable to maintain the initial high concentrations, which causes a decrease in the availability of this substrate in the skeletal muscles (Snow et al. 1988). This decrease may be due to the inhibition of lipolysis by

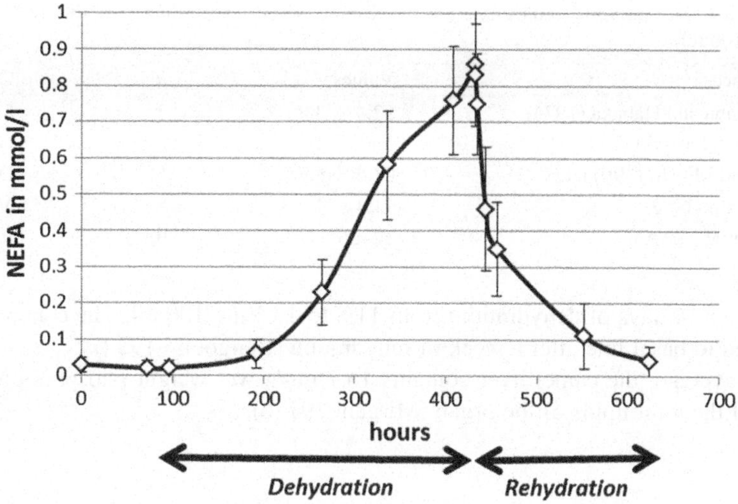

Fig. 3.5 Change in NEFA concentration during water deprivation and then rehydration period [calculated from Bengoumi (1992)]

acidosis (Issekutz and Miller 1962; Snow et al. 1988). Decrease in NEFA is associated with an increase in triglycerides (Snow et al. 1988), something that was not seen in other studies where there was an increase in both parameters and was explained by hepatic reesterification of NEFA (Poso et al. 1983).

3.2.3.4 Phospholipids

Phospholipids are a major component of all cell membranes, notably of the camel erythrocytes contributing to the remarkable properties of the camel RBC (Warda and Zeisig 2000). Its dosage is used in human clinical investigation of different diseases such as diabetes or cerebral pathologies. However, in animals, its dosage is rarely used. Serum phospholipid concentration is determined using phospholipase D and choline oxidase A synthase and acyl coenzyme A oxidase.

In camel, phospholipids were the major constituent of VLDL (10.6 ± 1.2 mg/100 ml), LDL (24.7 ± 3.1 mg/100 ml), and HDL (38.1 ± 0.08 mg/100 ml). Low-density lipoprotein, VLDL, and HDL were important plasma lipoprotein carriers for cholesterol (67.9 ± 9.5%), triglyceride (55.8 ± 7.8%), and phospholipid (51.9 ± 1.6%), respectively (Asadi et al. 2008).

Few references are available in camel (Table 3.4).

Neither age effect nor sex effect was described (Asadi et al. 2009), although a lower value in female (21.3 ± 5.9) than in male (27.1 ± 8.2 mg/100 ml) was reported by Faye and Mulato (1991). As for other lipidic parameters, the phospholipids increased significantly during water privation, the values in plasma being multiplied

Table 3.4 Phospholipid concentration in camel serum according to different authors (in mg/100 ml)

References	Values	*n*	Country
Abu-Sinna and Habaka (1974)	52.4 ± 3.2	7	Egypt
Leat and Northrop (1975)	36.1	–	UK
Faye and Mulato (1991)	22.4 ± 6.7	52	Djibouti
Bengoumi (1992)	11.4 ± 1.8	8	Morocco
Asadi et al. (2009)	12–26	93	Iran

by 4 after 14 days of dehydration: from 11.8 to 41.5 mg/100 ml. The concentration returned to basal line after a week of rehydration (Bengoumi 1992).

On average, the camel liver contains 16.7 mg/g wet weight phospholipids, i.e., 54% of the total lipids of the organ (Mirgani 1977b).

3.2.3.5 Total Lipids

In humans, the clinical interest of the total lipid determination is the investigation of hyperlipidemia in arteriosclerosis, diabetes, and cardiac diseases. Total lipids are obviously correlated to other lipid parameters and are not used in clinical investigation for animals in general. Few references are available in camel regarding total lipids: for example, 3.15 ± 0.08 g/l for Nazifi et al. (2009b). The total lipids in camel appeared in higher concentration in camel serum (4–8 g/l on average) than in other ruminants (Christie 1981): 2.2–4.8 g/l in cattle, 1.6–2.2 g/l in sheep, and 2.6–3.15 g/l in goat. However, lower values are sometimes reported in camel, for example, 1.04 ± 0.15 g/l (Al-Ali et al. 1988).

The total lipid content in camel serum was higher in adult (6.7 ± 1.2 g/l) than in camel baby (5.24 ± 1.1 g/l) and camel yearling (5.50 ± 1.05 g/l), but no sex difference was observed (Mohamed 2008; Nazifi et al. 2000). A seasonal variation was also mentioned. The values were reported to be 3.0 ± 1.57 g/l in summer vs 5.12 ± 1.35 g/l in winter (Nazifi and Gheisari 1999).

With a range of 1.65–2.01 g/l, no significant change occurred with pregnancy (Saeed and Khan 2012).

In Sudan, Anafi breed presented higher total lipids than Arabi breed, 6.69 ± 1.10 g/l and 4.42 ± 1.09 g/l, respectively (Mohamed 2008). In Saudi Arabia, Al-Qarawi (1999) showed also a breed effect, with higher values in white-coat color animals (Waddah breed) both in summer (9.45 ± 0.98) and in winter (11.9 ± 1.23 g/l) than in black-coat camel (Majaheem breed) in both seasons (4.08 ± 0.42 in summer and 6.72 ± 0.53 g/l in winter). The brown-coat color camel (Homor breed) had intermediate values: 7.42 ± 0.91 in summer and 8.38 ± 0.63 g/l in winter. Reverse to the observation of Nazifi and Gheisari (1999), the values were higher in summer (Al-Qarawi 1999).

The total lipids could be modulated by the energy level of the diet (Adel and El-Metwaly 2012), the values varying between 8.73 and 9.32 g/l according to the type of diet.

3.3 Conclusion

Energetic metabolism in camels is different from that in other ruminants with higher glycemia. For most of the energetic parameters, the blood serum concentrations were sensitive to the feeding time and obviously by the energy level of the diet. In consequence, comparisons in the literature have to be careful if data on those aspects are not reported in the methodology. Elsewhere, camel is remarkable by its ability to well manage the alternation of good and bad feeding periods where hump fat is mobilized or stored with a higher inertia than the other species.

References

Abd El-Hag SED, Shaddad SA, Hassan T (2005) Status of some chemical and biochemical parameters of camel blood in the rainy season in the Sudan. J Anim Vet Adv 4(8):713–715

Abd El-Salaam AM, Zeidan AEB, Abdalla EB, Allam MWA (2011) Thermoregulation and blood biochemical changes in male dromedary camels during hot-humid and hot-dry environments under Egyptian conditions. J Camel Pract Res 18(2):297–304

Abdelgadir SE, Wahbi AA, Idris OF (1979) Biochemical studies of some blood and plasma constituents on camel (*Camelus dromedarius*). In: Cockrill WR (ed) Camels and camelids. Scandinavian Institute of African Studies, Uppsala, pp 438–443

Abdelgadir SE, Wahbi AA, Idris OF (1984) Some blood and plasma constituents of the camel. In: Cockrill WR (ed) Proceedings of "the camelid, an all-purpose animal, Khartoum workshop on camels, December 1979", vol 1. SIDA, Uppsala, pp 438–441

Abdin-Bey MR (2001) Metabolic and endocrine changes during detomidine-HCl sedation in dromedary camel. J Camel Pract Res 8(2):181–183

Abu Damir H, Eldirdiri NI, Adam SEI, Howarth JA, Salih YM, Idris OF (1993) Experimental copper poisoning in the camel (*Camelus dromedarius*). J Comp Pathol 108:191–208

Abu-Sinna GME, Habaka SM (1974) The cholesterol and phospholipid contents of the serum in the camel (*Camelus dromedarius*). Ann Zool 10(3):57–60

Abu-Tarboush HM, Dawood AA (1993) Cholesterol and fat contents of animal adipose tissues. Food Chem 46:89–93

Adel EM, El-Metwaly H (2012) Effect of feed additive "exogenous enzymes" on growth performance of Maghraby camels. Life Sci J 9(4):4830–4835

Ahmed SH, Omer SAA, Gameel AA (2009) Some normal constituents in serum and cerebrospinal fluid in Sudanese camels (*Camelus dromedarius*). Assiut Vet Med J 55(123):163–170

Aichouni A, Jeblawi R, Dellal A, Hammou H, Aggad H (2010) Breed variation in blood constituents of the one-humped camel (*Camelus dromedarius*) in Algeria. J Camelid Sci 3:19–25

Aichouni A, Belhadia M, Kebir N, Aggad H (2013) Season influence on serum organic parameters of dromedarius (*Camelus dromedarius*) in Algeria. Biochem Biotechnol Res 1(1):8–12

Al-Ali AK, Husayni HA, Power DM (1988) A comprehensive biochemical analysis of the blood of the camel (*Camelus dromedarius*). Comp Biochem Physiol 89B(1):35–37

Al-Ani FK, Al-Azzawi WARA, Jermukly MS, Razzaq KK (1992) Studies on some hematological parameters of camel and lama in Iraq. Bull Anim Health Prod Afr 40:103–106

Al-Azraqi AA (2007) Relationship of serum leptin concentration to fat deposition in slaughtered young camels. J Anim Vet Adv 6:16–19

Albomohsen H, Mamouei M, Tabatabaei S, Fayazi J (2011) Metabolite composition variations of follicular fluid and blood serum in Iranian dromedary camels during the peak breeding season. J Anim Vet Adv 10(3):327–331

Al-Harbi MS (2012) Some hematologic values and serum biochemical parameters in male camels (*Camelus dromedarius*) before and during rut. Asian J Anim Vet Adv 7:1219–1226

Ali S, Ahmad N, Akhtar N, Zia-ur-Rahman NDE (2008) Metabolite contents of blood serum and fluid from small and large sized follicles in dromedary camels during the peak and the low breeding seasons. Anim Reprod Sci 108:446–456

Ali A, Derar D, Tharwat M, Zeitounb MM, Alsobyil FA (2016) Dystocia in dromedary camels: prevalence, forms, risks and hematobiochemical changes. Anim Reprod Sci 170:149–156

Al-Qarawi A (1999) The chronobiological blood parameter changes as correlated to different traits of *Camelus dromedarius* in Saudi Arabia. J Camel Pract Res 6(1):45–48

Al-Rukibat R, Ismail ZB (2014) Biochemical analysis of serum and synovial fluid in clinically normal young camels (*Camelus dromedarius*). Vet Sci Dev 4(1). http://www.pagepress.org/journals/index.php/vsd/article/view/5371/6375

Alshamsi NS, Ksiksi TS, Ashraf SS (2015) Altered serum enzymes and biochemical levels in Arabian racing camels with bone fractures. J Anim Plant Sci 25(4):1072–1080

Al-Sobayil FA, Mousa HM (2009) Clinical and biochemical studies on wry-neck syndrome in camels in Saudi Arabia. J Camel Pract Res 16(2):179–182

Al-Suhaimi E, Abu-se'da K, Akbar A (2009) Relationship of leptin hormone with insulin and glucose in Arabian camel (*Camelus dromedarius*). J Camel Pract Res 16:201–207

Al-Sultan SI (2003) Studies of some normal biochemical parameters of Majaheem breed of camel (*Camelus dromedarius*) in Saudi Arabia. J Camel Pract Res 10(1):79–80

Amin ASA, Abdoun KA, Abdelatif AM (2007) Seasonal variation in blood constituents of one-humped camel (*Camelus dromedarius*). Pak J Biol Sci 10(8):1250–1256

Asadi F, Shahriari A, Asadian P, Pourkabir M, Samadaei M (2008) Composition and electrophoretic mobility of plasma lipoproteins of dromedary camels (*Camelus dromedarius*). Am J Vet Res 69 (7):880–885

Asadi F, Shahriari A, Asadian P, Pourkabir M, Sabzikar A, Ojaghee R (2009) Serum lipid, glucose, free fatty acids and liver triglyceride in sub-adult and adult camels (*Camelus dromedarius*). Rev Med Vet 160(12):552–556

Ateeq G, Kouider S, Kolb E (1984) Haematology of dromedaries, cell counts, haemoglobin, proteins, urea, cholesterol, AST and ALT. Age and sex variations (in German). Archiv Exp Vet Med 38(5):664–675

Ayoub MA, Saleh AA (1998) A comparative physiological study between camels and goats during water deprivation. In: Alhadrami G, Ayoub MA (eds) Proceedings of the 3rd annual meeting for animal production under arid conditions, vol 1, 2–3 May 98, Al-Ain, UAE, pp 71–87

Ayoub MA, El-Khouly AA, Mohamed TM (2003) Some hematological and biochemical and steroid hormones levels in one-humped camel during different physiological conditions. Emir J Agric Sci 15(1):44–55

Azwai SM, Saltani H, Thomas PC, Shareha AM, El-Gammoudi F, Mohamed SO (1990) Note on cholesterol, glucose, urea and total protein concentration in serum of normal camels. Camel Newsl 7:94

Badakhshan Y, Mirmahmoudi R (2016) Blood metabolites of one-humped camel (*Camelus dromedarius*) versus sheep during summer heat stress. Iran J Vet Med 10(1):65–71

Badawy MT, Gawish M, Khalifa MA, El-Nouty FD, Hassan GA (2008) Seasonal variations in hemato-biochemical parameters in mature one humped she-camels in the north-western coast of Egypt. Egypt J Anim Prod 45(2):155–164

Badryyah R, Al-Suwaigh E, Al-Suhaimi A (2005) Comparative study on some biochemical constituents of plasma in male camels and goats. J Camel Pract Res 12(2):141–143

Barakat MZ, Abdel Fattah M (1971) Seasonal and sexual variations of certain constituents of normal camel blood. Zentrabl Veterinarmed A 18:174–176

Barakat MZ, Abdel-Fattah M (1970) Biochemical analysis of normal camel blood. Zentrabl Veterinarmed A 17:550–557

Bargaâ R, Lektib I, Farh M, El Abbadi N, Belhouari A, El Khasmi M (2016) Seasonal variations of biochemical profile and its correlations with thyroid and adrenal gland hormones in Moroccan camel (*Camelus dromedarius*). Int J Agric Environ Res 2(6):1858–1872

Barsham MA, Elbagir NM, Barri MES (2005) Serum biochemical changes associated with experimentally-induced hypothyroidism in one-humped camels (*Camelus dromedarius*). J Camel Pract Res 12(2):117–121

Bartha T, Sayed-Ahmeda A, Rudas P (2005) Expression of leptin and its receptors in various tissues of ruminants. Domest Anim Endocrinol 29:193–202

Bengoumi M (1992) Biochimie clinique du dromadaire et mécanismes de son adaptation à la déshydratation, Thèse de doctorat ès Sciences Agronomiques. I.A.V. Hassan II, rabat, Maroc, 184 p

Bengoumi M, Faye B, De La Farge F (1998) Clinical enzymology in the dromedary camel (*Camelus dromedarius*). Part III: Effect of dehydration on serum AST, ALT, GGT, AP, LDH and urine GGT activities. J Camel Pract Res 5(1):119–122

Bengoumi M, Moutaouakil F, De La Farge F, Faye B (2004) Seasonal variations of the plasma thyroid hormone concentrations and the body temperature in the dromedary camel. In: Gahlot TK (ed) Selected research on camelid physiology and nutrition. Camelid, Bikaner, pp 238–244

Bengoumi M, Faulconnier Y, Tabarani A, Sghiri A, Faye B, Chilliard Y (2005) Effect of feeding level on body weight, hump size, lipid content and adipocyte volume in the dromedary camel. Anim Res 54:383–393

Bizzetti M, Farah ME, Gugliucci B, Demi S (1988) Glycolipid metabolism in dromedaries reared in Somalia (in Italian). Ann Fact Med Vet Pisa 41:373–376

Chandrasena LG, Emmanuel B, Gilanpour H (1979a) A comparative study of glucose metabolism between the camel and the sheep. Comp Biochem Physiol 62A:837–840

Chandrasena LG, Emmanuel B, Hamar DW, Howard BR (1979b) A comparative study of ketone body metabolism between the camel (*Camelus dromedarius*) and the sheep (*Ovis aries*). Comp Biochem Physiol 64(1):109–112

Chavanne P, Boué A (1950) Taux normaux de l'urée et du glucose sanguine chez le dromadaire nord-africain. Rev Elev Méd Vét Payx Trop 4(4):183

Chehma A, Faye B, Bastianelli D (2010) Valeurs nutritionnelles des plantes vivaces des parcours sahariens algériens pour dromadaires. Fourrages 204:263–268

Chiericato GM, Warfa AA, Schiapelli KP (1986) Influence of sex of the dromedary on some constituents of the blood. Rivista Di Zootecniae Veterinaria 14(3):196–199

Chilliard Y (1987) Bibliographic review: Quantitative variations and metabolism of lipids in adipose tissue and liver during the gestation-lactation cycle. 2: in the ewe and the cow [in French]. Reprod Nutr Develop 27(2A):327–398

Chilliard Y (1989) Particularités du métabolisme des lipides et du métabolisme énergétique chez le dromadaire. Options méditerranéennes, série séminaires no 2, pp 101–110

Chilliard Y, Ferlay A, Mansbridge RM, Doreau M (2000) Ruminant milk fat plasticity: nutritional control of saturated, polyunsaturated, trans and conjugated fatty acids. Ann Zootech 49:181–205

Chilliard Y, Delavaud C, Bonnet M (2005) Leptin expression in ruminants: nutritional and physiological regulations in relation with energy metabolism. Domest Anim Endocrinol 29 (1):3–22

Christie WW (1981) Lipid metabolism in ruminant animals, vol VIII. Pergamon Press, Oxford, 460 p

Congiu S (1953) Dressing percentage, weight distribution, and correlations between different parts of the body in the Somali dromedary. Zootecnica Veterinaria 8:188–191

Dahlborn K, Benlamlih S, Wallsten C, Hossaini-Hilali J (1992) Effect of food deprivation on fluid balance and glucose regulation in pregnant and lactating camels. In: Allen WR, Higgins AJ, Mayhew IG, Snow DH, Wade JF (eds) Proceedings of the 1st international camel conference, 2–6 Feb 92. R&W, Dubai, pp 275–277

Delavaud C, Bengoumi M, Faye B, Levieux D, Chilliard Y (2013) Plasma leptin, glucose and non-esterified fatty acid variations in dromedary camels exposed to prolonged periods of underfeeding or dehydration. Comp Biochem Physiol A 166:177–185

Durand M, Kchouk M (1958) Etude de quelques constants hématologiques et chimiques chez le dromadaire. Bull Acad Vet France 31:197–198

El Khasmi M, Chakir Y, Riad F, Safwate A, Tahri EH, Farh M, El Abbadi N, Abouhafs R, Faye B (2013) Effects of transportation stress during the hot-dry season on some haematological and physiological parameters in Moroccan dromedary camels (*Camelus dromedarius*). J Life Sci 7 (1):13–25

El Khasmi M, Chakir Y, Bergaâ R, Barka K, Lektib I, El-Abbadi N, Belhouari A, Faye B (2015) Impact of transport distance on stress biomarkers levels in dromedary camel (*Camelus dromedarius*). Emir J Food Agric 27(6):507–512

El-Bahrawy KA, El-Hassanein EE (2011) Seasonal variations of some blood and seminal plasma biochemical parameters of male dromedary camels. Am-Euras J Agric Environ Sci 10 (3):354–360

El-Boshy M, Abbas H, El-Khodery S, Osman S (2009) Cytokine response and clinicopathological findings in Brucella infected camels (*Camelus dromedarius*). Vet Med 54(1):25–32

El-Harairy MA, Zeidan AEB, Afify AA, Amer HA, Amer AM (2010) Ovarian activity, biochemical changes and histological status of the dromedary she-camel as affected by different seasons of the year. Nature Sci 8(5):54–65

Elias E, Yagil R (1984) Haematological and serum biochemical values in lactating camels (*Camelus dromedarius*) and their newborn. Refuah vet, vol 41, pp 7–13

Elmahdi B, Sallmann HP, Fuhrmann H, von Engelhardt W, Kaske M (1997) Comparative aspects of glucose tolerance in camels, sheep, and ponies. Comp Biochem Physiol A Physiol 118 (1):147–151

Elnahas A (2008) Ultrasonographical examination of one humped camels (*Camelus dromedarius*) liver with some haematological and biochemical aspects, DVM dissertation. University of Leipzig, Germany, 107 p

Elsanhoty RM, El-Gohery SS, Badr FH (2011) Cholesterol reduction in camel hump fat using b-cyclodextrin. J Verbr Lebensm 6:183–189

Emmanuel B (1979) Comparative biochemical studies between the camel and other ruminants. In: Camels. IFS Symposium, Sudan, 321–346.

Emmanuel B (1980) Oxidation of butyrate to ketone bodies and CO2 in the rumen epithelium, liver, kidney, heart and lung of camel (*Camelus dromedarius*), sheep (*Ovis aries*) and goat (*Capra hircus*). Comp Biochem Physiol 65B:699–704

Emmanuel B (1981) Further metabolic studies in the rumen of camel (*Camelus dromedarius*) and sheep (*Ovis aries*). Comp Biochem Physiol 63:155–158

Emmanuel B (1984) Comparative biochemical studies. In: Cockrill WR (ed) Proceedings of workshop "the camelids , an all-purpose animal", Khartoum, December 1979. Scandinavian Institute of African Studies, Uppsala, pp 449–478

Emmanuel B, Nahapetian A (1980) Fatty acid composition of depot fats, and rumen wall of the camel (*Camelus dromedarius*). Comp Biochem Physiol 67B:701–704

Faye B (1997) Guide de l'élevage du dromadaire. Ed. Sanofi, Libourne, 126 pp

Faye B, Mulato C (1991) Facteurs de variation des paramètres protéo-énergétiques, enzymatiques et minéraux dans le plasma chez le dromadaire de Djibouti. Rev Elev Med Vét Pays Trop 44:325–334

Faye B, Saint-Martin G, Cherrier R, Ruffa A (1992) The influence of high dietary protein, energy and mineral intake on deficient young camel: I. Changes in metabolic profiles and growth performance. Comp Biochem Physiol 102A:409–416

Faye B, Jouany JP, Chacornac JP, Ratovonanahary M (1995) L'élevage des grands camélidés. Analyse des initiatives réalisées en France. INRA Productions Animales 8(1):3–17

Faye B, Bengoumi M, Viateau E, Tourret M, Chilliard Y (2001a) Adipocyte patterns of adipose tissue in camel hump and kidney. J Camel Res Pract 8:29–33

Faye B, Bengoumi M, Messad S, Chilliard Y (2001b) Fat storage and adipocyte patterns in camel: a tool for management of reproduction. Adv Reprod 5(3):10c

Faye B, Alsharary FZ, Al-Rwaily SH (2012) Gestion et évaluation du statut énergétique du dromadaire. Revue des BioRessources 2(2):1–16

Faye B, Bengoumi M, Al-Masaud A, Konuspayeva G (2015a) Comparative milk and serum cholesterol content in dairy cow and camel. J King Saud Univ Sci 27:168–175

Faye B, Bengoumi M, Al-Masaud A, Konuspayeva G (2015b) Low milk cholesterol in camel milk: true or not? In: Almaty K, Konuspayeva G (eds) Proceedings of 4th conference of ISOCARD, "silk road camel: the camelids, main stakes for sustainable development", 8–12 June 2015. Special issue of scientific and practical journal Veterinariya #2 (42) 2015, oral communication, pp 189–191

Ghodsian I, Nowrouzian I, Schels HF (1978) A study of some hematological parameters in the Iranian camels. Trop Anim Health Prod 10:109–110

Gorban AM, Izzeldin OM (1999) Study on cholesteryl ester fatty acids in camel and cow milk lipid. Int J Food Sci Technol 34:229–234

Guerouali A, Filali RZ (1992) Maintenance energy requirements of the dromedary camel. In: Allen WR, Higgins AJ, Mayhew ID, Snow DH, Wade JF (eds) Proceedings of the 1st international camel conference. R & W Publications Newmarket, England, pp 251–254

Gupta AK, Vyas KK, Dwaraknath PK, Pareek PK (1979a) Effect of breeding season, castration and exogenous testosterone on blood glucose level and eosinophil count of male camels (*Camelus dromedarius*). Indian J Anim Sci 49(7):554–556

Gupta GC, Joshi BP, Rai P (1979b) Observations on haematology of camel. Indian Vet J 56:269–272

Gupta L, Kumar RA, Ghanshyam T, Rajesh D, Garg R (2012) Effect of feeding different proportions of groundnut haulms (*Arachis hypogaea*) and cluster bean straw (*Cyamopsis tetragonoloba*) on nutrient utilisation and serum biochemical parameters in dromedary camels. Trop Anim Health Prod 44:1689–1695

Gutierrez C, Corbera JA, Juste MC, Doreste F, Morales I (2005) An outbreak of abortions and high neonatal mortality associated with *Trypanosoma evansi* in dromedary camels in the Canary Islands. Vet Parasitol 130:163–168

Hamad B, Aggad H, Hadef L, Adaika A (2017) Effect of seasons on blood biochemical parameters in male dromedary camels in Algeria. Indian J Anim Res B 672:1–5

Hekmatimoghaddam S, Rasooli A, Sazmand A, Hamidinejat H, Jafari H, Nouri M (2011) Serobiochemical alternations in dromedary camels naturally infected with Theileria spp. In: Iran proceedings of XVth international congress on animal hygiene, July 2011, vol 1, pp 455–457

Heller R, Lechner M, Von Engelhardt W (1986) Forestomach motility in the camel (*Camelus dromedarius*). Comp Biochem Physiol A Physiol 84(2):285–288

Hussein MF, Salah MS, Mogawer HH, Gar ElNabi AR (1992) Effect of lactation on the haemogram and certain blood constituents of the dromedary camel. J Appl Anim Res 1:43–50

Hussein YA, Al-Eknah MM, AL-Shami SA, Mandour MA, Fouda TA (2012) Coat color breed variation in blood constituents among indigenous Saudi Arabia camel strains. Mansoura Vet Med J 14(1):191–204

Ismail ZB, Qureshi T, Levy M, Khamas W, Nour A (2003) Preliminary report on some physiological parameters in pregnant she-camel (*Camelus dromedarius*) in North America. J Camel Pract Res 10(2):125–129

Issekutz B, Miller H (1962) Plasma free fatty acids during exercise and the effect of lactic acid. Exp Biol Med 110(2):237–239

Kadim IT, Maghoub O, Al-Maqbaly RS, Annamalai K, Al-Ajmi DS (2002) Effects of age on fatty acid composition of the hump and abdomen depot fats of the Arabian camel (*Camelus dromedarius*). Meat Sci 62:245–251

Kadim I, Mahgoub O, Purchas RW (2008) A review of the growth, and of the carcass and meat quality characteristics of the one-humped camel (*Camelus dromedaries*). Meat Sci 80:555–569

Kamal AM, Salama OA (2009) Lipid fractions and fatty acid composition of colostrums, transitional and mature she-camel milk during the first month of lactation. Asian J Clin Nutr 1:23–30

Kamili A, Bengoumi M, Faye B (2006) Assessment of body condition and body composition in camel by barometric measurements. J Camel Pract Res 13(1):67–72

Kaneko JJ, Harvey JW, Bruss M (1997) Clinical biochemistry of domestic animals, 5th edn. Academic, New York, 932 pp

Kataria N, Sareen M, Kataria AK, Gahlot TK, Bhatia JS, Ghosal AK (1995) Studies on cerebrospinal fluid in healthy camels (*Camelus dromedarius*). J Camel Pract Res 2(1):29–32

Khadjeh GH, Mojabi A, Mayahi M (1997) Normal value of some biochemical constituents in blood of Iranian one-humped camels (*Camelus dromedarius*). Indian J Vet Sci 67:414–417

Khan AA, Kohli IS (1973) A note on variations in blood serum cholesterol in camel (*Camelus dromedarius*) before and during rut. Indian J Anim Sci 43(12):1094–1095

Khan AA, Kohli S (1981) Blood serum cholesterol in male camel (*Camelus dromedarius*) before and during rut. Indian Vet J 58:498

Konuspayeva G, Lemarie E, Faye B, Loiseau G, Montet D (2008) Fatty acid and cholesterol composition of camel's (*Camelus bactrianus*, *Camelus dromedarius* and hybrids) milk in Kazakhstan. Dairy Sci Technol 88:327–340

Konuspayeva G, Faye B, Mussaad A (2014) Some lipid components of the camel milk and blood in intensive farm in Saudi Arabia. Emir J Food Agric 26(4):349–353

Kouider S, Kolb E (1982a) Studies into diurnal glucose levels and enzyme activities (aspartate amino-transferase, alanine amino-transferase, acid alkaline phosphatases, glucose 6-phosphate dehydrogenase) in serum of camels. Archiv Exp Vet Med 26(4):601–610

Kouider S, Kolb E (1982b) Concentrations of glucose and urea in serum of camels. Monatsch Vet Med 37(4):140–142

Kouider S, Ateeq G, Kolb E (1988) Studies into levels of total proteins, urea, total fat, cholesterol and bilirubin in blood plasma of camels in the course of one year (in German). Vet Med 43:139–142

Kumar M, Banerjee S (1962) Biochemical studies on Indian camel (*Camelus dromedarius*). 3. Plasma insulin-like activity and glucose tolerance. J Sci Indust Res 21:291–292

Leat WMF, Northrop CA (1975) Plasma lipids and lipoproteins of some herbivorous mammals. Proc Nutr Soc 34:105A–107A

Mal G, Sena DS, Sahani MS (2006) Haemato-biochemical changes in camels infested with mange during winter and summer season. J Camel Pract Res 13(2):174–174

Maloiy GMO (1972) Comparative studies on digestion and fermentation rate in the fore-stomach of the one-humped camel and the zebu steer. Res Vet Sci 13:476–482

Maloiy GMO, Clemens ET (1980) Gastrointestinal osmolality electrolyte and organic acid composition in five species of east African herbivorous mammals. J Anim Sci 51(4):917–924

Mazur A, Bauchart D, Chilliard Y, Didier R, Rayssiguier Y (1986) Teneur en lipides du foie chez la vache en début de lactation : comparaison des différentes techniques d'estimation des triglycérides hépatiques. Reprod Nutr Devel 26(1B):355–356

Mehrotra V, Gupta ML (1989) Seasonal variations in certain blood constituents in camel. Indian J Anim Sci 59:1559–1561

Mills GL, Taylaur CE (1971) The distribution and composition of serum lipoproteins in eighteen animals. Comp Biochem Physiol 40B:489–501

Mirgani T (1977a) Fatty acid composition of hump triglycerides of the camel *Camelus dromedarius*. Comp Biochem Physiol B Comp Biochem 58(2):211–213

Mirgani T (1977b) Fatty acid composition of liver triglycerides of camel. World Rev Anim Prod 13 (4):57–59

Mirgani T (1981) Effect of fasting on camel tissue lipid. J Arid Environ 4:359–361
Mirgani T, Uro A, Sallmann HB (1987) Studies on lipogenic enzymes in camel tissues. I. Enzymes of citrate cleavage pathway, vol 1983. Scientific Reports University of Khartoum, Sudan, pp 71–77
Mohamed HE (2008) Factors affecting the plasma lipid status in camels (*Camelus dromedarius*). Res J Biol Sci 3(4):444–445
Mohamed HA, Hussein AN (1999) Studies on normal haematological and serum biochemical values of the 'Hijin' racing camels (*Camelus dromedarius*) in Kuwait. Vet Res Commun 23 (4):241–248
Mohamed MA, Mohamed A, Locatelli A (1984) Water deprivation effects on the hematological and hematochemical pictures of *Camelus dromedarius*. Rev Elev Med Vet Pays Trop 37:313–317
Mohammed AK, Sackey AKB, Tekdek LB, Gefu JO (2007) Serum biochemical values of healthy adult one humped camel (*Camelus dromedarius*) introduced into a sub-humid climate in Shika-Zaria, Nigeria. J Camel Pract Res 14(2):191–194
Momenah MA (2014) Some blood parameters of one humped she-camels (*Camelus dromedaries*) in response to parasitic infection. Life Sci J 11(5):118–123
Moolchandani A, Sareen M (2016) Blood biochemical study in Trypanosomiasis infected camel (*Camelus dromedarius*). Res Rev J Vet Sci 2(1):41–43
Mosaferi S, Niasari-Naslaji A, Abarghani A, Gharahdaghi A, Gerami A (2005) Biophysical and biochemical characteristics of Bactrian camel semen collected by artificial vagina. Theriogen 63 (1):92–101
Naeini AT, Nazifi S (2001) Biochemical and cytological properties of blood and peritoneal fluid in clinically healthy adult camels (*Camelus dromedarius*). J Camel Pract Res 8(2):123–126
Nagpal AK, Roy AK, Chirania BL, Patil NV (2011) Growth, nutrient utilization and serum profile in camel calves as affected by dietary protein levels. Indian J Anim Nutr 28(2):166–171
Narnawane SD, Nagarajan G, Dahiya SS (2015) Hemato-biochemical studies in Indian camel (*Camelus dromedarius*) affected by contagious ecthyma. Indian J Vet Pathol 39(2):168–170
Nazifi S, Gheisari HR (1999) The influences of thermal stress on serum lipids of camel (*Camelus dromedarius*). J Camel Pract Res 6(2):307–309
Nazifi S, Rezakhani A, Gheisari HR (1998) Physical, biochemical and cytologic properties of blood and synovial fluid in clinically normal adult camel (*Camelus dromedarius*). J Vet Med A 45:155–160
Nazifi S, Gheisari R, Poorabbas H (1999) The influence of thermal stress on serum biochemical parameters of dromedary camels and their correlation with thyroid activity. Comp Haematol Int 9:49–53
Nazifi S, Gheisari M, Poorkabir A, Saadatfar S (2000) Serum lipids and lipoproteins in clinically healthy male camels (*Camelus dromedarius*). Vet Res Commun 24:527–531
Nazifi S, Oryan A, Bahrami S, Razavi SM (2009a) Evaluation of haematological and serum biochemical parameters in Iranian camels (*Camelus dromedarius*) infected with haemotrophic mycoplasma (Eperythrozoon) spp. Comp Clin Pathol 18:329–332
Nazifi S, Nikahval B, Mansourian M, Razavi SM, Farshneshani F, Rahsepar M, Javdani M, Bozorgi H (2009b) Relationships between thyroid hormones, serum lipid profile and erythrocyte antioxidant enzymes in clinically healthy camel (*Camelus dromedarius*). Revue Méd Vét 160(1):3–9
NRC (2001) Nutrient requirement of dairy cattle, 7th Rev. edn. National Academy Press, Washington, 401 p
Nyang'ao JMN, Olaho-Mukani W, Maribei JM, Omuse JK (1997) A study of some haematological and biochemical parameters of the normal dromedary camel in Kenya. J Camel Pract Res 4 (1):31–33
Omidi A, Zh S, Montazer-Torbati MB, Mostafai M (2014a) Metabolic profile of pregnant, non-pregnant and male two-humped camels (*Camelus bactrianus*) of Iran. Iran J Vet Med 8 (4):235–242

Omidi A, Sajedi Z, Torbati MBM, Nik HA (2014b) Lipid profile and thyroid hormone status in the last trimester of pregnancy in single-humped camels (*Camelus dromedarius*). Trop Anim Health Prod 46:609–614
Orliac D (1980) Contribution à l'étude de la biochimie sanguine de dromadaires et de chèvres sahariens no 71, Thèse doct. Vét. Toulouse, France
Orlov VK, Servetnik-Chalaia GK, Zagibaĭlova NB (1985) Fractional and fatty-acid composition of the lipids of horse and camel meat (in Russian). Vopr Pitan 4:71–76
Osman TEA, Al-Busadah KA (2000) Effects of age and lactation on some biochemical constituents of camel blood in Saudi Arabia. J Camel Pract Res 7(2):149–152
Osman TEA, Al-Busadah KA (2003) Normal concentrations of twenty biochemical parameters of she-camels, cows and ewes in Saudi Arabia. Pak J Biol Sci 6(14):1253–1256
Perk K, Lobl K (1961) A study of the serum proteins and lipoproteins of the camel and their relation to its resistance to heat and thirst. Refuah Vet 18(3):168–172
Poso AR, Soveri T, Oksanen HE (1983) The effect of exercise on blood parameters in standardbred and Finnish-bred horses. Acta Vet Scand 24:170–184
Raiymbek G (2013) Composition, quality and histochemical properties of bactrian camel (*Camelus bactrianus*) individual muscles, PhD in biology. Kazakh National University Al-Farabi, Kazakhstan
Rana JS, Singh S, Singh P, Singh J (1998) Experimental studies on jejunal obstruction in the camel (*Camelus dromedarius*). J Camel Pract Res 5(2):281–286
Rawdah TN, El-Faer MZ, Koreish SA (1994) Fatty acid composition of the meat and fat of the one-humped camel (*Camelus dromedarius*). Meat Sci 37:149–155
Raziq K, Younas M, Kakar MA (2008) Camel—a potential dairy animal in difficult environments. Pak J Agric Sci 45(2):263–267
Rezakhani A, Habibabadi SN, Ghojogh MM (1997) Studies on normal haematological and biochemical parameters of Turkmen camel in Iran. J Camel Pract Res 4(1):41–44
Röttcher D, Schillinger D, Zweygarth E (1987) Trypanosomiasis in the camel (*Camelus dromedarius*). Rev Sci Tech Off Int Epiz 6(2):463–470
Rutagwenda T, Lechner-Doll M, Schwartz HJ, Schultka W, Von Engelhardt W (1990) Dietary preference and degrability of forage on a semi-arid thornbush savannah indigenous ruminants, camels and donkeys. Anim Feed Sci Technol 35:179–192
Saeb M, Baghshani H, Nazifi S, Saeb S (2010) Physiological response of dromedary camels to road transportation in relation to circulating levels of cortisol, thyroid hormones and some serum biochemical parameters. Trop Anim Health Prod 42:55–63
Saeed A, Khan IA (2012) Alterations in serum lipids and lipoproteins profile of pregnant camels (*Camelus dromedarius*) at term. Comp Clin Pathol 21:1019–1021
Saeed A, Hussein MM, Khan IA, Chand G, El-Yousuf RA (2004a) Effect of sex and age on blood biochemical profile in camel. J Camel Pract Res 11(1):73–77
Saeed A, Hussain MM, Khan IA, El-Yousuf RJ (2004b) Effect of age on some serum constituents of camels in United Arab Emirates. Indian J Anim Sci 74(3):278–280
Saeed A, Khan IA, Hussein MM (2009) Change in biochemical profile of pregnant camels (*Camelus dromedarius*) at term. Comp Clin Pathol 18(2):139–143
Sahraoui N, Doudou A, Douadji O, Babelhadj B, Hornick JL (2016) Impact of natural vegetation on some biochemical parameters of the Arabian camel (*Camelus dromedarius*) in Algeria. J Camelid Sci 9:62–71
Sarwar A, Majeed MA (1997) Interrelationships between 30 parameters of blood in normal one-humped camel in summer. J Camel Pract Res 4(1):35–39
Sarwar A, Chaudhry MN, Khan IR, Abbas S, Majeed MA (1991) Eight serum biochemical values of one-humped camel (*Camelus dromedarius*): effect of sex, age in males and lactation and/or pregnancy in females. Pak Vet J 11:62–68
Sayed-Ahmed A, Elmorsy SE, Rudas P, Bartha T (2003) Partial cloning and localization of leptin and leptin receptor in the mammary gland of the Egyptian water buffalo. Domest Anim Endocrinol 25:303–314

Sazmand A, Rasooli A, Nouri M, Hamidinejat H, Hekmatimoghaddam S (2011) Serobiochemical alterations in subclinically affected dromedary camels with *Trypanosoma evansi* in Iran. Pak Vet J 31(3):223–226

Shaheen HM (2001) The effect of feed and water deprivation on ingestive behavior and blood constituents in camels: comparison with sheep and goats. J Camel Pract Res 8(2):153–162

Shukla MK, Khan MJZ, Pathan MM, Siddiquee GM, Latif A (2009) Biochemical profile of uncastrated male Kutchi camel during the breeding season. J Camel Pract Res 16(1):213–215

Snow DH, Billah A, Ridha A (1988) Effects of maximal exercise on the blood composition of the racing camel. Vet Rec 123:311–312

Soliman MK, Shaker M (1967) Cytological and biochemical studies on the blood of adult she-camels. Indian Vet J 44:989–995

Souilem O, Chine O, Alguemi C, Gogny M (1999) Etude de la glycémie chez le dromadaire (*Camelus dromedarius*) en Tunisie : résultats préliminaires. Rev Méd Vét 150(6):537–542

Tajik J, Sazmand A, Moghaddam SHH, Rasooli A (2013) Serum concentrations of thyroid hormones, cholesterol and triglyceride, and their correlations together in clinically healthy camels (*Camelus dromedarius*): effect of season, sex and age. Vet Res Forum 4(4):239–243

Valleras A, Stevens CE (1971) *Motility of the llama* and guanaco stomach. Am J Physiol 220:275–282

Warda M, Zeisig R (2000) Phospholipid- and fatty acid-composition in the erythrocyte membrane of the one-humped camel (*Camelus dromedarius*) and its influence on vesicle properties prepared from these lipids. Deutsche Tierärz Woch 107(9):368–373

Wasfi IA, Hafez AM, El-Tayeb FMA, El-Taher AY (1987) Thyroid hormones, cholesterol and triglyceride levels in the camel. Res Vet Sci 42:418

Wasfi IA, Gadir FA, Abdulla OM (1989) The acute effects of dexamethasone on some haematological parameters and serum biochemistry in the camel (*Camelus dromedarius*). Rev Élev Méd Vét Pays Trop 42(4):479–483

Wensvoort J, Kyle DJ, Orskov ER, Bourke DA (2001) Biochemical adaptation of camelids during periods where feed is withheld. Rangifer 21(1):45–48

Wilson RT (1984) The camel. Longman, London

Wood JD (1984) Fat deposition and the quality of fat tissue in meat animals. In: Wiseman JW (ed) Fats in animal nutrition. Butterworth, London, pp 406–435

Yagil R (1985) The desert camel. Comparative physiological adaptation. Karger, London, 162 p

Yagil R, Berlyne GM (1977) Glucose loading and dehydration in the camel. J Appl Physiol 42 (5):690–693

Yousif HS, Omer SA, Ahmed SH (2016) Influence of thermal environment change on blood metabolites, leukocytic and erythrocytic indices and clinical parameters of in-door camel (Camelus dromedaries). Glob J Biol Agric Health Sci 5(1):20–24

Zaher H, El-Zahar H, Al Sharifi S, Shety T (2017) Alterations in hematological and biochemical parameters affecting the reproductive performance in female camels (*Camelus dromedaries*). Int J Vet Health Sci Res 5(1):155–160

Zeidan AEB, Abd El-Salam AM, El-Malky OM, Ahmadi EA, Sarhan DMA, Daader AH (2008) Biochemical and histological changes of the ovary of the dromedary camel during breeding and non-breeding seasons. Egypt J Basic Appl Physiol 7:287–308

References

Raymond A, Bazoin A, Nouri M, Hanifehnezhad H, Heidari [illegible] S (2011) Serological survey of infections in [illegible] (Camelus dromedarius) [illegible] Trop Anim Health Prod [illegible] 43(1):223–[illegible]

Shalash MR (1983) The effect of season and water deprivation on [illegible] in the camel (Camelus dromedarius) [illegible] J Camel Pract Res [illegible]

[illegible] (2016) [illegible] during the breeding season. J Camel Pract Res [illegible]

[illegible] and the [illegible] of the [illegible] Vet Res [illegible]

Skidmore JA, Billah M, [illegible] (1996) [illegible] and [illegible] studies on the [illegible] camels. [illegible]

[illegible] (1999) [illegible] [illegible] Reprod [illegible]

Tariq [illegible] A, [illegible] (2016) Serum concentrations of thyroid hormones, cortisol and [illegible] in clinically healthy camels (Camelus dromedarius): effect of season, sex and age. Vet [illegible]

[illegible] and [illegible]

Wang [illegible] and [illegible] of the [illegible] and its [illegible] Camelus [illegible]

[illegible] (2012) [illegible] Vet Sci [illegible]

Wani [illegible] (1992) [illegible] effects of [illegible] the camel (Camelus dromedarius) [illegible]

[illegible] (2011) [illegible] of [illegible] camel [illegible] Reprod [illegible]

Wilson RT (1984) The camel. Longman, London

Wood AT (1995) [illegible] quality of [illegible] Wilson RM (ed) [illegible]

[illegible] (2001) [illegible] and [illegible] in the camel [illegible]

Yousif HS, [illegible] SA, Ahmed NM (2016) [illegible] of [illegible] and chemical [illegible] camel (Camelus dromedarius) [illegible] Sci [illegible]

[illegible] (2017) [illegible] and [illegible] parameters affecting the [illegible] performance in [illegible] (Camelus dromedarius) [illegible] Appl Anim [illegible]

[illegible] (2005) [illegible] the dromedary camel [illegible] and [illegible] Appl Physiol 2:28–[illegible]

Chapter 4
Nitrogen and Protein Parameters

Proteins are an essential component of the blood. The usual parameters analyzed in the blood for clinical and nutritional purpose are (except hormones and metalloenzymes which already discussed) **total proteins**, **albumin**, and **globulins**. In addition, **haptoglobin** is also investigated as an indicator of inflammation.

Regarding nonprotein nitrogen regularly determined, the **urea** is the most important, followed by **creatinine**, **uric acid**, **bilirubin**, and more rarely **ammonia**.

4.1 Nonprotein Nitrogen

4.1.1 Blood Urea Nitrogen (BUN)

Urea, the main nonprotein nitrogen (NPN), is the ultimate metabolite of the digestion of proteins in ruminant; it is synthetized in the liver (urea cycle) and excreted by the kidney in urine. Blood urea concentration within normal range can be interpreted as reflection of the protein intake, while the values out the normal range are an indication of renal for liver failures. However, urea metabolism in camel as in other ruminants is different than in monogastric because urea in blood returned partly to the rumen by transport across the rumen wall and via saliva. In rumen, the microflora can break down the NPN to ammonia and synthesize it into proteins that are later on hydrolyzed in the abomasum and intestine. When ammonia is produced too rapidly in the rumen or if its concentration becomes too high, appreciable amounts are absorbed directly into the bloodstream, reconverted to urea in the liver, and excreted through the kidneys in the urine (Fig. 4.1a, b).

However, urea excretion in camel appears very low compared to other species as it has been stated for long time (Schmidt-Nielsen et al. 1957). Blood urea is widely reused for protein synthesis with a higher efficiency in camel compared to other species especially in case of poor nitrogen content of the diet (Emmanuel et al. 1976). Urea recycling is finally more efficient in camel than in sheep. It has been

© Springer International Publishing AG, part of Springer Nature 2018
B. Faye, M. Bengoumi, *Camel Clinical Biochemistry and Hematology*,
https://doi.org/10.1007/978-3-319-95562-9_4

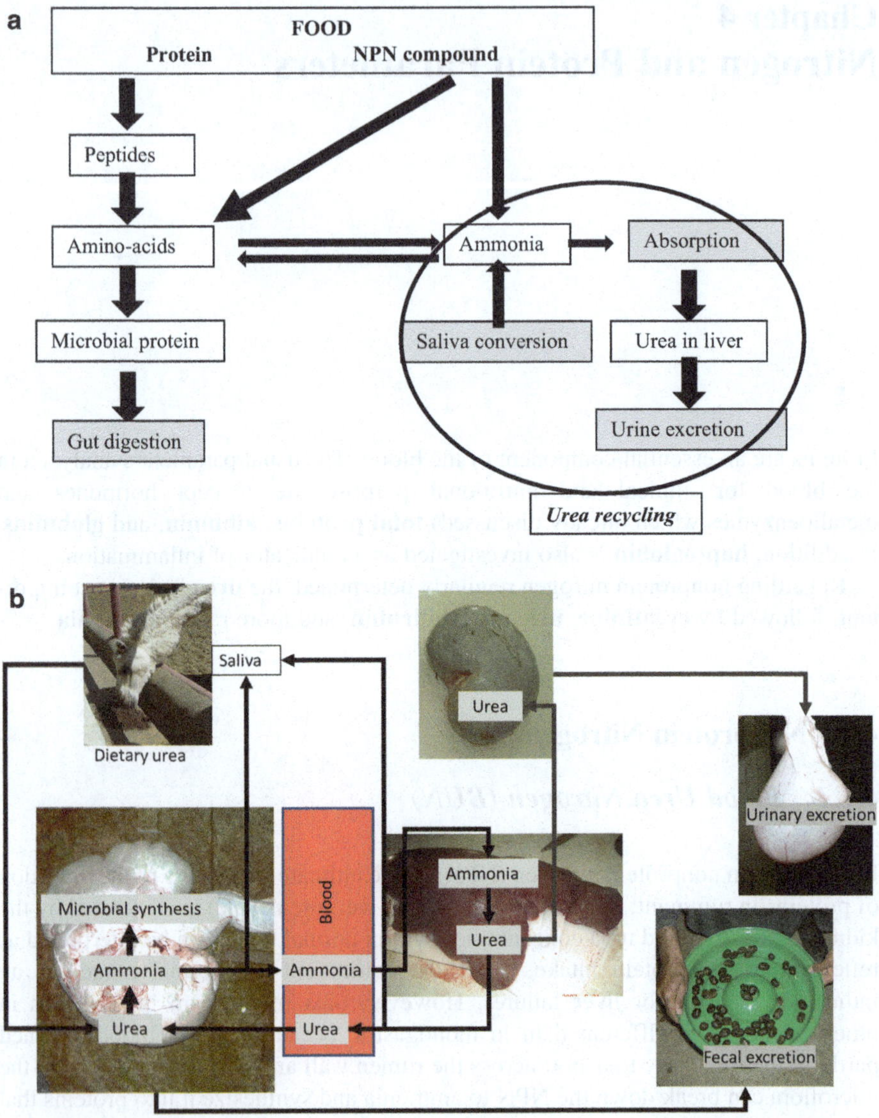

Fig. 4.1 (**a**) Schema of the digestion and absorption of nitrogenous compounds in ruminants. (**b**) Urea flow in camel (photo B. Faye)

stated that urea recycling in normally watered camel was 94%, while it was 75% in desert sheep and 79% in desert goat (Mousa et al. 1983). In case of water deprivation, this percentage could increase to 97% in camel.

This particularity of urea metabolism in camel could explain its higher sensitivity to urea intoxication (e.g., with molasses-urea blocks). Nitrogen recycling could improve the nitrogen balance (Gihad et al. 1989), but at reverse, in intensive systems, with diet rich in highly degradable nitrogen, the risk of alkalosis with hyperuremia is high (Kayouli et al. 1995).

Blood urea is determined by using enzymatic kits based on the kinetic method with urease (Orsonneau et al. 1992). Urea in plasma or serum is stable throughout 9 days at room temperature (Saeed et al. 1995). However, it is recommended to perform the analysis the same day or store samples at +4 °C for 3 or 4 days maximum.

4.1.1.1 Normal Values of Uremia

The range of normal urea in blood is 8–30 mg/100 ml (Table 4.1) but could reach wider values between 5 and 40 mg/100 ml (Bengoumi 1992). As for glucose, all values reported in the literature in mmol/l are converted in mg/100 ml to make easier the comparison. The conversion unit is 1 mmol/l = 2.8 mg/100 ml. In the review of Yadav and Bissa (1998), the blood urea ranged from 11.8 ± 0.3 (Elias and Yagil 1984) to 78.12 mg/100 ml (Azwai et al. 1990). Those values are comparable to the range observed in other ruminants.

The oldest reference regarding normal blood urea in North African camel gave mean values between 32 and 55 mg/100 ml (Chavanne and Boué 1950), but lower values were given for Indian camel: 6.3 ± 1.8 mg/l (Mathur et al. 1981). The variability between animals is high, but the main cause of this variability is the NPN compounds of the diet. Kataria and Kataria (2004) did not observe significant effect of drought on the uremia, the mean values being 23.9 ± 1.1 in non-affected area vs 24.1 ± 1.1 mg/100 ml in drought area which would be linked to same watering or NPN intake.

4.1.1.2 Physiological Factors of Variation

Hyperuremia

Hyperuremia is observed during renal failure or acute dehydration or nitrogen intoxication. Mild increase of uremia is observed during high nitrogen or protein intake and final stage of starvation with acute proteolysis.

Hypouremia

Hypouremia is observed during liver disease and during nitrogen and protein deficiency (Bengoumi 1992).

Table 4.1 Uremia in camel according to different authors (in mg/100 ml)

References	Values (mg/100 ml)	*n*	Country
Durand and Kchouk (1958)	32–57	–	Tunisia
Schmidt-Nielsen et al. (1957)	11.5–28.3	2	Algeria
Soliman and Shaker (1967)	3.10 ± 0.56	80	Egypt
Barakat and Abdel-Fattah (1970)	12 ± 3.8	260	Egypt
Idris and Tartour (1970)	2–45	142	Sudan
Emmanuel et al. (1976)	19.6–52.6	2	Iran
Abdelgadir et al. (1979)	14.6 ± 3.9	96	Sudan
Orliac (1980)	14 ± 4.2	102	Algeria
Mathur et al. (1981)	6.29 ± 1.77	20	India
Kouider and Kolb (1982)	25.7 ± 4.9	6	Syria
Mousa et al. (1983)	13.3 ± 1.5	3	Sudan
Ateeq et al. (1984)	15.4–26.0	88	Syria
McGrane and Kenyon (1985)	15.6–48.4	–	Sudan
Chiericato et al. (1986)	17.0–18.4	24	Somalia
Abdalla et al. (1988)	11.9 ± 2.6	23	UAE
Kouider et al. (1988)	12.3 ± 1.9	6	Syria
Snow et al. (1988)	39.7–41.6	9	UAE
Azwai et al. (1990)	31.7 (21.8–78.2)	142	Libya
Faye and Mulato (1991)	35.9 ± 17.5	52	Djibouti
Faye et al. (1992)	15.9 ± 4.3	32	Djibouti
Faye et al. (1995)	30 ± 14.8	82	France
Abu-Damir et al. (1993)	5.3 ± 1.3	5	Sudan
Rezakhani et al. (1997)	14.7–18.9	83	Iran
Sarwar and Majeed (1997)	12.04 ± 0.36	56	Pakistan
Ayoub and Saleh (1998)	4.7 ± 0.7	3	UAE
Baraka et al. (2000)	11.06 ± 1.0	38	Egypt
Bogin (2000)	30 ± 9	–	Israel
Chaudhary and Iqbal (2000)	15.06 ± 1.73	16	UAE
Osman and Al-Busadah (2000)	19.9 ± 2.24	6	Saudi Arabia
Tabatabaei Naeini and Nazifi (2001)	14.9 ± 5.35	50	Iran
Ben-Romdhane et al. (2003)	34.2 ± 9.6	165	Tunisia
Osman and Al-Busadah (2003)	49.8 ± 5.5	5	Saudi Arabia
Kataria and Kataria (2004)	23.89 ± 1.12	83	India
Al-Busadah (2007)	14.2 ± 1.7	60	Saudi Arabia
Mohammed et al. (2007)	13.8 ± 1.5	11	Nigeria
Ahmed et al. (2009)	17.93 ± 0.27	30	Egypt
Al-Sobayil and Mousa (2009)	22.8 ± 1.7	5	Saudi Arabia
Nazifi et al. (2009)	15.8 ± 0.8	20	Iran
Aichouni et al. (2010)	15.65 ± 5.9	48	Algeria
Patodkar et al. (2010)	18.99 ± 0.17	16	India
Sazmand et al. (2011)	26.9 ± 0.98	93	Iran
Abd El-Baky and Salem (2011)	26 ± 2	70	Egypt

(continued)

Table 4.1 (continued)

References	Values (mg/100 ml)	*n*	Country
Hekmatimoghaddam et al. (2011)	26.8 ± 0.95	92	Iran
Muhammad et al. (2011)	15.4 ± 16.9	12	Nigeria
Nagpal et al. (2011)	38.70 ± 3.49	5	India
Adel and El-Metwaly (2012)	34.85 ± 7.64	6	Iran
Al-Haj Ali et al. (2012)	15.7 ± 1.3	30	UAE
Al-Harbi (2012)	30.5 ± 0.16	10	Saudi Arabia
Deen (2013)	19.5–22.5	250	India
Osman et al. (2014)	12.36 ± 1.48	20	Egypt
Alshamsi et al. (2015)	14.8 ± 3.9	60	UAE
El-Deeb and Buczinski (2015)	9.6–12.8	15	Saudi Arabia
Narnaware et al. (2015)	36.7 ± 1.4	6	India
Ali et al. (2016)	8.79 ± 1.18	6	Saudi Arabia
Al-Jassim et al. (2016)	10.05 ± 0.53	12	Australia
Badakhshan and Mirmahmoudi (2016)	34.26 ± 1.17	18	Iran
Moolchandani and Sareen (2016)	15.04 ± 1.26	6	India
Sahraoui et al. (2016)	7.3 ± 2.02	22	Algeria
Hamad et al. (2017)	34.8 ± 8.3	30	Algeria

Sex Effect

No sex effect was reported by Idris and Tartour (1970), Chiericato et al. (1986), Bengoumi (1992), Al-Busadah (2007), Mohammed et al. (2007), or Omidi et al. (2014) contrary to Faye and Mulato (1991) who found lower values in males (23.3 ± 13.5) than in females (39 ± 17.2 mg/100 ml) in Djibouti, but this effect could be linked to a regional variation and imbalance in the sexual distribution between regions. Lower values in male (29.4 mg/100 ml on average) compared to female (55.6 mg/100 ml) was also reported in Tunisia (Ben-Romdhane et al. 2003) and in India (18.97 ± 1.61 vs 22.29 ± 0.76 mg/100 ml, respectively) by Deen (2013).

With average uremia between 30.2 and 30.8 mg/100 ml, no really significant changes were observed during the rutting season in male (Al-Harbi 2012). At reverse, the lactation stage could have a significant effect. Uremia was highest the first day of lactation (72 mg/100 ml on average) and decreased up to 41.4 mg/100 ml on average (Elias and Yagil 1984) probably in relation with the increase of protein catabolism at the beginning of lactation to satisfy the increased demand in nitrogen (Bengoumi 1992).

Age Effect

Contradictory results are reported on the effect of age: no variation (Faye and Mulato 1991; Rezakhani et al. 1997; Deen 2013), lower values in young camel (range 2–30) than in adult (15–45 mg/100 ml) according to Idris and Tartour (1970), and slightly higher values in young camels (mean, 37.6) than in adult (mean, 33 mg/100 ml) according to Ateeq et al. (1984). Osman and Al-Busadah (2000) found also higher values of uremia in camel calves (22.4 ± 1.1) compared to lactating dams (18.5 ± 1.2 mg/100 ml) or non-lactating she-camels (19.9 ± 2.2 mg/100 ml).

The lower values in young camel (less than 4 years) were confirmed by Ben-Romdhane et al. (2003) who found 30.6 mg/100 ml, while it was 34.8 between 4 and 10 years old camel and 36.0 mg/100 ml above 10 years old camels. There was no difference between suckling camel calves (26.78 ± 1.77) and weaned calves (25.29 ± 2.2 mg/100 ml) as well as in their lactating dams (Omer et al. 2010).

Pregnancy Effect

The uremia was higher in pregnant females compared to non-pregnant ones: the median value was 14.5 for pregnant vs 10.98 mg/100 ml in non-pregnant. Moreover, uremia increased significantly in the last trimester of gestation (Omidi et al. 2015) probably in relationship with the higher requirement for energy at the last stage of pregnancy increase of the metabolism of the fetus. Similar significant difference between pregnant (43.8 mg/100 ml) and non-pregnant (34.2 mg/100 ml) was revealed by Ben-Romdhane et al. (2003) and also by Ayoub et al. (2003): 15.5 ± 1.5 in pregnant camel vs 7.9 ± 1. 5 mg/100 ml in non-pregnant ones. However, Saeed et al. (2009) did not find significant difference: 19.6 ± 4.6 in pregnant vs 22.1 ± 4.0 mg/100 ml in non-pregnant camels. The absence of the effect of pregnancy on uremia was reported on Bactrian camel (Omidi et al. 2014) and in dromedary camel in Nigeria (Muhammad et al. 2011).

Breed Effect

No breed variability was found, for example, in Algeria, the values varying between 15.2 and 16.8 mg/100 ml according to the described phenotypes (Aichouni et al. 2010).

Similar results were reported for the Indian camels with values varying between 19. 5 ± 1.0 and 22.5 ± 1.5 mg/ml according to the camel breed (Deen 2013). A lack of breed variability was also observed by Al-Busadah (2007) in Saudi Arabia, with mean values between 28.8 and 32.4 mg/ml. At reverse, a slight breed effect (comparison of camel phenotypes based on their coat color) was recorded by Al-Qarawi (1999), and a seasonal variation was observed for each breed with higher values in summer than in winter (Fig. 4.2), contrary to most of the data of the literature (see below). Based on coat color difference of camels in Saudi Arabia, urea nitrogen concentration of the serum was higher for black and white, 16.6 ± 0.8 and 14.9 ± 0.6, respectively, than brown 14.7 ± 0.4 mg/100 ml (Hussein et al. 2012).

4.1.1.3 Effect of the Diet and Dehydration

In young camel receiving mineral and/or concentrate supplementation, the uremia increased significantly in control and only mineral supplemented groups (up to

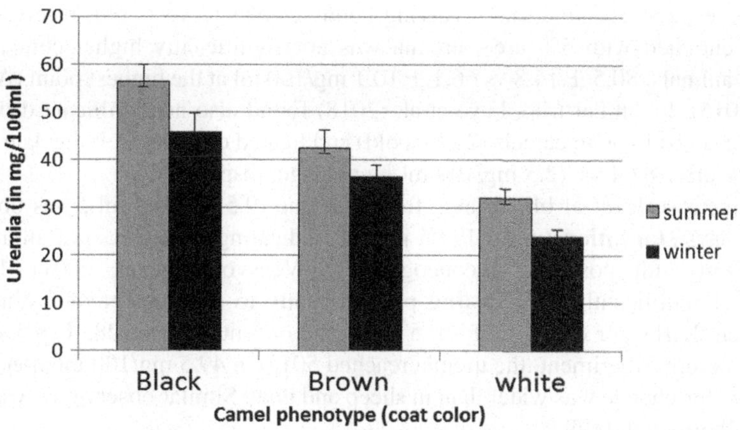

Fig. 4.2 Seasonal and camel breed changes of uremia in Saudi Arabia [calculated from Al-Qarawi (1999)]

45 mg/100 ml in control group), while groups receiving concentrates maintained their uremia at a lower level, between 13.5 and 14.1 mg/100 ml (Faye et al. 1992). The increase of uremia appeared after a 2-month period of low-protein diet indicating an excessive catabolism, lightly attenuated in mineral supplemented group (up to 36 mg/100 ml). Elsewhere, in their regional analysis at Djibouti, Faye and Mulato (1991) confirmed that uremia was significantly higher in peri-urban camel farms (55.6 mg/100 ml on average) compared to arid regions with low feeding resources (e.g., 17.5 mg/100 ml in Obock region and 21.3 mg/100 ml in Rift Valley).

The seasonal effect reported by some authors is linked to feeding resources availability (Bengoumi 1992). In dry season, camel uremia was quite lower (15.85 ± 0.84) than during the green season (25.7 ± 0.8 mg/100 ml) in Pakistan (Amin et al. 2007). Similar observations were done in Algeria (Aichouni et al. 2013) with significant higher uremia in winter (25.1 ± 4.2) than in summer (14.9 ± 3.9 mg/100 ml), while Hamad et al. (2017) found the reverse in Algeria also. In Nigeria Mohammed et al. (2007) found higher uremia in wet season (14.9 ± 4.0) than in dry season (12.7 ± 3.3 mg/100 ml). In Morocco, Chakir (2016) found higher uremia in May/June (32.5 ± 2. 6) than in January (29.7 ± 1.8 mg/100 ml).

In a trial including the distribution of three types of diets with increasing percentage of crude protein, Nagpal et al. (2011) did not observe a significant change of uremia. Indeed, those concentrations were nonsignificantly higher in low-protein diet (38.7 ± 3.5 mg/100 ml for CP = 9.1%) than in medium-protein diet (31.5 ± 2.4 mg/100 ml for CP = 11.5%) and high crude protein diet (33.1 ± 3.2 for CP = 14.0%). At reverse, Saini et al. (2007) observed higher urea levels in camels following the dietary nitrogen supplementation through urea and Khejri leaves. Gupta et al. (2012) comparing also three types of diet including 66.3 ± 2.3, 61.8 ± 2.2, and 58.5 ± 1.6% of crude protein, respectively, observed a significant correlation with uremia, 43.7 ± 3.6, 37.5 ± 7.3, and 28.7 ± 1.6 mg/

100 ml, respectively. In a trial involving young camels (2 years old) receiving date blocks enriched with 5% urea, uremia was not significantly higher compared to control animals: 80.5 ± 14.8 vs 66.1 ± 10.1 mg/100 ml at the highest point (Alafaliq et al. 2015). In similar trial, Faye et al. (2018) found also no significant difference between control young camels (2 years old) and treated camels receiving date blocks with 5% urea: 67.4 vs 72.3 mg/100 ml on average, respectively.

An increase level of blood urea, from baseline (9.5 mg/100 ml), was observed after 5 days **starvation** (up to 19.05 mg/ml) indicating an increase of amino acid catabolism with possible gluconeogenesis (Wensvoort et al. 2001). Uremia increased significantly after **fasting** proportionally to the number of fasting days (Shaheen 2001): for example, from a baseline concentration of 28. 1 ± 13.9 mg/100 ml before experiment, the uremia reached 501.0 ± 49.5 mg/100 ml after 4 days fasting. This change was wider than in sheep and goat. Similar observation was done by Dahlborn et al. (1992).

The variation of uremia with the **hydration** status was regularly studied in the camel for a long time. In the experiment of Mousa et al. (1983), uremia increased from 13.3 ± 1.5 to 21.7 ± 2.5 mg/100 ml after water deprivation. After 20 days of water restriction, Al-Haj Ali et al. (2012) reported an increase of uremia from 13.3 ± 1.4 as baseline level to 34.9 ± 3.1 mg/100 ml on dehydrated camels. However, no difference was reported between dehydrated camels and rehydrated ones (Etzion and Yagil 1986).

After 14 days of water restriction, Bengoumi (1992) recorded a high hyperuremia passing from a baseline of 37.2 ± 6.2 to a maximum of 122.5 ± 20.4 mg/100 ml at the end of restriction period and returned to normal levels after a week of rehydration (Fig. 4.3).

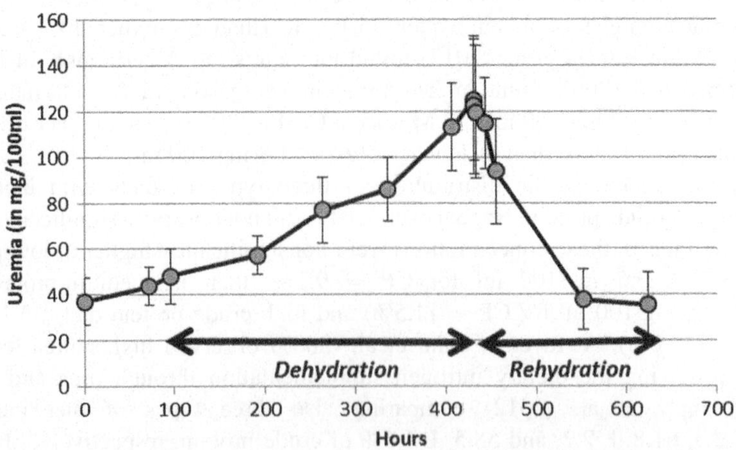

Fig. 4.3 Change in uremia of camel all along a cycle of dehydration (14 days) and rehydration [calculated from Bengoumi (1992)]

4.1.1.4 Disease Effect

Uremia is an important parameter in clinical investigation, and many references are available in the literature on camel. For example, comparing sick camels suffering from anorexia, decreased body gain, indigestion, and lowered reproductive performance expressed as delayed ovulation, anestrus, and repeat breeder, Zaher et al. (2017) observed a higher uremia in affected animals (20.6 ± 0.4) than in healthy ones (14.6 ± 0.4 mg/100 ml).

In case of **trypanosomosis**, a slight decrease depending on the stage of the disease was observed due to the reduction of appetite of the affected camel. When camels lost all its fat storage, the increase of protein catabolism could provoke a hyperuremia (Faye et al. 1992). Thus, some authors did not find significant effect (Chaudhary and Iqbal 2000; Sazmand et al. 2011; Abd El-Baky and Salem 2011; Moolchandani and Sareen 2016). At reverse Kamal (2008) found an important increase from 5.9 ± 0.9 (healthy camels) to 13.16 ± 1.2 mg/100 ml in affected animals. Gutierrez et al. (2005) also found hyperuremia in camels affected by abortion associated with *Trypanosoma evansi* infection (17.1 ± 3.6), while healthy camels exhibit quite lower values (8.12 ± 1.9 mg/100 ml).

Camels suffering from **mange** expressed higher uremia in winter season (36.9 ± 4.1 vs 31.7 ± 1.7 mg/100 ml in healthy camel) but not in summer, 29.66 ± 1.8 and 27.7 ± 0.35 mg/100 ml, respectively (Mal et al. 2006). At reverse, no effect on uremia was observed in camel affected by gastrointestinal parasites (Osman et al. 2014).

In camels affected by **theileriosis**, serum urea increased significantly (52.0 ± 5.9 mg/100 ml), the baseline value of healthy animals being 6.8 ± 2.1 mg/100 ml only (Ismael et al. 2014) contrary to Hekmatimoghaddam et al. (2011) who observed no significant difference.

According to the same author, **pasteurellosis** would also induce an increase of uremia in similar proportion, while for Kamal (2008), urea concentration is only multiplied by 2 in case of pasteurellosis. Uremia of camels with liver abscess (21.2 ± 3.4) is higher than non-affected camels (12.5 ± 0.3 mg/100 ml) (El-Deeb and Fouda 2013). In case of **paratuberculosis**, similar change occurred: 12.5 ± 0.3 in healthy camel vs 21.2 ± 3.4 mg/100 ml in sick camel (El-Deeb et al. 2014).

Although there are no clear biological relationships, some authors found higher uremia in case of infectious diseases as respiratory syndrome (El-Naser and Khamis 2009), noninfectious disorders as bone fracture (Alshamsi et al. 2015), or dystocia (Ali et al. 2016). Such drop in BUN was attributed by the authors by tissues' trauma.

Even if there is no evident relationship, ruminal acidosis should provoke also an increase of uremia (11.5 ± 1.1 mg/100 ml) compared to normal level of healthy animals (Kamal 2008).

Contrary to **monensin toxicosis** (Al-Jassim et al. 2016), the **narasin poisoning** provokes a toxic hyperuremia (Abu-Damir et al. 2013) with an increase from 8.7 ± 3.0 in control group to a maximum of 44.0 ± 20.0 mg/100 ml in young poisoned camel.

4.1.1.5 Urea in Milk

Urea is a part of the nitrogen fraction of the milk. In cow milk, urea represents 48% of the nonprotein nitrogen fraction. The determination of urea in milk gives information on the protein status of the animal, and positive correlations were reported between uremia and urea in milk (DePeters and Ferguson 1992; Roseler et al. 1993). Reverse to cow milk, few references are available in camel milk. On 102 samples from Bactrian and dromedary camel in Kazakhstan, a mean value of urea concentration in milk was reported to be 81.6 ± 60.4 mg/l (Faye et al. 2010). In cow milk, the concentration is generally comprised between 80 and 400 mg/l (Vérité et al. 1995). In camel, Faye et al. (2010) estimated the normal level of urea in milk between 30 and 120 mg/l. Surprisingly, some milk samples showed nil values in urea concentration that was never reported in cow or goat milk. The high recycling capacity of camel as recalled above could explain this particularity.

In the same reference of Faye et al. (2010), a seasonal effect was reported with higher urea concentration in milk in spring (111.5 ± 77.6 mg/l) than in autumn (46.4 ± 38.0 mg/l) corresponding to the higher nutritive values of feeding resources in spring. Indeed, 92% of milk samples with high quantity of urea (more than 130 mg/l) were collected at spring.

At reverse, no species effect (Bactrian or dromedary) or regional effect was observed.

4.1.1.6 Urea in Other Substrates

The **cerebrospinal fluid (CSF)** has higher urea concentration than blood of the same camel: 30.5 ± 1.4 vs 15.4 ± 1.42 mg/100 ml, respectively (Ahmed et al. 2009). At reverse, Nazifi and Maleki (1998) recorded double value in serum (30.5 ± 1.4) than in CSF (15.4 ± 1.4 mg/100 ml).

In **ruminal fluid**, urea concentration in heathy camels was 5.7 ± 0.4 mg/100 ml. This value decreased in case of indigestion, acidosis, or bloat but increased in case of skin necrosis, pasteurellosis, mange, or trypanosomosis (Kamal 2008). In healthy camels, Baraka et al. (2000) found lower concentrations (2.4 ± 0.2 mg/100 ml), and this value increased significantly in case of trypanosomosis (3.6 ± 0.3), frothy bloat (3.8 ± 0.1), and overall rumen acidosis (4.7 ± 1.0 mg/100 ml). At reverse, ruminal urea concentration decreased significantly in case of skin necrosis (1.41 ± 0.22) and caseous lymphadenitis (1.48 ± 0.2 mg/100 ml).

In different joints, synovial fluid contained between 12.4 ± 5.5 and 19 ± 7.4 mg/100 ml (Al-Rubikat and Ismail 2014).

In **peritoneal fluid**, urea concentration was lower and more variable than in the blood of the same camel 12.7 ± 8.9 vs 14.9 ± 5.4 mg/100 ml in the blood (Tabatabaei Naeini and Nazifi 2001).

In **synovial fluid,** urea concentration was lower (16.4 ± 6.93) than in the serum (22.9 ± 8.25 mg/100 ml) as stated by Al-Rubikat and Ismail (2014).

4.1.2 Uric Acid

Uric acid is a heterocyclic compound, product of the metabolic breakdown of purine nucleotides. It is a normal component of urine. In human, high blood concentrations of uric acid can lead to gout. The references in camel are not common and the normal values are not clearly defined. Uric acid concentration in plasma or serum is measured using uricase method.

Compared to human where the serum uric acid is comprised between 3.4–7.2 mg/100 ml (200–430 μmol/l) for men and 2.4–6.1 mg/100 ml for women (140–360 μmol/l), the concentration in camel appeared lower. As for herbivorous especially ruminant, serum uric acid concentration is very low because it is catabolized by uricase in the liver into allantoin which is excreted in urine. Two groups of values are reported in the literature. A first group of references indicated acid uric concentrations around 2 mg/100 ml: for example, Badakhshan and Mirmahmoudi (2016) found 2.2 ± 0.2 in a comparative study with sheep which showed higher significant values (3.2 ± 0.2 mg/100 ml). In a second group of references, the uric acid concentration in camel serum was between 0.1 and 0.2 mg/100 ml, for example, 0.16 ± 0.03 mg/100 ml in the study of Al-Ali et al. (1988) or from 0.11 ± 0.2 and 0.20 ± 0.04 according to the diet (Nagpal et al. 2011). Values of 0.14 ± 0.04 mg/100 ml were also reported by Barakat and Abdel-Fattah (1970). For Bogin (2000), the mean values in camel serum were 0.5 ± 0.2 mg/100 ml.

No sexual difference was reported: 0.14 ± 0.03 mg/100 ml in both sexes (Barakat and Abdel Fattah 1971). With quite higher mean values (8.47 mg/100 ml on average), Omidi et al. (2015) did not observe difference between pregnant and non-pregnant camel. However, higher concentrations were reported in weaned calves (more than one year) compared to suckling ones: 3.04 ± 0.49 vs 2.75 ± 0.38 mg/100 ml (Omer et al. 2010) although no difference was reported between camel calf (0.25 ± 0.06) and lactating adult (0.21 ± 0.02) or dry adult (0.24 ± 0.02 mg/100 ml) (Osman and Al-Busadah 2000).

A seasonal variation was observed with quite higher average in dry hot summer season (2.99 ± 0.08 mg/100 ml) in comparison to the other seasons, between 0.34 ± 0.05 in winter season and 0.24 ± 0.07 mg/100 ml in rainy season (Babeker et al. 2011).

The effect of diseases was investigated by Kamal (2008): comparatively to control camel (1.68 ± 0.13 mg/100 ml), the uric acid in serum significantly increases in camels affected by trypanosomosis (2.50 ± 0.17), pasteurellosis (2.61 ± 0.21), inflammation of the soft palate (2.55 ± 0.20), and ruminal acidosis (2.21 ± 0.14 mg/100 ml). At reverse, there is no effect of wryneck syndrome of the uric acid in blood (0.12 ± 0.05) compared to normal camel (0.14 ± 0.02 mg/100 ml) (Al-Sobayil and Mousa 2009).

The uric acid in cerebrospinal fluid is in lower concentration (2.17 ± 0.04) than in blood serum of the same camel: 3.05 ± 0.2 mg/100 ml (Ahmed et al. 2009).

4.1.3 Creatinine

Creatinine is synthesized in the muscles within a complex biological system involving creatinine. The daily production of creatinine is constant and linked to the muscular mass and not physical exercise. As creatinine is excreted by the kidneys, essentially by glomerular filtration, a deficient filtration in the kidney provokes an increase of creatinine blood level. Creatinine levels in blood but also in urine may be used to calculate the creatinine clearance which is correlated approximately with the glomerular filtration rate (GFR). However, the determination of urinary creatinine is tedious and time-consuming and requires collection of urine for 24 h which is a heavy procedure that could be implemented only during experiments. Therefore, determination of serum or plasma creatinine, an important indicator of kidney integrity, is preferred. In camel, the role of the kidney being highly important in the adaptation to water restriction and electrolytes' balance, the dosage of creatinine was regularly achieved to investigate the kidney functions (Yagil 1993; Kamili et al. 2013). Plasma or serum creatinine concentration is determined using a kinetic colorimetric method, using picric acid (Jaffe) method, or an enzymatic technique using iminohydrolase.

4.1.3.1 Normal Values

In the camel normally watered, the creatininemia is around 0.8–2 mg/100 ml (Bengoumi 1992). For example, Sarwar and Majeed (1997) reported mean values of 0.86 ± 0.02 mg/100 ml, while Shukla et al. (2009) reported 1.02 ± 0.28 mg/100 ml. Other values are reported in the Table 4.2.

4.1.3.2 Physiological Variability

With recorded concentrations of 1.36 ± 0.66 in male and 1.7 ± 0.24 mg/100 ml in female camel, no significant sexual difference was observed (Al-Sultan 2003). No sexual difference was also observed neither by Mohammed et al. (2007), 1.0 ± 0.12 in male camel and 0.91 ± 0.01 mg/100 ml in female camel, nor by Saeed et al. (2004), 1.7 ± 0.3 mg/100 ml in both sexes. However, Patodkar et al. (2010) revealed a slight but significant higher creatininemia in male (2.37 ± 0.14) than in female (1.87 ± 0.21 mg/100 ml). Barakat and Abdel Fattah (1971) had also found higher values in male (2.04 ± 0.07 and 2.37 ± 0.05 according to green and dry season) than in female (1.83 ± 0.05 and 1.95 ± 0.05 mg/100 ml, respectively).

A significant difference according to the physiological status of the camel was observed by Osman and Al-Busadah (2000): 1.61 ± 0.22 in lactating she-camels vs 2.4 ± 0.37 mg/100 ml in non-lactating ones. Ayoub et al. (2003) reported a higher creatininemia in pregnant camel (1.42 ± 0.04) than in non-pregnant (1.20 ± 0.06 mg/100 ml) contrary to Omidi et al. (2014) who stated no significant

Table 4.2 Creatinine concentration in camel serum according to different authors

References	Values (mg/100 ml)	*n*	Country
Soliman and Shaker (1967)	0.73 ± 0.02	80	Egypt
Barakat and Abdel-Fattah (1970)	2.05 ± 0.1	260	Egypt
Abdelgadir et al. (1979)	1.89 ± 0.3	96	Sudan
Orliac (1980)	2.0 ± 0.3	102	Algeria
Mathur et al. (1981)	0.96 ± 0.14	20	India
Abdalla et al. (1988)	1.5 ± 0.22	23	UAE
Bengoumi (1992)	1.34 ± 0.18	240	Morocco
Abu-Damir et al. (1993)	1.43 ± 0.08	5	Sudan
Ayoub and Saleh (1998)	1.6 ± 0.06	3	UAE
Nazifi and Maleki (1998)	2.02 ± 0.05	21	Iran
Chaudhary and Iqbal (2000)	1.86 ± 0.25	16	UAE
Osman and Al-Busadah (2003)	1.5 ± 0.1	5	Saudi Arabia
Salman and Afzal (2004)	1.90 ± 0.15	28	UAE
Mohammed et al. (2007)	0.97 ± 0.1	11	Nigeria
Kamal (2008)	0.85 ± 0.05	20	Egypt
Ahmed et al. (2009)	1.56 ± 0.12	30	Egypt
Al-Sobayil and Mousa (2009)	1.54 ± 0.14	5	Saudi Arabia
Nazifi et al. (2009)	1.41 ± 0.04	20	Iran
Saeed et al. (2009)	1.34 ± 0.22	60	UAE
Aichouni et al. (2010)	1.09 ± 0.03	48	Algeria
Patodkar et al. (2010)	2.13 ± 0.18	16	India
Abd El-Baky and Salem (2011)	0.47 ± 0.10	70	Egypt
Nagpal et al. (2011)	1.45 ± 0.06	5	India
Adel and El-Metwaly (2012)	0.97 ± 0.05	6	Iran
Al-Harbi (2012)	1.53 ± 0.47	10	Saudi Arabia
Al-Rubikat and Ismail (2014)	1.8 ± 0.15	100	Jordan
Alshamsi et al. (2015)	1.57 ± 0.32	60	UAE
Narnaware et al. (2015)	0.96 ± 0.11	6	India
Ali et al. (2016)	1.23 ± 0.37	6	Saudi Arabia
Al-Jassim et al. (2016)	1.25 ± 0.06	12	Australia
Badakhshan and Mirmahmoudi (2016)	1.38 ± 0.08	18	Iran
Sahraoui et al. (2016)	1.48 ± 0.27	22	Algeria
Hamad et al. (2017)	2.00 ± 0.29	30	Algeria

difference between pregnant (1.62 ± 0.17) and non-pregnant (1.38 ± 0.07 mg/100 ml) female camel.

No age effect was observed, creatinine in camel calf (2.6 ± 0.19 mg/100 ml) being comparable to dry adult. Similar observation was done by Saeed et al. (2004) and Deen (2013). However, light difference between young suckling camel (1.33 ± 0.20) and young weaned camel (1.21 ± 0.13 mg/100 ml) was related (Omer et al. 2010). On the other hand, creatinine concentration increased at the last trimester of gestation passing from 0.73 to 1.0 mg/100 ml (Omidi et al. 2015). In male, a slight variation was observed according to the breeding season, the values varying between 1.45 ± 0.65 at rutting period and 1.57 ± 0.65 at the post-rut season (Al-Harbi 2012).

The blood creatinine being independent of the feed composition (Nagpal et al. 2011; Adel and El-Metwaly 2012; Chakir 2016), no seasonal variation is observed (Bengoumi 1992; Salman and Afzal 2004; Mohammed et al. 2007) even in case of drought (Kataria and Kataria 2004), the creatininemia being similar in affected area (1.45 ± 0.04) than in non-affected one (1.43 ± 0.09 mg/100 ml) whatever the sex of the animal. However, Babeker et al. (2013) found higher creatinine concentration in serum during dry hot summer (1.45 ± 0.13) compared to the other seasons (0.97 ± 0.10 mg/100 ml). Amin et al. (2007) found also a quite higher creatininemia at the green season (1.54 ± 0.04) compared to dry season (0.84 ± 0.05 mg/100 ml). In Algeria, Aichouni et al. (2013) observed significant higher creatininemia in winter (1.3 ± 0.26) than in summer (1.05 ± 0.02 mg/100 ml). Moreover, Gupta et al. (2012) found a positive correlation of creatinine concentration in serum with the nitrogen intake in camels, values being 2.2 ± 0.3, 1.9 ± 0.2, and 1.7 ± 0.3 mg/100 ml with nitrogen intake of 159.4 ± 9.9, 128.4 ± 7.8, and 96.0 ± 4.2 g/day, respectively.

4.1.3.3 Water Restriction Effect

The decrease of the glomerular filtration during severe dehydration leads to a decrease of creatinine clearance. After 10 days water deprivation in summer, blood creatinine increased by 60% and the urine creatinine by 147%, and the clearance decreased by 72% (Yagil and Berlyne 1977). After 14 days dehydration, Bengoumi (1992) observed an increase of the blood creatinine by 276% passing from 1.31 to 4.93 mg/100 ml on average. Concentrations remained high after 24 h after watering (3.34 mg/100 ml) and then decreased regularly to reach 1.5 mg/100 ml after a week. At the same time, creatinine clearance decreased from 281 to 110 ml/min after dehydration period and reached 294 ml/min after 7 days rehydration (Bengoumi 1992). In the experiment of Kataria et al. (2003), creatinine clearance was divided by 3 after 24 days dehydration in winter and by 5 after 12 days' dehydration in summer (Fig. 4.4). Reverse to these last authors, Ayoub and Saleh (1998) observed a nonsignificant increase of creatinine after 72 h dehydration (from 1.6 ± 0.06 to 1.67 ± 0.03 mg/100 ml) and a significant decrease after rehydration (1.2 ± 0.06 mg/100 ml). That means that camels were dehydrated before.

After 34 days of water deprivation during cold season, creatinine was increased by 30% passing from 1.18 ± 0.28 to 1.53 ± 0.14 mg/100 ml, while the glomerular filtration rate (GFR) decreased by 20% passing from 1.33 ± 0.23 to 1.06 ± 0.21 ml/min/kg (Kamili et al. 2013). The GFR was rehabilitated rapidly after rehydration, in less than 2 h (Etzion and Yagil 1986). After 20 days dehydration, Al-Haj Ali et al. (2012) observed also an increase of serum creatinine from 1.3 ± 0.1 to 2.2 ± 0.1 mg/100 ml. These changes were less important than the study of Bengoumi (1992), but the seasonal conditions of the experiment (summer vs winter) could have an important impact explaining the observed differences. For example, Yagil and

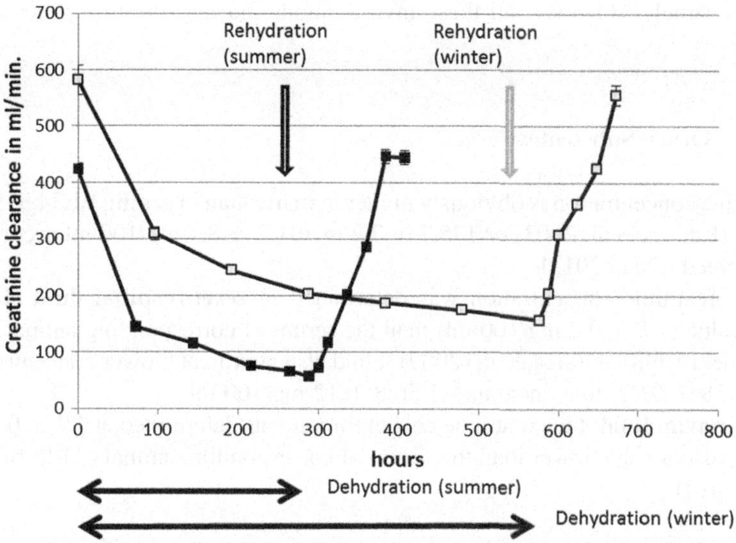

Fig. 4.4 Changes (mean and SD) in creatinine clearance (in ml/min) in camel during dehydration and rehydration periods in summer (black square) and in winter (open square) [calculated from Kataria et al. (2003)]

Berlyne (1977) stated that moderate dehydration did not affect plasma creatinine concentrations. Globally, GFR is lower in camel than in sheep and goat (Ouajd and Kamel 2009).

4.1.3.4 Disease Effect

In case of **bone fracture**, a slight but significant increase of creatinine was observed (1.7 ± 0.04 mg/100 ml) (Alshamsi et al. 2015). Contrary to Chaudhary and Iqbal (2000) or Abd El-Baky and Salem (2011), Kamal (2008) observed a significant increase of creatinine in case of **trypanosomosis** (1.9 ± 0.08) and also in case of **pasteurellosis** (1.95 ± 0.05), **inflammation of the soft palate** (1.96 ± 0.08), and **ruminal acidosis** (1.93 ± 0.4 mg/100 ml) compared to healthy camel (0.85 ± 0.05 mg/100 ml). At reverse, Gutierrez et al. (2005) did not find significant effect of *Trypanosoma evansi* outbreak on the creatinine concentration in serum: 1.38 ± 0.19 and 1.32 ± 0.33 mg/100 ml in control and affected group, respectively.

In diseased camels suffering from anorexia, decrease body gain, indigestion, and lowered reproductive performance expressed as delayed ovulation, anestrus, and repeat breeder (Zaher et al. 2017), the creatinine concentration in serum was almost double (2.17 ± 0.07) than in healthy animals (1.12 ± 0.05 mg/100 ml). Camels affected by **theileriosis** have a higher creatininemia (Ismael et al. 2014): the values changed from 1.15 ± 0.12 in normal serum to 2.38 ± 0.33 mg/100 ml in serum of affected camels. However, all these investigations did not link this effect to renal failure.

4.1.3.5 Other Substrates

Creatinine concentration is obviously higher in **urine** than in serum: 91.34 ± 5.8 mg/100 ml (Kataria et al. 2003) or 175.7 ± 7.2 to 191.2 ± 8.1 mg/100 ml according to camel breed (Deen 2013).

The creatinine concentration was determined in **cerebrospinal fluid (CSF)** at lower value (1.5 ± 0.2 mg/100 ml) than the serum of corresponding animal (Nazifi and Maleki 1998). Ahmed et al. (2009) found also significant lower concentration in CSF (1.38 ± 0.02) than in serum (1.56 ± 0.12 mg/100 ml).

In **synovial fluid**, the creatinine concentration was determined at 1.1 ± 0.42 mg/100 ml, also a value lower than the serum of corresponding animal (Al-Rubikat and Ismail 2014).

4.1.4 Ammonia

Ammonia is one of the compounds participating in the urea cycle. It is a toxic molecule normally detoxified in mammals mainly by its conversion to urea in the liver (Visek 1984). Some studies involving dairy cattle concluded that high protein feeding could probably lead to elevated ammonia concentrations in tissues (Jordan and Swanson 1979). Moreover, addition of urea in the diet would cause hyperammonemia, particularly in camel characterized by its high urea recycling efficiency as mentioned above. If ammonia concentration is too elevated in the rumen, blood, cerebrospinal fluid, and other tissues could lead to poisoning of the animal (Ntiranyibagira et al. 2015). Notably, liver damage could lead to an elevated amount of ammonia in the blood (hyperammonemia) associated to a decrease of urea. However, the severity of ammonia intoxication being poorly correlated with its blood concentration, the determination of ammonemia is of low clinical interest (Stahl 1963).

Few references are available in camel, especially since **blood ammonia** is normally low and not easily traceable (Mathur et al. 1981). Usually ammonia determination is limited to other substrates like ruminal fluid or milk during experiment trials.

Regarding **camel urine**, an old reference (Read 1925) stated that the camel excreted no ammonia, but it was not confirmed later by Smith and Silvette (1928) who found that ammonia represented 1.7 to 19.1% of the total urine nitrogen, except in pregnant Bactrian camel. The recent investigations on camel urine did not mention ammonia as an essential component (El-Nagi and Torki 2007; Ahamad et al. 2017).

In **ruminal fluid** of healthy camel, ammonia concentration was 4.33 ± 0.16 mg/100 ml (Baraka et al. 2000) and 4.9 ± 0.4 mg/100 ml (Kamal 2008). This value increased significantly in case of trypanosomosis, skin disease, lymphadenitis, manga, pasteurellosis, and other diseases with a maximum of 6.3 mg/100 ml.

The presence of ammonia in **milk** is an indicator of contamination, and its variability is very high as it is depending of the degree of contamination. In consequence, its clinical interest is quite low. In Kazakhstan, the range was between 0 and 46 mg/L with a mean of 5 ± 10 mg/l with no significant difference between camel species (Bactrian, dromedary, and hybrids) or regions but a significant seasonal variation, with higher values occurring in summer when the risk of contamination is higher (Konuspayeva 2007).

4.2 Proteins

The present chapter will include serum proteins, **albumin** and **globulins**, and **haptoglobin**. Total serum protein concentration is determined using a colorimetric method (biuret) or densitometry.

4.2.1 Serum Proteins

The concentration of the total protein in serum is mainly linked to the nitrogen level in feed and to the dehydration status of the animal. Also, it could increase with an increase in specific proteins as albumin or immunoglobulin (see the chapter on protein indicators). The serum protein (or total protein) concentration in the camel is between 6.3 and 8.3 g/100 ml, as cited in the literature (Table 4.3). Generally, authors do not agree on the effect of age because the difference is mainly observed before 6 months. While some concluded that there is no effect (Ghodsian et al. 1978), others observed major variations according to age classes (Chartier et al. 1986). These differences can be explained by the little comparable observation protocols in terms of age, animal feeding, as well as the experimental or seasonal conditions during blood sampling (Ghosal et al. 1973). These values are comparable to those of cattle in similar conditions (Hassan et al. 1968) and more generally to those of ruminants (Jatkar et al. 1962; Soliman and Shaker 1967; Kataria et al. 2000).

4.2.1.1 Seasonal Variation

The seasonal variations identified by some authors (Ghosal et al. 1973; Mehrotra and Gupta 1989; Kataria et al. 2002; Salman and Afzal 2004) are attributed to dietary changes. Indeed, the decline in the protein rate in summer (6.3 ± 0.2 vs 6.8 ± 0.12 in winter) is associated with the decline in food intake, the decline in the quality of diet, and the lower availability of forage during the summer period. However, reverse trend was reported with significant higher values in summer than in winter, namely,

Table 4.3 Observed values of total serum proteins according to different authors (in g/100 ml)

References	*n*	Mean/SD (or range)
Kumar et al. (1961)—India	–	6.40 ± 0.55
Perk and Lobl (1961)—Israel	15	6.98 ± 5.4 (M)
	15	6.66 ± 4.1 (F)
Jatkar et al. (1962)—India	20	6.6 ± 0.27
Soliman and Shaker (1967)—Egypt	8	7.0 ± 0.2
Hassan et al. (1968)—Sudan	37	8.2 ± 0.6
Barakat and Abdel-Fattah (1970)—Egypt	260	6.42 ± 0.4
Ghosal et al. (1973)—India	43	6.6 ± 0.01
Pegram and Scott (1976)—Ethiopia	16	7.72 ± 0.85
Ghodsian et al. (1978)—Iran	99	6.8 (5.1–9.3)
Boid et al. (1980)—Sudan	20	6.48 ± 0.71
Orliac (1980)—Algeria	102	6.8 ± 0.4
Mathur et al. (1981)—India	20	6.46 ± 1.58
Biagi (1983)—Somalia	200	6.24 ± 0.8
Hassan et al. (1983)—India	20	6.5 ± 0.5
Sellaouati (1984)—Tunisia	107	6.5 ± 0.36
Chartier et al. (1986)—Mauritania	114	6.3–8.3
Chiericato et al. (1986)—Somalia	24	6.9 ± 0.5
Abdo et al. (1987)—Saudi Arabia	80	5.8 ± 0.3
Abdalla et al. (1988)—UAE	23	6.1 ± 0.37
Yagoub (1988)—Sudan	97	7.1 ± 1.6
Bengoumi et al. (1989)—Morocco	62	6.52 ± 0.4
Azwai et al. (1990)—Libya	142	7.62 (4.9–12.3)
Al-Ani et al. (1992)—Iraq	15	6.6 ± 0.45
Knight et al. (1994)—UAE	52	6.32 ± 0.54
Sarwar and Majeed (1997)—Pakistan	56	7.78 ± 0.22
Ayoub and Saleh (1998)—UAE	3	7.5 ± 0.1
Dalvi et al. (1998)—India	–	7.42 ± 0.54
Baraka et al. (2000)—Egypt	38	8.05 ± 0.34
Osman and Al-Busadah (2000)—Saudi Arabia	6	9.83 ± 1.03
Chaudhary et al. (2003)—UAE	30	5.7 ± 0.15
Liu (2003)—China[a]	25	6.8 ± 1.01
Osman and Al-Busadah (2003)—Saudi Arabia	5	7.1 ± 0.3
Saeed et al. (2004)—UAE	167	6.1 ± 0.4
Salman and Afzal (2004)—UAE	28	6.33 ± 0.27
Abd El-Hag et al. (2005)—Sudan	26	7.61 ± 1.1
Sadiek and Saleh (2006)—Egypt	8	6.85 ± 0.65
Al-Busadah (2007)—Saudi Arabia	60	7.74 ± 0.68
Mohammed et al. (2007)—Nigeria	11	6.5 ± 0.15
Ali et al. (2008)—Pakistan	92	5.53 ± 0.3
Al-Sultan (2008)—Saudi Arabia	50	5.2 ± 0.8
Mohri et al. (2008)—Iran	11	6.2 ± 0.6

(continued)

Table 4.3 (continued)

References	n	Mean/SD (or range)
Al-Sobayil and Mousa (2009)—Saudi Arabia	5	7.3 ± 2.3
Faye et al. (2009)—UAE	44	6.1 ± 0.9
Nazifi et al. (2009)—Iran	20	7.03 ± 0.2
Baghshani et al. (2010)—Iran	10	6.08 ± 0.61
Patodkar et al. (2010)—India	16	7.49 ± 0.37
Albomohsen et al. (2011)—Iran	50	7.16 ± 0.33
Hekmatimoghaddam et al. (2011)—Iran	92	7.40 ± 0.08
Adel and El-Metwaly (2012)—Egypt	18	9.21 ± 0.15
Elkhair and Hartmann (2012)—Sudan	14	5.8 ± 0.4
Hussein et al. (2012)—Saudi Arabia	200	7.5 ± 0.28
Ahmadi-Hamedani et al. (2014)—Iran	9	8.1 ± 0.31
Ismael et al. (2014)—Saudi Arabia	23	5.78 ± 0.17
Li and Hai (2014)—China[a]	20	6.3 ± 0.37
Ali et al. (2015)—Saudi Arabia	10	7.49 ± 0.96
Al-Saiady et al. (2015)—Saudi Arabia	18	6.23 ± 0.13
Alshamsi et al. (2015)—UAE	60	6.5 ± 0.35
Omidi et al. (2015)—Iran	20	6.07 (median)
Badakhshan and Mirmahmoudi (2016)—Iran	18	7.52 ± 0.12
Yousif et al. (2016)—Sudan	15	5.89 ± 0.62
Ali et al. (2016)—Saudi Arabia	6	4.81 ± 3.2

[a]Bactrian camel

7.9 vs 7.6 g/100 ml (Abdoun et al. 2012) or 6.3 vs 6.16 g/100 ml (Chakir 2016). Babeker et al. (2013) reported a quite higher proteinemia in the rainy season (9.2 ± 0.78 in July and 9.3 ± 0.15 g/100 ml in September) than that in the dry season (6.35 ± 0.12 in October and 2.16 ± 0.12 g/l in April). Surprisingly, Amin et al. (2007) found a lower proteinemia in the green season (7.1 ± 0.1) than in the dry season (8.4 ± 0.1 g/100 ml), while the reverse could be expected. In Sudan, Yousif et al. (2016) noticed that total proteins are lower in summer (5.3 ± 0.3 g/100 ml) than in autumn (6.5 ± 0.4 g/100 ml), with the concentration in winter being intermediate (5.8 ± 0.484 g/l).

4.2.1.2 Physiological Variation

The influence of sex or age is not commonly mentioned. Serum protein concentration would be lower among suckling camels than in adult animals (Elias and Yagil 1984; Chartier et al. 1986; Al-Sultan 2008). For example, Bengoumi et al. (1989) reported a value of 5.3 ± 0.6 g/100 ml in the camel calf versus 6.5 ± 0.4 in their dam. Comparing the non-lactating adult she-camel to the camel calves, Osman and Al-Busadah (2003) found a similar total protein concentration (9.8 ± 1.0 and 9.8 ± 3.6 g/100 ml), while the lactating camel had a significant lower value

(7.9 ± 0.95 g/l). However, the optimum value would be between 1 and 5 years (Yagoub 1988). In few cases, no difference was reported between suckling, weaned, and adult camels (Ali et al. 2008; Omer et al. 2016). In the same vein, no sex effect was also reported (Biagi 1983; Bengoumi et al. 1989; Liu 2003).

A significant lower value was reported in pregnant camel (7.0 g/100 ml) compared to the non-pregnant one 6.4 g/100 ml (Saeed et al. 2009), which was not confirmed by Khadjeh (2001) who found a value of 6.7 g/100 ml in the pregnant camels versus 6.8 g/100 ml in the non-pregnant ones. There was no clear effect of the breeding season as Zeidan et al. (2008) reported a value of 6.2 ± 0.2 at breeding season, which is similar to the value found in the hot-dry months of the non-breeding season (6.8 ± 0.4) but significantly lower than that obtained in the hot-humid months of the non-breeding season (7.2 ± 0.3 g/100 ml). In males, with mean values between 7.2 and 7.4 g/100 ml, no effect of rutting season occurred (Al-Harbi 2012). The concentrations in total proteins seemed to be lower in the lactating camel than in the non-lactating one (6.9 ± 0.1 vs 7.3 ± 0.1 g/100 ml, respectively) (Hussein et al. 1992).

The range of values in the Bactrian camel (5.5–7.0 g/100 ml) seems to be lower than those in the dromedary (6.3–8.7 g/100 ml) (Higgins and Kock 1984). A significant difference was also reported between camel breeds in Saudi Arabia, with higher values in the black-coat camel (Majaheem) (7.0 g/100 ml) compared to the white-coat camel (Waddah breed) 6.3 g/100 ml on average (Hussein et al. 2012).

4.2.1.3 Effect of Effort

Total proteins increase slightly after intense physical effort (Snow et al. 1988). This change is due to a slight decrease in plasma volume during physical activity, as already pointed out for hematocrit. However, this increase is not confirmed in the camel after 30 min of trotting (Auer et al. 2015).

4.2.1.4 Effect of Dehydration

After 6 days of water deprivation, serum protein concentration would increase almost 15% and up to 25% after 8 days dehydration (Ghosal et al. 1975) and even by 40% after 2 weeks (Bengoumi 1992). These are rather low values in comparison to other domestic species. In sheep, for example, the serum protein concentration increases by 60% after 5 days of deprivation of water. However, some authors observe no change (Mahmud et al. 1984) or even a slight decrease in the concentration of serum proteins in the dehydrated camel (Hassan 1971). Because of their high molecular weight, the serum proteins cannot pass through the vascular endothelium. So, their concentration increases with the decrease of plasma volume as for other mammals.

4.2.1.5 Effect of Diseases

Serum protein would also be affected by parasitic infestation of the animal. Thus, Athar Khan et al. (1992) observed a significant decrease in serum proteins in animals presenting **coccidian oocysts** in their feces, in comparison with free animals. However, camels that are positive to **trypanosomosis** diagnosis tests did not show any difference compared with healthy camels (Chaudhary and Iqbal 2000; Abd El-Baky and Salem 2011; Sazmand et al. 2011). But, Ahmadi-Hamedani et al. (2014) found a decrease by 13% when they compared affected camels to negative ones, and Hussain et al. (2016) reported a decrease from 7.4 ± 0.1 in control ones to 5.1 ± 0.1 g/100 ml in infected camels. In Iran, Karimi et al. (2015) observed a similar decrease in total proteins in affected camels (6.4 ± 0.1 g/100 ml) compared to unaffected ones (6.6 ± 0.1 g/100 ml). Kamal (2008) also revealed a significant decrease in total proteins in blood in case of trypanosomosis, namely, from 7.8 ± 0.7 in control camels to 5.3 ± 0.7 g/l in affected camels. Surprisingly, Corbera et al. (2003) observed a reverse effect in an outbreak of abortion related to trypanosomosis where the total protein passed from 6.3 ± 0.4 in the control group to 7.9 ± 0.7 g/100 ml in the outbreak group. This increase is noted and would be linked to dehydration.

A decline of serum albumin was described in case **of ruminal acidosis** or **pasteurellosis**, but not with mange (Kamal 2008). **Liver abscess** was also linked to total proteins decline in serum, from 6.3 ± 0.4 to 5.1 ± 0.5 g/100 ml (El-Deeb and Fouda 2013).

A similar trend was reported in the camel affected by **theileriosis** (Youssef et al. 2015). At reverse, there is no observed effect of the trace element status (Faye et al. 2009; Li and Hai 2014). The **wryneck** syndrome is related to a slight but significant decrease in total proteins (Al-Sobayil and Mousa 2009) as well as paratuberculosis (El-Deeb et al. 2014). A strong inexplicable decline of total proteins occurred in the camel affected by **respiratory diseases** (El-Naser and Khamis 2009).

Gastrointestinal parasitism also provokes a significant decline in total proteins as stated by Osman et al. (2014) who reported that the proteinemia was 4.8 ± 0.61 vs 6.7 ± 1.26 g/l in the control camel.

4.2.2 Protein Fractions

Albumin and **globulins** are the main proteins of blood plasma where they bind many components as cations, fatty acids, hormones, bilirubin, thyroxine, and pharmaceutical compounds. Percentage and concentration of different fractions are determined using cellulose acetate electrophoresis. Serum albumin contributes to the regulation of blood volume by maintaining the oncotic pressure. Serum globulins are not hydrosoluble and include a wide variety of molecules as α1-globulines (α1-antitrypsin, α1-antichymotrypsin, HDL, prothrombin, etc.), α2-globulines (ceruloplasmin, antithrombin II, haptoglobin, cholinesterase, plasminogen, etc.), β-globulins (LDL, transferrin, fibrinogen, transcobalamin, etc.), and γ-globulins (IgG, IgM, IgE, etc.).

Hypoalbuminemia is caused by liver damage, nephrotic syndrome, enteropathy, and malnutrition, while hyperalbuminemia is linked to water restriction. Hypoglobinemia is linked to specific deficiencies. The ratio albumin/globulin (A/G) is also used as indicator of inflammatory process or immunological disturbances. Camel albumin was isolated and characterized recently (Malik et al. 2013).

4.2.2.1 Normal Values

In camel, serum albumin concentration varies between 25 and 45 g/l (Table 4.4), while it is 35–50 in human, 26–37 in horse, 30–36 in cattle, 24–30 in sheep, 27–39 in goat, and 36–48 in llama (Bengoumi 1992; Kaneko et al. 1997).

Total globulins in camel range from 20 to 50 g/l (Table 4.5). In proportion, α-globulins represent 15–20%, β-globulins 25–45%, and γ-globulins 50–65% of the total globulins. These values vary according to animal species (Fig. 4.5).

4.2.2.2 Physiological Variations

Age Effect

Albumin and γ-globulins increased with age (Ghodsian et al. 1978; Elias and Yagil 1984; Höhne and Niepage 1985; Chartier et al. 1986; Khadjeh 1998; Saeed et al. 2004; Elkhair and Hartmann 2011), and the β-globulins decreased, while α-globulins did not change significantly (Abdo et al. 1987; Bengoumi et al. 1989). For example, in Morocco, Bengoumi et al. (1989) reported values for concentration of albumin ranging from 30.0 ± 2.3 in young camel to 37.4 ± 3.0 g/l in adult, while β-globulin was higher in young camels (6.8 ± 1.8) than in adults (8.0 ± 0.9 g/l). Serum γ-globulins were higher in adults (14.1 ± 1.2) than in camel calves (6.8 ± 1.2 g/l).

For albumin, Faye and Mulato (1991) in Djibouti found a slight but significant difference between young camels (31.3 ± 3.5) and adult camel (33.5 ± 3.3 g/l). In Emirates, Chaudhary et al. (2003) found a significant lower serum albumin concentration in camel calves (23.7 ± 0.8) than in adults (30.6 ± 0.8 g/l). Among the components of globulin, β1-globulin was quite higher in calf (14.2 ± 0.2) compared to adult (9.7 ± 0.3 g/l), while it was at reverse for γ-globulin (4.1 ± 0.2 and 8.6 ± 0.3 g/l in calf and adult, respectively). However, there was no difference for total globulins. In consequence, the ratio A/G was reverse in camel calf (0.9 ± 0.04) compared to adult (1.2 ± 0.0). For Ahmadi-Hamedani et al. (2014), if total proteins are effectively higher in adult than in young camel (respectively, 80.9 ± 3.1 and 66.8 ± 2.9 g/l), there was no significant difference for albumin (42.9 ± 3.1 and 40.2 ± 2.4 g/l, respectively), while globulin concentrations were higher in adult

Table 4.4 Serum albumin concentration in camel according to different authors (mean and SD or range)

References	Values (g/l)	n	Country
Durand and Kchouk (1958)	41.5 ± 8	–	Tunisia
Perk and Lobl (1961)	40	30	Israel
Ghosal et al. (1975)	22.5–24	5	India
Pegram and Scott (1976)	26.0 ± 8.4	16	Ethiopia
Abdelgadir et al. (1979)	38 ± 3.2	96	Sudan
Orliac (1980)	38	102	Algeria
Biagi (1983)	27.5 ± 4.1	200	Somalia
Chiericato et al. (1986)	34–36.4	24	Somalia
Abdo et al. (1987)	36	80	Saudi Arabia
Abdalla et al. (1988)	45 ± 0.35	23	UAE
Snow et al. (1988)	28.6–31	9	UAE
Faye and Mulato (1991)	32.7 ± 3.5	52	Djibouti
Bengoumi (1992)	37 ± 5	5	Morocco
Faye et al. (1995)	36.4 ± 4.7	182	France
Sarwar and Majeed (1997)	43.4 ± 1.1	56	Pakistan
Ayoub and Saleh (1998)	32.3 ± 0.7	3	UAE
Chaudhary et al. (2003)	30.6 ± 0.8	30	UAE
Osman and Al-Busadah (2003)	37 ± 3	5	Saudi Arabia
Saeed et al. (2004)	39.0 ± 3.0	167	UAE
Salman and Afzal (2004)	42;2 ± 1.9	28	UAE
Abd El-Hag et al. (2005)	40.5 ± 2.0	26	Sudan
Sadiek and Saleh (2006)	32.6 ± 4.6	8	Egypt
Al-Busadah (2007)	42.9 ± 4.0	60	Saudi Arabia
Mohammed et al. (2007)	33.98 ± 0.98	11	Nigeria
Ali et al. (2008)	36.3 ± 2.0	92	Pakistan
Mohri et al. (2008)	34.0 ± 4.0	11	Iran
Zeidan et al. (2008)	36.7–44.6	67	Egypt
Al-Sobayil and Mousa (2009)	30.0 ± 0.4	5	Saudi Arabia
Momenah (2014)	40.3 ± 3.0	8	Egypt
Nazifi et al. (2009)	28.98 ± 2.5	20	Iran
Baghshani et al. (2010)	38.3 ± 2.10	10	Iran
Patodkar et al. (2010)	41.3 ± 2.1	16	India
Albomohsen et al. (2011)	38.8 ± 1.4	50	Iran
Al-Mujalli et al. (2011)	38.02 ± 1.13	20	Saudi Arabia
Hekmatimoghaddam et al. (2011)	37.3 ± 0.5	92	Iran
Nagpal et al. (2011)	35.4 ± 0.7	5	India
Adel and El-Metwaly (2012)	55.8 ± 1.8	18	Egypt
Hussein et al. (2012)	32 ± 2.4	200	Saudi Arabia
Al-Rubikat and Ismail (2014)	44.0 ± 2.8	100	Jordan
Ismael et al. (2014)	34.2 ± 1.5	23	Saudi Arabia
Ali et al. (2015)	61.08 ± 8.11	10	Saudi Arabia
Omidi et al. (2015)	30.35 (median)	20	Iran

(continued)

Table 4.4 (continued)

References	Values (g/l)	n	Country
Ali et al. (2015)	61.1 ± 8.1 (M)	10	Saudi Arabia
Ali et al. (2016)	29.5 ± 15.6	6	Saudi Arabia
Badakhshan and Mirmahmoudi (2016)	51.5 ± 0.8	18	Iran

Table 4.5 Serum globulin concentration in camel according to different authors (mean and SD or range)

References	Values (g/l)	*n*	Country
Pegram and Scott (1976)	51.2 ± 10.3	16	Ethiopia
Faye et al. (1995)	32.7 ± 5.1	182	France
Sarwar and Majeed (1997)	35.1 ± 1.5	56	Pakistan
Ayoub and Saleh (1998)	42.7 ± 1.4	3	UAE
Chaudhary et al. (2003)	25.5 ± 0.6	30	UAE
Saeed et al. (2004)	23.0 ± 3.0	167	UAE
Sadiek and Saleh (2006)	35.9 ± 3.5	8	Egypt
Al-Busadah (2007)	34.6 ± 1.8	60	Saudi Arabia
Al-Sultan (2008)	24.0 ± 0.3	50	Saudi Arabia
Ali et al. (2008)	18.6 ± 1.6	92	Pakistan
Zeidan et al. (2008)	24.9–30.0	67	Egypt
Momenah (2014)	52.8 ± 3.9	8	Egypt
Nazifi et al. (2009)	41.3 ± 1.7	20	Iran
Saeed et al. (2009)	27.6 ± 3.6	28	UAE
Patodkar et al. (2010)	33.6 ± 2.0	16	India
Albomohsen et al. (2011)	32.8 ± 2.7	50	Iran
Al-Mujalli et al. (2011)	33.07 ± 0.74	20	Saudi Arabia
Adel and El-Metwaly (2012)	36.3 ± 2.3	18	Egypt
Hussein et al. (2012)	28 ± 2.2	200	Saudi Arabia
Ismael et al. (2014)	36.4 ± 1.3	23	Saudi Arabia

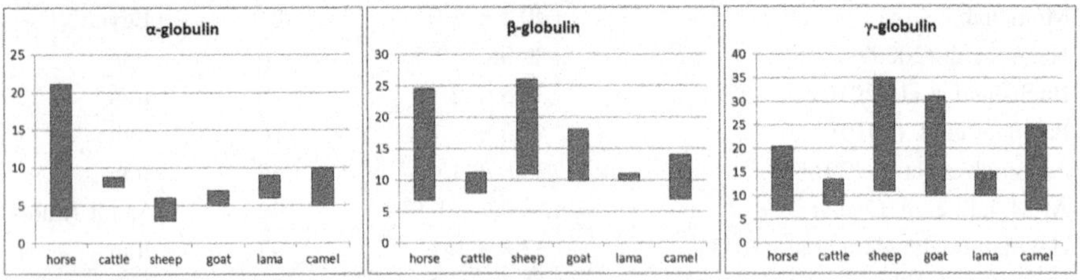

Fig. 4.5 Ranges of the different globulins (in g/l) according to species [calculated from Bengoumi (1992)]

camels, essentially due to the important increase of γ-globulins, in double quantity in adults (22.6 ± 0.2) than in young animals (12.3 ± 1.2 g/l).

The effect of age on serum globulins could be explained by the immature lymphoid system in young camel (Ahmadi-Hamedani et al. 2014). This difference is mainly observed during the first 6 months. The same albumin concentration (33.4 ± 2.1 g/l) was found in suckling and weaned camel calf (Omer et al. 2010).

At reverse, Osman and Al-Busadah reported higher albumin concentration in camel calves (44.6 ± 1.7) compared to adults (between 31.1 ± 3.9 and 39.0 ± 3.8 g/l according to their physiological stage).

Sex Effect

Usually, there is no effect of the sex on the total proteins and their fractions (Perk and Lobl 1961; Höller and Hassan 1966; Biagi 1983; Chartier et al. 1986; Chiericato et al. 1986; Bengoumi et al. 1989; Saeed et al. 2004; Mohammed et al. 2007). However, a significantly higher concentration in γ-globulin was observed in female than in males, 19.9 ± 1.8 and 14.3 ± 2.2 g/l, respectively (Ahmadi-Hamedani et al. 2014). The difference could be related to the physiological status of the female because serum albumin appeared significantly lower in lactating females compared to males, 50.0 ± 4.9 and 60.0 ± 2.2 g/l, respectively (Elkhair and Hartmann 2011). Regarding globulins, there was difference, except for γ-globulin fraction which was proportionally higher in female than in male (26.3 ± 5.2 vs 17.7 ± 1.7%).

Pregnancy-Lactation Effect

There was no change in total proteins during pregnancy of camel (Bhargava et al. 1964; Eltohamy et al. 1986), but an important export of proteins to colostrum provokes a decrease after the parturition (Elias and Yagil 1984). In fact, lower albumin concentration was observed in lactating camels compared to dry ones, 31.1 ± 3.9 vs 39.0 ± 3.8 g/l, respectively (Osman and Al-Busadah 2000). At reverse, significant lower serum albumin concentration was reported in pregnant she-camels at term (36.2 ± 2.0) compared to non-pregnant (4.1 ± 3.7 g/l), while no difference was revealed on serum globulin (Saeed et al. 2009). Ayoub et al. (2003) reported opposite results: serum albumin was higher in pregnant camels (27.0 ± 0.5) than in non-pregnant ones (24.9 ± 0.2 g/l), while no difference occurred for globulin. If a slight significant difference occurred between camel in lactation and non-lactating animal (38.2 ± 0.9 vs 41.1 ± 1.0 g/l, respectively), no difference was observed for globulin (Hussein et al. 1992).

Breeding season did not seem to have significant effect both on albumin and globulins (Ali et al. 2008; Zeidan et al. 2008).

There is no effect of castration (Kchouk and Durand 1958). For example, albumin was reported at 34.8 ± 2.0 in whole males and 35.4 g/l in castrated ones (Bengoumi et al. 1989).

Seasonal and Feeding Effect

Serum proteins are influenced by food intake (Faye and Mulato 1991) and, in consequence, are decreasing at the dry season (Barakat and Abdel Fattah 1971; Ghosal et al. 1973; Chartier et al. 1986; Kouider et al. 1988). Mohammed et al. (2007) did not report significant difference between seasons, 34.5 ± 2.9 and 33.6 ± 3.7 g/l in dry and rainy season, respectively.

In fact, the seasonal variation is not clear and available results of the literature are contradictory. Indeed the effect of season is mainly due to the difference in availability of feeding resources and watering. For example, Amin et al. (2007) did not observe seasonal difference in serum albumin concentration, while they reported high globulin in dry season (58.3 ± 3.9) than in green season (40.0 ± 3.8 g/l) like Abd El-Hag et al. (2005). Values in summer were higher than in winter at least in Indian context (Ghosal et al. 1975), but opposite observations were done in the same country with higher albumin in winter (35.1 ± 2.5) than in summer (29.0 ± 1.5 g/l), while the same values are observed for globulin (20.9 ± 5.0 vs 19.5 ± 7.6 g/l in winter and summer, respectively) although no seasonal variability for total proteins and the ratio A/G occurred (Mal et al. 2006). In India also, Kataria et al. (2002) found significant higher albumin in hot season (11.5 ± 0.8) than in cold one (9.5 ± 0.3 g/l) and a reverse tendency for globulin which was lower in hot season than in winter (23.2 ± 2.2 and 24.1 ± 2.3, respectively). In the United Arab Emirates, Salman and Afzal (2004) did not reveal significant seasonal difference: 4.22 ± 0.19 in winter vs 4.14 ± 0.14 g/l in summer. In Morocco, Bengoumi (1992) reported higher total proteins and albumin during the green period from February to May but no changes in globulins and their fractions. In Sudan, Babeker et al. (2011) observed higher serum albumin during the dry humid season (October, 38.2 ± 0.3) and rainy hot summer (September, 33.0 ± 6.0) and low at rainy season (July, 20.8 ± 2.1) and overall at the dry hot summer (April, 17.8 ± 0.6 g/l). In fact, autumn is the season where albumin was in higher concentration in serum (37.7 ± 4.5), summer being lower (28.2 ± 4.7) and winter being intermediate (Yousif et al. 2016). Serum globulin was also lower in summer (24.3 ± 6.6) than in autumn and winter (27.7 ± 5.0 g/l).

Testing two types of diet on young camel with, respectively, 13.1 and 12.4% crude protein and 10.2 and 15.3% crude fiber, Al-Saiady et al. (2015) did not find significant effect neither on serum albumin (45.2 ± 1.2 and 42.4 ± 1.2 g/l) nor on serum globulins (17.1 ± 1.0 and 17.6 ± 1.0 g/l). Similar absence of effect on protein pattern was observed by Adel and El-Metwaly (2012) and Nagpal et al. (2011). At reverse, Poonia et al. (2015) observed a significant effect of feeding after testing three types of diets. Concentrations varied from 61.5 ± 1.0 to 69.2 ± 0.4 g/l for total proteins, 30.6 ± 0.4 to 33.7 ± 0.1 g/l for albumin, and 30.9 ± 0.5 to 35.5 ± 0.2 g/l for globulins.

Breed Effect

No breed effect was revealed in the dromedary camel from Saudi Arabia neither on total proteins nor on albumin and globulin (Al-Busadah 2007; Asadi et al. 2009) although Hussein et al. (2012) found some differences in albumin concentration according to the coat color. Al-Busadah (2007) did not observe difference of globulin fractions in three main camel breeds of Saudi Arabia.

Water Restriction Effect

Serum protein concentration is a good indicator of hydration status because it is directly linked to volemia increasing during dehydration and decreasing during hyperhydration. No effect of cycle dehydration/rehydration on total proteins, globulin and albumin, was observed by Ayoub and Saleh (1998) with a short water restriction period (72 h) which is the normal watering period for camels. Similar observations were done by Ghosal et al. (1975) after 10 and 16 days of water restriction. Yet, reversely, Bengoumi (1992) got an increase of the total proteins after 14 days dehydration, the concentration passing from 61 ± 6 before the experiment to 85 ± 6 g/l and back to normal value 4 days after rehydration. This increase involved albumin (37 ± 5 and 52 ± 5 g/l before dehydration and after 14 days dehydration) and globulins (24 ± 3 and 35 ± 2 g/l, respectively). In all the case, the return to baseline values occurred 4 days after rehydration. The increase of globulins was observed for all components, from 3.9 ± 0.8 to 6.1 ± 1.0 (α-), from 6.6 ± 0.7 to 11.8 ± 1.3 (β-), and from 13.8 ± 2.2 to 17.0 ± 1.5 g/l (γ-globulin) before and after 14 days dehydration. At the same time, the ratio A/G was constant (1.5 approximatively).

4.2.2.3 Disease Effect

Generally albumin, α- and β-globulins decrease during hepatic failure, and γ-globulins increase during infections and allergy and decrease during immunodeficiencies. In camel affected by **mange**, serum albumin decreased (from 35.1 ± 2.5 to 32.0 ± 3.1 in winter but not in summer), while the globulins increased in both seasons, in winter from 20.9 ± 5.0 in healthy camel to 30.5 ± 4.0 in affected ones and in summer from 19.5 ± 7.6 to 26.2 ± 3.0 g/l, respectively, probably in relation to elevated immunoglobulin levels, the decrease of albumin content being an osmoregulatory response (Mal et al. 2006). These changes, both for albumin and globulin, were not confirmed by Kamal (2008). At reverse, Momenah (2014) observed a decrease of total proteins, albumin and globulins, in camel affected by parasitic

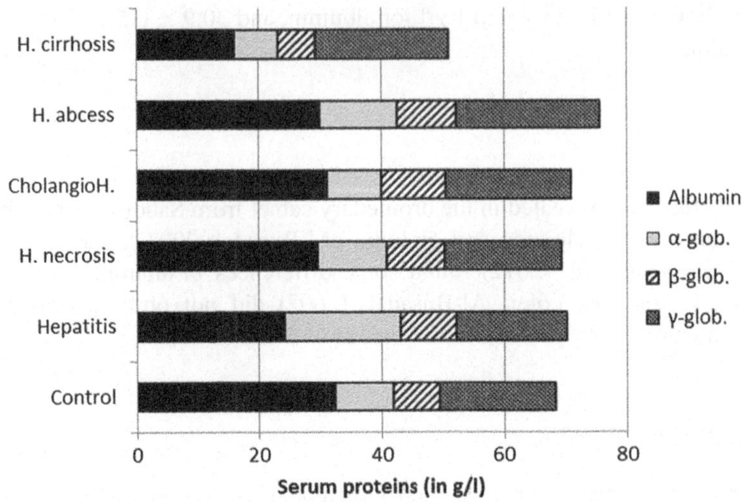

Fig. 4.6 Changes in protein patterns according to the type of liver damage [from after Sadiek and Saleh (2006)]

infection. Albumin decreased also in camel affected by cystic echinococcosis (Heidarpour et al. 2012; Abdel Aziz et al. 2016).

Opposite effect for albumin and globulins was also reported in case of camel **trypanosomosis**: while serum albumin concentration decreased from 30.8 ± 1.4 to 23.8 ± 1.2 g/l, globulins concentration increased from 32.7 ± 1.8 to 42.3 ± 1.6 g/l (Karimi et al. 2015). Similar figure was reported formerly by Abd El-Baky and Salem (2011): albumin declined from 37.0 ± 3.6 to 27.6 ± 2.5 when globulin increased from 40.5 ± 1.6 to 48.8 ± 2.8 g/l. The increase of globulins was due to the higher proportion of γ-globulin in affected camel. Indeed, γ-globulins increased from 11.6 ± 1.5 to 25.5 ± 1.3 g/l, while α1-globulin decreased significantly from 4.7 ± 0.5 to 1.0 ± 0.4 g/l. At the same time, α2-globulin did not change significantly 8.4 ± 1.2 and 7.2 ± 1.1 in healthy and affected camels, respectively, as well as β-globulin, 6.9 ± 0.8 and 8.4 ± 0.7 g/l, respectively (Karimi et al. 2015). Consequently, the ratio A/G decreased from 0.96 ± 0.07 to 0.58 ± 0.06. The increase of γ-globulins in camel affected by *Trypanosoma evansi* was formerly reported by Boid et al. (1980): the percentage of γ-globulin increased from 25.8 ± 3.3% in control camel to 41.1 ± 6.0% in infected animals, while α- and β-globulins did not change significantly. At the same time, albumin percentage declined from 48.8 ± 2.7 to 36.3 ± 4.9% in control and infected camels, respectively. A similar pattern was described by Pegram and Scott (1976) in Ethiopia. A slight decrease (−6.8%) of albumin in camels affected by trypanosomosis was also reported by Moolchandani and Sareen (2016). Kamal (2008) found also an important decline of albumin in affected camels (from 31.8 ± 3.2 to 10.5 ± 3.5 g/l) but no significant change for globulin.

In case of **liver damage**, the protein pattern in serum changed significantly (Fig. 4.6) with an increase of albumin and total globulin (especially α- and β-globulins) except in case of cirrhosis (Sadiek and Saleh 2006).

In camel affected by **low reproductive performances** (delayed ovulation, anestrus, repeat breeding, etc.), serum albumin was lower in concentration (21.5 ± 0.3) than in control animals (38.2 ± 1.4 g/l) (Zaher et al. 2017).

At reverse, El-Deeb and Fouda (2013) observed an albumin decrease in camel affected by liver abscess (37.4 ± 1.2 to 30 ± 4 g/l). **Paratuberculosis** in camel is also linked to similar albumin decrease from 37.4 ± 2.0 to 30.0 ± 2.0 g/l (El-Deeb et al. 2014). If the total proteins decreased significantly in camel affected by **contagious ecthyma**, it is essentially due to globulin that declined from 33.2 ± 3.6 in control animal to 20.5 ± 1.2 g/l, while albumin did not change significantly (Narnaware et al. 2015). Albumin as well as total protein decreased in case of **brucellosis** (El-Boshy et al. 2009): in control camel, albumin concentration in serum was 41.2 ± 3.1 g/l, while lower values occurred in camel infected by *Brucella abortus* (31.4 ± 2.5) or *Brucella melitensis* (34.2 ± 2.9 g/l). In case of **babesiosis**, the increase of globulin is more important (from 27.6 ± 2.3 to 46.0 ± 4.1 g/l) than albumin, from 34.9 ± 1.6 to 38.4 ± 4.1 g/l (Swelum et al. 2014).

In racing camel suffering from bone fractures, the albumin was significantly higher (31.2 ± 3.3) in comparison with control camel (29.5 ± 2.2 g/l) (Alshamsi et al. 2015).

Contrary to the other species, there is no effect of transportation stress although road transportation has been reported to cause dehydration and may manifest itself as a hyperproteinemia and hyperalbuminemia (Baghshani et al. 2010). The resistance of camel to dehydration is probably the main explanation of this difference.

The decrease of albumin in camel serum was also observed in case of acidosis and pasteurellosis, while no effect on globulin was revealed (Kamal 2008).

Total proteins and albumin fell down (from 59.0 ± 1.11 to 48.2 ± 2.5 and from 29.83 ± 0.76 to 17.3 ± 1.9 g/l, respectively) while globulin was stable in case of monensin toxicity (Al-Jassim et al. 2016).

4.2.2.4 Other Substrates

The **synovial fluid** contained fewer albumins than the serum: 13.0 ± 5.5 g/l (Al-Rubikat and Ismail 2014).

In **cerebrospinal fluid**, the concentration of albumin is very low, 0. 263 ± 0.02 g/l only, while in serum of the same camels, the concentration was 37.63 ± 1.5 g/l (Ahmed et al. 2009). Similar findings were revealed by Nazifi and Maleki (1998) who found 0.32 ± 0.04 g/l of total proteins in CSF vs 53 ± 1.6 g/l in serum.

In **follicular fluid**, albumin concentration was 42.8 ± 1.3 in small-size follicle vs 41.8 ± 1.1 g/l, and the difference was significant. At reverse, the concentration in globulin was higher in large follicle than in small one: 24.3 ± 1.4 vs 19.4 ± 2.6 g/l (Albomohsen et al. 2011). However, no significant difference related to the size of follicle appeared both for albumin and globulin to Ali et al. (2008) although a great difference occurred between the peak of breeding season (24.6 ± 0.6) and non-breeding season (15.6 ± 0.6 g/l). This higher globulin contents in the follicular fluid during the peak breeding season might be related to the higher level of follicular activity.

4.2.3 Bilirubin

Bilirubin is the normal pigment of the catabolism cycle of hemoproteins of which 80% originates from hemoglobin during breakdown of the red blood cells. Bilirubin is mainly synthetized in the spleen, transported by blood to the liver where it is conjugated with glucuronic acid and excreted in bile. In the intestine lumina, conjugated bilirubin is reduced to urobilinogen, urobilin, and stercobilin. Around 10% of urobilinogen is reabsorbed into the bloodstream, filtered in the kidney, and excreted in urine. Therefore, blood total bilirubin exists under two components: nonconjugated also called indirect or free bilirubin and conjugated one also called direct bilirubin (McGilvery and Goldstein 1983). Bilirubin concentration is measured in plasma or serum using diazoreaction method. The increase of plasma bilirubin causes jaundice or icterus characterized by yellow coloration of tissues and mucosa due to the accumulation of bilirubin. There are three types of icterus, prehepatic, hepatic, and posthepatic. To identify the origin of icteria, there is a need to measure concentration of direct bilirubin. Prehepatic icterus is mainly due to the accumulation of indirect bilirubin caused by an excess of bilirubin mainly during hemolytic diseases like babesiosis and copper poisoning. Hepatic icterus is due to the accumulation of indirect bilirubin caused by hepatic diseases that reduce glucoronoconjugation, while posthepatic icterus is due to the lack of excretion of conjugated bilirubin in bile during cholestasis.

In camel, the bilirubinemia is varying between 0.5 and 8.6 mg/l, i.e., 0.8–15 μmol/l (Table 4.6), while the normal ranges in other species are 4.15–2 mg/l (horse), 0.11–5.0 mg/l (cattle), 1–5 mg/l (sheep), 0–1 mg/l (goat), and 0–10 mg/l (llama). The range appears higher in camelids than in other species (Kaneko et al. 1997). The limit level of hyperbilirubinemia in human is around 100 mg/l and critical value at 200 mg/l.

The physiological factors of variation were not frequently studied. Soliman and El-Aroumsi (1965) found higher values in males (7.2 ± 0.23) than in females (6.5 ± 0.35 mg/l), but this sexual difference was not confirmed by Sellaouati (1984).

Table 4.6 Bilirubin concentration in camel serum according to different authors

References	Values (mg/l)	*n*	Country
Soliman and Shaker (1967)	1.22 ± 0.17	80	Egypt
Abdelgadir et al. (1979)	3.2 ± 1.6	96	Sudan
Orliac (1980)	1.1 ± 0.6	102	Algeria
Al-Ali et al. (1988)	0.49 ± 0.0	20	Saudi Arabia
Abd El-Hag et al. (2005)	1.12 ± 1.37	26	Sudan
Bengoumi (2008)	1.85 ± 0.3	10	Morocco
Al-Tayib (2014)	0.04	1	Saudi Arabia
Narnaware et al. (2015)	8.6 ± 2.0	6	India
Al-Jassim et al. (2016)	1.7 ± 0.06	12	Australia
Badakhshan and Mirmahmoudi (2016)	13.0 ± 3.2	18	Iran
Moolchandani and Sareen (2016)	8.2 ± 0.12	6	Iran

No significant age or physiological status effect was observed (Osman and Al-Busadah 2000): 3.0 ± 0.17 in camel calf, 2.57 ± 0.17 in lactating she-camel, and 3.27 ± 0.46 mg/l in non-lactating ones. In spite of variation all along the lactation (from 7.5 ± 2.0 at the first stage of lactation to 31 ± 4 mg/l at the peak of lactation), there was no difference between lactating, on average 10.6 ± 1.0 mg/l, and non-lactating, 12.0 ± 1.0 mg/l (Hussein et al. 1992).

According to Bengoumi (1992), the values of total bilirubin were below the detection limit of 1.16 mg/l (2.0 μmol/l) by colorimetric method. Exceptionally, after seasonal monitoring, concentrations of 1.28 mg/l in castrated adult camels and 1.46 mg/l in young camels were observed in September. According to the same author, the total bilirubin increased in serum after 14 days dehydration (up to 1.87 mg/l) and then returned to baseline 4 days after rehydration. This change is probably related to the hemoconcentration.

A seasonal variation was also observed by Kouider et al. (1988) but with a different pattern than Bengoumi (1992), the maximum value being observed in summer (Fig. 4.7). However, a high standard deviation occurred.

There was no effect of contagious ecthyma on the concentration of bilirubin: 7.1 ± 1.6 mg/l in affected camel compared to 8.6 ± 2.0 mg/l in healthy camel (Narnaware et al. 2015). In case of trypanosomosis in camel, the bilirubinemia did not change also significantly (Moolchandani and Sareen 2016). No effect of monensin intoxication was observed, the slight decrease of bilirubinemia (from 1.7 ± 0.4 to 1.26 ± 0.6 mg/l being nonsignificant (Al-Jassim et al. 2016). At reverse, total bilirubin in serum increased in case of copper poisoning (Abu-Damir et al. 1993). It increased also in camel affected by parasitic infection, the concentration reaching 19.8 ± 0.5 in animals with general symptoms vs 13.8 ± 0.7 in healthy camels, but the values are expressed in g/l which is quite higher than expected normal values (Momenah 2014). Heidarpour et al. (2012) observed an increase of

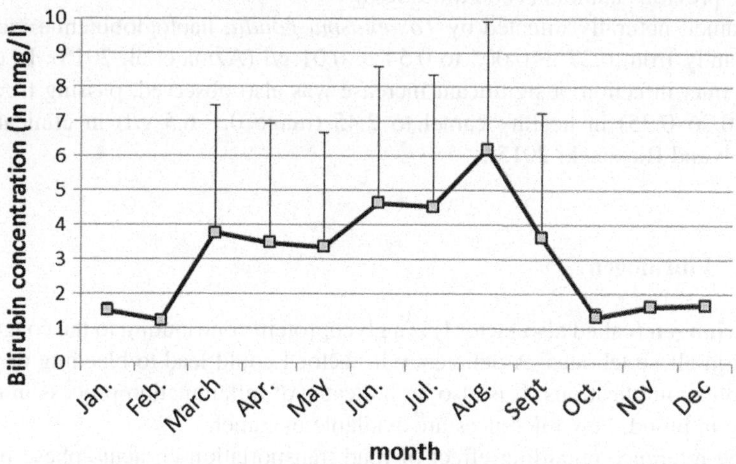

Fig. 4.7 Seasonal change of bilirubinemia in camel [calculated from Kouider et al. (1988)]

total bilirubin in camel affected by cystic echinococcosis: from 2.2 ± 0.06 to 3.0 ± 0.09 mg/l.

Babesiosis provoked in camel a significant increase of total bilirubin from 2.7 ± 0.2 mg/l in healthy camel to 5. 5 ± 0.6 in camel affected by babesiosis and up to 7.7 ± 2.0 mg/l when the animals are affected by babesiosis and other pathogens (Swelum et al. 2014).

4.2.4 Haptoglobin (Hp), Fibrinogen, and Pepsinogen

4.2.4.1 Haptoglobin (Hp)

The **haptoglobin** is a α2-globulin synthesized in the liver, sometimes used in clinic to investigate intravascular hemolytic anemia that provokes a hypohaptoglobinemia. In veterinary clinic, it is mainly monitored as indicator of inflammatory process. There are few references in camel. The mean serum basal concentration of Hp in apparently healthy dromedary camels was 0.27 ± 0.04 g/l with a range 0.11–0.61 g/l (Nazifi et al. 2012).

Testing the impact of stress related to road transportation on acute phase proteins, the mean concentrations of haptoglobin in the basal pre-transport conditions were 0.3 ± 0.04 g/l and ranged from 0.11 to 0.50 g/l with no difference between male and female (Baghshani et al. 2010). Moreover, no effect of duration in transportation was observed.

Tharwat and Al-Sobayil (2015) investigated camel haptoglobin to evaluate the effect of racing on the inflammation biomarkers. In their observations, they found a highly significant increase of Hp after race, but with a value of 3.7 ± 1.4 mg/l compared to a value of 0.3 ± 0.1 mg/l before race, they reported quite lower values than the previous authors (1000 times less).

In camel, naturally infected by *Toxoplasma gondii*, haptoglobinemia increased significantly from 0.27 ± 0.005 to 0.54 ± 0.01 g/l (Azma et al. 2015). In case of urinary tract infection, a significant increase was also observed, passing from 0.31 (range 0.26–0.35) in healthy camel to 2.45 (range 0.1–6.5 g/l) in clinical cases (El-Deeb and Buczinski 2015).

4.2.4.2 Fibrinogen

The **fibrinogen** (called also factor I) is a glycoprotein contributing to the coagulation process in all vertebrates. A deficiency in factor I could lead to bleeding or thromboembolic complications. It is also an indicator of inflammation process in case of increase in blood. Few references are available in camel.

In the reference regarding effect of road transportation on acute-phase proteins (Baghshani et al. 2010), the mean concentrations of fibrinogen in the basal pre-transport conditions were 3.10 ± 0.47 and ranged from 2.2 to 3.6 g/l.

A significant increase was observed in clinical cases or urinary tract infection, from 3.27 to 4.29 g/l on average (El-Deeb and Buczinski 2015).

4.2.4.3 Pepsinogen

The **pepsinogen** is a protein secreted by the stomach as precursor of the pepsin, one the main enzyme of the digestion. An elevation in blood in veterinary clinic is linked to gastrointestinal parasitism.

In the reference of Kataria and Kataria (2006), plasma pepsinogen estimations were made in healthy and affected camels in three different conditions (camels infected with haemonchosis, with drought, and with pica). The overall mean values of plasma pepsinogen in healthy camels were 151.61 ± 14.17 mU tyrosine. In the affected camels, a significant increase was observed in the mean values of this parameter when compared with those of healthy stock. The highest values were observed in pica-affected dromedaries.

4.3 Conclusion

The protein metabolism in camel is marked by the high level of nitrogen recycling. One of the most important consequences of this particularity is the sensitivity of camel to urea intoxication. The proportion and the quantity of proteins in the serum are within the range observed in other species but with a lower variability except for γ-globulin. As for energetic parameters, the nitrogen parameters have a high interest for nutritional investigations. Regarding some proteins presenting clinical interest (like haptoglobin, fibrinogen, or pepsinogen), the references are scarce.

References

Abdalla OM, Wasfi IA, Gadir FA (1988) The Arabian race camel normal parameters-I. Haemogram, enzymes and minerals. Comp Biochem Physiol 90A(2):231–239

Abdel Aziz AR, El Meghanawy RA, Ibrahim FF, El Zahaby DI (2016) Occurrence of hydratic cyst in camels (*Camelus dromedarius*) and their effect on meat quality. Egypt J Chem Environ Health 2(1):48–63

Abd El-Baky AA, Salem SI (2011) Clinicopathological and cytological studies on naturally infected camels and experimentally infected rats with *Trypanosoma evansi*. World Appl Sci J 14:42–50

Abdelgadir SE, Wahbi AA, Idris OF (1979) Biochemical studies of some blood and plasma constituents on camel (*Camelus dromedarius*). In: Cockrill WR (ed) Camels and camelids. Scandinavian Institute of African Studies, Uppsala, pp 438–443

Abd El-Hag SED, Shaddad SA, Hassan T (2005) Status of some chemical and biochemical parameters of camel blood in the rainy season in the Sudan. J Anim Vet Adv 4(8):713–715

Abdo MS, Hassanien MM, Manna ME, Hamed M (1987) Electrophoretic pattern of serum proteins in the Arabian camels. Indian Vet J 64:841–844

Abdoun KA, Samara EM, Okab AB, Al-Haidary AI (2012) A comparative study on seasonal variation in body temperature and blood composition of camels and sheep. J Anim Vet Adv 11 (6):769–773

Abu-Damir H, Eldirdiri NI, Adam SEI, Howarth JA, Salih YM, Idris OF (1993) Experimental copper poisoning in the camel (*Camelus dromedarius*). J Comp Pathol 108:191–208

Abu-Damir H, Al-Haj Ali M, Abbas TA, Omer EA, Al-Fihail AM (2013) Narasin poisoning in the dromedary camel (*Camelus dromedarius*). Comp Clin Pathol 22:305–311

Adel EM, El-Metwaly H (2012) Effect of feed additive "exogenous enzymes" on growth performance of Maghraby camels. Life Sci J 9(4):4830–4835

Ahamad SR, Alhaider AQ, Raish M, Shakeel F (2017) Metabolomic and elemental analysis of camel and bovine urine by GC–MS and ICP–MS. Saudi J Biol Sci 24:23–29

Ahmadi-Hamedani M, Ghazvinian K, Kokhaei P, Barati M, Mahdavi A (2014) Comparison of effects of age and sex on serum protein electrophoretic pattern in one-humped camels (*Camelus dromedarius*) in Semnan, Iran. Open Vet J 4(1):4–8

Ahmed SH, Omer SAA, Gameel AA (2009) Some normal constituents in serum and cerebrospinal fluid in Sudanese camels (*Camelus dromedarius*). Assiut Vet Med J 55(123):163–170

Aichouni A, Belhadia M, Kebir N, Aggad H (2013) Season influence on serum organic parameters of dromedarius (*Camelus dromedarius*) in Algeria. Bioch Biotech Res 1(1):8–12

Aichouni A, Jeblawi R, Dellal A, Hammou H, Aggad H (2010) Breed variation in blood constituents of the one-humped camel (*Camelus dromedarius*) in Algeria. J Camelid Sci 3:19–25

Alafaliq A, Ben Abdallah A, Almasaud A, Faye B (2015) The use of date blocks for supplementary feeding of growing camels. In: Konuspayeva G (ed) Proceedings of 4th conference of ISOCARD,"silk road camel: the Camelids, main stakes for sustainable development", June 8–12, 2015, Almaty, Kazakhstan [special issue of Scientific and Practical Journal Veterinariya #2 (42) 2015], poster session, p 438

Al-Ali AK, Husayni HA, Power DM (1988) *A comprehensive biochemical analysis of the blood of the camel (Camelus dromedarius)*. Comp Biochem Physiol 89B(1):35–37

Al-Ani FK, Al-Azzawi WARA, Jermukly MS, Razzaq KK (1992) Studies on some haematological parameters of camel and llama in Iraq. Bull Anim Prod Afr 40:103–106

Albomohsen H, Mamouei M, Tabatabaei S, Fayazi J (2011) Metabolite composition variations of follicular fluid and blood serum in Iranian dromedary camels during the peak breeding season. J Anim Vet Adv 10(3):327–331

Al-Busadah K (2007) Some biochemical and haematological indices in different breeds of camels in Saudi Arabia. Sci J King Faisal Univ (Basic Appl Sci) 8(1):131–142

Al-Haj Ali M, Adem A, Chandranath IS, Benedict S, Pathan JY, Nagelkerke N, Nyberg F, Lewis LK, Yandle TG, Nicholls GM, Frampton CM, Kazzam E (2012) Responses to dehydration in the one-humped camel and effects of blocking the renin-angiotensin system. PLoS One 7(5): e37299

Al-Harbi MS (2012) Some hematologic values and serum biochemical parameters in male camels (*Camelus dromedarius*) before and during rut. Asian J Anim Vet Adv 7:1219–1226

Ali S, Ahmada N, Akhtar N, Zia-ur-Rahman NDE (2008) Metabolite contents of blood serum and fluid from small and large sized follicles in dromedary camels during the peak and the low breeding seasons. Anim Reprod Sci 108:446–456

Ali A, Refaat D, Tharwat M, Al-Sobayil F (2015) Impotentia generandi in male dromedary camels: breeding soundness, haematology, biochemistry and testosterone level. Comp Clin Pathol 24:727–731

Ali A, Derar D, Tharwat M, Zeitounb MM, Alsobyil FA (2016) Dystocia in dromedary camels: prevalence, forms, risks and hematobiochemical changes. Anim Reprod Sci 170:149–156

Al-Jassim RAM, Shahsavari A, Owen H, Khamas W (2016) Gross pathology, biochemistry and histology of selected organs of camels suffering from suspected monensin toxicosis in Australia. J Vet Sci Technol 7(3):1–5

Al-Mujalli AM, Al-Naeem AA, Al-Ghamdi G, Al-Swailem A, Al-Yamani E, Shehata AM, Al-Dubaib MA, Hashad M, El-Lithy DA, Mahmoud OM, Alfayez M (2011) Cellular and biochemical blood profile in camels suffering from dubduba syndrome. Sci J King Faisal Univ (Basic Appl Sci) 12(2):165–172

Al-Qarawi A (1999) The chronobiological blood parameter changes as correlated to different traits of *Camelus dromedarius* in Saudi Arabia. J Camel Pract Res 6(1):45–48

Al-Rubikat R, Ismail ZB (2014) Biochemical analysis of serum and synovial fluid in clinically normal young camels (*Camelus dromedarius*). Vet Sci Develop 4(1):40–42

Al-Saiady MY, Mogawer HH, Al-Mutairi SE, Bengoumi M, Musaad A, Gar-Elnaby A, Faye B (2015) Factors affecting feed intake, body weight, testicular size and testosterone, follicle stimulating hormone (FSH) and luteinizing hormone (LH) serum concentrations in peri-pubertal male camels. Afr J Agric Res 10(14):1709–1713

Alshamsi NS, Ksiksi TS, Ashraf SS (2015) Altered serum enzymes and biochemical levels in Arabian racing camels with bone fractures. J Anim Plant Sci 25(4):1072–1080

Al-Sobayil FA, Mousa HM (2009) Clinical and biochemical studies on wry-neck syndrome in camels in Saudi Arabia. J Camel Pract Res 16(2):179–182

Al-Sultan SI (2003) Studies of some normal biochemical parameters of Majaheem breed of camel (*Camelus dromedarius*) in Saudi Arabia. J Camel Pract Res 10(1):79–80

Al-Sultan SI (2008) Studies on hematological and certain serum biochemical values in young Magaheim dromedary camel at Al-Ahsa province. Vet Res 2(3–4):34–37

Al-Tayib O (2014) Zoonotic balantidiasis in camel from Saudi Arabia. Sch Acad J Biosci 2 (7):445–447

Amin ASA, Abdoun KA, Abdelatif AM (2007) Seasonal variation in blood constituents of one-humped camel (*Camelus dromedarius*). Pak J Biol Sci 10(8):1250–1256

Asadi F, Shahriari A, Asadian P, Pourkabir M, Sabzikar A, Ojaghee R (2009) Serum lipid, glucose, free fatty acids and liver triglyceride in sub-adult and adult camels (*Camelus dromedarius*). Rev Med Vet 160(12):552–556

Ateeq G, Kouider S, Kolb E (1984) Haematology of dromedaries, cell counts, haemoglobin, proteins, urea, cholesterol, AST and ALT. Age and sex variations (in German). Archiv Exp Vet Med 38(5):664–675

Athar Khan M, Sarwer Khan M, Ashfaq-Ul-Rehman (1992) The incidence and effects of coccidiosis on certain blood parameters in camels of Pakistan. In: International camel conference, Dubaï (UAE), 2–6 Feb 1992. R&W, England, p 406

Auer R, Gleiß A, Windberger U (2015) Towards a basic understanding of the properties of camel blood in response to exercise. Emir J Food Agric 27(3):302–311

Ayoub MA, Saleh AA (1998) A comparative physiological study between camels and goats during water deprivation. In: Alhadrami G, Ayoub MA (eds) Proceedings of the third annual meeting for animal production under arid conditions, 2–3 May 98, vol 1. Al-Ain, Dubai, pp 71–87

Ayoub MA, El-Khouly AA, Mohamed TM (2003) Some hematological and biochemical and steroid hormones levels in one-humped camel during different physiological conditions. Emir J Agric Sci 15(1):44–55

Azma F, Razavi SM, Nazifi S, Rakhshandehroo E, Sanati AR (2015) A study on the status of inflammatory systems in camels naturally infected with *Toxoplasma gondii*. Trop Anim Health Prod 47:909–914

Azwai SM, Saltani H, Thomas PC, Shareha AM, El-Gammoudi F, Mohamed SO (1990) Note on cholesterol, glucose, urea and total protein concentration in serum of normal camels. Camel Newsl 7:94

Babeker EA, Elmansoury YHA, Suleem AE (2011) The influence of seasons on blood constituents of dromedary camel (*Camelus dromedarius*). Online J Anim Feed Res 3(1):1–8

Babeker EA, Elmansoury YHA, Suleem AE (2013) The influence of seasons on blood constituents of dromedary camel (*Camelus dromedarius*). Online J Anim Feed Res 3(1):1–8

Badakhshan Y, Mirmahmoudi R (2016) Blood metabolites of one-humped camel (*Camelus dromedarius*) versus sheep during summer heat stress. Iranian J Vet Med 10(1):65–71

Baghshani H, Nazifi S, Saeb M, Saeb S (2010) Influence of road transportation on plasma concentrations of acute phase proteins, including fibrinogen, haptoglobin, serum amyloid A, and ceruloplasmin, in dromedary camels (*Camelus dromedarius*). Comp Clin Pathol 19:193–198

Baraka TA, El-Sherif MT, Kubesy AA, Illek J (2000) Clinical studies of selected ruminal and blood constituents in dromedary camels affected by various diseases. Acta Vet Brno 69:61–68

Barakat MZ, Abdel-Fattah M (1970) Biochemical analysis of normal camel blood. Zentrabl Veterinarmed A 17:550–557

Barakat MZ, Abdel Fattah M (1971) Seasonal and sexual variations of certain constituents of normal camel blood. Zentrabl Veterinarmed A 18:174–176

Bengoumi M, Kessabi M, Hamliri A (1989) Teneurs et fractionnement des protéines sériques chez le dromadaire: effet de l'âge et du sexe. Maghreb Vet 20:31–33

Bengoumi M (1992) Biochimie clinique du dromadaire et mécanismes de son adaptation à la déshydratation, Thèse de doctorat ès Sciences Agronomiques. I.A.V. Hassan II, rabat, Maroc, 184 p

Bengoumi M (2008) Tolerance study of tiludronate in the dromedary camel. J Camelid Sci 1:24–31

Ben-Romdhane SM, Romdhan N, Feki M, Sanhagi H, Kaabachi N (2003) Blood biochemistry in dromedary (*Camelus dromedarius*). Rev Med Vet 154:695–702

Bhargava AK, Mehrotra PN, Banerjee S (1964) Biochemical studies on Indian camel (*Camelus dromedarius*). V. Serum proteins and their variations with age, sex, pregnancy, rut and infection. Indian J Exp Biol 2:52–58

Biagi G (1983) [further investigations on the seroprotein observations of Somalian *Camelus dromedarius*] (in Italian). Ann Fac Med Vet Pisa 35:193–200

Bogin E (2000) Clinical pathology of Camelids: present and future. Rev Méd Vét 151(7):563–568

Boid R, Luckins AG, Rae PF, Gray AR, Mahmoud MM, Malik KH (1980) Immunoglobulin levels and electrophoretic patterns of serum proteins in camels infected with Trypanosoma evansi. Vet Parasitol 6:333–345

Chakir Y (2016) Réponses physiologiques au stress de pré-abattage chez le dromadaire (*Camelus dromedarius*), Thèse no CED00/2016. Université Hassan de Casablanca, Maroc, 217 p

Chaudhary ZI, Iqbal J (2000) Incidence, biochemical and haematological alterations induced by natural trypanosomosis in racing dromedary camels. Acta Trop 77:209–213

Chaudhary ZI, Iqbal J, Rashid J (2003) Serum protein electrophoretic pattern in young and adult camels. Aust Vet J 81(10):625–626

Chartier C, Chartier F, Lepers J, Pesce JL (1986) Etude préliminaire de quelques paramètres sanguins usuels du dromadaire mauritanien (*Camelus dromedarius*). Rev Elev Méd Vét Pays Trop 39:395–401

Chavanne P, Boué A (1950) Taux normaux de l'urée et du glucose sanguins chez le dromadaire nord-africain. Rev Elev Méd Vét Pays Trop 4(4):183

Chiericato GM, Warfa AA, Schiapelli KP (1986) Influence of sex of the dromedary on some constituents of the blood. Rev Zootec Vet 14(3):196–199

Corbera JA, Morales M, Pulido M, Montoya JA, Gutierrez C (2003) An outbreak of nutritional muscular dystrophy in dromedary camels. J Appl Anim Res 23:117–122

Dalvi SH, Mantri AM, Talvelkar BA, Kulkarni BA, Patankar DD, Walawalkar M (1998) Blood metabolic profiles of Indian camel (*Camelus dromedarius*) under hot humid climate of Konkan region. Indian Vet J 75:217–220

Dahlborn K, Benlamlih S, Zine-Filali R, Gerouali A, Hossani-Hilali I, Oukessou M (1992) Food deprivation and refeeding in the camels (*Camelus dromedarius*). Am J Physiol 262:R1000–R1005

Deen A (2013) Serum creatinine, urea nitrogen and endogenous creatinine clearance based glomerular filtration rate in camels to evaluate renal functions. Camel Int J Vet Sci 1(1):1–12

DePeters EJ, Ferguson JD (1992) Non-protein nitrogen and protein distribution in the milk of cows. J Dairy Sci 75:3192–3209

Durand M, Kchouk M (1958) Etude de quelques constants hématologiques et chimiques chez le dromadaire. Bull Acad Vet France 31:197–198

El-Boshy M, Abbas H, El-Khodery S, Osman S (2009) Cytokine response and clinicopathological findings in Brucella infected camels (*Camelus dromedarius*). Vet Med 54(1):25–32
El-Deeb WM, Fouda TA (2013) Liver abscesses in dromedary camels (Camelus dromedaries): oxidative stress biomarkers and pro-inflammatory cytokines. J Vet Sci Technol 4:1–4
El-Deeb WM, Fouda TA, El-Bahr SM (2014) Clinico-biochemical investigation of paratuberculosis of dromedary camels in Saudi Arabia: proinflammatory cytokines, Acute phase proteins and oxidative stress biomarkers. Pak Vet J 34(4):484–488
El-Deeb WM, Buczinski S (2015) The diagnostic and prognostic importance of oxidative stress biomarkers and acute phase proteins in urinary tract infection (UTI) in camels. Peer J 3:e1363 18 p
Elias E, Yagil R (1984) Hematological and serum biochemical values in lactating camels and their newborns. Refuah Vet 41:7–13
Elkhair NM, Hartmann H (2011) Serum protein capillary electrophoretic patterns in camels (*Camelus dromedarius*): influence of age and sex. In: Proceedings of the 3rd conference on the international society of camelid research and development (ISOCARD), 29 Jan–1 Feb 2012, Muscat, Oman, pp 49–52
El-Nagi AH, Torki AI (2007) Chemical and biochemical composition of pregnant camel urine (*Camelus dromedarius*). Int J Biol Biotechnol 4(4):433–435
El-Naser EMA, Khamis GFA (2009) Some hematological and blood serum biochemical indices associated with respiratory affections by camels. Assiut Vet Med J 55(123):154–162
Eltohamy MM, Salama A, Youssef AEA (1986) Blood constituents in relation to the reproduction state in the camel (*Camelus dromedarius*). Beit Trop Land Vet Med 24:425–430
Emmanuel B, Howard BR, Emady M (1976) Urea degradation in the camel. Can J Anim Sci 56:595–601
Etzion Z, Yagil R (1986) Renal function in camels (*Camelus dromedarius*) following rapid rehydration source. Physiol Zool 59(5):558–562
Faye B, Mulato C (1991) Facteurs de variation des paramètres protéo-énergétiques, enzymatiques et minéraux dans le plasma chez le dromadaire de Djibouti. Rev Elev Med Vét Pays Trop 44:325–334
Faye B, Saint-Martin G, Cherrier R, Ruffa A (1992) The influence of high dietary protein, energy and mineral intake on deficient young camel: I. Changes in metabolic profiles and growth performance. Comp Biochem Physiol 102A:409–416
Faye B, Jouany JP, Chacornac JP, Ratovonanahary M (1995) L'élevage des grands camélidés. Analyse des initiatives réalisées en France. INRA Prod Anim 8:3–17
Faye B, Seboussi R, Alhadrami G (2009) Meta-analysis of the interactions between selenium status and haematological and mineral parameters in camel blood. Trends Comp Biochem Physiol 14:25–34
Faye B, Konuspayeva G, Almasaud A, Alafaliq A, Ben Abdallah A (2018) The effect of date-urea blocks as supplementary feeding on growth of young camels. Emr J Food Agric 30(4):320–325
Faye B, Konupayeva G, Loiseau G (2010) Variability of urea concentration in camel milk in Kazakhstan. Dairy Sci Technol 90:707–713
Ghodsian I, Nowrouzian I, Schels HF (1978) A study of some hematological parameters in the Iranian camels. Trop Anim Health Prod 10:109–110
Ghosal AK, Apanna TC, Dwaraknath PK (1973) Studies on the seasonal variation in the blood constituents of Indian camel (*Camelus dromedarius*). Indian J Anim Sci 43:642–644
Ghosal AK, Apanna TC, Dwaraknath PK (1975) A note on the effect of short-term water deprivation on certain blood characteristics in the camel (*Camelus dromedarius*). Indian J Anim Sci 45:105–108
Gihad EA, El-Gallad TT, Sooud AE, El-Nasr A, Farid MFA (1989) Feed and water intake, digestibility and nitrogen utilization by camels compared to sheep and goats fed low protein desert by products. In: Tisserand JL (ed) Actes du "Séminaire sur la Digestion, la Nutrition et l'Alimentation du Dromadaire". Ouargla, Algérie, 28 Février–1 Mars 1988, Options Méditerranéennes A(2), pp 75–81

Gupta L, Kumar RA, Ghanshyam T, Rajesh D, Garg R (2012) Effect of feeding different proportions of groundnut haulms (*Arachis hypogaea*) and cluster bean straw (*Cyamopsis tetragonoloba*) on nutrient utilisation and serum biochemical parameters in dromedary camels. Trop Anim Health Prod 44:1689–1695

Gutierrez C, Corbera JA, Juste MC, Doreste F, Morales I (2005) An outbreak of abortions and high neonatal mortality associated with *Trypanosoma evansi* in dromedary camels in the Canary Islands. Vet Parasitol 130:163–168

Hamad B, Aggad H, Hadef L, Adaika A (2017) Effect of seasons on blood biochemical parameters in male dromedary camels in Algeria. Indian J Anim Res On-Line B 672:1–5

Hassan YM, Hoeller H, Hassan IM (1968) Observations on the blood constituents of camels in the Sudan. Sudan J Vet Sci Anim Husb 9:464–476

Hassan YM (1971) A note on the effect of dehydration on a camel. Sudan J Vet Sci Anim Husb 12:111–112

Hassan AB, Shawkat ME, Elatrash SM, Attah AH (1983) Comparative study on the effect of some trypanocidal drugs on serum constituents and enzymatic activity in camels. Indian Vet J 60:518–526

Hekmatimoghaddam S, Rasooli A, Sazmand A, Hamidinejat H, Jafari H, Nouri M (2011) Serobiochemical alternations in dromedary camels naturally infected with Theileria spp. In: Iranian proceedings of the XVth international congress on animal hygiene, vol 1, 7/2011, pp 455–457

Heidarpour M, Mohri M, Borji H, Moghdass E (2012) Oxidative stress and trace elements in camel (*Camelus dromedarius*) with liver cystic echinococcosis. Vet Parasitol 187:459–463

Higgins AJ, Kock RA (1984) The camel in health and disease. I. A guide to the clinical examination, chemical restraint and medication of the camel. Br Vet J 140:485–504

Höhne HK, Niepage H (1985) Studies on the blood components of dromedaries (*Camelus dromedarius*) (in German). Dtsch Tierärz Wschr 92:322–327

Höller HK, Hassan YM (1966) Determination of bleeding blood components of camel in Sudan (in German). Dtsch Tierärz Wschr 73:553–556

Hussain M, Saeed Z, Gulsher M, Shaikh RS, Ali M, Ahmad AN, Hussain I, Akhtar M, Iqbal F (2016) A report on the molecular detection and seasonal prevalence of *Trypanosoma brucei* in dromedary camels from Dera Ghazi Khan District in southern Punjab (Pakistan). Trop Biomed 33(2):268–275

Hussein MF, Salah MS, Mogawer HH, Gar ElNabi AR (1992) Effect of lactation on the haemogram and certain blood constituents of the dromedary camel. J Appl Anim Res 1:43–50

Hussein YA, Al-Eknah MM, AL-Shami SA, Mandour MA, Fouda TA (2012) Coat color breed variation in blood constituents among indigenous Saudi Arabia camel strains. Mansoura Vet Med J 14(1):191–204

Idris OF, Tartour G (1970) Serum urea in camels of the Sudan: its variation with age, sex and region. In: Proceedings of the 5th Vet. conference. Sudan Veterinary Association, Khartoum, Sudan

Ismael AB, Swelum AA, Khalaf AF, Abouheif MA (2014) Clinical, haematological and biochemical alterations associated with an outbreak of theileriosis in dromedaries (*Camelus dromedarius*) in Saudi Arabia. Pak Vet J 34(2):209–213

Jatkar PR, Kohli RN, Bhatt PL (1962) Quantitative biochemical studies on camel's blood. Part III. Total serum-protein content. Indian Vet J 39(10):548–550

Jordan ER, Swanson LV (1979) Effect of crude protein on reproductive efficiency, serum total protein and albumin in the high producing dairy cow. J Dairy Sci 62:58–63

Kamal AM (2008) Some biochemical, hematological and clinical studies of selected ruminal and blood constituents in camels affected by various diseases. Res J Vet Sci 1(1):16–27

Kamili A, Bengoumi M, Oukessou M, Faye B, Lefebvre HP (2013) Assessment of glomerular filtration rate in normally hydrated and dehydrated dromedary camel by plasma exogenous creatinine clearance test. Emir J Food Agric 25(4):314–319

Kaneko JJ, Harvey JW, Bruss M (1997) Clinical biochemistry of domestic animals, 5th edn. Academic, New York, p 932

Karimi A, Rahbari S, Yousefi A (2015) Blood parasites of camels from central regions of Iran: comparative evaluation of various detection techniques and serum protein components. J Adv Parasitol 2(1):1–4

Kataria AK, Kataria N, Sharma KN (2000) A differential study of serum proteins profile in camel (*Camelus dromedarius*) Vis-à-Vis some other species. J Immunol Immunopathol 2:52–55

Kataria AK, Kataria N, Sharma KN (2002) Serum protein and immunoglobulin profile in camel (*Camelus dromedarius*). Indian J Anim Hlth 41(1):5–8

Kataria N, Kataria AK, Agarwal VK, Garg SL, Sahani MS (2003) Effect of seasonal dehydration on creatinine clearance in Indian dromedary camels. J Camel Pract Res 10(2):91–97

Kataria N, Kataria AK (2004) Use of blood analytes in assessment of stress due to drought in camel. J Camel Pract Res 11(2):129–133

Kataria N, Kataria AK (2006) Plasma levels of gastrin and pepsinogen in camel (*Camelus dromedarius*). J Camel Pract Res 13(2):141–144

Kayouli C, Jouany JP, Dardillat C, Tisserand J-L (1995) Particularités physiologiques du dromadaire: conséquences pour son alimentation. Options Méditerranéennes : Série B Etudes et Recherches 13:143–155

Kchouk M, Durand M (1958) Quelques dosages chimiques dans le sang des dromadaires en Tunisie. Arch Inst Pasteur Tunis 35:3–57

Khadjeh GH (1998) Electrophoretic pattern of serum proteins in the Iranian one-humped camel (*Camelus dromedarius*). Indian J Anim Sci 68:476–477

Khadjeh GH (2001) Concentration of serum proteins in pregnant and non-pregnant Iranian one-humped camels (*Camelus dromedarius*). J Camel Pract Res 8(1):89–90

Knight PK, Cluer D, Manefield GW, Gorde AK (1994) Haematology in the racing camel at rest: seasonal and training variations. In: Saltin B, Rose RJ (eds) The racing camel. Physiology, metabolic functions and adaptations. Acta Physiol Scand Suppl 150:19–23

Konuspayeva GS (2007) Variabilité physico-chimique et biochimique du lait des grands camélidés (*Camelus dromedarius, Camelus bactrianus et hybrides*) au Kazakhstan, PhD Université de Montpellier II, France, Sciences des aliments, 253 p

Kouider S, Kolb E (1982) Concentrations of glucose and urea in serum of camels. Monatsch Vet Med 37(4):140–142

Kouider S, Ateeq G, Kolb E (1988) Studies into levels of total proteins, urea, total fat, cholesterol and bilirubin in blood plasma of camels in the course of one year (in German). Mh Vet Med 43:139–142

Kumar M, Ghosh P, Banerjee S (1961) Biochemical studies on Indian camel. I. Blood proteins and lipids. J Sci Ind Res 20C:236–238

Li WX, Hai WJ (2014) Studies of sulphur-induced copper deficiency in the Bactrian camel. J Anim Vet Adv 13(14):886–890

Liu ZP (2003) Studies of the haematology and trace element status of adult Bactrian camels (*Camelus bactrianus*) in China. Vet Res Commun 27:397–405

Mahmud HM, Abdulhamid HM, Locatelli A (1984) Water deprivation effects on the hematological and hematochemical pictures of *Camelus dromedarius*. Rev Elev Med Vet Pays Trop 37:313–317

Mal G, Sena DS, Sahani MS (2006) Haemato-biochemical changes in camels infested with manage during winter and summer season. J Camel Pract Res 13(2):174–174

Malik A, Al-Senaidy A, Skrzypczak-Jankun E, Jankun J (2013) Isolation and characterization of serum albumin from *Camelus dromedarius*. Exp Ther Med 6:519–524

McGilvery R, Goldstein G(1983) Biochemistry: a functional approach, III edn. W.B. Saunders, Philadelphia, 909 pp

McGrane JJ, Kenyon SJ (1985) Laboratory diagnosis manual for field veterinarians in the Sudan. Overseas Development Administration Publications, London, pp 76–78

Mathur GN, Ghosal AK, Bhatia JS (1981) Note on certain blood constituents in the Indian camel. Indian J Anim Sci 51(12):1179–1180

Mehrotra V, Gupta ML (1989) Seasonal variations in certain blood constituents in camel. Indian J Anim Sci 59:1559–1561

Mohammed AK, Sackey AKB, Tekdek LB, Gefu JO (2007) Serum biochemical values of healthy adult one humped camel (*Camelus dromedarius*) introduced into a sub-humid climate in Shika-Zaria. Nigeria J Camel Pract Res 14(2):191–194

Mohri M, Moosavian HR, Hadian MJ (2008) Plasma biochemistry of one-humped camel (*Camelus dromedarius*): effects of anticoagulants and comparison with serum. Res Vet Sci 85:554–558

Momenah MA (2014) Some blood parameters of one humped she-camels (*Camelus dromedaries*) in response to parasitic infection. Life Sci J 11(5):118–123

Moolchandani A, Sareen M (2016) Blood biochemical study in Trypanosomiasis infected camel (*Camelus dromedarius*). Res Rev J Vet Sci 2(1):41–43

Mousa HM, Ali KE, Hume ID (1983) Effects of water deprivation on urea metabolism in camels, desert sheep and desert goats fed dry desert grass. Comp Biochem Physiol 74A(3):715–720

Muhammad BF, Aliyu D, Njidda AA, Madigawa IL (2011) Some haematological, biochemical and hormonal profile of pregnant and non-pregnant she-camels (*Camelus dromedarius*) raised in a Sudan savanna zone in Nigeria. J Camel Pract Res 18(1):73–77

Nagpal AK, Roy AK, Chirania BL, Patil NV (2011) Growth, nutrient utilization and serum profile in camel calves as affected by dietary protein levels. Indian J Anim Nut 28(2):166–171

Narnaware SD, Nagarajan G, Dahiya SS (2015) Hemato-biochemical studies in Indian camel (*Camelus dromedarius*) affected by contagious ecthyma. Indian J Vet Pathol 39(2):168–170

Nazifi S, Maleki K (1998) Biochemical analysis of serum and cerebrospinal fluid in clinically normal adult camels (*Camelus dromedarius*). Res Vet Sci 65:83–84

Nazifi S, Oryan A, Bahrami S, Razavi SM (2009) Evaluation of haematological and serum biochemical parameters in Iranian camels (*Camelus dromedarius*) infected with haemotrophic mycoplasma (Eperythrozoon) spp. Comp Clin Pathol 18:329–332

Nazifi S, Oryan AM, Ansari-Lari M, Tabandeh MR, Mohammadalipour A, Gowharnia M (2012) Evaluation of sialic acids and their correlation with acute-phase proteins (haptoglobin and serum amyloid a) in clinically healthy Iranian camels (*Camelus dromedarius*). Comp Clin Pathol 21:383–387

Ntiranyibagira E, Patil NV, Bhagwat SR, Lateef A, Xu K, Liu H (2015) Effects of different levels of urea supplementation on nutrient intake and growth performance in growing camels fed roughage based complete pellet diets. Anim Nutr 1:356–361

Omer SA, Salawa ME, Khougali H, Gussey AA, Samad HA (2010) Studies on some biochemical and haematological indices of Sudanese camels (*Camelus dromedarius*). Sudan Univ. Sci. Technol. Bull., pp 1–7

Omer SA, Khougali SME, Agab H, Samad GHA (2016) Studies on some biochemical and haematological indices of Sudanese camels (*Camelus dromedarius*). http://www.sustech.edu/staff_publications/20090615152717900.pdf

Omidi A, Zh S, Montazer-Torbati MB, Mostafai M (2014) Metabolic profile of pregnant, non-pregnant and male two-humped camels (*Camelus bactrianus*) of Iran. Iran J Vet Med 8 (4):235–242

Omidi A, Fathi MH, Asiaban M (2015) Elevated levels of blood urea nitrogen and creatinine in the last trimester of pregnancy of dromedary camels (*Camelus dromedarius*). Iran J Vet Med 9 (4):249–255

Orliac D (1980) Contribution à l'étude de la biochimie sanguine de dromadaires et de chèvres sahariens, Thèse doct. Vét., Toulouse, France, No 71

Orsonneau JL, Massoubre C, Cabanes M, Lustenberger P (1992) Simple and sensitive determination of urea in serum and urine. Clin Chem 38(5):619–623

Osman TEA, Al-Busadah KA (2000) Effects of age and lactation on some biochemical constituents of camel blood in Saudi Arabia. J Camel Pract Res 7(2):149–152

Osman TEA, Al-Busadah KA (2003) Normal concentrations of twenty biochemical parameters of she-camels, cows and ewes in Saudi Arabia. Pak J Biol Sci 6(14):1253–1256

Osman FA, Gaadee H, Sayed GA (2014) Clinico-hematological and biochemical changes in camels affected with gastro-intestinal parasites. Int J Adv Vet Med Sci 8(5):154–161

Ouajd S, Kamel B (2009) Physiological particularities of dromedary (*Camelus dromedarius*) and experimental implications. Scand J Lab Anim Sci 36(1):19–29

Patodkar VR, Somkuwar AP, Parekar S, Khade N (2010) Influence of sex on certain biochemical parameters in nomadic camels (*Camelus dromedarius*) nearby Pune, in Maharashtra. Vet World 3(3):115–117
Pegram RG, Scott JM (1976) The prevalence and diagnosis of Trypanosoma evansi infection in camels in southern Ethiopia. Trop Anim Health Prod 8:20–27
Perk K, Lobl K (1961) A study of the serum proteins and lipoproteins of the camel and their relation of its resistance to heat and thirst. Refuah Vet 18:163–168
Poonia R, Srivastava A, Sena S, Srivastava M (2015) A study on serum proteins of camel (*Camelus dromedaries*) maintained on different diets. Int J Basic Appl Sci 4(1):1–4
Read BE (1925) Chemical constituents of camel's urine. J Biol Chem 64:615–617
Rezakhani A, Nazifi HS, Maghrebi GM (1997) Studies on normal haematological and biochemical parameters of Turkmen camel in Iran. J Camel Pract Res 4(1):41–44
Roseler DK, Ferguson JD, Sniffen CJ, Herrema J (1993) Dietary protein degradability effects on plasma and milk urea nitrogen in Holstein cows. J Dairy Sci 76:525–534
Sadiek AH, Saleh MA (2006) Implications of serum proteins as a diagnostic and prognostic marker of hepatic diseases in camels. In: Proceedings of the international scientific conference on camels. Part II, Al-Qassim, Saudi Arabia, 10–12 May 2006, pp 904–914
Saeed A, Afzal M, Saeed A (1995) Effect of storage on some constituents of camel serum. Aust Vet J 72:212–215
Saeed A, Hussein MM, Khan IA, Chand G, El-Yousuf RA (2004) Effect of sex and age on blood biochemical profile in camel. J Camel Pract Res 11(1):73–77
Saeed A, Khan IA, Hussein MM (2009) Change in biochemical profile of pregnant camels (*Camelus dromedarius*) at term. Comp Clin Pathol 18(2):139–143
Sahraoui N, Doudou A, Douadji O, Babelhadj B, Hornick JL (2016) Impact of natural vegetation on some biochemical parameters of the Arabian camel (*Camelus dromedarius*) in Algeria. J Camelid Sci 9:62–71
Saini N, Singh GP, Nagpal AK (2007) Nutrient utilization from cluster bean straw, supplemented with urea and *Prosopis cineraria* leaves in growing camel. Indian J Dairy Sci 60:342–344
Salman R, Afzal M (2004) Seasonal variations in hematological and serum biochemical parameters in racing camels. J Camel Sci 1:63–65
Sarwar A, Majeed MA (1997) Interrelationships between 30 parameters of blood in normal one-humped camel in summer. J Camel Pract Res 4(1):35–39
Sazmand A, Rasooli A, Nouri M, Hamidinejat H, Hekmatimoghaddam S (2011) Serobiochemical alterations in subclinically affected dromedary camels with *Trypanosoma evansi* in Iran. Pak Vet J 31(3):223–226
Schmidt-Nielsen B, Schmidt-Nielsen K, Houpt TR, Jarnum SA (1957) Urea excretion in the camel. Am J Physiol 188:477–484
Sellaouati K (1984) Contribution à l'établissement des paramètres biochimiques sanguins du dromadaire, Thèse Méd. Vét. Sidi Thabet, Tunisie
Shaheen HM (2001) The effect of feed and water deprivation on ingestive behaviour and blood constituents in camels: comparison with sheep and goats. J Camel Pract Res 8(2):153–162
Shukla MK, Khan MJZ, Pathan MM, Siddiquee GM, Latif A (2009) Biochemical profile of uncastrated male Kutchi camel during the breeding season. J Camel Pract Res 16(1):213–215
Smith HW, Silvette H (1928) Note on the nitrogen excretion of camels. J Biol Chem 78:409–411
Snow DH, Billah A, Ridha A (1988) Effects of maximal exercise on the blood composition of the racing camel. Vet Rec 123:311–312
Soliman MK, El-Aroumsi S (1965) Serum bilirubin levels in Egyptian buffaloes and camels. Vet Rec 77:633
Soliman MK, Shaker M (1967) Cytological and biochemical studies on the blood of adult she-camels. Indian Vet J 44:989–995
Stahl J (1963) Studies of the blood ammonia in liver disease. Its diagnostic, prognostic, and therapeutic significance. Ann Intern Med 58:1–24

Swelum AA, Ismael AB, Khalaf AF, Abouheif MA (2014) Clinical and laboratory findings associated with naturally occurring babesiosis in dromedary camels. Bull Vet Inst Pulawy 58:229–233

Tabatabaei Naeini A, Nazifi S (2001) Biochemical and cytological properties of blood and peritoneal fluid in clinically healthy adult camels (*Camelus dromedarius*). J Camel Pract Res 8:123–126

Tharwat M, Al-Sobayil F (2015) The impact of racing on serum concentrations of acute-phase proteins in racing dromedary camels. Comp Clin Pathol 24:575–579

Vérité R, Rétif S, Faverdin P, (1995) Milk urea as an index for nutritive protein balance urinary N excretion in dairy cows on conserved diets. In: Proceedings of the VII symposium on protein metabolism nutrition, Estaeion Zooteeniea Nacional, Vale de Santazem, 24–27 May 1995, pp 33–48

Visek WJ (1984) Ammonia: its effects on biological systems, metabolic hormones, and reproduction. J Dairy Sci 67:481–498

Wensvoort J, Kyle DJ, Orskov ER, Bourke DA (2001) Biochemical adaptation of camelids during periods where feed is withheld. Rangifer 21(1):45–48

Yadav SB, Bissa UK (1998) Factors affecting some blood constituents in camels—a review. In: Proceedings of the 3rd annual meeting for animal production under arid conditions, vol 2. UAE University Publications, Al-Ain, pp 32–48

Yagil R, Berlyne GM (1977) Glucose loading and dehydration in the camel. J Appl Physiol 42 (5):690–693

Yagil R (1993) Renal function and water metabolism in the dromedary. In: Bourke E, Mallick NP, Pollak VE (eds) Moving points in nephrology. Contributions to neprology, vol 102. Karger, Basel, pp 161–170

Yagoub IA (1988) Studies on haematological parameters and some biochemical blood constituents of Sudanese camels (*Camelus dromedarius*). Acta Vet Brno 38:99–106

Yousif HS, Omer SA, Ahmed SH (2016) Influence of thermal environment change on blood metabolites, leukocytic and erythrocytic indices and clinical parameters of in-door camel (*Camelus dromedaries*). Glob J Biol Agric Health Sci 5(1):20–24

Youssef SY, Yasien S, Ali Mousa WH, Nasr SM, El-Kelesh EAM, Mahran KM, Abd-El-Rahman AH (2015) Vector identification and clinical, hematological, biochemical, and parasitological characteristics of camel (*Camelus dromedarius*) theileriosis in Egypt. Trop Anim Health Prod 47:649–656

Zaher H, El-Zahar H, Al Sharifi S, Shety T (2017) Alterations in hematological and biochemical parameters affecting the reproductive performance in female camels (*Camelus dromedaries*). Int J Vet Health Sci Res 5(1):155–160

Zeidan AEB, Abd El-Salam AM, El-Malky OM, Ahmadi EA, Sarhan DMA, Daader AH (2008) Biochemical and histological changes of the ovary of the dromedary camel during breeding and non-breeding seasons. Egypt J Basic Appl Physiol 7:287–308

Faye B, Konuspayeva G, Almasaud A, Alafaliq A, Ben Abdallah A (2018) The effect of date-urea blocks as supplementary feeding on growth of young camels. Emirates Journal of Food and Agriculture 30 (4):320-325

Rezakhani A, Nazifi Habibabadi S, Maghrebi Ghojogh M (1997) Studies on normal haematological and biochemical parameters of Turkmen camel in Iran. J Camel Pract Res 4(1): 41-44Aichouni A, Belhadia M, Kebir N, Aggad H (2013) Season influence on serum organic parameters of dromedarius (Camelus dromedarius) in Algeria. Bioch Biotech Res 1(1): 8-12Abd El-Baky A A, Salem S I (2011) Clinicopathological and cytological studies on naturally infected camels and experimentally infected rats with Trypanosoma evansi. World appl Sci J 14: 42-50

Chapter 5
Clinical Enzymology

Enzymes are proteins catalyzing biological reactions. Their activities in blood and tissues are indicators of cells suffering and are used to explore expecially liver and kidney damages. The most used in clinical investigations are ASAT, ALAT, GGT, CPK, ALP and GLDH. In addition, metalloenzymes can be investigated. They are linked to some microminerals' metabolism: ceruplasmin for copper, superoxide-dismutase for zinc and glutathione-peroxidase for selenium.

In clinical enzymology, the concentration of the enzyme in blood is estimated using an activity measurement assessed by the speed of the reaction that it is catalyzed. It corresponds to the amount of substrate converted or the amount of product formed per unit of time. Several factors influence enzyme activity, including the concentration of the enzyme, the temperature, the pH, and the concentration of substrates or some activators or inhibitors. In order to have a good correlation between the enzyme concentration and its activity, it is necessary that all these factors are at optimal level.

Clinical enzymology represents one of the para-clinical examinations most appropriate for early diagnosis, to confirm clinical suspicions and to follow the evolution of many diseases in humans and animals. The principles of the use of enzymes in semiotics are widely treated elsewhere (Kaneko et al. 1997). However, before addressing the clinical enzymology in the camel, it seems useful to recall some of these principles.

© Springer International Publishing AG, part of Springer Nature 2018
B. Faye, M. Bengoumi, *Camel Clinical Biochemistry and Hematology*,
https://doi.org/10.1007/978-3-319-95562-9_5

5.1 General Principles of Clinical Enzymology

The most important principles of clinical enzymology will be presented here, especially some general criteria for interpreting the results of laboratory:

1. Generally, the increase in serum enzyme activity indicates cell suffering. The cells of multicellular organisms are particularly rich in enzymes. During tissue injury, these enzymes pass in the interstitial media and then in the general bloodstream. The increase in plasma enzyme activity indicates tissue damage. However, in the case of kidney damage, this increase is observed mainly in urine.
2. In some cases, serum enzyme activity decreases during some tissue damage or poisoning. In fact, some enzymes, bio-synthetized in tissues, including the liver tissue, and released into the blood where they are active, have their plasma activity decreasing during damage of these tissues or poisoning. For example, the cholinesterase activity decreases during liver failure or poisoning by organophosphates. Moreover, certain enzymes are genetically deficient in some individuals.
3. The plasma or red blood cell activity of some enzymes may decrease during mineral deficiencies or increase during an excessive intake. Indeed, some micronutrients are cofactors for many enzymes like copper for ceruloplasmin, copper and zinc for superoxide dismutase, or selenium for glutathione peroxidase. Correlation between the enzyme activity and the provided trace elements is essential to assess the nutritional status in these minerals.
4. Reverse to principle (2), the biosynthesis of certain enzymes can be increased during tissue damage. It is the case, for example, for gamma-glutamyl transferase (GGT) and the alkaline phosphatases (PAL) during cholestasis. We called this a phenomenon of induction. For example, in humans, alcohol dehydrogenase and GGT are induced by excessive consumption of alcohol.

Among these principles, the first one is the more commonly used in clinical biochemistry for cellular damage diagnosis, severity assessment, and monitoring of their evolution. However, the detection of cell damage using measurements of enzyme activity requires knowledge of enzyme tissue distribution. Indeed, the increase in plasma activity of a specific enzyme from a tissue indicates that a cytolysis is occurring in this tissue. There are numerous specific enzymes to a particular tissue but their dosages are not always practical. Enzymes that are easy to determine often are ubiquitous. However, the combination of several enzymes and other blood parameters generally allows clarifying a diagnosis. In addition, the tissue distribution of enzymes varies from one species to another and leads to difficulty in the interpretation of results and on the convenient choice of enzymatic markers in animal clinical biochemistry.

On this point, the tissue distribution of enzymes in camel studies shows that this animal differs significantly from other ruminants. Indeed, most of the enzymes commonly used in clinical biochemistry are, in camel, most concentrated in the kidney than in other tissues unlike most other domestic species in which the liver plays a decisive role. This feature would be in relation to the significant metabolic role of the kidney in the dromedary camels, particularly during dehydration (Bengoumi et al. 1997a, b, 1998a, b).

The severity of a cell lesion depends on its intensity and extent. For the last parameter, it is accepted that the increase in serum enzyme activity is proportional to the number of injured cells. Therefore, the degree of increase in serum enzyme activity allows estimating the extent of the cell damage. Consequently, the intensity of the cytolysis can be appreciated from the intracellular distribution of enzymes. Indeed, cytoplasmic enzymes are released into serum even if cell injury is slight, in particular, during an increase in cell membrane permeability. Conversely, enzymes contained in the cell organelles are released only if the damage is serious, affecting the organelles' integrity. So the increase in serum activity of mitochondrial enzymes indicates that the cytolysis is serious and that the damaged cells are lost.

The follow-up of the changes in damaged tissue by measurement of enzyme activities should also take into account the serum half-life of these enzymes. This last being defined as the time required to make the serum activity of the released enzymes cut in half. Serum activity of the enzymes with short half-life such as transaminases diminishes quickly if the disease's evolution is favorable and the cellular lesion rapidly restored, while the serum activity of enzyme with long half-life as gamma-glutamyl transferase (GGT) persists even if the lesion disappeared and if healing is established. In general, transaminases, lactate dehydrogenase (LDH), and creatinine kinase (CK) have a serum half-life less than 48 h, while alkaline phosphatases (PAL) and GGT may exceed 10 days.

5.1.1 Peculiarities of Blood Sampling for Enzyme Activity Determination

5.1.1.1 The Choice of Anticoagulant

Sampling for enzyme activity measurement should take in account the nature of the anticoagulant used. Indeed, some anticoagulants are likely to alter the plasma activity of some enzymes (Friedel et al. 1975; André 1981; Jones 1985a, b). In this respect, several examples can be cited: (1) the enzymes with mineral cofactors such as alkaline phosphatases, arginase, and sorbitol dehydrogenase are inhibited by anticoagulants chelating minerals such as EDTA (ethylene diamine tetra-acetic); (2) oxalates and fluorides decrease the plasma activity of GGT but without effect on transaminases (ASAT and ALAT); and (3) the activity of GGT in serum is lower than in the plasma obtained on lithium heparinate, while the reverse is observed for the transaminases (ASAT and ALAT).

Serum appears the best substrate for enzymatic assays. However, the delay of serum collection after blood clotting, which is about 3 h, does not allow the measurement of the activity of certain enzymes with very short half-life such as sorbitol dehydrogenase. Furthermore, the red blood cells being particularly rich in enzymes, hemolysis, even slight, may strongly influence the plasma or serum enzyme activity. Therefore, it is useless to perform measurements of enzyme activity on hemolyzed samples. It is possible to use sampling tubes for serum with silicone gel that fits between the red blood cells and serum during centrifugation, preventing collection of red blood cells with serum.

5.1.1.2 Samples Conditioning

Temperature and storage duration of blood samples have a significant effect on enzyme activity. This effect varies according to the enzymes and species. For example, in cattle, plasma ALT activity decreases by 3% per day at 20 °C, while GGT activity decreases by only 0.5%. The stability of enzymes also varies according to the species. At −20 °C, plasma sheep ASAT activity is more stable than that of cattle.

In camel, few specific data on enzymes' stability are available. Saeed et al. (1995) showed, for example, that storage of camel serum samples at room temperature (23–25 °C) ensured the stability of LDH during 1 day. This time was 3 days for the ASAT, 6 days for ALAT and GGT, and 8 days for AP. CK activity is declining after 4 h only. At refrigeration temperature (4–5 °C), stability of most camel enzymes is guaranteed for 9 days. However a decrease in enzyme activity of CK, LDH, and ALAT is observable after 1, 6, and 7 days, respectively (Saeed et al. 1995). Our observations showed that freezing at −20 °C allows enzymes' stability for at least 3 months. However, the enzymes, being proteins, are very sensitive to heat shock, including the cycle freezing-thawing. Therefore, it is recommended to avoid measurement of enzyme activity in samples thawed and then refrozen (Clifton 1985).

In view of the risk of misinterpretation from hemolyzed samples, it is unnecessary to perform measurements of enzyme activity in such samples. So, conditioning of the samples at +4 °C just after collection followed by a quick centrifugation has to be ensured to avoid hemolysis. Although the camel RBCs are resistant to hemolysis (see Chap. 2), the remoteness of the laboratory from sampling place and the carrying conditions to labs including shaking on rough roads can cause destruction of red blood cells.

5.1.2 *Interpretation of the Results of Enzyme Activity Measurement*

5.1.2.1 Units

Enzyme concentrations are very low in cells and even more in the plasma. So, the determination of this concentration is very difficult in practice. In a tissue or a fluid, the enzyme concentration is estimated from the rate of the reaction that it catalyzes. However, this estimate is valid only if the speed of the reaction is maximal, i.e., when all the active enzyme sites are occupied. Indeed, the enzyme activity depends on not only the amount of the enzyme in the milieu but also of the environmental conditions, including pH, temperature, concentration of substrates and products, and other factors. Different units have been used in the past which made the comparison of results rather difficult. The need for standardization led the International Society of Clinical Biology (ISCB) to propose an international unit (U/l) corresponding to the amount of enzyme able to catalyze the conversion of 1 micromole of substrate or the release of 1 micromole of product per minute in optimal reactive conditions of temperature, pH, substrate concentration, and other factors (Kaneko et al. 1997). Most of the plasma enzyme activity measurements use ready-made kits containing substrates and a buffer for the pH stability. However, temperature measurement of enzyme activity can vary from one country to another or from one laboratory to the other depending on the method chosen. Certainly, there are formulas for conversion of enzymatic activities according to the temperature used in the analyses, but it requires that the authors should indicate the conditions of analysis which is not always the case. Currently, most laboratories use a temperature of 37 °C which is close to the natural conditions of enzymatic reactions in animals.

5.1.2.2 Comparison with Usual Values

Interpretation of the results of enzyme activity measurement in clinical biochemistry classically passes through their comparison with usual or reference values. In camel, the reference values are rarely available because of the difficulty to obtain it and the relative limited number of bibliographic references. However, in case of disorders characterized by tissue damage, enzyme activity generally follows the law of "all or nothing." Indeed, any slight cell lesion causes the release of enzymes in plasma in very large quantities. There is therefore a very significant "quantitative jump." Usually, it is admitted that there is really cell damage if enzyme activity equals at least the double of the usual values.

After these instructions on the clinical enzymology, we will address the main enzymes used in animal clinical and nutritional biochemistry.

5.2 The Main Enzymes Used in Camel

5.2.1 Aspartate Aminotransferase: ASAT (E.C. 2.6.1.1)

Aspartate aminotransferase formerly called transaminase glutamic oxaloacetic (TGO) is one of the most commonly used enzymes in clinical biochemistry. ASAT intervenes in the metabolism of amino acids. It catalyzes the transfer of amine (NH2) from α-amino acid (glutamate) to α-keto acid (oxaloacetate) to give another α-amino acid (aspartate) and another α-keto acid (α-ketoglutarate). It is therefore a reaction of biosynthesis of one amino acid from another. The sense of in vitro reaction depends on the concentration of the substrate, but in vivo glutamic acid is the NH_2 donor:

$$\text{Glutamate} + \text{Oxalo-acetate} \xleftrightarrow{\text{ASAT}} \text{Aspartate} + \alpha\text{-ketoglutarate}$$

5.2.1.1 Principle of Dosage

The activity of the ASAT is determined from the rate of disappearance of NADH,H^+, which, added to the product of catalysis of aspartate and of the α-ketoglutarate (oxaloacetate), is degraded in NAD^+ according to the catalytic reactions below:

$$\text{Aspartate} + \alpha\text{-ketoglutarate} \xleftrightarrow{\text{ASAT}} \text{Glutamate} + \text{Oxalo-acetate}$$

$$\text{Oxalo-acetate} + \text{NADH}, H^+ \xleftrightarrow{\text{Malate dehydrogenase}} \text{Malate} + NAD^+$$

The speed of NADH, H + degradation, is measured by spectrophotometry at 340 nm.

5.2.1.2 Tissue Localization and Clinical Significance

ASAT is a mixed enzyme, i.e., both cytoplasmic and mitochondrial. Unlike other ruminants where it is concentrated in skeletal and heart muscles, and secondarily in the liver (Braun 1985), in camel ASAT is an enzyme present firstly in the kidney and the muscle, liver, and myocardium (Bengoumi et al. 1997a). The kidney with 29 $\pm$ 2.4 U/g has roughly double of the liver activity (13 $\pm$ 1.5 U/g) and muscle (15 $\pm$ 1.8 U/g), while the ASAT activity in the heart is relatively low (4 $\pm$ 0.9 U/g). However, Afzal and Saeed (1995) observe a tissue location in camel comparable to that of cattle. The increase of its plasma activity is rather a consequence of muscle or liver damage, while the increase in urinary activity indicates renal damage. Moreover, ASAT has a strong activity in red blood cells (0.9 $\pm$ 0.2 U/g of hemoglobin) which is roughly three times the plasma activity. Therefore, any hemolysis can strongly influence the results of the analyses.

5.2.1.3 Usual Values

In camel, serum mean usual values for ASAT activity range from 37 to 131 U/l (Table 5.1), but lower values were reported regularly in apparently healthy animals (Kouider and Kolb 1982). These values are close to those of the cattle (78–132 U/l) but below those of sheep (264–360 U/l), goats (157–513 U/l), horse (226–366-U/l), and llama (216–378-U/l) (Kaneko 1989). ASAT activity in serum appeared significantly higher in summer than in winter, probably in relationship with the heat stress: 29 $\pm$ 1 vs 15 $\pm$ 6 U/l (Nazifi et al. 1999). Similar pattern was reported by Kataria et al. (1991) who observed an increase in ASAT in camels exposed to hot temperature

Table 5.1 Normal values in the main enzymes (ASAT, ALAT, ALP, and LDH) in camel according to different authors

References	Enzyme activities (U/l)			
	ASAT	ALAT	ALP	LDH
Ghosal and Dwaraknath (1971)	11 ± 1	–	–	–
Adam et al. (1974)	25	11	–	–
Boid (1980)	38 ± 11.9	16 ± 4	31 ± 9	–
Orliac (1980)	94 ± 42	15 ± 3	54 ± 22	–
Halabi et al. (1982)	–	–	34 ± 17	–
Al-Amrousi and Wasfi (1984)	44 ± 6	6 ± 1	–	–
Sellaouti (1984)	79 ± 13	8 ± 2	51 ± 1	–
Abdo et al. (1985)	93 ± 1	35 ± 1	1 ± 0	–
Chiericato et al. (1986)	74–93	8.6–10	58–72	605–616
El Dirdiri et al. (1987)	35 ± 10	19 ± 7	–	345 ± 96
Abdalla et al. (1988)	96 ± 24	–	–	–
Al-Ali et al. (1988)	81 ± 37	–	63 ± 11	2620 ± 510
Bizzeti et al. (1988)	41 ± 14	6 ± 2	–	–
Ali et al. (1989)	71–79	–	–	–
Ben-Zvi et al. (1989)	65 ± 4	–	–	–
Kataria and Bhatia (1991)	36 ± 0	5 ± 0	27 ± 0	479 ± 7
Abu Damir et al. (1993)	–	–	166 ± 23	377 ± 84
Haroun (1994)	12 ± 5	9 ± 4	43 ± 21	–
Knight et al. (1994a)	87 ± 58	–	–	850 ± 350
Faye et al. (1995a)	48 ± 14	–	–	–
Nyang'ao et al. (1997)	6 ± 2	3 ± 1	11 ± 4	142 ± 25
Sarwar and Majeed (1997)	48 ± 3	4 ± 0	–	–
Nazifi and Maleki (1998)	15 ± 3	19 ± 2	109 ± 9	1680 ± 23
Nazifi et al. (1999)	149 ± 6	19 ± 3	23 ± 3	684 ± 13
Chaudhary and Iqbal (2000)	99.37 ± 4.17	10.92 ± 1.54	77.23 ± 3.88	923.0 ± 36.9
Bogin (2000)	105 ± 17	11 ± 3	82 ± 13	392 ± 64
Tabatabaei Naeini and Nazifi (2001)	117.3 ± 46.9	31.24 ± 16.4	115.6 ± 10.4	322.1 ± 24.2
Khadjeh (2002)	106.3 ± 2.54	10.93 ± 0.56	–	–
Ayoub et al. (2003)	53.8–77.1	13.1–18.5	44.3–73.5	297–428
Osman and Al-Busadah (2003)	164.6 ± 39.9	17.2 ± 3.6	60.0 ± 7.2	455 ± 75.9
Seleim et al. (2003)	31.4 ± 1.65	19.87 ± 1.13	95 ± 3	–
Ahmad et al. (2004)	51.97 ± 3.71	14.59 ± 1.86	59.65 ± 7.26	–
Sarwar et al. (2004)	47.8 ± 2.6	4.33 ± 0.12	–	–
Seboussi et al. (2004)	90.2 ± 4.6	18.0 ± 4.0	–	–
Barsham et al. (2005)	–	–	27.61 ± 2.8	–
Gutierrez et al. (2005)	42.2 ± 24.6	11.1 ± 2.9	–	592.8 ± 187
Mal et al. (2006)	45.6 ± 7.5	6.7 ± 1.5	78.8 ± 10.1	–

(continued)

Table 5.1 (continued)

References	Enzyme activities (U/l)			
	ASAT	ALAT	ALP	LDH
Al-Busadah (2007)	29.9 ± 3.1	11.23 ± 1.6	–	253 ± 26
Elnahas (2008) (Egypt)	138.3 ± 39.4	15.3 ± 5.9	55.5 ± 18.9	–
Elnahas (2008) (Germany)	139.3 ± 36	6.5 ± 1.9	149.8 ± 70.1	–
Mohri et al. (2008)	73.8 ± 22.5	13.3 ± 4.8	279.4 ± 198	–
Bengoumi (2008)	83 ± 10	14 ± 3	232 ± 34	–
Al-Sobayil and Mousa (2009)	43 ± 3	17.2 ± 5.42	31.4 ± 12.2	–
Faye et al. (2009)	81.4 ± 57.7	11.1 ± 5.7	88.8 ± 34.7	463 ± 189.4
Nazifi et al. (2009b)	118.4 ± 6.8	31.05 ± 2.5	–	–
Saeed et al. (2009)	84.2 ± 20.6	10.4 ± 2.12	179.6 ± 23.2	822.7 ± 111
Aichouni et al. (2010)	73.3 ± 17.0	4.4 ± 2.6	–	331 ± 77
Shen and Li (2010)[a]	39.5 ± 4.9	13.9 ± 2.9	48.9 ± 13.5	192.6 ± 81.6
Bengoumi et al. (1997b)	76–183	7–28	51–178	815–3122
Abd-El-Baky and Salem (2011)	30 ± 2.0	11.3 ± 1.5	85.0 ± 5.0	–
Babeker and Suleem (2013)	20.35 ± 2.1	7.7 ± 0.96	83.65 ± 11.5	–
Al-Rukibat and Ismail (2014)	162.1 ± 156.3	8.3 ± 0.49	134.3 ± 74.03	894.8 ± 539.4
Ismael et al. (2014)	97.8 ± 4.6	9.2 ± 0.8	–	807.5 ± 84.6
Momenah (2014)	80.11 ± 2.2	56.11 ± 1.65	–	–
Omidi et al. (2014)[a]	151.8 ± 24.5	14.5 ± 2.0	345.6 ± 68.4	–
Tharwat and Al-Sobayil (2014)	90 ± 30	–	10 ± 9	–
Alshamsi et al. (2015)	99.9–121.4	–	191 ± 89	452 ± 53
Badakhshan and Mirmahmoudi (2016)	101.3 ± 4.22	20.9 ± 1.5	57.7 ± 18.16	710.8 ± 38.9
Hussain et al. (2016)	32.72 ± 1.77	13.09 ± 1.12	92.77 ± 1.23	–
Hamad et al. (2017)	111.4 ± 34.0	27.8 ± 18.0	108.3 ± 30.9	928 ± 447
Zaher et al. (2017)	74.92 ± 2.93	8.5 ± 0.52	–	434.2 ± 12.8

[a]Bactrian camel (non-pregnant)

(83 ± 1) compared to those under cold conditions (68 ± 9 U/l). No breed effect has been reported (Seboussi et al. 2004; Aichouni et al. 2010).

5.2.1.4 Physiological Variations

Age

As in other domestic animals, age does not seem to influence plasma activity of ASAT (Bengoumi et al. 1997b; Sarwar et al. 2004). However, according to some authors, this activity decreases in the dromedary between 7 and 16 years (Salutini and Biagi 1983) or from 4 years (Kataria and Bhatia 1991) with mean values ranging from 44 ± 1 before 4 years old, 36 ± 1 between 4 and 10 years old, and 28 ± 1 U/l after 10 years old. However, values seem to be higher in camel calves than in she-camels. Osman and Al-Busadah (2000) reported 217 ± 25 U/l in calves vs 80 ± 8 U/l in lactating he-camels and 123 ± 10 U/l in non-lactating she-camels. Irregular pattern according to age was reported by Seboussi et al. (2004).

Sex

In most species, the sex has no significant influence on plasma activity of ASAT. It is the same in the dromedary camel according to a majority of authors (Salutini and Biagi 1983; Ateeq et al. 1984; Chiericato et al. 1986; El Dirdiri et al. 1987; Bengoumi et al. 1997a; Sarwar et al. 2004). However, for Adam et al. (1974) and Kataria and Bhatia (1991), serum ASAT activity is higher in males than in females (39 ± 1 vs 331 ± 1 U/l, respectively). Similar feature is recorded by Seboussi et al. (2004): 104 ± 8 in male vs 82 ± 1 in female. But this difference could be due to the rutting season. Indeed, in male camel, ASAT is significantly higher in non-breeding season: 62 ± 3 vs 41 ± 4 U/l in rutting season (Zeidan and Abbas 2003).

Pregnancy-Lactation

In general, no effect of the physiological stage of pregnancy and lactation (or castration in male) on camel plasma activities of ASAT was reported (Elias and Yagil 1984; Ateeq et al. 1984; Sarwar et al. 1992; Bengoumi et al. 1997b; Osman and Al-Busadah 2000; Khadjeh 2002; Saeed et al. 2009; Omidi et al. 2014). However, on a large number of animals (240 camels), a slight difference was reported by Seboussi et al. (2004) between pregnant (85 ± 2 U/l) and non-pregnant camel (82 ± 2 U/l). Sarwar et al. (2004) found higher ASAT activity in camel heifers (66 ± 7 U/l) than in non-lactating camel (38 ± 3 for non-pregnant and 44 ± 6 U/l for pregnant camels). During pregnancy, ASAT activity seems to increase at the end (after 360 days of gestation), i.e., just before parturition, with values passing from 93 ± 6 U/l to 158 ± 11 U/l (El-Belely et al. 1988).

A slight breed effect was described, ASAT activity being higher in brown-coat and white-coat camel compared to black-coat camel: 61 ± 2 and 59 ± 2 vs 46 ± 2 U/l, respectively (Hussein et al. 2012).

ASAT activity is sensitive to selenium poisoning in camel: between two levels of supplementation (0–4 mg vs 8–16 mg), ASAT is increased by 47% (Faye et al. 2009).

Physical Effort

In well-trained racing camels, physical effort has no significant effect on plasma activity of ASAT (Beaunoyer 1992; Rose et al. 1994). However, in the grazing dromedary in pasture not accustomed to race, a gallop over a distance of 4 km caused an increase in the activity of ASAT passing from 95 to 162 U/l (Bengoumi et al. 1998b).

Dehydration

During a total water deprivation-induced dehydration by more than 25%, the plasma activity of ASAT does not seem to be influenced (Mohamed et al. 1984). The observed increase comes solely from the hemoconcentration (Bengoumi et al. 1998a). In normal hydrated camels, Ben-Zvi et al. (1989) reported ASAT values at 65 ± 4 U/l, while it was 114 ± 20 U/l after 10 days of dehydration.

Diseases

Globally, ASAT is used in liver or kidney disease investigations in camel (Belina et al. 2015), but some important diseases in camel as trypanosomosis, internal parasites, and diarrhea could have an effect on the enzyme activity.

Anemia is the main symptom in camel. Its development and persistence in the course of the disease induce anoxic conditions which manifest signs of dysfunction in various organs as a result of vascular damage. This is followed by the release of large quantities of cytoplasmic and mitochondrial enzymes, especially ASAT and ALAT, into serum (Enwezor and Sackey 2005). After inoculation of *Trypanosoma evansi*, ASAT activity increased from 39 U/l on day 6 after contamination to 236 U/l on day 27 (Boid 1980), while in camel treated with quinapyramine, the values fell back to normal level. In Suratex-positive camel, the mean ASAT values passed from 99 ± 4 (negative camels) to 116 ± 4 U/l. Similar effect was reported in smear-positive camels (105 ± 4) in Chaudhary and Iqbal (2000)). The increase was more pronounced in the study of Hussain et al. (2016), from 33 ± 2 in healthy camels to 55 ± 3 U/l in infected camels.

At reverse, no significant difference was observed in healthy camels compared to sick ones in an outbreak of abortion associated with *Trypanosoma evansi* in Canary Islands (Gutierrez et al. 2005): 42 ± 25 vs 47 ± 17 in sick and control camels, respectively.

The presence of gastrointestinal parasites contributes to the increase of enzyme activity due to cell suffering. In affected camels, ASAT values were reported to be 159 ± 8.61, while it was 105 ± 11.31 in healthy camels (Osman et al. 2014), but reverse results were reported in camel infected with *Haemonchus contortus* (68.17 ± 2.15 U/l) or with mixed gastrointestinal parasites (77.5 ± 2.87 U/l) compared to non-infected camel (93.0 ± 1.0). At reverse, mange has no effect on ASAT concentrations except in winter (Mal et al. 2006).

Similar figure occurs in young camel (Abdo et al. 1985). Change in ASAT activity occurred also in camel affected by cystic echinococcosis affecting the liver, with values passing from 79 ± 10 to 156 ± 16 U/l (Heidarpour et al. 2012).

The tick infestation is linked to a high increase of AST activity, from 90 ± 30 in control camel to 396 ± 206 U/l in infested camel and 413 ± 206 U/l in camel that died because of heavy infestation (Tharwat and Al-Sobayil 2014). Similar increase was reported by Momenah (2014).

In case of theileriosis, ASAT activity is double than healthy camels passing from 97.9 ± 4.7 to 193.7 ± 18.3 U/l (Ismael et al. 2014). It is the same with babesiosis: from 93.35 ± 4.61 to 175.46 ± 27.35 (Swelum et al. 2014). Camels affected by pasteurellosis have on average a higher plasma ASAT activity: 40.73 ± 2.63 vs 31.4 ± 1.65 in healthy camels (Seleim et al. 2003). Similar figure is reported regarding camels affected by arthritis due to *Mycoplasma*: 105.7 ± 2.2 in positive camels vs 78.16 ± 1.3 U/l in negative ones (Shoieb and Sayed-Ahmed 2016).

In camels affected by wryneck syndrome (paralysis of muscles of the neck), ASAT increases significantly: 132.8 ± 10.3 vs 43 ± 3 U/l in healthy camels (Al-Sobayil and Mousa 2009). A significant increase of ASAT activity is reported in racing camel affected by bone fracture, with a mean value passing from 99.9 to 300.2 U/l (Alshamsi et al. 2015).

During experimental copper poisoning, it was observed an increase in ASAT (Abu Damir et al. 1993) that might be associated with liver and kidney lesions due to poisoning. The maximum value (more than 100 U/l) was observed on the third day after poisoning.

At reverse, a quite important increase was observed in camels affected by narasin poisoning (Abu Damir et al. 2013), with values passing from 76 ± 16 in control camels to 1467 ± 998 in young affected camels to 1142 ± 717 U/l in adult ones. Monensin toxicity has similar effect, the ASAT activity reaching 3500 to 8200 U/l in affected camels (Chaudhary et al. 1998).

Serum ASAT is also increased linearly with the CMT score in case of clinical mastitis (Ali et al. 2016a, b) passing from 33 ± 1 U/l in healthy camel to 52 ± 1 U/l in camel with CMT score +++ (Fig. 5.1). ASAT activity is linearly increasing with the duration of dystocia, passing from 3.20 ± 2.92 U/l in control group to 10.5 ± 5.7 U/l in group with dystocia duration of 72 h (Ali et al. 2016a, b). In the young camel, diarrhea is not associated with a significant change of ASAT values in plasma (Bengoumi et al. 1998c). ASAT is higher in camel infected by brucellosis, up to 116.4b ± 0.12 U/l, while control camel was 81.5 ± 6.15 U/l only (El-Boshy et al. 2009).

5.2.1.5 ASAT in Other Substrates

In camel milk, ASAT activity is lower than in serum: 8–9 U/l (Yadav et al. 2015). This value increases in case of mastitis: from 9 ± 0 U/ l in camel with CMT = 0 to 12 ± 0 U/l with CMT = +++ (Ali et al. 2016a, b).

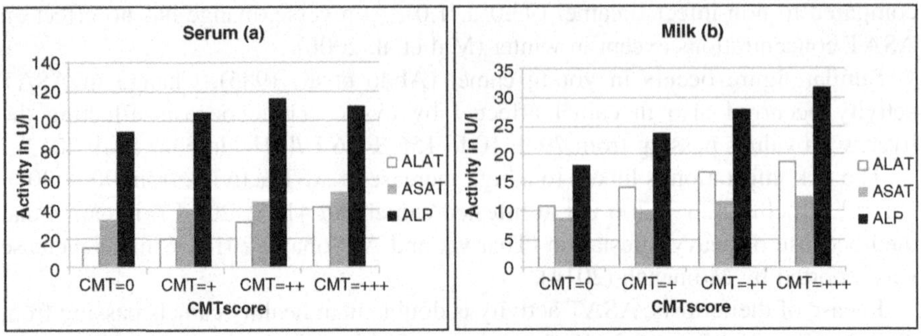

Fig. 5.1 Changes in serum (**a**) and milk (**b**) ASAT, ALAT, and ALP activities according to the CMT score of camel [calculated from Ali et al. (2016a)]

In peritoneal fluid of healthy camels, ASAT concentrations are lower than in blood: 45 ± 32 U/l (Tabatabaei Naeini and Nazifi 2001). In cerebrospinal fluid, ASAT activity has been determined at 4 ± 0 U/ l (Nazifi and Maleki 1998) and 21 ± 0 U/l (Ahmed et al. 2009). In synovial fluid, values of 22 ± 24 U/l (Al-Rukibat and Ismail 2014) and 22 ± 2 U/l (Omer and Gameel 2009) were reported. Few references are regarding the AST concentration in camel urine (Nazifi et al. 2005).

5.2.2 *Alanine Aminotransferase: ALAT (E.C. 2.6.1.2)*

Alanine aminotransferase, also called transaminase glutamo-pyruvique (TGP), is often investigated in clinical biochemistry in association with ASAT.

5.2.2.1 Catalyzed Reaction

ALAT intervenes in the metabolism of amino acids. It catalyzes the transfer of the amine (NH_2) from an α-amino acid (glutamate) to an α-keto acid (pyruvate) to synthetize a new α-amino acid (alanine) as well as another α-keto acid (α-ketoglutarate). It is therefore a reaction for biosynthesis of one amino acid using another one. The sense of in vitro reaction depends on the concentration of the substrate, but in vivo, glutamic acid is the main NH_2 donor:

$$\text{Glutamate} + \text{Pyruvate} \xleftrightarrow{\text{ALAT}} \text{Alanine} + \alpha\text{-ketoglutarate}$$

5.2.2.2 Principle of Dosage

The activity of ALAT is determined from the rate of disappearance of NADH,H$^+$, which, added to pyruvate, the product of catalysis of alanine is changed into lactate and NAD$^+$ (under the effect of catalyze of lactate dehydrogenase. Thus, the dosage of ALAT is based on the following catalyzed reactions:

$$\text{Alanine} + \alpha\text{-ketoglutarate} \xleftrightarrow{\text{ALAT}} \text{Glutamate} + \text{Pyruvate}$$

$$\text{Pyruvate} + \text{NADH} \cdot \text{H}^+ \xleftrightarrow{\text{Lactate dehydrogenase}} \text{Lactate} + \text{NAD}^+$$

The disappearance of NADH·H$^+$ is measured by spectrophotometry at 340 nm.

5.2.2.3 Tissue Localization and Clinical Significance

ALAT is a cytoplasmic enzyme. In camels, ALAT is mainly present in the kidney (5 ± 0 U/g) and then in liver (4 ± 0 U/g), in skeletal muscle (3 ± 0 U/g), and myocardia (3 ± 1 U/g) (Bengoumi et al. 1997a), reverse to other ruminants where it is present mainly in the muscle and heart and secondly in the liver and kidney (Braun 1985). According to Afzal and Saeed (1995), the enzyme activity of ALAT should be more important in the skeletal muscle (10 ± 3 U/g) than in the kidney (2 ± 1). By the way, the increase of plasmatic activity is in priority an indicator of the muscular or liver failure, while kidney failure is characterized by an increase of urinary activity. As for ASAT, ALAT is relatively concentrated in red blood cells (0.4 ± 0.1 U/g Hb). In consequence, hemolysis can influence the enzyme activity of ALAT and could lead to misinterpretation of results. In peritoneal fluid of healthy camels, similar values than in plasma were reported: 31 ± 23 U/l (Tabatabaei Naeini and Nazifi 2001).

5.2.2.4 Usual Values

In camel, the mean usual values of serum or plasma activity for ALAT are varying from 6 to 25 U/l (Table 5.1). These values are close to those of horse (3–23 U/l), higher than in llama (6–14 U/l) but lower than in other domestic ruminants: 14–38 U/l in cattle, 26–38 U/l in sheep, and 24–83 U/l in goat. No breed effect was reported in the different camel types from Algeria (Aichouni et al. 2010).

5.2.2.5 Physiological Variations

Age

As for other domestic animals, the age seems to have no significant effect on serum activity of ALAT (Sarwar et al. 2004). However, for some authors, the activity is lower between 7 and 16 years old (Salutini and Biagi 1983) or from 4 years old (Kataria et al. 1991) or from 5 to 7 years (Seboussi et al. 2004). A slight significant difference was reported between camel calves (15 $\pm$ 2 U/l) and non-lactating adult camels (19 $\pm$ 2 U/l), but not with lactating ones (Osman and Al-Busadah 2000).

Sex

In most of the species, no evident effect of the sex has been reported on serum ALAT activity in camels (Salutini and Biagi 1983; Ateeq et al. 1984; Chiericato et al. 1986; El Dirdiri et al. 1987; Bengoumi et al. 1997b; Sarwar et al. 2004). Some references found a higher activity in males compared to females (Adam et al. 1974; Kataria et al. 1991; Seboussi et al. 2004). ALAT is slightly but significantly lower during the rutting season in male camel: 46 $\pm$ 2 vs 51 $\pm$ 2 U/l (Zeidan and Abbas 2003). Similar seasonal variation was reported by Al-Harbi (2012).

Gestation, Lactation, and Castration Effect

In camel, no effect of physiological status during pregnancy or lactation, as well as of castration, is reported (Elias and Yagil 1984; Ateeq et al. 1984; Sarwar et al. 1992; Bengoumi et al. 1997b; Osman and Al-Busadah 2000; Khadjeh 2002; Saeed et al. 2009; Omidi et al. 2014). However, a significant lower value was reported in pregnant camel (17 $\pm$ 0 U/l) compared to non-pregnant camel (19 $\pm$ 1 U/l) although these values are not biologically different (Seboussi et al. 2004).

No breed effect was reported on ALAT activity (Seboussi et al. 2004; Hussein et al. 2012). Seasonal effect is sometimes reported (Hamad et al. 2017).

Physical Effort

Due to the low concentration in the skeletal muscle, physical effort has no significant effect on serum ALAT activity (Bengoumi et al. 1998b).

Dehydration Effect

As for ASAT, water deprivation provoking a dehydration of more than 25% has no impact on plasma activity of ALAT (Mohamed et al. 1984). The increase observed in some references is coming exclusively from the decrease in plasma volume (Ben-Zvi et al. 1989; Bengoumi et al. 1998a).

Disease Effect

Camels affected by *Trypanosoma evansi* have significantly higher values of ALAT (Sazmand et al. 2011; Hussain et al. 2016; Chaudhary and Iqbal 2000). In experimentally infected camel with *Trypanosoma evansi*, ALAT activity increased from 16 U/l to 38 U/l on day 27 and then returned to 13 U/l in camel treated with trypanocide (Boid 1980).

According to Osman et al. (2014), gastrointestinal parasites provoke also an increase of serum values of ALAT activity passing from 113 ± 2 in healthy camels to 163 ± 4 in affected camels (see note 1 above). Similar pattern is observed by Abdo et al. (1985), with values passing from 35 ± 1 in healthy camel to more than 38 U/l in camel infested by *Strongylus* or *Haemonchus* sp. (Abdo et al. 1985). In camel affected by mange (Mal et al. 2006), no significant difference was observed with healthy camels in summer (6 ± 1 in healthy camel vs 5 ± 0 U/l in affected camel), but a slight difference occurred in winter, the affected camels having a significant higher value (9 ± 1 U/l) than healthy camels (7 ± 2 U/l).

The camels affected by *Theileria camelensis* had their ALAT activity multiplied by 3 passing from 9 ± 1 in healthy animals to 27 ± 3 U/l in affected ones (Ismael et al. 2014). ALAT increased in camel affected by babesiosis, from 12 ± 1 to 16 ± 3 U/l (Swelum et al. 2014). Contrary to ASAT, ALAT activity is not changing in camel calf affected by arthritis due to *Mycoplasma* (Shoieb and Sayed-Ahmed 2016).

The calf diarrhea also has no effect on ALAT activity as for ASAT: 21 ± 6 vs 20 ± 8 U/l in diarrheic and non-diarrheic camel, respectively (Bengoumi et al. 1998c). In case of pasteurellosis, higher plasma ALAT activity is observed in affected camels compared to healthy ones: 25 ± 2 vs 2 ± 1, respectively (Seleim et al. 2003).

No change in ALAT activity was reported in the experimental copper toxicity trial described by Abu Damir et al. (1993). At reverse, an explosive effect was reported in camels suffering from narasin poisoning (an antibiotic used in chicken feed) with values growing from 13 ± 2 in control animals to 228 ± 153 in young affected camels and to 427 ± 345 U/l in adult affected camels (Abu Damir et al. 2013). In case of monensin toxicity, ALAT activity can reach 5700 to 6900 U/l (Chaudhary et al. 1998).

As for ASAT, ALAT is increasing in mastitic camel (Fig. 5.1) passing from 12.99 ± 0.12 to 41.55 ± 0.99 in healthy and CMT score +++, respectively (Ali et al. 2016a, b).

5.2.2.6 ALAT in Other Substrates

In camel milk, the range of ALAT activity is 9–11 U/l according to Yadav et al. (2015) and 11 ± 0 U/l according to Ali et al. (2016a, b). ALAT activity in cerebrospinal fluid of camel was reported at 7 ± 0 U/l (Ahmed et al. 2009) and 10 ± 1 U/l (Nazifi and Maleki 1998). In synovial fluid, it was reported at 36 ± 26 U/l (Al-Rukibat and Ismail 2014).

5.2.3 *Alkaline Phosphatases: ALP (E.C. 3.1.3.1)*

Alkaline phosphatases are among the first enzymes used in clinical biochemistry. In 1920, it had already discovered that serum activity of ALP increased in case of bone or liver damage.

5.2.3.1 Catalyzed Reaction and Principles of Dosage

ALP are non-specific esterases which catalyze hydrolysis of phosphorous links of several substrates including ATP. The name "alkaline" refers to the alkaline pH (pH = 10) for optimal in vitro activity (Kaneko 1989).

Measurement of the catalytic activity of ALP is based on the hydrolysis of *p*-nitrophenyl phosphate into *p*-nitrophenol in alkaline milieu (pH = 10) with the presence of magnesium as cofactor:

$$p\text{-nitrophenyl-phosphate} \xleftrightarrow{\text{ALP+Mg}^{2+}} p\text{-nitrophenol} + H_3PO_4$$

The kinetic of appearance of *p*-nitrophenol is determined by spectrophotometry at 400 nm.

5.2.3.2 Tissue Localization and Clinical Significance

ALP are group of isoenzymes present in several cells. In general, their activity is associated with the cell microvilli absorption and secretion, such as the epithelium of the biliary canaliculi, intestines, kidney tubules, and placenta. ALP are highly concentrated in osteoblasts responsible for ossification (Whitby et al. 1984). In camel, ALP are concentrated especially in the kidney (28 ± 3 U/g), while they are low in the liver (6 ± 1 U/g) and less than the limit of detection in the heart and skeletal muscle (Bengoumi et al. 1997a). However, according to Afzal and Saeed (1995), the highest amount of ALP is observed in the small intestine of the camel (226 ± 51 U/g).

The main source of serum ALP is the bone (Kaneko 1989), and their plasma activity increases in bone disorders (Rico et al. 1975). Moreover, this activity also increases in hepatobiliary disorders including cholestasis (Kaneko 1989). The activity of ALP in red blood cells is nil which excludes any effect of hemolysis on plasma activity of the ALP (Bengoumi et al. 1997a).

In peritoneal fluid of healthy camel, Tabatabaei Naeini and Nazifi (2001) reported quite lower values than in the bloodstream: 30 ± 4 vs 116 ± 10 U/l.

5.2.3.3 Usual Values

In adult camel, the mean ALP activity varies between 32 and 110 U/l (Bengoumi et al. 1997a), but a larger range is reported (Yadav and Bissa 1998; Hussein et al. 1982) especially when it includes young animals because of the higher values among 1- to 6-month-old camel (172–386 U/l).

These values are comparable to those of other ruminants (Table 5.1): from 30 to 488 U/l in cattle, 68 to 387 U/l in sheep, 93 to 387 U/l in goat, and 41 to 92 U/l in llama. However, they are lower than those of equines (143 to 395 U/l) (Kaneko 1989).

ALP usually determined in serum is the total alkaline phosphatases (TAP). The TAP levels in serum consist of several enzyme isoforms produced by the bone, liver, intestine, kidney, spleen, and placenta, but the main part in serum is liver and bone isoforms. Bone-specific alkaline phosphatase (BAP) has been determined in camel blood: the mean concentration was 86 ± 7 U/L with a slight nonsignificant fluctuation over 24 h with minimum and maximum concentrations at 75 ± 7 and 97 ± 4 U/L, respectively (Al-Sobayil 2010).

5.2.3.4 Effect of Physiological Factors

ALP being closely associated with mechanisms of bone metabolism, their enzyme activity is related to lactation and growth.

Age Effect

As for other ruminants, plasma activity of the ALP in camel is higher among young animals and decreases with age (Halabi et al. 1982; Kaneko 1989; Kataria and Bhatia 1991; Bengoumi et al. 1997b): for example, from 199 U/l in young camel to 64 U/l on average in adult (Bengoumi et al. 1997b). Similar observation was done by Osman and Al-Busadah (2000): 48 ± 47 U/l in camel calves vs 35 ± 6 U/l in lactating adults.

This decrease is most important during the first months. Thus, in the newborn, serum activity decreases from 289 U/l at birth to 190 U/l at the end of the first month (Elias and Yagil 1984). In camel, values remain high even after 18 months

(Bengoumi et al. 1997b). Indeed, the catalytic activity of the ALP is related to the osteogenesis action of osteoblasts, which is very active in growing young camels, and it continues beyond 18 months in this species.

Sex Effect

According to El Dirdiri et al. (1987), the plasma activity of the ALP is higher in females than in males, while Kataria and Bhatia (1991) or Babeker and Suleem (2013) reported the opposite: 124 ± 12 in male vs 43 ± 7 U/l in female. Chiericato et al. (1986) observed no variation related to sex. Other more recent studies showed that plasma ALP activity is not different between females and castrated males. Elsewhere no difference was reported between non-pregnant females (346 ± 68 U/l) and males (382 ± 52 U/l) (Omidi et al. 2014). At reverse, the entire males have higher activity in ALP plasma (88 U/l) than females (60 U/l) and castrated males (54 U/l), especially during the rutting season, which can be linked to hyperactivity of the entire males during this time (Bengoumi et al. 1997b).

Contrary to ASAT and ALAT, alkaline phosphatases seem significantly higher during rutting season: 53 ± 2 vs 49 ± 2 U/l (Zeidan and Abbas 2003). Regarding breed effect, no significant difference was reported, with values varying between 129 ± 2 and 137 ± 1 U/l according to the types of camels (Hussein et al. 2012).

Gestation Effect

The serum activity of ALP varies slightly during gestation. From 23 U/l for a camel at the beginning of pregnancy, it increases to 34 U/l in the sixth month and then decreases in late gestation. This increase can be attributed to the secretory activity of the placenta (Eltohamy et al. 1986). This pregnancy effect does not seem be very important (Bengoumi et al. 1997b).

According to some other authors, there is no effect of gestation on the serum concentration in ALP: on average, 37 in pregnant vs 35 U/l in non-pregnant camel (Muhammad et al. 2011). At reverse, Khadjeh (2002) reported a slight higher activity in pregnant camel (73 ± 4 U/l) than in non-pregnant camel (61 ± 3 U/l), and Saeed et al. (2009) found 192 ± 10 in pregnant camel vs 179 ± 23 U/l in non-pregnant camel. Omidi et al. (2014) reported a reverse significant difference with higher activity in non-pregnant (346 ± 68 U/l) than in pregnant camel (109 ± 10 U/l).

Lactation Effect

According to Elias and Yagil (1984), serum activity of the ALP is higher among lactating camels. This effect would be linked to the bone demineralization resulting from the leakage of plasma calcium to the udder.

However, reverse results were reported (Osman and Al-Busadah 2000) with higher values in non-lactating she-camels (87 ± 21 U/l) than in lactating ones (35 ± 6 U/l).

Dehydration Effect

Dehydration does not seem to influence plasma activity of the ALP, despite a slight osteodystrophy due to a low urinary excretion of phosphorus. Indeed, during dehydration, a decline in plasma concentration of the osteocalcin indicates a slow-down in the activity of osteoblasts, and consequently a reduction in plasma activity of the ALP would be masked by the decrease of plasma volume (Bengoumi et al. 1996). In their experiment, Bengoumi et al. (1998a) found no significant changes during water deprivation (from 67 to 90 U/l), but a slight nonsignificant increase was observed after starting rehydration.

Physical Effort Effect

Given their low concentration in muscle tissue, physical effort has no significant influence on plasma activity of the ALP (Snow et al. 1988), in spite of a slight nonsignificant increase (up to 95 U/l), 3 h after the end of the race (Bengoumi et al. 1998b; Beaunoyer 1992).

Disease Effect

Results of literature are contradictory (Chaudhary and Iqbal 2000); it is indeed difficult to establish a link between the change in ALP activity and the disease. Regarding trypanosomosis, there is no effect on ALP activity according to Sazmand et al. (2011): 94 ± 7 vs 118 ± 9 U/l in infected and non-infected camels, respectively. Similar results were reported by Gutierrez et al. (2005) and Ahmad et al. (2004): 59 ± 7 U/l in normal camel serum and 55 ± 12 IU/l in hemoparasitized ones.

At reverse, Hussain et al. (2016) found significant difference with higher values in infected camels (122 ± 1 U/l) compared to non-infected (92 ± 1 U/l) camels. Such results were formerly reported by Abd-El-Baky and Salem (2011): 85 ± 5 vs 134 ± 21 U/l in negative and positive camels, respectively. In contrast, in 1980, Boid observed in experimentally infected camel a decrease from 46 U/l before contamination to 13 U/l on day 41. On treated camel, the ALP activity increased from 13 to 26 U/l (Boid 1980).

Gastrointestinal parasites as *Strongylus* increase ALP activity, with values passing from 1.28 ± 0.16 in healthy camel to 1.66 ± 0.15 in camels infected by *Camellostrongylus* and to 2.30 ± 0.12 U/l in case of mixed infestation (Abdo et al. 1985) although the values appear very low compared to other data from the literature (Table 5.1).

The mange has no effect on the plasma values of ALP in serum, whatever the season (Mal et al. 2006). In camel affected by wryneck syndrome, ALP values are slightly higher (43 ± 17.7 U/l) than in healthy camels (31.4 ± 12.2 U/l), but this difference is nonsignificant (Al-Sobayil and Mousa 2009). In case of tick infestation, ALP activity increased strongly passing from 10 ± 9 U/l to 107 ± 54 U/l (Tharwat and Al-Sobayil 2014).

In young camels affected with diarrhea, a significant decrease in plasma activity of the ALP was observed for the sickest animals passing from 476 ± 159 U/l in healthy animals to 324 ± 167 U/l in sick camels (Bengoumi et al. 1998c). This could be linked to growth decrease induced by the disease. Indeed, osteoblasts, which are heavily involved in bone growth, are rich in ALP, and any slowdown in this growth translates into a decrease in enzyme activity. The increasing of serum ALP activity is also observed in case of mastitis (Ali et al. 2016a, b): 92.6 ± 1.54 with CMT = 0 to 110.15 ± 2.50 U/l with higher CMT scoring (Fig. 5.1).

ALP activity is also decreasing in racing camel affected by lameness or bone fracture, with values passing from 191 ± 89 to 138 ± 36 and 112 ± 51 U/l, respectively (Alshamsi et al. 2015). At reverse ALP activity is increasing in camel affected by *Pasteurella multocida* (Seleim et al. 2003).

ALP concentrations in adult camels affected by induced hypothyroidism increased significantly after sodium thiocyanate injection inducing the thyroid disorder, with values of 27 ± 3 U/l on the first month, 36 ± 6 U/l on the second month, and 26 ± 4 U/l on the third month postinjections (Barsham et al. 2005). ALP is affected also by copper poisoning, the concentrations increasing 2–3 days after intoxication (Abu Damir et al. 1993). An increase of ALP, up to 164–260 U/l, is also observed in camels intoxicated by pyrethroid linked to the liver damage induced by the poison (Abubakr et al. 2012). In camel affected by monensin toxicity, the values can reach 2250–4300 U/l (Chaudhary et al. 1998). In Bactrian camel affected by copper deficiency provoking a syndrome of emaciation, ALP was increased significantly passing from 49 ± 13 U/ l in healthy animals to 130 ± 29 U/l in affected ones (Shen and Li 2010).

5.2.3.5 ALP in Milk

Usually, ALP concentration (not activity) is determined in milk in order to control the pasteurization because alkaline phosphatases disappear after heat treatment in cow milk. However, camel ALP being thermoresistant, its interest as a marker of milk pasteurization cannot be used (Loiseau et al. 2001; Wernery et al. 2008). In raw milk, the average alkaline phosphatase activity was lower in camel milk (21.31 μg/ ml) than in cow and buffalo milk (Yoganandi et al. 2014). This low value is confirmed by Yadav et al. (2015): 16.04–24.93 U/l. But after pasteurization, the average alkaline phosphatase activities is disappearing in pasteurized cow milk while is still ranging from 6 to 10 U/l for pasteurized camel milk (Lorenzen et al. 2011). The ALP activity in milk is influenced by mastitis (Ali et al. 2016a, b): the minimum is observed in camel with CMT = 0 (18 ± 0 U/l), while the maximum occurred in camel with the highest CMT scoring (32 ± 0 U/l).

5.2.3.6 ALP in Other Substrates

ALP activity in cerebrospinal fluid is lowest: 12 ± 3 U/l (Achaaban et al. 2009). However, higher values were reported by other authors: 33 ± 4 U/l (Nazifi and Maleki 1998) and 76 ± 0 U/l (Ahmed et al. 2009). In synovial fluid, high activity was observed: 116 ± 74 U/l (Al-Rukibat and Ismail 2014) and 79 ± 0 U/l (Omer and Gameel 2009).

5.2.4 *Lactate Dehydrogenase: LDH (E.C. 1.1.1.27)*

5.2.4.1 Catalyzed Reaction and Principle of Dosage

LDH is an enzyme that catalyzes the reversible conversion of pyruvate to lactate. The meaning of the reaction is determined by the availability of oxygen and energy (NADH, H+) (Kaneko 1989):

$$\text{Pyruvate} + \text{NADH} \cdot \text{H}^{+} \xleftrightarrow{\text{LDH}} \text{Lactate} + \text{NAD}^{+}$$

The catalytic activity of the LDH is determined from the catalyzed reaction by measuring the speed of oxidation of NADH H^+ at 340 nm.

5.2.4.2 Tissue Localization and Clinical Interest

In blood, the catalytic activity of LDH is about 150 times higher in red blood cells than in plasma, and so, even a slight hemolysis can alter its plasma activity (Kaneko 1989; Bengoumi et al. 1997a). In camel, LDH activity is especially concentrated in the myocardium (1978 ± 136 U/g), followed by the muscle (1045 ± 118 U/g). The liver (49 ± 3 U/g) and kidney (6 ± 1 U/g) have a low LDH activity (Bengoumi et al. 1997a). However, according to Afzal and Saeed (1995), LDH activity is significantly higher in the skeletal muscles (227 ± 71 U/g) than in the heart (174 U/g ± 15 U/g).

In other domestic animals, LDH is concentrated in several tissues but mainly in muscles, the myocardium, and the liver (Braun 1985). However, serum LDH activity increases especially during muscle or myocardium damage. The electrophoretic separation of different isoenzymes of LDH can specify the affected organ. In practice, the increase in this activity was observed during nutritional myopathy in ruminants and in the case of the myoglobinuria in horses (Benjamin 1984).

5.2.4.3 Usual Values

In camel, the usual average values of serum LDH activity range from 337 to 2620 U/l (Table 5.1). They are similar to those of cattle (692–1445 U/l) but higher than those of equines (162–412 U/l), sheep (238–440 U/l), goats (123–392 U/l), and the llama (88–487 U/l).

5.2.4.4 Physiological Factors of Variation

Age

Age influences serum LDH activity which is higher among young camels than adults: on average 1824 U/l vs 1179 U/l in young and adult camels, respectively (Bengoumi et al. 1997b), or 597 ± 14 in less than 4-year-old camels, 537 ± 16 in 4- to 10-year-old camels, and 302 ± 6 U/l in more than 10-year-old camels (Kataria and Bhatia 1991). This age effect would be related to the sensitivity of young animals to the muscular effort during herd movements in pastures. Indeed, the camel is an animal that is characterized by an ambulatory grazing that can get him walk between 20 and 30 km daily (Faye 1997).

Sex, Castration, and Pregnancy-Lactation

Serum LDH activity doesn't change neither according to gender, castration, gestation, nor lactation stage (Chiericato et al. 1986; El Dirdiri et al. 1987; Al-Busadah 2007; Saeed et al. 2009). However, due to the higher activity in male camels during rutting season, plasma LDH activity is quite higher (Zeidan and Abbas 2003), 360 ± 14 IU/l, compared to 213 ± 13 IU/l in non-breeding season. This difference was also reported by Bengoumi et al. (1997b) with higher values in whole males (1421 U/l) than in castrated ones (1146 U/l). The effect of rutting could explain the seasonal variation observed in male camel with a significant increase of LDH in February (380 ± 1 U/l) compared to the pre-rutting month (357 ± 2 U/l), post-rutting month (350 ± 1 U/l), or non-rutting month (355 ± 2 U/l) (Al-Harbi 2012).

In female racing camel, the lower values in LDH (Fig. 5.2) as well as other enzymes as AST and CK were reported in October–November and the maximum in August and December (Knight et al. 1994a). Maximal values were also reported in winter in Algeria (Hamad et al. 2017).

With a range of 303–561 U/l in three camel breeds from Algeria, no significant breed effect was observed by Aichouni et al. (2010). Similar observations were done in Saudi camel breeds (Al-Busadah 2007).

As for AST, LDH increased by 45% when selenium supplementation overpasses 4 mg/day in camel (Faye et al. 2009).

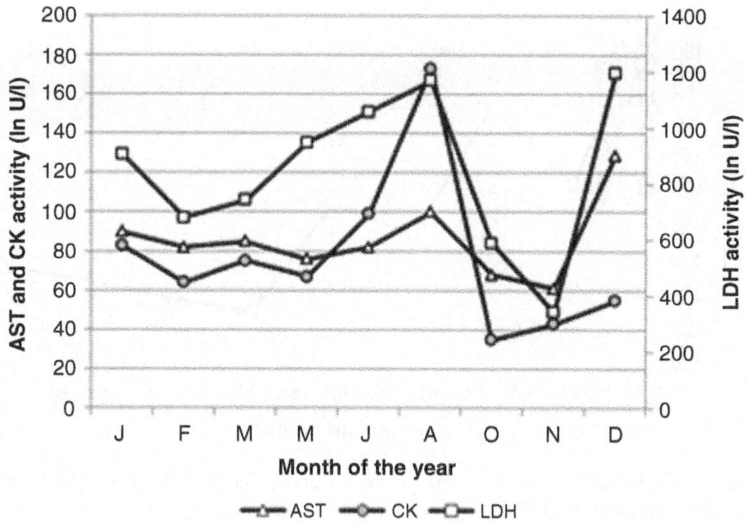

Fig. 5.2 Monthly changes in LDH, AST, and CPK in female racing camel [calculated from Knight et al. (1994a, b)]

Dehydration

A medium dehydration does not seem to influence plasma LDH activity. LDH is an indicator of the integrity of the muscle cell or heart attack. The stability of its activity during a 14-day water privation shows that these tissues are not suffering. However, LDH peak occurs after 2 weeks; LDH activity increased suddenly from 1334 to 2227 U/l but decreased rapidly after watering (Bengoumi et al. 1998a).

Physical Effort

Given its high concentration in the muscle and heart tissue, physical effort strongly influences the plasma activity of LDH (Snow et al. 1988; Beaunoyer 1992; Bengoumi et al. 1998b; Knight et al. 1994a) as shown in Fig. 5.3.

In case of intense physical effort (race), plasma LDH activity remains higher even after 24 h: from a basal value of 1200 U/l, LDH activity increased up to 2000 U/l just after racing and then declined up to 1500 U/l after a day (Bengoumi et al. 1998b). A peak of LDH activity is observed between 2 and 4 h after the end of the race (Knight et al. 1994a; Bengoumi et al. 1998b), but the highest plasma LDH activity is observed 5 minutes after the end of the exercise (Knight et al. 1994b). Thus, interpretation of the results of the LDH dosage must take into account the effort made by the animals that move in the pasture, as they always have higher values than the average.

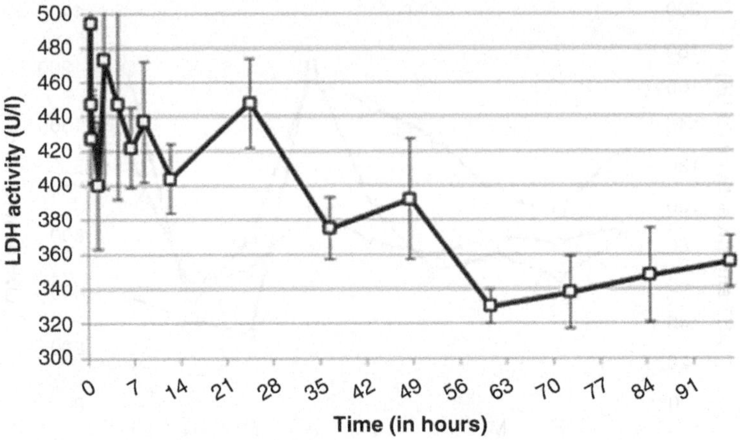

Fig. 5.3 Change in LDH activity in pre- (time 0) and postexercise up to 96 h after racing [calculated from Beaunoyer (1992)]

5.2.4.5 Pathological Variations

With values of LDH activities varying from 983 ± 36 in control camels, 1128.2 ± 48 in Suratex-positive animals, and 1089 ± 47 U/l in smear-positive camels, no significant effect of trypanosomosis is observed (Chaudhary and Iqbal 2000). This lack of effect was confirmed later by Sazmand et al. (2011): 479 ± 46 in infected camels vs 505 ± 40 U/l in healthy camels. It is confirmed also by Gutierrez et al. (2005) on camel population affected by an outbreak of abortion due to trypanosomosis.

A strong effect of the presence of theilerioses is also observed, LDH activity in plasma passing from 807 ± 85 in healthy camel to 1743 ± 134 U/l in affected animals (Ismael et al. 2014). Babesiosis in camel is also linked to a significant increase of LDH, from 727 ± 24 to 1219 ± 137 and even 2134 ± 359 U/l in case of association between babesiosis and other pathogens as gastrointestinal nematodes (Swelum et al. 2014).

LDH activity is increasing in racing camel affected by bone fracture: from 452.1 ± 52.6 to 674.9 ± 199.2 U/l in healthy and affected camels, respectively (Alshamsi et al. 2015).

LDH activity is also sensitive to copper poisoning and tends to increase, but a high individual variability occurs (Abu Damir et al. 1993). The narasin poisoning has a particular strong impact on LDH activity which is multiplied by 6, one week after contamination (Abu Damir et al. 2013). The effect is still higher in case of monensin poisoning: 13,000 to 21,000 U/l (Chaudhary et al. 1998).

No significant difference was observed in diarrheic young camels (2735 ± 1244 U/l) compared to non-diarrheic ones (2513 ± 1252 U/l) (Bengoumi et al. 1998c).

In Bactrian camel from China, affected by the syndrome "emaciation ailment" linked to copper deficiency, LDH increased significantly in affected animals, with values being multiplied by 1.7 (Shen and Li 2010).

5.2.4.6 LDH in Other Substrates

In camel milk, the LDH activity is 132–168 U/l (Yadav et al. 2015).

Contrary to serum activity, LDH activity in synovial fluid is very low: 12.7 ± 14.81 U/l (Al-Rukibat and Ismail 2014). LDH activity was also determined in camel hump. The result has shown activity comparable to that in serum: 420 ± 150 U/l (Al-Rehaimi et al. 1989). LDH activity is high in camel urine (Nazifi et al. 2005).

5.2.5 Gamma-Glutamyl Transferase: GGT or γ-GT (E.C. 2.3.2.2)

GGT is a carboxypeptidase transferring γ-glutamyl group to a peptide or convenient acceptor (Kaneko 1989). It is usually used in clinical biochemistry investigations.

5.2.5.1 Dosage

The catalytic activity of GGT is determined using the reaction below:

$$\text{l-}\gamma\text{-glutamyl-}p\text{-nitroanilide} + p\text{-nitroaniline} \overset{\text{GGT}}{\longleftrightarrow} \text{Glycylglycine} + \text{l-}\gamma\text{-glutamylglycylglycine}$$

The rate of appearance of *p*-nitroaniline is measured by spectrophotometry at 405 nm.

5.2.5.2 Tissue Localization and Clinical Interest

GGT is an enzyme associated with cell membrane. It is mainly distributed in the epithelial cells of the proximal tube in the kidney, biliary duct and mammary glands (Meister 1974), and intestine (Kaneko 1989). In the kidney and liver of sheep, a part of the GGT (5%) is soluble in cell; the remaining part (>95%) is related to cellular structures such as membranes and mitochondria (Braun et al. 1978). As in domestic mammals, the camel kidney is the richest organ of GGT (65 ± 4 U/g) followed far away by the liver (3 ± 1 U/g) (Afzal and Saeed 1995; Bengoumi et al. 1997a). Skeletal and cardiac muscles have GGT activities below the limit of detection (Bengoumi et al. 1997a). Furthermore, GGT activity is nil in camel erythrocytes, explaining the absence of the effect of hemolysis on plasma activity of this enzyme (Bengoumi et al. 1997a). Nazifi et al. (2005) explored GGT activity in camel organs showing a high activity in the liver (202 ± 9 IU/g) and kidney (122 ± 14 in the medulla and 109 ± 26 IU/g in the cortex) followed by the spleen (84 ± 5 IU/g) and abomasal pylori (58 ± 10 IU/g). The values ranged between 10 and 16 IU/g in the lymph node, lung, muscle, and ovary. In all other tissues, the activity is below 5 IU/g (Nazifi et al. 2002). High activity is also described in camel urine.

Despite its high concentration in the kidney, the increase in the catalytic activity of the GGT in blood indicates more hepatobiliary disorders such fascioliasis and metabolic or toxic hepatitis (Meyer 1982; Hasim and Braun 1989). In fact, because of its high molecular weight, GGT does not pass into the general bloodstream during renal damage. In this case, its activity increases in urine (Hasim and Braun 1989; Bengoumi et al. 1990). Besides, in camels suffering from intense dehydration, urinary GGT activity increased significantly (Bengoumi 1992).

5.2.5.3 Usual Values

Usual values of serum activity of GGT in dromedary camel vary from 8 to 28 U/l (Table 5.2). They are similar to those of the llama (7–29 U/l), horses (4–14 U/l), and cattle (6–18 U/l) but lower than those of small ruminants (20–52 U/l). The threshold indicating a liver cell suffering is probably lower in camel (approximately 15 U/l according to Faye et al. (1995a) than in dairy cow (22 U/l according to Barnouin and Paccard (1988)). Common urinary GGT activity values are lower than 5 U/l.

5.2.5.4 Physiological Variation Factors

Age

Age has no significant influence on camel plasma activity of GGT from 45 days (Bengoumi et al. 1997b). In newborn calves, serum GGT activity rises strongly after ingestion of colostrum, which is very rich in this enzyme (Hasim and Braun 1989), but such analysis was not performed in the camel. Faye et al. (1992a) observed values similar to in less than 2 years old camel compared to adult ones.

Sex Effect

As in other ruminants, GGT does not vary with sex (Braun et al. 1978; El Dirdiri et al. 1987; Faye and Mulato 1991; Bengoumi et al. 1997b). In Bactrian camel, Omidi et al. (2014) observed quite comparable values in pregnant (22 ± 2) and non-pregnant female (22 ± 1) and male (20 ± 1 U/l).

However, in male, a slight seasonal difference was observed with lower GGT activity at the rutting period (February, 16 ± 1 U/l) compared to pre-rutting

Table 5.2 Normal values of the main enzymes (GGT, CK, and LDH) in camel according to different authors

	Enzymatic parameters		
References	GGT	CPK	GLDH
Chiericato et al. (1986)	–	124–128	–
El Dirdiri et al. (1987)	19.6 ± 5.01	34.2 ± 114	–
Abdalla et al. (1988)	18 ± 7.4	129 ± 77.2	–
Bizzeti et al. (1988)	8.43 ± 2.27	–	–
Faye and Mulato (1991)	8.4 ± 3.1	–	14.1 ± 21.3
Faye et al. (1992a)[a]	6.7 ± 1.6	–	14.1 ± 7.5
Knight et al. (1994a)	–	72 ± 69	–
Faye et al. (1995a)	10.1 ± 5.8	–	5.8 ± 10.8
Chaudhary and Iqbal (2000)	15.41 ± 1.85	113.21 ± 5.38	–
Bogin (2000)	16 ± 6	41 ± 30	–
Khadjeh (2002)	16.93 ± 0.95	76.06 ± 8.41	–
Ayoub et al. (2003)	8.8–11.1	23.2–55.9	–
Osman and Al-Busadah (2003)	25.6 ± 7.8	408 ± 127	–
Seboussi et al. (2004)	–	86.1 ± 43	–
Gutierrez et al. (2005)	12.6 ± 4.8	56.5 ± 12.4	–
Bengoumi (2008)	18 ± 4	80 ± 17	–
Elnahas (2008) (Egypt)	12.4 ± 7.07	–	12.02 ± 6.8
Elnahas (2008) (Germany)	5.9 ± 4.07	–	9.5 ± 3.3
Mohri et al. (2008)	17.7 ± 4.1	192.0 ± 59.0	–
Nazifi et al. (2009b)	–	179.45 ± 1.09	–
Saeed et al. (2009)	14.15 ± 3.2	82.2 ± 21.3	–
Aichouni et al. (2010)	–	44.8 ± 26	–
Shen and Li (2010)[b]	19.3 ± 3.7	–	–
Abd-El-Baky and Salem (2011)	24.33 ± 4.04	–	–
Hussein et al. (2012)	–	124.8 ± 1.36	–
Ismael et al. (2014)	8.23 ± 1.03	–	–
Abu Damir et al. (2013)	–	92 ± 56	–
Tharwat and Al-Sobayil (2014)	13 ± 6	–	–
Alshamsi et al. (2015)	17.2 ± 2.9	105.1–113.1	–
Hamad et al. (2017)	–	292 ± 118	–
Zaher et al. (2017)	15.17 ± 1.17	84.42 ± 2.43	–

(November, 25 ± 2 U/l), post-rutting (May, 21 ± 1 U/l), and non-rutting periods (August, 25 ± 2 U/l) (Al-Harbi 2012).

Other Physiological Factors

The physiological stage of gestation and lactation, castration, dehydration, and exercise does not seem to influence serum GGT activity in camel (Bengoumi et al.

1997b, 1998a). However, Saeed et al. (2009) reported a slight higher activity in non-pregnant camel (14 ± 3) than in pregnant camel (11 ± 3 U/l). In camel receiving high selenium supplementation (more than 4 mg/day), GGT activity increased by 62% (Faye et al. 2009).

Disease Effect

The effect of trypanosomosis is unclear. According to Chaudhary and Iqbal 2000, GGT activity was 18.15 ± 2.11 in Suratex-positive animals and 16.83 ± 1.95 in smear-positive animals, compared to 15.41 ± 1.85 U/l in healthy camels, i.e., without significant effect. With a value of 14.2 ± 9.1 U/l, no effect was also observed in an outbreak of abortion linked to trypanosomosis reported in Canary Islands (Gutierrez et al. 2005). At reverse, for Abd-El-Baky and Salem (2011), the GGT activity is double in positive animals (48.33 ± 7.64) compared to negative ones (24.33 ± 4.04 U/l). The effective impact on liver is depending on the exact status of the camel: in phase of true parasitemia or seropositive only. According to such status, the effect on GGT could be different.

In two studies conducted in Djibouti and France (Faye and Mulato 1991; Faye et al. 1995a), there was a geographical variation of GGT values in camel probably due to the parasitic status of animals depending on environmental conditions. Indeed GGT is an indicator of hepatobiliary disorders in which the worms with biliary tropism of the genus *Fasciola* play a major role. However, fascioliasis is uncommon in the dromedary camel, except in animals living in wetlands as the Nile Delta, Rajasthan Canal, or oases (Faye 1997). There is no significant impact of tick infestation on GGT activity (Tharwat and Al-Sobayil 2014).

Contrary to ASAT, GGT did not change in camels affected by wryneck syndrome (Al-Sobayil and Mousa 2009). At reverse, theileriosis contributes to a strong increase of GGT activity. The values are reported to be 8 ± 1 U/l in healthy animals vs 39 ± 7 in affected camels (Ismael et al. 2014). Babesiosis is also a cause of GGT elevation in serum, from 9 ± 1 to 22 ± 6 U/l (Swelum et al. 2014).

GGT activity does not significantly increase in racing camel affected by lameness compared to healthy animals, but the difference is very slight: 20 ± 3 vs 17 ±. U/l, respectively, in female and 20 ± 3 vs 21 ± 3 U/l, respectively, in male (Alshamsi et al. 2015). GGT activity increases significantly in camels affected by monensin poisoning, with values passing from 3–25 to 200–390 U/l (Chaudhary et al. 1998).

No difference was observed in case of diarrhea in young camel: 16 ± 10 U/l in non-diarrheic animals compared to 19 ± 11 U/l in diarrheic ones (Bengoumi et al. 1998c). In case of other diseases associated with reproductive failure, GGT is increased by 800% passing from 15 ± 1 to 127 ± 8 U/l (Zaher et al. 2017).

Other Substrates

In camel milk, GGT is inactivated between 10 and 20 min at 72 °C and thus, contrary to ALP, can be used as indicator for the pasteurization of camel milk (Loiseau et al. 2001). In camel milk, the range in raw milk is 16–25 U/l (Yadav et al. 2015). These values are not corresponding to that reported by Lorenzen et al. (2011), 342 ± 37 U/l and 329 ± 22 U/l in two different pools of raw milk, while the remaining activity after heat treatment was 2 ± 4 and 0 U/l, respectively.

5.2.6 *Creatine Kinase: CK (E.C. 2.7.3.2)*

5.2.6.1 Catalyzed Reaction and Principle of Dosage

CK, also known as creatine phosphokinase (CPK), catalyzes the conversion of creatine phosphate or phosphocreatine to creatine. This reaction is important for the production of energy in the muscle fibers (Kaneko 1989). In consequence, CK is important in cells consuming ATP rapidly for their energy, especially the skeletal muscle and myocardium but also the brain, retina, and spermatozoa. CK catalyzes hydrolysis of the energy-rich bond synthesizing ATP by directly transferring released energy and phosphate to ADP. This reaction is reversible, so that when the muscle is rich in ATP, the energy is recovered to make phosphocreatine. On the contrary, when ATP becomes necessary, this direct transfer restores the energy stored:

$$\text{Creatine-Phosphate} + \text{ADP} \xleftrightarrow{\text{CK}} \text{Creatine} + \text{ATP}$$

$$\text{Glucose} + \text{ATP} \xleftrightarrow{\text{Hexokinase}} \text{Glucose-6-Phosphate} + \text{ADP}$$

$$\text{Glucose-6-Phosphate+}$$
$$\text{NADP} \xleftrightarrow{\text{Glucose-6-Phosphate Dehydrogenase}} \text{Gluconate-6Phosphate} + \text{NADPH} \cdot \text{H}^+$$

The catalytic activity of CK is determined from the reaction catalyzed by measuring the rate of appearance of NADPH, H^+, at 240 nm.

5.2.6.2 Tissue Location and Clinical Interest

The tissue distribution of the CK has been described in the camel by Afzal and Saeed (1995). Given its role, it is especially active in the skeletal muscles (6540 ± 1259 U/g) and myocardium (4135 ± 930 U/g). It can be found also in the brain (482 ± 71 U/g), small intestine (357 ± 27 U/g), and spleen (178 ± 14). Other organs (liver, kidney, pancreas, and lungs) contain less than 30 U/g on average. In other domestic species, and without exception, CK is also concentrated primarily in the skeletal muscle and myocardium with some activity in the brain.

In domestic animals, the serum activity of the CK increases especially during muscle or myocardium damage. Electrophoretic separation of different isoenzymes of CK allows to identify the affected organ. Special kits to determine the CK of cardiac origin (CK-MB) in humans suffering from heart disease are available. In practice, the increase in serum activity of the CK is observed during muscle or heart attack (Benjamin 1984).

5.2.6.3 Usual Values

In camel, the average usual plasma serum activity of CK values is between 40 and 120 U/l (Table 5.2). They are closed to those of cattle and other domestic animals.

No breed effect was reported (Aichouni et al. 2010). However, Hussein et al. (2012) reported very slight but significant lower CK activities in white-coat camels (118 ± 2.0 U/l) compared to brown-coat (125 ± 1 U/l) and black-coat camels (129 ± 2 U/l). Obviously, even if these values are statistically different, it has no biological significance.

5.2.6.4 Physiological Variations

Apart from the physical effort, the effects of other physiological factors of variation (age, sex, castration, gestation, lactation, dehydration) on the CK activity were not commonly studied in the camel. Some authors reported effects of some physiological factors but with no clinical signification nor biochemical interpretation.

El Dirdiri et al. (1987) have no observed sexual difference in CK activity: 32 ± 10 vs 36 ± 12 U/l in male and female camel, respectively. Seboussi et al. (2004) also have found quite similar values in male camel (83 ± 4.0 U/l) and female camel (87 ± 4 U/l).

There was no effect of pregnancy and breed also (Seboussi et al. 2004; Saeed et al. 2009). However, old camels (more than 8 years old) had a significant lower CK activity (58 ± 4 U/l) than 5- to 7-year-old camels (84 ± 5 U/l) and young camels (<4 years old): 98 ± 4 U/l (Seboussi et al. 2004). Seasonal effect with significant higher values in summer was reported in Algeria (Hamad et al. 2017).

Effect of Exercise

Physical effort increases rapidly plasma CK activity with a peak at 6 h after the race and remains significantly high at more than 24 h in trained animals (Beaunoyer 1992; Mohamed and Hussein 1999) (Fig. 5.4).

According to Bengoumi et al. (1998b), CK activity is multiplied by 4.7 after exercise, passing on average from 39 to 183 U/l. Surprisingly, Snow et al. (1988) did

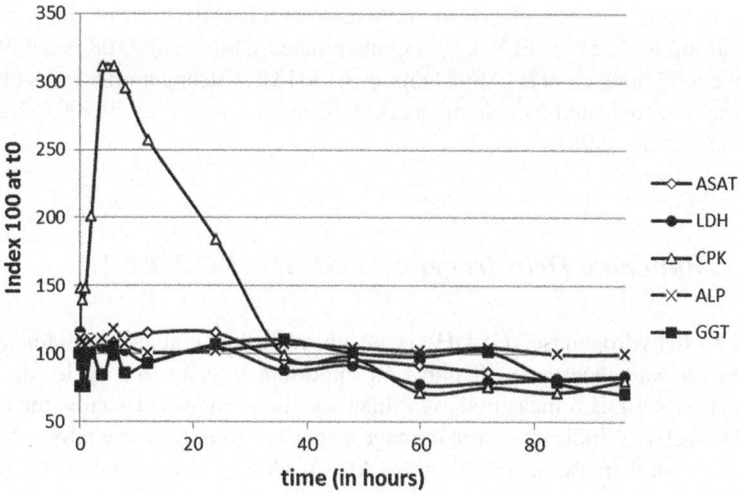

Fig. 5.4 Changes in main plasma enzyme activities before (t0) and after maximal exercise (from 10 min to 96 h). For comparison, all values were set to 100 at t0 [calculated from Beaunoyer (1992)]

not observe a significant difference in CK activity before and after the exercise, but they did not mention the sampling time post-racing. According to Rose et al. (1994), an exponential increase is observed in CK activity during exercise and is positively correlated to the duration of the race and training degree of camels. Myocardial CK (CK-MB), an isoenzyme of CPK, is not a good indicator for cardiac suffering because its values (range 0.19–0.60 ng/ml) did not change significantly during exercise (Tharwat et al. 2013). According to Abdalla et al. (1988), CK-MB activity in camel serum is on average 12 ± 8 U/l.

Dehydration Effect

A significant but unexplained decrease of CK activity in case of dehydration was described by Mohamed et al. (1984).

Disease Effect

No effect of trypanosomosis was reported by Chaudhary and Iqbal (2000), Gutierrez et al. (2005), and Sazmand et al. (2011).

The wryneck syndrome is marked by a spectacular increase of CK (1314 ± 58 U/l) compared to the activity in healthy camels (64 ± 23 U/l). This effect is due to the muscle injury which is the main lesion observed in this syndrome (Al-Sobayil and Mousa 2009). CK is also strongly increased in racing camels affected by bone fracture with values multiplied by more than 10 (Alshamsi et al. 2015).

Nasarin poisoning provokes a sharp increase in CK activity: from 92 ± 56 U/l in control group to 5121 ± 3271 U/l in contaminated adults and 4508 ± 2959 U/l in contaminated young camels (Abu Damir et al. 2013). A stronger effect was observed in monensin-intoxicated camels on serum CK activity reaching 30,500–51,300 U/l (Chaudhary et al. 1998).

5.2.7 *Glutamate Dehydrogenase: GLGH (EC 1.4.1.2)*

Glutamate dehydrogenase (GLDH) is an enzyme that converts glutamate to α-ketoglutarate and vice versa. It plays an important role in urea cycle. In clinical investigation, GLDH is measured to evaluate the liver function. Usually, the increase of GLDH activity indicates liver damage and liver toxicity especially when it is associated with increase of ASAT and/or ALAT. However, in camel, its use was not very common.

5.2.7.1 Usual Values

There are very few data in the literature about the GLDH activity in camel. In a study achieved at Djibouti, Faye and Mulato (1991) found values ranging from 0 to 97 U/l with an average of 14.1 U/l, i.e., substantially higher values than in cattle (5–11 U/l). The values observed in females are on average higher than in males (Faye and Mulato 1991), but there is no effect of age.

5.2.7.2 Disease Effect

GLDH is usually a marker of nonparasitic liver damage. At Djibouti, camels from some regions (east and southwest) have presented digestive disorders marked by hypermagnesemia, hyperuricemia, and a rise of the GLDH (up to 27.4 U/l on average) that might lead to hepatorenal toxicity. Such impairment is related to the consumption of almost exclusively bushes named *Salvadora persica* and *Cadaba rotundifolia*.

5.3 The Main Metalloenzymes

The metalloenzymes are enzymes having a metal ion in their molecule. In consequence, their activities in blood are well correlated to the trace element level. In the present document, three main metalloenzymes are presented: ceruloplasmin, glutathione peroxidase, and superoxide dismutase.

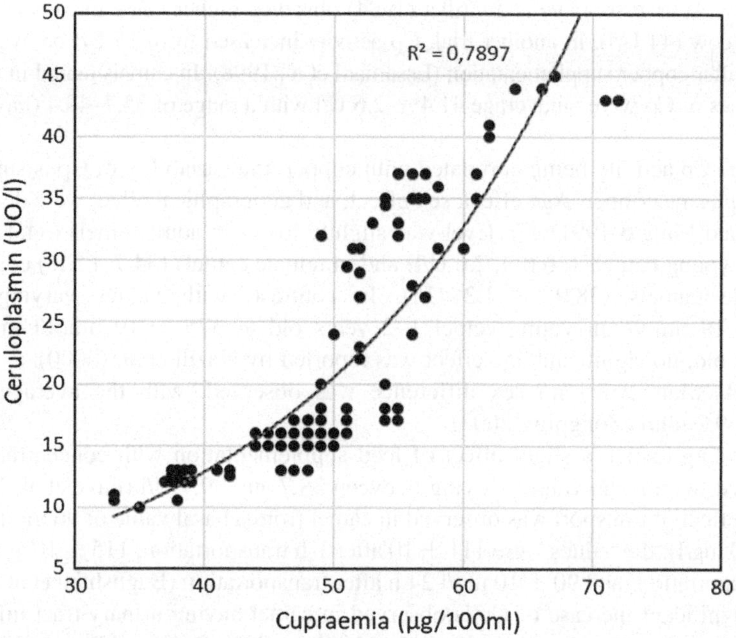

Fig. 5.5 Nonlinear relationship between ceruloplasmin and plasma copper in camel calculated in different experiments [recalculated from Essamadi et al. (1998)]

5.3.1 Ceruloplasmin or Cp (EC 1.16.3.1)

The ceruloplasmin (or caeruloplasmin) is a ferroxidase and the main copper-carrying protein in the bloodstream. Synthetized in the liver, Cp contains six atoms of copper and carries 95% of the total copper plasma. So, Cp activity is highly correlated to the copper status of the animals including camel (Abdel Rahim 1983; Faye et al. 1986; Faye and Mulato 1991) although it appears that the coefficient of correlation in field conditions is lower in camel than in other ruminants' species (Faye and Grillet 1984) and that a nonlinear regression model (Fig. 5.5) improved the observed correlation (Essamadi et al. 1998).

The camel ceruloplasmin was purified and characterized showing on average 5.8 ± 0.3 atoms of copper per molecule (Essamadi et al. 2002) with no difference with Cp of the young camel (Essamadi et al. 2000).

Several units were used in the literature and, as such, the comparisons are sometimes difficult. Most of the references are using oxidase units/l (U/l), but also µmol/l or mg/l are used.

The activity in camel was reported as lower than in cow (11 vs 34 U/l, respectively) but with higher sensitivity to copper supplementation (22 vs 39 U/l, respectively) in a comparative experiment (Bengoumi et al. 1998d). Similar difference was reported by Srivastava and Dwaraknath (1971) with values of 25 and 69 U/l, respectively, for camel and cow. At reverse, Faye and Grillet (1984) observed higher value in camel (51 U/l) than in cow (44 U/l). In another trial, Cp activity increased from 15 U/l on average to 45 U/l after copper supplementation (Essamadi et al. 1998). In camels reared in France, the values of Cp were on average 41.4 ± 2.6 U/l with a range of 35.7–49.4 (Faye et al. 1995b).

In the Cp activity being correlated with copper, the variability of Cp is similar to that of plasma copper. Age effect, sex effect, and geographical effect were described (Faye and Mulato 1991): Cp level was slightly lower in adult camels (34.4 ± 2.8) than in young camels (36.6 ± 2.6 U/l) and in female camels (34.7 ± 2.6) compared to male camels (38.9 ± 1.2 U/l). In contrast, with values varying from 353 ± 19 μmol/l in young camel 1–2 years old to 375 ± 19 μmol/l in camel 4 years old, no significant age effect was reported by Nazifi et al. (2000).

In Bactrian camel no sex difference was observed, with the average being 54.4 ± 9.2 mg/l (Zongping 2003).

In young camel, a slight effect of feed supplementation with concentrates was observed, with mean values varying between 38.7 and 39.7 U/l (Faye et al. 1992b).

No effect of transport was observed in camel from a basal value of 96 mg/l (range 41–130 mg/l): the values were 111 ± 10 after 1-h transportation, 115 ± 10 at the end of transportation, and 90 ± 10 mg/l 24 h after transportation (Baghshani et al. 2010).

A significant increase of Cp is observed in camel having urinary tract infection, with mean values passing from 90 mg/l to 1060 mg/l (El-Deeb and Buczinski 2015). In Bactrian camel affected by sway disease in China, a low quantity of Cp was revealed: 34.4 ± 19.6 mg/l (Liu et al. 1994). Similarly, Bactrian camels affected by "emaciation ailment" had on average a lower Cp value than unaffected ones: 135 ± 29 mg/l vs 232 ± 34 mg/l (Shen and Li 2010).

5.3.2 *Glutathione Peroxidase or GSH-Px (EC 1.11.1.9)*

The glutathione peroxidase (GSH-Px) includes several isoenzymes linked to different tissues of the organism. It belongs to the peroxidase enzyme family and plays consequently an important role to protect the organism from oxidative damage. In fact, it is one of the most important antioxidant enzymes in humans and animals. Because the enzyme includes four atoms of selenium in its molecule, it is regarded as a selenoprotein. The purified enzyme contained 16 ng of selenium per mg of protein (Chafik et al. 2018).

5.3.2.1 Usual Values

So, GSH-Px is generally correlated with the selenium status of animals, including camel (El-Magawry et al. 1988; Bengoumi et al. 1998e; Faye and Seboussi 2009). This correlation is higher in camel than in cow (Bengoumi et al. 1998e). The determination of GSH-Px activity is done in red blood cells, and the results are expressed in IU/g Hb.

The first reference regarding GSH-Px in camel is probably El-Magawry et al. (1988) who have found no difference between mean values of 25 ± 0.64 in male and 23.7 ± 0.17 IU/g Hb in female camel, which is apparently healthy. With a range of 15 to 36 IU/g Hb without sex or age effect, similar figure was reported by Hamliri et al. (1990). No sex and age effect was also reported by Taha et al. (2010).

At reverse, GSH-Px activity measured in whole blood was reported to be higher in females than males (Abdel Rahim 2005) with value three times higher in female (18.64 ± 0.03 EU/ml) than in male (6.32 ± 0.02 EU/ml).

5.3.2.2 Se Supplementation Effect

However, the variability of GSH-Px activity in RBC being in relation with selenium status and the reference values could change considerably according to selenium supplementation. Thus, the level of GSH-Px activity varied from 51.63 ± 10.4 in non-supplemented camels to 79.26 ± 10.4 during selenium supplementation and to 131.7 ± 17.2 IU/g Hb after supplementation, showing that maintenance of activity after selenium supply is linked to the longer lifespan of the RBC compared to cow (Bengoumi et al. 1998e).

In camel receiving 2 mg or 4 mg Se supplementation per day, GSH-Px activity was significantly higher (54.1 ± 22.5 and 61.9 ± 25.9 IU/g Hb, respectively) than in non-supplemented animals (28.0 ± 8.3 IU/g Hb), and the mean was 48.01 ± 25.07 IU/g Hb (Seboussi et al. 2008). In another trial involving pregnant camels (Seboussi et al. 2009a), GSH-Px activity was, on average, significantly higher in supplemented she-camels with 2 mg Se (47.5 ± 25.6) compared to non-supplemented camels (18.1 ± 8.7 IU/g Hb). In these females before and after parturition, a difference in GSH-Px activity was also reported: the values decreased significantly after parturition passing from 21.2 ± 7.5 to 14.8 ± 8.8 IU/g Hb in non-supplemented camels and from 58.1 ± 23.9 to 33.4 ± 20.7 IU/g Hb in supplemented camels. At parturition, the camel calves born from supplemented dams had threefold GSH-Px values higher than the control calves: 73.8 ± 2.9 vs 25.0 ± 3.2 IU/g Hb (Seboussi et al. 2009a). However, no age affect, both in male and female camels, was observed (Taha et al. 2010).

In an experiment involving young camels receiving very high selenium (Se) supplementation, up to 8 mg/day (Seboussi et al. 2010), the activity of erythrocyte glutathione peroxidase oscillated between 25.1 and 127.1 IU/g Hb with an average value of 72.30 ± 23.3 IU/g Hb. The values in 8 mg/day Se-supplemented camels (76.1 IU/g Hb) and 4 mg Se-supplemented camels (65.9 IU/g Hb) were significantly higher than in 2 mg Se-supplemented animals (50.5 IU/g Hb) and non-supplemented ones (51.0 IU/g Hb).

In camel receiving very high dose of selenium (from 8 to 16 mg/day) in order to evaluate the threshold limit for selenosis, the GSH-Px activity varied between 26.85 and 174.16 IU/g Hb with a mean value of 79.32 ± 30.94 IU/g Hb (Seboussi et al. 2009b). In the meta-analysis published by Faye et al. (2009), the GSH-Px activity

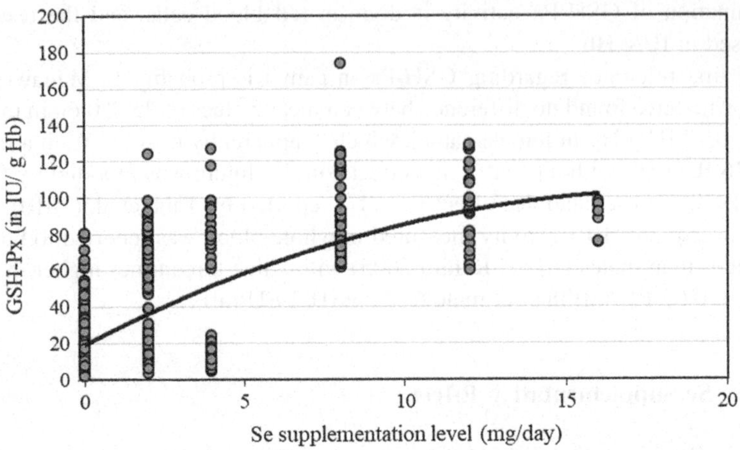

Fig. 5.6 Mean GSH-Px activity in camel RBC according to the level of Se supplementation in the diet [from Faye et al. (2009)]

reached a plateau from 8 mg/day Se supplementation, indicating a "saturation" of the activity in RBC (Fig. 5.6).

5.3.2.3 Disease Effect

In camel affected by nutritional muscular dystrophy (NMD), Corbera et al. (2003) found values of 33.6 ± 9.2 IUl/g Hb for young animals and 71.1 ± 22.8 IU/g Hb for their dams. The authors considered that these values were deficient based on former observations that GSH-Px activity was on average 288.5 ± 157.5 IU/g Hb in camel (Corbera et al. 2001).

5.3.3 Superoxide-Dismutase or SOD (EC 1.15.1.1)

SOD is an antioxidant enzyme catalyzing the dismutation of the radical O_2- into oxygen molecule O_2 or hydrogen peroxide (H_2O_2). Thus, SOD is a powerful antioxidant defense of the cells exposed to oxygen. Different isoenzymes are described according to their location: SOD_1 is located in the cytoplasm, SOD_2 in the mitochondria, and SOD_3 in extracellular fluids. SOD_1 and SOD_3 are copper and zinc dependent, while SOD_2 is manganese dependent, including in camel (Fouad 2015). Accordingly, a positive relationship is generally found between plasma level of copper/zinc and SOD activity in animals, but the correlations are not constantly observed. It is difficult to have coherent references in camel, with values being highly variable according to the different authors.

On average SOD activity is lower in camel compared to cow in similar feeding conditions: between 1474 ± 252 and 1813 ± 352 IU/100 gHb in camel according to the level of mineral supplementation vs 2254 ± 205 to 2436 ± 237 IU/100 gHb in cow (Bengoumi et al. 1998f). Different values with low daily variability were reported in Iran (Nazifi et al. 2009a): from 1323.4 ± 37.2 to 1412.2 ± 41.6 U/gHb (i.e., 100 more times than the previous reference).

The copper-zinc supplementation has no significant effect on camel SOD activity contrary to cow. Moreover, the copper-zinc competition on the binding sites of red blood cells could have different patterns in camel than in cow due to the low efficiency of zinc supplementation in the first species (Bengoumi et al. 1998f).

In camel affected by urinary tract infection, the SOD activity decreased significantly passing on average from 4.98 in healthy camel to 3.73 U/gHb in affected ones, with values returning to normal after treatment (El-Deeb and Buczinski 2015), but the published values are still quite different than all the other authors. Lower SOD activity was also revealed in Bactrian camel affected by emaciation ailment (Shen and Li 2010): 14.3 ± 1.9 vs 18.5 ± 2.3 μmol/l in sick camels.

5.4 Other Enzymes

Some other enzymes are sometimes investigated less for their clinical interest than in the frame of specific experiment. In the following table, some references regarding camel are listed (Table 5.3).

Other enzymes were determined in camel, but they provide any information on their use in clinical pathology. They are just mentioned as the only references available in camel.

5.4.1 Acid Phosphatase (EC 3.1.3.2) or ACP

ACP is an enzyme used to free attached phosphoryl groups from other molecules during digestion. In human, the determination of its activity in serum is used as a marker of prostate cancer. It could be used also to assess enzymatic damage caused by a kidney disease, liver disease, or heart attack. Few references are available in camel. The ACP serum values are lower in cold season (6.21 ± 0.37 U/l) than in hot season (8.20 ± 0.21 U/l) and higher in young camels (9.12 ± 0.32 U/l) than in adult camels (5.40 ± 0.16 U/l), but there is no sex difference (Kataria and Bhatia 1991).

5.4.2 Isocitrate Dehydrogenase (EC 1.1.1.42) or IDH

The IDH enzyme is included in Krebs cycle and catalyzes the reaction:

Table 5.3 Reference values of some minor enzymes in camel

Enzyme (U/l)[a]	Biological interest	Values (U/l)	References
Acid phosphatase (ACP)	Mainly prostate cancer	7.18 ± 0.21	Kataria and Bhatia (1991)
Isocitrate dehydrogenase (IDH)	Hepatocellular damage	7.74 ± 0.17	Kataria and Bhatia (1991)
Sorbitol dehydrogenase	Liver-kidney damage	0.50 ± 0.64	Boid (1980)
α-Amylase	Pancreatitis	2199-2197	Chiericato et al. (1986)
		881.4 ± 15.8	Khadjeh (2002)
		2325 ± 629	Mura et al. (2005)
Arginase	Liver damage	16.55 (0–28)	Adam et al. (1974)
G6PDH	Anemia	29.33 ± 5.51	Abd-El-Baky and Salem (2011)
Glutathione reductase (GSH)	Cellular stress	115.33 ± 12.22	Abd-El-Baky and Salem (2011)
		2300–3900	Kataria et al. (2010)
Catalase (in kU/l)	Oxidative stress	40.8–52.2	Lektib et al. (2016)
		72–92.3	Kataria et al. (2010)
		25.1 ± 2.9[b]	Shen and Li (2010)[c]
		60.1 ± 3.18	El-Khasmi et al. (2015)
Xanthine oxidase (mU/l)	Oxidative stress	50–80	Kataria et al. (2010)
	Liver damage	40–90	Mohamed (2006)
Paraoxonase (U/l)	Protection lipid oxidation	25.4 ± 1.3	Taha et al. (2010)
Glutathione reductase (U/l) (kU/l)	Oxidative stress	115.3 ± 12.2	Abd-El-Baky and Salem (2011)
		2.3–3.8	Kataria et al. (2010)
Glucose-6-phosphate dehydrogenase (U/l)	Protection oxidative damage	29.3 ± 5.5	Abd-El-Baky and Salem (2011)
Arginase (U/l) (IU/g)	Hepatotoxicity	0.28 (plasma)	Adam et al. (1974)
		567 ± 12 (liver)	Razmi et al. (2003)
		24 ± 10 (kidney)	
Maltase (U/g)	Pancreatic enzyme	14.4 ± 0.5	Mohamed et al. (2005)
Cellobiase (U/g)	Pancreatic enzyme	0.8 ± 0.01	Mohamed et al. (2005)
Trehalase (U/g)	Pancreatic enzyme	0.8 ± 0.04	Mohamed et al. (2005)
Lactase (U/g)	Pancreatic enzyme	0.6 ± 0.03	Mohamed et al. (2005)
Sucrase (U/g)	Pancreatic enzyme	0.8 ± 0.04	Mohamed et al. (2005)

(continued)

Table 5.3 (continued)

Enzyme (U/l)[a]	Biological interest	Values (U/l)	References
Gluco-amylase (U/g)	Pancreatic enzyme	61.4 ± 93.3	Mohamed et al. (2005)

[a]Except mention of other units
[b]In μmol/l
[c]Bactrian camel

$$\text{Isocitrate} + \text{NAD}^{+} \rightarrow \text{2-oxoglutarate} + \text{CO}_2 + \text{NADH, H+}$$

It includes different isoenzymes playing a role in the defense against cellular oxidative stress.

There is no sex or seasonal variability, but higher activity is observed in young camel (9.75 ± 0.03 U/l) compared to old adult camel (5.47 ± 0.22 U/l) (Kataria and Bhatia 1991).

5.4.3 *Sorbitol Dehydrogenase (EC. 1.1.1.14) or SDH*

SDH is an enzyme involved in carbohydrate metabolism. It concerts sorbitol, the sugar alcohol form of glucose, into fructose. Sorbitol dehydrogenase uses NAD+ as a cofactor. The zinc-dependent catalyzed reaction is:

$$\text{Sorbitol} + \text{NAD+} \rightarrow \text{Fructose} + \text{NADH} + \text{H}^{+}$$

In tissues where sorbitol dehydrogenase is low or absent, sorbitol can accumulate under conditions of hyperglycemia and consequently provokes impairment of those tissues as the retina, liver, or kidney.

5.4.4 *α-Amylase (EC 3.2.1.1)*

α-Amylase is a carbohydrase enzyme mainly present in saliva and in pancreatic secretion, and its role is to catalyze the hydrolysis of starch into simple sugars. The increase of amylase in serum may be linked to the inflammation of the pancreas. The activity of alpha-amylase of the camel serum was found to be higher (2325 U/l) than in cattle serum with 77 U/l only (Mura et al. 2005).

No significant difference was observed between pregnant and non-pregnant camels: 881 ± 15.8 vs 916.9 ± 21.5 U/l, respectively (Khadjeh 2002). No sex difference was also observed in camel (Chiericato et al. 1986). In camel pancreas, *α*-amylase activity is 3360 ± 127 U/g (Mohamed et al. 2005).

5.4.5 Lactoperoxidase or LPO (EC 1.11.1.7)

The lactoperoxidase is mainly secreted by the mammary gland but also by the salivary glands and some other mucosal glands. LPO catalyzes the oxidation of many organic and inorganic substrates as iodide, bromide, or thiocyanate by hydrogen peroxide. Through the production of those oxidized products, LPO has a strong antibacterial activity. So, LPO is used as an antimicrobial agent for reducing the bacterial microflora in milk: the activation of the LPO system by addition of hydrogen peroxide contributes to the extension of shelf life of cooled raw milk especially in southern countries (Puspitarini et al. 2013). A potential antiviral activity against hepatitis C of the camel milk LPO was also reported (Redwan et al. 2015).

Due to its main secretion in udder and its role in bacterial protection of raw milk, LPO is determined in milk. At reverse, no reference of LPO in serum is available. In different camel milk samples, Lorenzen et al. (2011) have quantified LPO activity and reported values between 888.3 ± 36.8 and 1172 ± 93.3 U/l. Camel LPO is less heat resistant than bovine LPO. It is almost disappearing in pasteurized camel milk and could be used as a marker of pasteurization (Tayefi-Nasrabadi et al. 2011; Wernery et al. 2008, 2013).

5.4.6 Catalase or CAT (EC 1.11.1.6)

Catalase is a common enzyme catalyzing the decomposition of hydrogen peroxide to water and oxygen. It plays a pivotal role in protecting the cell from oxidative damage by reactive oxygen species (ROS). The catalase activity in camel was determined in the liver (Al-Bar 2012) which contains a high level (32,225 U/g). In camel milk, catalase activity was also very high compared to cow and buffalo milk: 6 ± 0 U/l vs 1 ± 04 and 1 ± 0.0 U/l, respectively (Yoganandi et al. 2014).

In camel blood, CAT activity was lower in winter (41 ± 3) than in summer (52 ± 5 kU/l) (Lektib et al. 2016). Similar figure was reported formerly by Kataria et al. (2010). As a marker of stress, catalase activity is significantly increased with the distance of transportation passing from 60 ± 3 kU/l after 72–80 km transportation in truck to 94 ± 4 after 350–360 km (El-Khasmi et al. 2015).

In red blood cells, catalase activity is slightly lower in young camel than in adult camel: 3 ± 0 vs 4 ± 0 U/mgHb (Mousa et al. 2006). Camel catalase was recently purified (Chafik et al. 2017).

5.4.7 Xanthine Oxidase or XO (EC 1.17.3.2)

Xanthine oxidase generates reactive oxygen species (ROS) and catalyzes the oxidation of hypoxanthine to xanthine and can further catalyze the oxidation of xanthine to uric acid. Consequently it is an enzyme marker of oxidative stress and increases in case of liver damage. In camel, XO activity in plasma appears higher in hot or cold stressful environment (80 ± 4 and 74 ± 4 mU/l, respectively, in female camels with no difference with male camels) compared to moderate environmental temperature: 51 ± 3 and 50 ± 2 mU/l in male and female camel, respectively (Kataria et al. 2010). In camel plasma, XO activity was reported to be lower than in other species like cattle, sheep, or buffalo (Mohamed 2006). Such lower XO activity in camel serum compared to many other species was already reported by Al-Khalidi and Chaglassian (1965). A camel breed difference was reported by Mohamed (2006): 40 ± 9 in Anafi breed vs 90 ± 10 mU/l in Arabi breed.

In camel milk, XO activity is thermostable contrary to that observed in cow milk (Loiseau et al. 2001).

5.4.8 Lipase or LIP (EC 3.1.1.3)

Lipase is an esterase enzyme catalyzing the hydrolysis of fats. It contributes to the catabolism of dietary or tissue lipids. The increase of lipase in serum is a marker of pancreas damage. Pancreatic lipase was purified in camel (Khafagy et al. 2000), but no reference is available regarding serum level. In camel raw milk, lipase activity is 71 ± 24 U/l and 12 ± 2 U/l only after heat treatment (Lorenzen et al. 2011).

Other enzymes can be determined in camel blood, but the references are not common and their clinical interest is debatable (see Table 5.3).

More enzymes are analyzed in milk, especially to test their heat resistance (Loiseau et al. 2001).

5.5 Conclusion

The blood enzymes in camel are following similar patterns than in other species. The physiological variations are finally of low interest. Due to the law "all or nothing," it is rather the exceptional changes in their concentrations in relationship with cell disorders or stress than the variation within the normal range which must be taken into account in the clinical investigation.

References

Abdalla OM, Wasfi IA, Gadir FA (1988) The Arabian race camel normal parameters. I. Haemogram, enzymes and minerals. Comp Biochem Physiol 90A:237–239

Abdel Rahim AG (1983) The relationship between the concentrations of serum copper and zinc and the activities of serum enzymes copper oxidase and alkaline phosphates of dromedary camel. J Anim Environ 6:265–268

Abdel Rahim AG (2005) The relationship between whole blood selenium (se) concentration and the activity of the seleno-enzyme, glutathione peroxidase (GSH-Px, E.C.I.11.1.9) in camel (*Camelus dromedarius*). J Arid Environ 62:359–362

Abd-El-Baky AA, Salem SI (2011) Clinicopathological and cytological studies on naturally infected camels and experimentally infected rats with *Trypanosoma evansi*. World Appl Sci J 14:42–50

Abdo MS, Kawasmeh ZA, El-Manna MM (1985) Some blood parameters in male Arabian camels (*Camelus dromedarius*) naturally infested with intestinal parasites in Al-Hasa (Saudi Arabia). Indian Vet J 62:948–950

Abu Damir H, Eldirdiri NI, SEI A, Howarth JA, Salih YM, Idris OF (1993) Experimental copper poisoning in the camel (*Camelus dromedarius*). J Comp Pathol 108:191–208

Abu Damir H, Al-Haj Ali M, Abbas TA, Omer EA, Al-Fihail AM (2013) Narasin poisoning in the dromedary camel (*Camelus dromedarius*). Comp Clin Pathol 22:305–311

Abubakr MI, Nayel MN, Abdelrahman AO, Abuobida SA, Ahmed AT, Mirghani EF, (2012) Pyrethroid (lambda cyhalothrin) poisoning in camels. In: Proceedings of the 3rd conference on ISOCARD, 29 Jan–1 Feb 2012, Muscat, Oman, pp 58–61

Achaaban MR, Baiss M, El-Allali K, Oukessou M, Piro M (2009) The cerebrospinal fluid in the dromedary camel (*Camelus dromedarius*): sampling and composition. In: Proceedings of the 2nd conference on ISOCARD, 12–14 Mar 2009, vol 96, Djerba, Tunisia, abstr. 121

Adam SEI, Obeid HM, Ashour N, Tartour G (1974) Serum enzyme activities and haematology of normal and diseased ruminants in Sudan. Acta Vet Brno 43:225–231

Afzal M, Saeed A (1995) Distribution of enzymes between different organs of the camel (*Camelus dromedarius*). Aust Vet J 72:195

Ahmad S, Butt AA, Muhammad G, Athar M, Khan MZ (2004) Haematobiochemical studies on the haemoparasitized camels. Int J Agric Biol 6:331–334

Ahmed SH, Omer SAA, GAMEEL AA (2009) Some normal constituents in serum and cerebrospinal fluid in Sudanese camels (*Camelus dromedarius*). Assiut Vet Med J 55(123):163–170

Aichouni A, Jeblawi R, Dellal A, Hammou H, Aggad H (2010) Breed variation in blood constituents of the one-humped camel (*Camelus dromedarius*) in Algeria. J Camelid Sci 3:19–25

Al-Ali AK, Al-Husayni HA, Power D (1988) A comprehensive biochemical analysis of the blood of the camel (*Camelus dromedarius*). Comp Biochem Physiol 89B:35–37

Al-Amrousi AM, Wasfi IA (1984) Some biochemical parameters of mature female camels in eastern province of Saudi Arabia. Assiut Vet Med J 13:121–124

Al-Bar OAM (2012) Characterization of partially purified catalase from camel (*Camelus dromedarius*) liver. Afr J Biotechnol 11(40):9633–9640

Al-Busadah KA (2007) Some biochemical and haematological indices in different breeds of camel in Saudi Arabia. Sci J King Faisal Univ 8(1):131–142

Al-Harbi MS (2012) Some hematologic values and serum biochemical parameters in male camels (*Camelus dromedarius*) before and during rut. Asian J Anim Vet Adv 7(11):1219–1226

Ali BH, El-Sanhouri AA, Musa BE (1989) Some clinical, haematological and biochemical effects of four tranquilizers in camel (*Camelus dromedxarius*). Rev Elev Méd Vét Pays Trop 42 (1):13–17

Ali F, Hussain R, Qayyum A, Gul ST, Iqbal Z, Hassan MF (2016a) Milk somatic cell counts and some hemato-biochemical changes in sub-clinical mastitic dromedary she-camels (*Camelus dromedarius*). Pak Vet J 36(4):405–408

Ali A, Derar D, Tharwat M, Zeitounb MM, Alsobyil FA (2016b) *Dystocia in dromedary camels: prevalence, forms, risks and hematobiochemical changes*. Anim Reprod Sci 170:149–156
Al-Khalidi UAS, Chaglassian TH (1965) The species distribution of xanthine oxidase. Biochem J 97:318–320
Al-Rehaimi AA, Al-Ali AK, Mutairy AR, Dissanayake AR (1989) A comparative study of enzyme profile of camel (*Camelus dromedarius*) hump and sheep (*Ovis aries*) tail tissues. Comp Biochem Physiol 93B(4):857–858
Al-Rukibat R, Ismail ZB (2014) Biochemical analysis of serum and synovial fluid in clinically normal young camels (*Camelus dromedarius*). Vet Sci Develop 4(5371):40–42
Alshamsi NS, Ksiksi TS, Ashraf SS (2015) Altered serum enzymes and biochemical levels in Arabian racing camels with bone fractures. J Anim Plant Sci 25(4):1072–1080
Al-Sobayil FA (2010) Circadian rhythm of bone formation biomarkers in serum of dromedary camels. Res Vet Sci 89:455–459
Al-Sobayil FA, Mousa HM (2009) Clinical and biochemical studies on wry-neck syndrome in camels in Saudi Arabia. J Camel Pract Res 16(2):179–182
André F (1981) Intérêt du dosage enzymatique sérique en pathologie des animaux de rente. Le Point Vét 12:47–53
Ateeq G, Kouider S, Kolb E (1984) Haematology of dromedaries: cell counts, haemoglobin, proteins, urea, cholesterol, AST and ALT. Age and sex variations. Arch Exp Vet Med 38:664–675
Ayoub MA, El-Khouly AA, Mohamed TM (2003) Some hematological and biochemical and steroid hormones levels in one-humped camel during different physiological conditions. Emir J Agric Sci 15(1):44–55
Babeker EA, Suleem AE (2013) Observation of certain hematological and biochemical parameters in nomadic camels (*Camelus dromedarius*) in the Sudan. Univ Bakht Alruda Sci J 6:167–174
Badakhshan Y, Mirmahmoudi R (2016) Blood metabolites of one-humped camel (*Camelus dromedarius*) versus sheep during summer heat stress. Iran J Vet Med 10(1):65–71
Baghshani H, Nazifi S, Saeb M, Saeb S (2010) Influence of road transportation on plasma concentrations of acute phase proteins, including fibrinogen, haptoglobin, serum amyloid a, and ceruloplasmin, in dromedary camels (*Camelus dromedarius*). Comp Clin Pathol 19 (2):193–198
Barnouin J, Paccard P (1988) Facteurs de risque nutritionnels de la pathologie hépatique dans les troupeaux bovins laitiers en France. Can Vet J 29:915–920
Barsham MA, ElBagir NM, Barri MES (2005) Serum biochemical changes associated with experimentally induced hypothyroidism in one-humped camels (*Camelus dromedarius*). J Camel Pract Res 12(2):117–121
Beaunoyer DE (1992) Changes in serum enzyme activities after maximal exercise in camels. In: Proceedings of 1st international camel symposium, Dubaï, UAE, 2–7 Feb, pp 331–333
Belina D, Giro B, Ashenafi H, Demissie T, Muktar Y (2015) Review on camel liver pathology and its major diagnostic approaches. Glob J Vet Med Res 3(1):68–79
Bengoumi M (1992) Biochimie clinique du dromadaire et mécanismes de son adaptation à la déshydratation, Thèse de doctorat ès Sciences Agronomiques. I.A.V. Hassan II, Rabat, Maroc, 184 p
Bengoumi M (2008) Tolerance study of Tiludronate in the dromedary camel. J Camelid Sci 1:24–31
Bengoumi M, Kessabi M, Hamliri A (1990) Utilisation des enzymes en séméiologie hépatique des ruminants. Maghreb Vét 22:25–28
Bengoumi M, Robins SP, De La Farge F, Coxam V, Davicco MJ, Barlet JP (1996) Water restriction and bone metabolism in camels. Reprod Nutr Develop 36(5):545–554
Bengoumi M, Faye B, De La Farge F, Olson WG, Rico AG (1997a) Clinical enzymology in the dromedary camel (*Camelus dromedarius*). Part I. Enzyme activities and distributions of AST, ALT, GGT, AP and LDH in liver, kidney, muscle, myocardium and blood. J Camel Pract Res 4:19–23

Bengoumi M, Faye B, De La Farge F (1997b) Clinical enzymology in the dromedary camel (*Camelus dromedarius*). Part II. Effect of season, age, sex, castration, lactation and pregnancy on serum AST, ALT, GGT, AP and LDH activities. J Camel Pract Res 4:25–29

Bengoumi M, Faye B, De La Farge F (1998a) Clinical enzymology in the dromedary camel (*Camelus dromedarius*). Part III. Effect of dehydration on serum AST, ALT, GGT, AP, LDH and urine GGT activities. J Camel Pract Res 5(1):119–122

Bengoumi M, Faye B, De La Farge F (1998b) Clinical enzymology in the dromedary camel (*Camelus dromedarius*). Part IV. Effect of exercise on serum AST, ALT, GGT, AP, LDH and CK activities. J Camel Pract Res 5(1):123–126

Bengoumi M, Berrada J, Rochdi M, Hidane K, De Lafarge F, Faye B (1998c) Physiopathologie des diarrhées du chamelon au Maroc. Signes cliniques et perturbations métaboliques. Rev Elev Med Vét Pays Trop 51:277–281

Bengoumi M, Essamadi K, Tressol JC, Faye B (1998d) Comparative study of copper and zinc metabolism in cattle and camel. Biol Trace Elem Res 63:81–94

Bengoumi M, Essamadi AK, Tressol JC, Chacornac JP, Faye B (1998e) Comparative effect of selenium concentration and erythrocyte gluthatione peroxidase activity in cattle and camels. Anim Sci 67:461–466

Bengoumi M, Essamadi AK, Chacornac JP, Tressol JC, Faye B (1998f) Comparative relationship between copper-zinc plasma concentrations and superoxide dismutase activity in camels and cows. Vet Res 29:557–565

Benjamin MM (1984) Outline of veterinary clinical pathology, III edn. The Iowa State University Press, IOWA

Ben-Zvi Z, Vancreveld C, Yagil R (1989) Liver function and protein binding in camels. Comp Biochem Physiol 93A(2):349–352

Bizzeti M, Elmi FM, Gugliucci B, Demi S (1988) Enzymatic and mineral blood serum features in dromedary bred in Somaliland Note II. Ann Fac Med Vet Univ Pisa 41:377–380

Bogin E (2000) Clinical pathology of Camelids: present and future. Rev Méd Vét 151(7):563–568

Boid R (1980) Changes in the levels of some serum enzymes in dromedary camels infected with *Trypanosoma evansi*. Res Vet Sci 28:336–340

Braun JP (1985) Distribution des enzymes dans les organes des animaux. Int Soc Biol 11:322–329

Braun JP, Rico AG, Benard P (1978) Tissue and blood distribution of gamma-glutamyl transferase in the lamb and in the ewe. Res Vet Sci 28:37–40

Chafik A, Essamadia A, Çelikb SY, Mavid A (2017) Purification of camel liver catalase by zinc chelate affinity chromatography and pH gradient elution: an enzyme with interesting properties. J Chromatogr B 1070:104–111

Chafik A, Essamadia A, Çelikb SY, Solakc K, Mavid A (2018) Partial purification and some interesting properties of glutathione peroxidase from liver of camel (*Camelus dromedarius*). Russ J Bioorgan Chem 44(1):41–51

Chaudhary ZI, Iqbal J (2000) Incidence, biochemical and haematological alterations induced by natural trypanosomosis in racing dromedary camels. Acta Trop 77:209–213

Chaudhary ZI, Iqbal J, Raza M, Qandil MI (1998) Acute monensin toxicity in dromedary camels. J Camel Pract Res 5(2):271–273

Chiericato GM, Schiapelli MP, Warfa AA (1986) Caratteristiche del profilo enzimatico e minerale del dromedarius (*Camelus dromedarius*). Clin Vet (Milan) 109:155–158

Clifton B (1985) Storage stability of some bovine plasma enzymes. Am J Vet Res:1242–1244

Corbera JA, Gutierrez C, Morales M, Montel A, Montoya JA (2001) Assessment of glutathione peroxidase activity in the dromedary camel. Vet Res 32:185–191

Corbera JA, Morales M, Pulido M, Montoya JA, Gutierrez C (2003) An outbreak of nutritional muscular dystrophy in dromedary camels. J Appl Anim Res 23:117–122

El Dirdiri NI, Suliman HB, Shommein AM (1987) Normal serum activities of some diagnostic enzymes in dromedary camel in Sudan. Vet Res Commun 11:201–203

El-Belely MS, Abou-Ahmed MM, Hemeida NA, El-Baghdady YRM (1988) The effect of pregnancy on certain serum organic constituents in the camel. Assiut Vet Med J 20(40):148–159

El-Boshy M, Abbas H, El-Khodery S, Osman S (2009) Cytokine response and clinicopathological findings in Brucella infected camels (*Camelus dromedarius*). Vet Med 54(1):25–32

El-Deeb WM, Buczinski S (2015) The diagnostic and prognostic importance of oxidative stress biomarkers and acute phase proteins in urinary tract infection (UTI) in camels. Peer J 3:1363 18 p

Elias E, Yagil R (1984) Haematological and serum biochemical values in lactating camels (*Camelus dromedarius*) and their newborn. Refuah Vet 41:7–13

El-Khasmi M, Youssef C, Mohamed F, El- Hassane T, Fouad F, El-Abbadi N, Rachid A, Faye B (2015) Impact of transport distance on stress biomarkers levels in dromedary camel (*Camelus dromedarius*). In: Konuspayeva G (ed) Proceedings of the 4th conference of ISOCARD, "silk road camel: the Camelids, main stakes for sustainable development", June 8–12, 2015 Almaty, Kazakhstan, special issue of Scientific and Practical Journal Veterinariya #2(42) 2015, oral communication, pp 180–182

El-Magawry SM, Ibrahim IA, Dowidar MF, Yousif MA (1988) Gluthatione peroxidase activity in camels' erythrocytes (*Camelus dromedarius*) in relation to blood selenium concentration. Assiut Vet Med J 20(40):20–24

Elnahas A (2008) Ultrasonographical examination of one humped camels (*Camelus dromedarius*) liver with some haematological and biochemical aspects, DVM dissertation. University of Leipzig. Germany, 107 p

Eltohamy MM, Salama A, Yousef AEA (1986) Blood constituents in relation to thye reproduction state of the camel (*Camelus dromedarius*). Beitr Landw Vet Med 24:425–430

Enwezor FNC, Sackey AKB (2005) Camel trypanosomosis—a review. Veterinarski Archiv 75 (5):439–452

Essamadi K, Bengoumi M, Tressol JC, Chacornac JP, Faye B (1998) Comparative relationship of plasma copper concentration and ceruloplasmin activity of camel and cow. Trends Comp Biochem Physiol 5:211–220

Essamadi AK, Bengoumi M, Zaoui D, Faye B, Bellenchi GC, Musci G, Calabrese L (2000) Young camel ceruloplasmin: purification and characterisation. In: Atelier International sur le chamelon: "le Chamelon, futur de l'élevage camélin". Ouarzazate, 24–26 October 1999. Maroc Rev Elev Méd Vét Pays Trop 53:111–114

Essamadi K, Bengoumi M, Zaoui D, Faye B, Bellenchi C, Musci G, Calabrese L (2002) Purification and partial characterization of camel (*Camelus dromedarius*) ceruloplasmin. Comp Biochem Physiol B 131:509–517

Faye B (1997) Le guide de l'élevage du dromadaire. Ed. Sanofi, Libourne, 125 pp

Faye B, Grillet C (1984) La carence en cuivre chez les ruminants domestiques de la région d Awash. Rev Elev Med Vét Pays Trop 37:42–60

Faye B, Mulato C (1991) Facteurs de variation des paramètres protéo-énergétiques, enzymatiques et minéraux dans le plasma chez le dromadaire de Djibouti. Rev Elev Méd Vét Pays Trop 44:325–334

Faye B, Seboussi R (2009) Selenium in camel—a review. Nutrients 1:30–49

Faye B, Grillet C, Tessema A (1986) Teneur en oligo-éléments dans les fourrages et le plasma des ruminants domestiques en Ethiopie. Rev Elev Med Vét Pays Trop 39:227–237

Faye B, Saint-Martin G, Cherrier R, Ali RM (1992a) The influence of high dietary protein, energy and mineral intake on deficient young camel (*Camelus dromedarius*). II: Changes in metabolic profiles and growth performance. Comp Biochem Physiol 102A:409–416

Faye B, Saint-Martin G, Cherrier R, Ali RM (1992b) The influence of high dietary protein, energy and mineral intake on deficient young camel: II: Changes in mineral status. Comp Biochem Physiol 102A:417–424

Faye B, Ratovonanahary M, Chacornac JP, Soubre P (1995a) Metabolic profiles and risks of diseases in camels in temperate conditions. Comp Biochem Physiol 112A:67–73

Faye B, Jouany JP, Chacornac JP, Ratovonanahary M (1995b) L'élevage des grands camélidés. Analyse des initiatives réalisées en France. INRA Prod Anim 8:3–17

Faye B, Seboussi R, Alhadrami G (2009) Meta-analysis of the interactions between selenium status and haematological and mineral parameters in camel blood. Trends Comp Biochem Physiol 14:25–34

Fouad D (2015) Molecular modelling and phylogeny of the manganese superoxide dismutase from the camel, *Camelus dromedarius*. Pak J Zool 47(4):1015–1023

Friedel R, Trautschold I, Gärtner K, Helle-Feldmann M, Gaudssuhn D (1975) Effects of blood sampling on enzyme activities in the serum of small laboratory animals. Z Klin Chem Klin Biochem 13(11):499–505

Ghosal AK, Dwaraknath PK (1971) Normal serum transaminase activities in domestic animals. Indian J Anim Health 10:61–62

Gutierrez C, Corbera JA, Juste MC, Doreste F, Morales I (2005) An outbreak of abortions and high neonatal mortality associated with *Trypanosoma evansi* in dromedary camels in the Canary Islands. Vet Parasitol 130:163–168

Halabi M, Kouider S, Kolb E (1982) Untersuchungen über die aktivitat der suren und alkalischen phosphate im blut serum von mannlischen und weisslischen kamelen unterschiedilieschen alters. Mh Vet Med 37:143–145

Hamad B, Aggad H, Hadef L, Adaika A (2017) Effect of seasons on blood biochemical parameters in male dromedary camels in Algeria. Indian J Anim Res Online B 672:1–5

Hamliri A, Khallaayoune K, Johnson DW, Kessabi M (1990) The relationship between the concentration of selenium in the blood and the activity of glutathione peroxidase in the erythrocytes of the dromedary camel (*Camelus dromedarius*). Vet Res Commun 14(1):27–30

Haroun EM (1994) Normal concentrations of some blood constituents in young Najdi camels (*Camelus dromedarius*). Comp Biochem Physiol A Physiol 108(4):619–622

Hasim M, Braun JP (1989) *Importance diagnostique de la gamma glutamyl transférase chez les ruminants*. Point Vét 120:169–174

Heidarpour M, Mohri M, Borji H, Moghdass E (2012) Oxidative stress and trace elements in camel (*Camelus dromedarius*) with liver cystic echinococcosis. Vet Parasitol 187:459–463

Hussain R, Khan A, Zahid Abbas R, Ghaffar A, Abbas G, Rahman TU, Ali F (2016) Clinico-hematological and biochemical studies on naturally infected camels with trypanosomiasis. Pak J Zool 48(2):311–316

Hussein MF, Al-Sabayei AA, Hassan HA (1982) A study of some aspects of blood chemistry in Saudi Arabian camels (*Camelus dromedarius*). Sudan J Vet Sci Anim Husb 23:69–72

Hussein YA, Al-Eknah MM, AL-Shami SA, Mandour MA, Fouda TA (2012) Coat color breed variation in blood constituents among indigenous Saudi Arabia camel strains. Mansoura Vet Med J 14(1):191–204

Ismael AB, Swelum AA, Khalaf AF, Abouheif MA (2014) Clinical, haematological and biochemical alterations associated with an outbreak of theileriosis in dromedaries (*Camelus dromedarius*) in Saudi Arabia. Pak Vet J 34(2):209–213

Jones DG (1985a) Stability and storage characteristics of enzymes in cattle blood. Res Vet Sci 38:301–306

Jones DG (1985b) Stability and storage characteristics of enzymes in sheep blood. Res Vet Sci 38:307–311

Kaneko JJ (1989) Clinical biochemistry of domestic animals, 4th edn. Academic, New York

Kaneko JJ, Harvey JW, Bruss M (1997) Clinical biochemistry of domestic animals, 5th edn. Gulf Professional, New York, p 932

Kataria N, Bhatia JS (1991) Activity of some enzymes in the serum of dromedary camels. Res Vet Sci 51:174–176

Kataria N, Sareen M, Bhatia JS (1991) Effect of climatic condition, sex and age on serum ASAT and ALAT levels in dromedary camel. Indian Vet J 68:596–598

Kataria N, Kataria AK, Pandey N, Gupta P (2010) Serum biomarkers of physiological defense against reactive oxygen species during environmental stress in Indian dromedaries. HVM Int J Bioflux Soc 2(2):55–60

Khadjeh GH (2002) Concentration of serum enzymes in prezgnant and non-pregnant Iranian one-humped camels. Indian J Anim Sci 72(5):391–392

Khafagy EZ, Abd El-Fattah MM, Sabri GM, Abd-El-Kader AH (2000) Studies on camel pancreatic lipase and colipase. Biochem Soc Trans 28(5):1089

Knight PK, Cluer D, Manefield GW, Gorde AK (1994a) Haematology in the racing camel at rest: seasonal and training variations. In: Saltin B, Rose RJ (eds) The racing camel. Physiology, metabolic functions and adaptations (chap. III). Acta Physiol Scand Suppl 150:19–23

Knight PK, Rose RJ, Evans DL, Cluer D, Henckel P, Saltin B (1994b) Metabolic responses to maximal intensity exercise in the racing camel. In: Saltin B, Rose RJ (eds) The racing camel. Physiology, metabolic functions and adaptations (chap. VIII). Acta Physiol Scand Suppl 150:61–77

Kouider S, Kolb E (1982) Glucose levels and various enzyme activities (aspartate aminotransferase, alanine aminotransferase, acid and alkaline phosphatases, glucose-6-phosphate dehydrogenase) in camel serum during the day. Arch Exp Vet Med 36:601–610

Lektib I, Bargaâ R, Farh M, El Abbadi N, Tahri E, Riad F, Belhouari A, Hammoumi A, El Mzibri M, El Khasmi M (2016) Blood oxidant stress in relation to seasonal activity of thyroid and adrenal glands in *Camelus dromedarius*. Int J Res Environ Stud 3:101–108

Liu ZP, Ma Z, Zhang YJ (1994) Studies on the relationship between sway disease of bactrian camels and copper status in Gansu province. Vet Res Commun 18(4):251–260

Loiseau G, Faye B, Serikbaeva A, Montet D (2001) Enzymes ability to serve as markers of pasteurized camel milk. In: International conference on new horizons in biotechnology, 18–21 avril 2001, Trivandrum, Inde

Lorenzen PC, Wernery R, Johnson B, Jose S, Wernery U (2011) Evaluation of indigenous enzyme activities in raw and pasteurised camel milk. Small Rumin Res 97:79–82

Mal G, Sena DS, Sahani MS (2006) Haemato-biochemical changes in camels infested with mange during winter and summer season. J Camel Pract Res 13(2):173–174

Meister A (1974) Glutathione: metabolism and function via the gamma glutamyl cycle. Life Sci 15:177

Meyer DJ (1982) The liver part I: biochemical test for the evaluation of the hepatobiliary system. Compend Contin Educ 4:663–673

Mohamed HE (2006) Purine derivatives in the plasma and urine and tissue xanthine oxidase (XO) in Sudanese camels (*Camelus dromedarius*). J Anim Vet Adv 5(4):310–312

Mohamed HA, Hussein AN (1999) Studies on normal haematological and serum biochemical values of the "Hijin" racing camels (*Camelus dromedarius*) in Kuwait. Vet Res Commun 23 (4):241–248

Mohamed MH, Mohamed AH, Locatelli A (1984) Water deprivation effects on the hematological and hematochemical pictures of *Camelus dromedarius*. Rev Elev Méd Vét Pays Trop 37 (3):313–317

Mohamed SA, Fahmy AS, Mohamed TM (2005) Carbohydrases in camel (*Camelus dromedarius*) pancreas. Purification and characterization of glucoamylase. Comp Biochem Physiol B 140:73–80

Mohri M, Moosavian HR, Hadian MJ (2008) Plasma biochemistry of one-humped camel (*Camelus dromedarius*): effects of anticoagulants and comparison with serum. Res Vet Sci 85:554–558

Momenah MA (2014) Some blood parameters of one humped she-camels (*Camelus dromedaries*) in response to parasitic infection. Life Sci J 11(5):118–123

Mousa HM, Omer OH, Ali BH, Al-Wabel N, Ahmed SM (2006) Antioxidant levels in tissues of young and adult camels (*Camelus dromedarius*). J Physiol Biochem 62(3):213–218

Muhammad BF, Aliyu D, Njidda AA, Madigawa IL (2011) Some haematological, biochemical and hormonal profile of pregnant and non-pregnant she-camels (*Camelus dromedarius*) raised in a Sudan savanna zone in Nigeria. J Camel Pract Res 18(1):73–77

Mura U, Mohamed AS, Osman AM, Ipata PL (2005) A comparative study of serum and organ alpha-amylases of the camel (*Camelus dromedarius*). Comp Biochem Physiol A 82(1):141–144
Nazifi S, Maleki K (1998) Biochemical analysis of serum and cerebrospinal fluid in clinically normal adult camels (*Camelus dromedarius*). Res Vet Sci 65:83–84
Nazifi S, Gheisari R, Poorabbas H (1999) The influence of thermal stress on serum biochemical parameters of dromedary camels and their correlation with thyroid activity. Comp Haematol Int 9:49–53
Nazifi S, Saeb A, Alizadeh A (2000) Studies on the relationship between haemoglobin types of adult dromedary camels and the concentrations of haemoglobin, copper, ceruloplasmin and iron. Comp Haematol Int 10(3):122–125
Nazifi S, Aminlari M, Mhammadi AH (2002) Distribution of gamma glutamyltransferase activity in the tissues of the camel (*Camelus dromedarius*). J Camel Pract Res 9(2):97–99
Nazifi S, Saeb M, Dadvar L (2005) Activity of the urinary enzymes in Iranian dromedary camels. J Fac Vet Med Univ Tehran 60(3):235–239
Nazifi S, Mansourian M, Nikahval B, Razavi SM (2009a) The relationship between serum level of thyroid hormones, trace elements and antioxidant enzymes in dromedary camel (*Camelus dromedarius*). Trop Anim Health Prod 41(1):129–134
Nazifi S, Oryan A, Bahrami S, Razavi SM (2009b) Evaluation of haematological and serum biochemical parameters in Iranian camels (*Camelus dromedarius*) infected with haemotrophic Mycoplasma (Eperythrozoon) spp. Comp Clin Pathol 18:329–332
Nyang'ao JMN, Olaho-Mukani W, Maribei JM, Omuse JK (1997) A study of some haematological and biochemical parameters of the normal dromedary camel in Kenya. J Camel Pract Res 4 (1):31–33
Omer SA, Gameel AA (2009) Normal values of some synovial fluid constituents in Sudanese camels (*Camelus dromedarius*). In: Proceedings of the 2nd ISOCARD conference, vol 96, Djerba, 12–14 Mar 2009, Abstr. 119
Omidi A, Sajedi Z, Montazer-Torbati MB, Mostafai M (2014) Metabolic profile of pregnant, non-pregnant and male two-humped camels (*Camelus bactrianus*) of Iran. Iran J Vet Med 8 (4):235–242
Orliac DG (1980) Contribution à l'étude de la biochilie sanguine de dromadaires et de chèvres sahariens, Thèse Med. Vet. Ecole Nationale Vétérinaire de Toulouse, France
Osman TEA, Al-Busadah KA (2000) Effects of age and lactation on some biochemical constituents of camel blood in Saudi Arabia. J Camel Pract Res 7(2):149–152
Osman TEA, Al-Busadah KA (2003) Normal concentrations of twenty biochemical parameters of she-camels, cows and ewes in Saudi Arabia. Pak J Biol Sci 6(14):1253–1256
Osman FA, Gaadee H, Sayed GA (2014) Clinico-hematological and biochemical changes in camels affected with gastro-intestinal parasites. IJAVMS 8(5):154–161
Puspitarini OR, Al-Baarri AN, Legowo AM, Bintoro P, Hintono A (2013) The activation method of lactoperoxidase system to inhibit microbial activity in fresh milk. Anim Prod 15(2):119–126
Razmi N, Nazifi S, Motaalleh G (2003) Distribution of arginase in tissues of camel (*Camelus dromedarius*). J Camel Pract Res 10(2):135–137
Redwan EM, Almehdar HA, EL-Fakharany EM, Baiga AK, Uversky VN (2015) Potential antiviral activities of camel, bovine, and human lactoperoxidases against hepatitis C virus genotype 4. R Soc Chem Adv 5:60441–60452
Rico AG, Godfrain JC, Braun JP, Benard P, Burgat V (1975) Dosages enzymatiques en clinique bovine. Rev Méd Vét 126:53–68
Rose RJ, Evans D, Henkel P, Knight PK, Cluer D, Saltin B (1994) Metabolic responses to prolonged exercise in the racing camel. In: Saltin B, Rose RJ (eds) The racing camel. Physiology, metabolic functions and adaptations (chap. VII) Acta Physiol Scand Suppl 150:49–60
Saeed A, Afzal M, Saeed A (1995) Effect of storage on some constituents of camel serum. Aust Vet J 72:212–215
Saeed A, Khan IA, Hussein MM (2009) Change in biochemical profile of pregnant camel (*Camelus dromedarius*) at term. Comp Clin Pathol 18:139–143

Salutini E, Biagi G (1983) GOT, GPT e CPK nel *Camelus dromedarius* Somalo di diversa eta e sesso. Atti Soc Sci Vet 37:375–377

Sarwar A, Majeed MA (1997) Interrelationships between 30 parameters of blood in normal one-humped camel in summer. J Camel Pract Res 4(1):35–39

Sarwar A, Majeed MA, Hur G, Khan IR (1992) Studies on the serum transferases and electrolytes of normal one humped camel in summer. Pak Vet J 12:175–182

Sarwar A, Majeed MA, Hur G, Khan LR (2004) Two transferases and four electrolytes in normal onehumped camel serum. J Camel Sci 1:57–61

Sazmand A, Rasooli A, Nouri M, Hamidinejat H, Hekmatimoghaddam S (2011) Serobiochemical alterations in subclinically affected dromedary camels with Trypanosoma evansi in Iran. Pak Vet J 31(3):223–226

Seboussi R, Faye B, Alhadrami G (2004) Facteurs de variation de quelques éléments trace (sélénium, cuivre, zinc) et d'enzymes témoins de la souffrance musculaire (CPK, ALT et AST) dans le sérum du dromadaire (*Camelus dromedarius*) aux Emirats Arabes Unis. Rev Elev Med Vét Pays Trop 57(1–2):87–94

Seboussi R, Faye B, Alhadrami G, Askar M, Ibrahim W, Hassan K, Mahjoub B (2008) Effect of different selenium supplementation levels on selenium status in camel. Biol Trace Elem Res 123:124–138

Seboussi R, Faye B, Askar M, Hassan K, Alhadrami G (2009a) Effect of selenium supplementation on blood status and milk, urine and fecal excretion in pregnant and lactating camel. Biol Trace Elem Res 128:45–57

Seboussi R, Faye B, Alhadrami G, Askar M, Bengoumi M, Elkhouly A (2009b) Chronic selenosis in camels. J Camel Pract Res 16(1):25–38

Seboussi R, Faye B, Alhadrami G, Askar M, Ibrahim W, Mahjoub B, Hassan K, Moustafa T, Elkhouly A (2010) Selenium distribution in camel blood and organs after different level of dietary selenium supplementation. Biol Trace Elem Res 133(1):34–50

Seleim RS, Tos AR, Sahar MR, Nada HS, Gobran RA (2003) ELISA and other tests in the diagnosis of Pasteurella multocida in camels. http://www.priory.com/vet/camel.htm

Sellaouti K (1984) Contribution à l'établissement des paramètres biochimiques sanguins du dromadaire, Thèse Médecine vétérinaire. Ecole vétérinaire de Sidi Thabet, Tunisia

Shen X, Li X (2010) Studies of "emaciation ailment" in the Bactrian camel. Afr J Biotechnol 9 (49):8492–8497

Shoieb SM, Sayed-Ahmed M (2016) Clinical and Clinicopathological findings of arthritic camel calf associated with Mycoplasma infection (*Camelus dromedarius*). J Dairy Vet Anim Res 3 (1):00068

Snow DH, Billah A, Ridha A (1988) Effects of maximal exercise on the blood composition of the racing camel. Vet Rec 123:311–312

Srivastava KB, Dwaraknath PK (1971) Ceruloplasmin (copper oxidases) activity in the serum of animals. Indian J Anim Sci 41:1044–1046

Swelum AA, Ismael AB, Khalaf AF, Abouheif MA (2014) Clinical and laboratory findings associated with naturally occurring babesiosis in dromedary camels. Bull Vet Inst Pulawy 58:229–233

Tabatabaei Naeini A, Nazifi S (2001) Biochemical and cytological properties of blood and peritoneal fluid in clinically healthy adult camels (*Camelus dromedarius*). J Camel Pract Res:123–126

Taha N, Abou El-Noeman S, Korshom M, Mandour A, El-Feky M (2010) Serum glutathione peroxidase and paraoxonase-1 enzyme activities in camel (*Camelus dromedarius*): effect of age and sex differences. J Camel Pract Res 17(1):51–57

Tayefi-Nasrabadi H, Hoseinpour-fayzi MA, Mohasseli M (2011) Effect of heat treatment on lactoperoxidase activity in camel milk: a comparison with bovine lactoperoxidase. Small Rumin Res 99(2–3):187–190

Tharwat M, Al-Sobayil F (2014) The effect of tick infestation on the serum concentrations of the cardiac biomarker troponin I, acid-base balance and haematobiochemical profiles in camels (*Camelus dromedarius*). Trop Anim Health Prod 46:139–144

Tharwat M, Al-Sobayil F, Buczinski S (2013) Effect of racing on the serum concentrations of cardiac troponin I and creatine kinase myocardial band in racing camels (*Camelus dromedarius*). Vet Res Comm 37:139–144

Wernery U, Fischbach S, Johnson B, Jose S (2008) Evaluation of alkaline phosphatase (ALP), γ-glutamyl transferase (GGT) and lactoperoxidase (LPO) activities for their suitability as markers of camel milk heat inactivation. Milchwissenschaft 63(3):265–267

Wernery U, Wernery R, Masko O, Lorenzen PC (2013) Lactoperoxidase: a suitable enzymatic marker of camel milk pasteurization. J Camel Pract Res 20(1):35–38

Whitby LG, Percy-Robb IW, Smith AF (1984) Lecture notes on clinical chemistry, III edn. Blackwell Scientific, Oxford

Yadav SB, Bissa UK (1998) Factors affecting some blood constituents in camels—a review. In: Proceedings of the 3rd annual meeting for animal production under arid conditions, vol. 2. UAE University Publications, pp 32–48

Yadav AK, Kumar R, Priyadarshini L, Singh J (2015) Composition and medicinal properties of camel milk: a review. Asian J Dairy Food Res 34(2):83–91

Yoganandi J, Jain AK, Mehta BM, Wadhwani KN, Aparnathi KD (2014) Comparative studies on selected enzymes activities of camel, cow and buffalo milk. J Camel Pract Res 21(2):249–252

Zaher H, El-Zahar H, Al Sharifi S, Shety T (2017) Alterations in hematological and biochemical parameters affecting the reproductive performance in female camels (*Camelus dromedaries*). Int J Vet Health Sci Res 5(1):155–160

Zeidan AEB, Abbas HE (2003) Physio-biochemical changes in the male dromedary camels during breeding (rutting) and non-breeding season. J Camel Pract Res 1(2):183–190

Zongping L (2003) Studies on the haematology and trace element status of adult Bactrian camels (*Camelus bactrianus*) in China. Vet Res Commun 27:397–405

Chapter 6
Macro-minerals and Electrolytes

The macro-minerals, namely, calcium (Ca), phosphorus (P), and magnesium (Mg), and electrolytes, namely, sodium (Na), potassium (K), chloride (Cl), and bicarbonates (H_2CO_3), play different essential roles in organisms as fundamental constituents of skeletal structures (bones and teeth) and by their intervention in many functions contributing to the live metabolism. They are also involved in the mechanism of adaptation of camel in desert conditions, the electrolytes participating actively to the cycle dehydration/rehydration and to the water metabolism. Moreover, due to the strong skeleton of the animals (long legs, long neck, heavy head), the bone metabolism where calcium and phosphorus are essential, the macro-minerals investigation in this species was relatively important in the scientific literature. The variability of these parameters in camel are consequently highly under the dependance of the growing and watering status.

In the camel, occasionally submitted to severe dehydration in its natural environment, the electrolytes play an important role. The legendary resistance of the camel to dehydration and heat, one of the most studied aspects of its biology (Yagil 1985; Wilson 1989; Bengoumi et al. 1993; Bengoumi and Faye 2002), is linked to several mechanisms of adaptation especially (1) reduction of water losses and (2) maintenance of homeostasis by regulation of the concentration of vital parameters and maximum excretion of metabolic wastes. Hydration status has an important effect on the hydro-electrolytic balance.

6.1 Hydro-electrolytic Balance in Camel

Water is the major constituent of the body; it accounts for about 60% of body weight in most domestic mammals. This rate varies in the dromedary camel from 58% to 75% (MacFarlane et al. 1962; Ghosal et al. 1974) depending on fattening and hydration status, which explain seasonal variations with higher values in the dry season than in the wet season (Banerjee and Bhattacharjee 1963). The distribution of

© Springer International Publishing AG, part of Springer Nature 2018
B. Faye, M. Bengoumi, *Camel Clinical Biochemistry and Hematology*,
https://doi.org/10.1007/978-3-319-95562-9_6

water in the different sectors expressed as a percentage of body weight is comparable to that of other domestic ruminants.

One of the remarkable characteristics of the dromedary is its ability to resist to dehydration which can reach more than 30% of body weight, i.e., a loss of about 50% of body water. For comparison, most domestic animals cannot survive to water losses exceeding 12–15% of body weight (Yagil 1985; Wilson 1989). Water deprivation causes water losses three times higher in cattle than in dromedaries: 6% of the weight per day in cattle compared to 2% of the weight per day in dromedaries (Ghosal et al. 1974). These losses are unevenly distributed among the various hydric sectors. A total water loss of 20% of the body weight induces a decrease in water in all compartments but especially in the intracellular and digestive tract (MacFarlane et al. 1962). The expression of the results as a percentage of the weight shows an increase in the plasma sector and a decrease in the other sectors. Indeed, during dehydration, water is transferred from the intracellular and interstitial sectors and from the digestive cavities to the plasma (Ghosal et al. 1974).

The water turnover corresponds to the volume of water renewed every 24 h. In dromedary, the water turnover doubles in summer and reaches its maximum value in lactating females (Siebert and MacFarlane 1971). The average water turnover (61 ml/kg/24 h) is lower than those of sheep (110 ml/kg/24 h) and cattle (148 ml/kg/24 h). The increase of water turnover in summer seems to be due mainly to the frequency of watering and sweating (MacFarlane et al. 1962; Ghosal et al. 1974). The state of dehydration in the dromedary causes a drop of 80% of the water turnover (Etzion et al. 1984), which marks its resistance to the water deprivation. This decrease is due to reduced base metabolism following hypothyroidism (Yagil et al. 1978; Bengoumi 1992; Bengoumi et al. 2003) and decreased renal and fecal losses (Yagil 1985). Water maintenance requirements are estimated at 6 liters per 100 kg of body during the dry season and the half during the wet season.

Water metabolism is controlled by several hormones including antidiuretic hormone, renin-angiotensin system, aldosterone, atrial natriuretic factor, etc. (Bengoumi et al. 1993).

6.2 Biological Signification of Observed Values

The electrolytes play an important role in the osmolality. This aspect being discussed in the first chapter on hematology, the present part will focus on the biological signification of the main electrolytes in camel.

6.2.1 Sodium

Sodium is an electrolyte occurring almost entirely in the fluids of the body where it serves a vital function in controlling osmotic pressures and acid-base balance, maintenance of cell membrane electric potential, and transmission of nerve influx. In a species like camel, particularly well adapted to water restriction, its role in water metabolism is particularly important. Obviously, natremia (as kaliemia) increased significantly during water deprivation (Ghosal et al. 1975). Sodium ions (Na+) constitute 93% of the base ion found in the blood (145 mmol/l in the plasma vs 10–15 mmol/l in cells). It has also some effect on muscle irritability and absorption of carbohydrates. The values are expressed in mmol/l (= mmol/l).

Sodium is mainly absorbed in the small intestine by simple diffusion. Sodium maintenance requirements are estimated at 8 g/100 kg of body weight. Sodium metabolism is mainly regulated by the kidney where an important tubular reabsorption counterbalanced electrically by chloride excretion in the proximal part and with potassium in the distal part of the glomerulus under the control of aldosterone produced by the adrenal cortex which is sensitive to osmolarity occurs. Kidneys are also sensitive to the plasma lower blood pressure, which results in lower filtration rates and activates the renin-angiotensin system. Renin, a peptide hormone produced in the glomeruli, initiates cascade or reactions producing angiotensin II that stimulates production of aldosterone. Antidiuretic hormone (ADH) reduces indirectly sodium concentration by increasing water reabsorption in the kidneys. Plasma sodium concentration is positively correlated to the total sodium content in the body. However, its variations are mainly linked to water distribution in the different hydric sectors. Plasma sodium concentration is the main indicator of plasma osmolarity.

6.2.1.1 Usual Values

In camel, natremia in normal conditions varied between 140 and 178 mmol/l (Table 6.1). Those values are comparable to that of cattle (132–152), sheep (139–152), goat (142–155), and llama (148–155) but slightly higher than horse (132–146) (Kaneko 1989). Hyponatremia reflects sodium depletion and decreases plasma osmolarity. It is mainly due to important hydration, renal failure, salt deficiency, or decrease in plasma protein concentration. It is generally linked to an edema. Hypernatremia is observed mainly during dehydration, renal failure, or acute salt poisoning.

6.2.1.2 Physiological Variations

Despite high values reported in growing camel calves by Omer et al. (2008), no age or parity effect was observed by most of the authors (Höller and Hassan 1966; Gupta et al. 1979; Salib and Soahir 1984; Rezakhani et al. 1997; Ahmed et al. 2003) although a significant statistical difference was observed between suckling and weaned calves (Omer et al. 2010): respectively, 123 ± 3 and 127 ± 3 mmol/l which seems to have no biological value. In their first year of camel calf life, natremia presented a sharp increase by almost a twofold level within the first month of age, followed by a slower, almost linear rise with advancing age, giving an overall mean value of about 130 ± 4 mmol/l. At the age of 7 months, the camel

Table 6.1 Natremia in camel blood serum according to different authors (all the values are expressed in mmol/l)

References	Mean (mmol/l)	SD	Nb	Country
Höller and Hassan (1966)	148	± 9	50	Sudan
Hassan et al. (1968)	300–390	–	32	Sudan
Barakat and Fatah (1970)	148	± 2	200	Egypt
Little et al. (1970)	163	± 1	40	Australia
Ghosal et al. (1973b)	162	± 5	46	India
Abdelgadir et al. (1979)	147	± 4	96	Sudan
Orliac (1980)	158	± 5	102	Algeria
Musa and Mukhtar (1982)	178	± 13	174	Sudan
Sellaouti (1984)	149	± 7	107	Tunisia
Abbas and Musa (1986)	140	± 12	36	Sudan
Abdalla et al. (1988)	154	± 1.5	20	UAE
Bizzeti et al. (1988)	153	± 15	44	Somalia
Elkasmi (1989)	143	± 3	60	Morocco
Evans et al. (1992)	149	± 2	5	UAE
Rezakhani et al. (1997)	158	± 8	31	Iran
Sarwar and Majeed (1997)	178	± 3	56	Pakistan
Khadjeh (1998)	150	± 1	109	Iran
Mohamed and Hussein (1999)	148	± 2	100	Kuwait
Naeini and Nazifi (2001)	147	± 7	50	Iran
Ahmed et al. (2003) (young)	148	± 11	4	Sudan
Ismail et al. (2003)	155	± 3	9	USA
Liu (2003)	148	± 32	35	China
Osman and Al-Busadah (2003)	168	± 1	5	Saudi Arabia
Kataria and Kataria (2004)	160	± 5	83	India
Sarwar et al. (2004)	178	± 3	56	Pakistan
Gutierrez et al. (2005)	154	± 4	16	Spain
Al-Busadah (2007)	155	± 19	60	Saudi Arabia
Mohammed et al. (2007)	145	± 1	11	Nigeria
Bengoumi (1992)	160	± 6	240	Morocco
Bengoumi (2008)	166	± 3	10	Morocco
Omer et al. (2008) (young)	542	± 10	12	Sudan
Al-Shami (2009)	146	± 11	105	Saudi Arabia
Al-Sobayil and Mousa (2009)	153	± 3	5	Saudi Arabia
Nazifi et al. (2009)	146	± 1	20	Iran
Saini et al. (2009)	138	± 4	3	India
Wernery et al. (2009)	149	± 5	747	UAE
Aichouni et al. (2010)	159	±14	48	Algeria
Eltahir et al. (2010)	160	± 29	30	Oman
Shen and Li (2010)	147	± 30	15	China[a]
Hekmatimoghaddam et al. (2011)	168	± 1	92	Iran
Sazmand et al. (2011)	168	± 2	93	Iran
Singh et al. (2015)	161	± 1	28	India

(continued)

Table 6.1 (continued)

References	Mean (mmol/l)	SD	Nb	Country
Tajik et al. (2015)	169	± 6	180	Iran
Shawaf (2017)	155	± 1	7	Saudi Arabia
Zaher et al. (2017)	154	± 0	51	Egypt

[a]Bactrian camel

calves had attained sodium levels comparable with those cited for adult camels (Hussein et al. 1992b).

A slight difference was reported between low-yield camel (151 ± 6) and high-yield one (144 ± 10 mmol/l). No sexual difference also was reported (Tajik et al. 2015). In a lactating camel (Hussein et al. 1992a), natremia increased significantly at the end of lactation (172.6 ± 12.4) compared to the beginning (25.1 ± 8.5 mmol/l).

Although there is no clear biological explanation (unless differences in hydration or feeding status), plasma sodium concentration would decrease significantly both in fertile and infertile camels at breeding season (Saini et al. 2009). Indeed, the natremia was above 140 mmol/l in both groups before and after breeding season, while it decreased to 125 (fertile) and 123 mmol/l (infertile) at breeding season. There was no significant difference for females according to the breeding season (Zeidan et al. 2008), but the values expressed in mg/100 ml cannot correspond to reference values in mmol/l: for example, 132 mg/100 ml in breeding season, i.e., 57.4 mmol/l.

In male, contradictory results were also reported regarding the effect of breeding status which means that other factors mainly salt feeding intake and dehydration status are responsible for these effects. Zia-Ur-Rahman et al. (2007) found a significant higher concentration at the rutting season compared to non-rutting (135 ± 3 vs 110 ± 4 mmol/l). On the contrary, sodium in the blood serum seems to be lower at the rutting season compared to non-rutting, 192 ± 6 vs 207 ± 6 mmol/l, respectively (Zeidan and Abbas 2003), but this difference is more marked in the hot-humid months (Abdel-Salaam et al. 2011). Besides, Desalegn et al. (2012) found higher sodium in serum of the camel at wet season (192 ± 6) than in the dry one (170 ± 6 mmol/l). Similar figure was observed in the subhumid area in Nigeria (Sackey et al. 2007). At reverse, in Kenya, higher natremia was reported in dry season compared to wet one: 195 ± 3 vs 115 ± 3 mmol/l (Kuria et al. 2013). In Algeria, Aichouni et al. (2013) reported lower concentrations of sodium in winter than in summer, respectively, 146 ± 11 and 170 ± 12 mmol/l. Similar difference was observed in Iran: 143 ± 1 in December vs 140 ± 1 mmol/l in June (Nazifi et al. 1999). For El-Harairy et al. (2010), natremia was significantly higher in summer (137 ± 4) and lower in autumn (126 ± 2), while values in spring (129 ± 3) and winter (133 ± 2) were in between.

Exercise seems to have no impact on the natremia of racing camel (Evans et al. 1992), the values remaining at similar level before and during the effort at different treadmill velocity (Fig. 2.3).

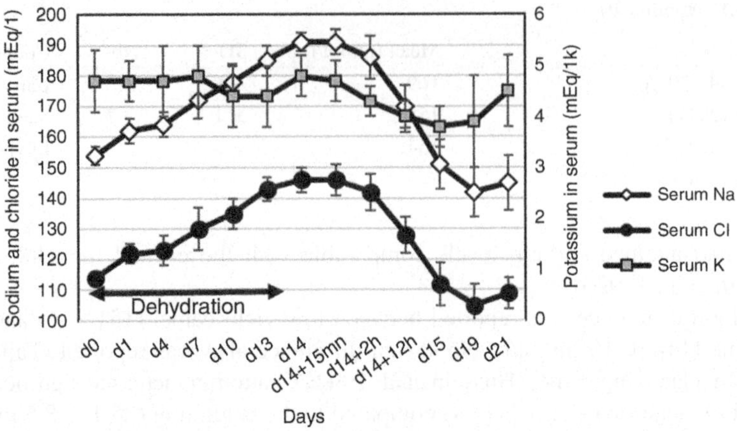

Fig. 6.1 Changes in electrolytes concentrations in camel blood serum during 14 days water deprivation and after rehydration (Bengoumi et al. 1993)

6.2.1.3 Effect of Dehydration

Sodium metabolism is strongly influenced by hydration status (Fig. 6.1). All authors report an increase in serum sodium concentration with water deprivation (Al-Qarawi 1997). For example, in a trial achieved in UAE, natremia passed from 164 ± 2 before dehydration to 173 ± 4 after 72 h water deprivation, then 149 ± 6 mmol/l after rehydration (Ayoub and Saleh 1998). Similar pattern was described by Al-Haj Ali et al. (2012): from 145 ± 1 to 175 ± 2 mmol/l after 20 days of water deprivation.

The overall effect of dehydration would be a simultaneous increase of plasma sodium concentration and urine excretion. During a 14-day dehydration, the natriuria passed from 20 ± 24 to 299 ± 213 with a peak at 408 ± 91 mmol/l on the fourth day (Bengoumi 1992). This excretion of sodium in the urine returned to baseline 24 h after rehydration. This increase during dehydration is due to decreased glomerular filtration and tubular sodium reabsorption. Indeed, the activation of antidiuretic hormone during dehydration induces an important reabsorption of water and a slight reabsorption of sodium from the digestive tract under the action of aldosterone (Bengoumi et al. 1993; Riad et al. 1994). During rehydration, plasma sodium concentration and urine excretion decrease rapidly. Increased plasma concentration of aldosterone, after rehydration, prevents a strong sodium renal depletion (Yagil and Etzion 1979; Bengoumi et al. 1993; Riad et al. 1994).

6.2.1.4 Clinical Variations

Reported data on effect of diseases on plasma sodium concentrations are contradictory. In their study regarding different diseases in the camel, Baraka et al. (2000) reported a significant hyponatremia in case of digestive disorders such as acidosis

(71 ± 9) and frothy bloat (26 ± 8), while the control camel showed also hyponatremia (110 ± 6 mmo/l). Similar findings were observed by Kamal (2008) including lower natremia in the camel affected by trypanosomosis in comparison to control camel: 132 ± 14 vs 160 ± 13 mmol/l. However, no change in natremia was observed in the camel affected by trypanosomosis (Baraka et al. 2000; Sazmand et al. 2011). Natremia and kalemia are not influenced by the stress status of the camel (El Khasmi et al. 2011).

The natremia decreased significantly in case of monensin intoxication from 156 ± 0 in control camel to 146 ± 1 mmol/l in affected one (Al-Jassim et al. 2016).

6.2.1.5 Sodium in Milk and Other Fluids

The Bactrian camel contained twice more sodium in its milk than the cow or goat (Wang et al. 2011). Globally, the range of sodium in camel milk is 220–690 mg/l (Yagil and Etzion 1980; Abu-Lehia 1987; Sawaya et al. 1984; Elamin and Wilcox 1992; Mostafidi et al. 2016) although Bengoumi et al. (1998) found a higher concentration, on average 902 ± 92 mg/l of sodium, as well as Hamed et al. (2016), 733 ± 22 mg/l. Significant differences were reported between 7 regions of Iran with values between 400 and 859 mg/l (Mostafidi et al. 2016). Shamsia (2009) reported higher values (580 ± 35 mg/l) than in human milk. Lower values were reported by Al-Wabel (2008) although the camel milk (116 ± 5) contained more sodium than the cow milk (91.6 ± 3.4), goat milk (101 ± 11), and sheep milk (95 ± 5 mg/l). Similar hierarchy was reported by Soliman (2005): 578 ± 12.2 in camel milk vs 517 ± 7 in buffalo milk, 497 ± 7 in cow milk, 503 ± 8 in goat milk, and only 160 ± 3 mg/l in human milk.

A significant variation in sodium content in milk was reported (Mal et al. 2007): 682 ± 12 in early lactation and 816 ± 21 mg/l in late lactation. In desert camel, higher sodium concentration in milk was reported (69 ± 1) compared to the milk from intensive farm (43 ± 9.77 mg/l) (Alwan and Zwaik 2014). Similar differences were reported by Elhassan et al. (2016) with higher value in desert camel (201.48 ± 10.9) than in intensive farm (130.31 ± 8.6 mg/100 ml), but the values were expressed by 100 ml and not by liter. The grazing of desert plant containing usually more salt (especially in halophytic plant area) could explain such differences. Effectively, camel milk from desert camel has more salty taste.

The concentration of sodium in peritoneal fluid is comparable to that in the blood serum: 146.45 ± 19.39 mmol/l (Naeini and Nazifi 2001). In cerebrospinal fluid, a similar concentration was reported also: 150.3 ± 1.70 mmol/l (Shawaf 2017). In synovial fluid, lower concentration in sodium than in the blood serum was reported, respectively, 152.4 ± 16.83 and 199.3 ± 40.49 mmol/l (Al-Rukibat and Ismail 2014). The camel urine is normally rich in sodium. In a normal camel, the concentration reached 57.04 ± 14.76 mmol/l (Kamalu et al. 2003). Higher concentrations in normally hydrated camels were reported by Bengoumi (1992): 154 ± 2 mmol/l.

The concentration of sodium in sweating fluid is similar to that of blood serum but could increase up to double according to the environment (El-Zeiny 2011).

Sodium concentration was analyzed all along the digestive tract showing values between 89 ± 6.8 in the rumen and 43 ± 14 mmol/l in the colon (Maloiy and Clemens 1980). In rumen fluid, Kamalu et al. (2003) reported a value of 36.44 ± 3.28 mmol/l, concentration significantly lower than in rumen fluid cattle. Higher concentration was reported in a healthy camel (108.38 ± 9.75 mmol/l), and such concentration increased in case of different digestive disorders like acidosis (127.9 ± 10.8) or frothy bloat (121.65 ± 11.5 mmol/l; Kamal 2008).

6.2.1.6 Sodium in Camel Meat and Wool

In camel meat, Mahmud et al. (2011) reported concentration of sodium at 252 mg/100 g. Lower values were reported by Siham and Daoud (2015), but on average camel meat was richer in sodium than beef or goat meat (respectively, 114.40 ± 4.98, 89.08 ± 6.40, and 76.0 ± 3.54 mg/100 g). Similar range was reported by Kadim et al. (2006): 104.7–257 mg/100 g. With a range of 139–150 mg/100 mg, there was no significant variability between the muscle and age (Ibrahim et al. 2017).

The concentration of the testis in sodium at the breeding season appeared quite higher (235.8 ± 108.0) than at non-rutting one (192.3 ± 20.0 μg/g), while no difference occurred in the epididymis, respectively, 1073.33 ± 75.5 and 1180.0 ± 80.0 μg/g (Zia-Ur-Rahman et al. 2007).

Dromedary camel raw hair contained on average 588 μg/g, but lower concentrations were observed in specific fiber, on average 194–196 μg/g (Helal 2015).

6.2.2 Potassium

Potassium is the main cation (K+) of intracellular fluid (98% of total potassium). It regulates intracellular osmotic pressure and acid-base balance. Its metabolism is mainly regulated by its passive absorption in the gastrointestinal tract and its excretion mainly in the urine and secondly in the feces and skin. The body content of potassium depends on the balance between its intake (the main source of potassium is vegetables) and excretion. Plasma potassium concentration is mainly regulated by the kidney with high absorption in the proximal tube and excreted in the urine in the distal tube of the glomerulus. Potassium has an important role in cell membrane potential, transmission of nerve influx, acid-base balance, and neuromuscular excitability with direct effect on muscular activity and cardiac functions. Absorption and excretion of potassium are electrically balanced by sodium or hydrogen protons. Plasma potassium concentration is not a good marker of total potassium content in the body.

6.2.2.1 Usual Values

The normal values of serum potassium in the camel are expected to be 5.1 ± 0.4 mmol/l with a range of 3.6–6.0 (Bogin 2000). Bengoumi (1992) reported a range of 3.5–6.3 mmol/l (Table 6.2), i.e., higher values than the horse (2.4–4.7) but the same with cattle (3.9–5.8), sheep (3.9–5.4), and goat (3.5–6.1). With a range of 4.6–7.1 mmol/l, the llama has an average higher kalemia (Kaneko 1989). Hypokalemia is observed during potassium depletion due to gastrointestinal disorders (diarrheas) and hyperkalemia in renal failure or some metabolic disorders (cellular lesions, adrenal cortical insufficiency, dehydration).

6.2.2.2 Physiological Variations

Plasma potassium concentration is mainly stable and not affected by physiological factors. There is no effect of age on kalemia (Höller and Hassan 1966; Gupta et al. 1979; Salib and Soahir 1984; Chiericato et al. 1983; Rezakhani et al. 1997; Omer et al. 2010), but the differences according to parity are unclear (Ahmed et al. 2003) with minimum values in second and fourth parity (mean: 4.4 mmol/l) and the maximum at the first and third lactation (mean: 5.2 and 5.7 mmol/l, respectively). No difference was observed according to the milk yield (Ahmed et al. 2003), and no difference was observed between lactating and non-lactating camel, although changes were observed throughout the lactation (Hussein et al. 1992a), with a maximum at the ninth month of lactation (6.4 ± 0.6) and a minimum at the first stage of lactation (4.8 ± 0.4 mmol/l). For Tajik et al. (2015), kalemia increased with age: 5.5 ± 0.2 in less than 5-year-old camel, 5.8 ± 0.1 in 5–10-year-old camel, and 6.4 ± 0.2 mmol/l in oldest animals. Potassium level in 1-year-old camel calves increased steadily, from about 1.34 ± 0.15 mmol/l at birth to a peak of 9.33 ± 0.09 mEq/l at 7 months. Thereafter, its concentration decreased gradually (Hussein et al. 1992b).

A slight sex difference was reported by Mohammed et al. (2007), the male showing a value of 5.3 ± 0.5 compared to 4.8 ± 0.2 mmol/l in female. Similar results were given by Tajik et al. (2015): 5.9 ± 0.1 in males vs 5.5 ± 0.2 in females. At reverse no sexual difference was reported by Babeker and Suleem (2013).

Also, no difference was observed in male according to the rutting season (Zeidan and Abbas 2003) although a difference was noted by Abdel-Salaam et al. (2008) that kalemia is decreasing significantly in Egypt during the hot-humid season.

Potassium in the blood serum changed significantly according to the breeding season: from 5.26 ± 0.13 before breeding season (September–November) to 4.66 ± 0.22 at breeding season (December) and to 3.9 ± 0.25 mmol/l in January–February (Abdel-Salaam et al. 2011). Similar pattern was observed in infertile camels (Saini et al. 2009). Zia-Ur-Rahman et al. (2007) found also a lower concentration in serum potassium in rutting season (2.78 ± 0.3) than in non-rutting one

Table 6.2 Kalemia in camel blood serum according to different authors (all values are expressed in mmol/l)

References	Mean (mmol/l)	SD	nb	Country
Höller and Hassan (1966)	3.5	± 0.4	50	Sudan
Barakat and Fatah (1970)	4.7	± 0.01	200	Egypt
Little et al. (1970)	6.3	± 0.3	40	Australia
Ghosal et al. (1973a)	5.1	± 0.3	46	India
Abdelgadir et al. (1979)	5.3	± 0.4	96	Sudan
Orliac (1980)	4.9	± 0.4	102	Algeria
Al-Amrousi et al. (1984)	4.4	± 0.13	40	Saudi Arabia
Sellaouti (1984)	5.1	± 0.3	107	Tunisia
Abbas and Musa (1986)	6.3	± 1.2	36	Sudan
Abdalla et al. (1988)	4.6	± 0.43	20	UAE
Bizzeti et al. (1988)	5.1	± 1.3	44	Somalia
Elkasmi (1989)	5.4	± 0.8	60	Morocco
Rezakhani et al. (1997)	5.08	± 0.41	31	Iran
Sarwar and Majeed (1997)	5.41	± 0.11	56	Pakistan
Khadjeh (1998)	5.04	± 0.12	109	Iran
Mohamed and Hussein (1999)	3.0–4.7	–	100	Kuwait
Naeini and Nazifi (2001)	5.85	± 0.22	50	Iran
Ahmed et al. (2003) (young)	6.0	± 1.5	4	Sudan
Ismail et al. (2003)	4.67	± 0.63	9	USA
Liu (2003)	4.23	± 0.66	35	China[a]
Osman and Al-Busadah (2003)	4.0	± 0.2	5	Saudi Arabia
Kataria and Kataria (2004)	5.81	± 0.23	83	India
Sarwar et al. (2004)	5.41	± 0.01	56	Pakistan
Gutierrez et al. (2005)	4.82	± 0.62	16	Spain
Al-Busadah (2007)	4.3	± 0.61	60	Saudi Arabia
Mohammed et al. (2007)	5.03	± 0.42	11	Nigeria
Bengoumi (2008)	4.5	± 0.32	10	Morocco
Omer et al. (2008) (young camel)	9.7	± 0.2	12	Sudan
Al-Sobayil and Mousa (2009)	5.2	± 0.6	5	Saudi Arabia
Nazifi et al. (2009)	5.46	± 0.07	20	Iran
Saini et al. (2009)	4.61	± 0.22	3	India
Wernery et al. (2009)	4.2	± 0.4	736	UAE
Aichouni et al. (2010)	5.72	± 0.8	48	Algeria
Eltahir et al. (2010)	5.75	± 1.14	30	Oman
Shen and Li (2010)	3.97	± 0.76	15	China[a]
Hekmatimoghaddam et al. (2011)	6.25	± 0.11	92	Iran
Sazmand et al. (2011)	6.27	± 0.11	93	Iran
Babeker and Suleem (2013)	3.97	± 0.66	101	Sudan
Singh et al. (2015)	4.31	± 0.07	10	India
Tajik et al. (2015)	5.82	± 0.4	180	Iran
Zaher et al. (2017)	6.4	± 0.19	51	Egypt

[a]Bactrian camel

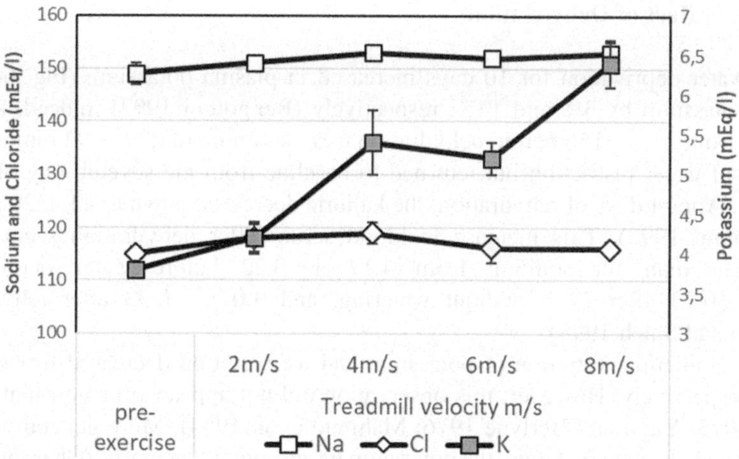

Fig. 6.2 Changes of electrolytes in serum blood of camel during exercise (calculated from Evans et al. 1992)

(4.19 ± 0.6 mmol/l). In Egypt, Zeidan et al. (2008) found significant higher values in female camel during breeding season than in non-breeding season but as for sodium with concentrations not corresponding to references of the literature, less than 1 mmol/l.

For Tajik et al. (2015) and Aichouni et al. (2013), kalemia was higher in summer (6.35 ± 0.12 and 5.79 ± 1.0, respectively) than in winter (5.29 ± 0.05 and 5.55 ± 1.3 mmol/l, respectively), while for Desalegn et al. (2012), it was higher during the dry season (5.6 ± 0.23) than during the wet season (4.79 ± 0.23 mmol/l). Nazifi et al. (1999) found also a slight higher value in summer (5.2 ± 0.02) than in winter (4.9 ± 0.02 mmol/l). For El-Harairy et al. (2010), the kalemia is quite higher in spring (6.6 ± 0.42) than in other seasons (4.43–4.93 mmol/l)[1] as shown in Fig. 1. For Babeker et al. (2011), the higher kalemia was observed in October (7.19 ± 2.60) and the lowest in July during the rainy season (2.04 ± 0.15 mmol/l).

Such seasonal difference was confirmed by Ghosal et al. (1973b) in India and Kuria et al. (2013) in Kenya. Kalemia would be higher in free-grazing camel than in indoor animals: 6.8 ± 0.31 vs 4.2 ± 0.1 mmol/l (Al-Shami 2009).

Contrary to natremia, exercises are linked to an increase of kalemia in racing camel serum (Evans et al. 1992). The serum concentrations were 3.8 ± 0.1 during the pre-exercise and 6.4 ± 0.3 mmol/l with a treadmill velocity of 8 m/s (Fig. 6.2).

[1]In the publication, the unit was in mg/100 ml, but the reported values corresponded to mmol/l.

6.2.2.3 Effect of Dehydration

Total water deprivation for 10 days increased in plasma potassium (Fig. 6.1) and urine potassium by 3% and 14%, respectively (Bengoumi 1992). Indeed, kaliuria passed from 245 ± 156 before dehydration to a maximum of 422 ± 97 mmol/l after 4 days of water restriction but returned to baseline from the seventh day of dehydration. After 4 days of rehydration, the kaliurie decreased down to 44 ± 29 mmol/l (Bengoumi 1992). This increase in blood serum after dehydration seems more important than for sodium: from 4.17 ± 0.22 before water deprivation, 7.60 ± 0.23 after 72 h without watering, and 4.07 ± 0.33 after rehydration (Ayoub and Saleh 1998).

The glomerular filtration of potassium and its excretion decreased by 66% and 57%, respectively. However, this observation did not appear to be constant (Yagil et al. 1975; Yagil and Berlyne 1976; Mahmud et al. 1984). More generally, in the dehydrated dromedary camel, the potassium tissue content increases (Charnot 1961), while the glomerular filtration and excretion of potassium decrease simultaneously. The metabolism of potassium depends on aldosterone, which stimulates its tubular excretion in exchange with reabsorbed sodium. However, in dehydrated dromedary where the action of aldosterone is weak, antidiuretic hormone regulates this metabolism (Charnot 1958; Yagil and Etzion 1979). After rehydration, plasma potassium concentration decreases, and its tubular excretion increases (Yagil and Berlyne 1976; Bengoumi et al. 1993; Riad et al. 1994).

6.2.2.4 Clinical Variations

Reverse to sodium, potassium in the blood increased significantly from 4.93 ± 0.05 to 5.9 ± 0.2 mmol/l in case of monensin toxicosis (Al-Jassim et al. 2016). In consequence, the Na/K ratio changes from 31.56 ± 0.36 in non-affected camel to 25.4 ± 0.9 mmol/l in intoxicated animals. No effect of trypanosomosis was reported by Sazmand et al. (2011) contrary to Baraka et al. (2000) who reported hyperkalemia in a camel affected by trypanosomosis (35.5 ± 4.13) compared to a healthy camel (25.75 ± 2.35 mmol/l). However, these values are out of the normal range for kalemia, and such results are doubtful. Surprisingly, Kamal (2008) found also a hyperkalemia in an affected camel, while it was the reverse for the other electrolytes: 6.15 ± 0.39 in a sick animal vs 4.58 ± 0.53 mmol/l in a healthy camel. In case of pasteurellosis, skin necrosis, frothy bloat, pasteurellosis, and indigestion, the potassium increased also significantly (Kamal 2008).

Potassium in the blood serum decreased significantly in the camel receiving injection (IM or IV) of dexamethasone (Wasfi et al. 1989).

6.2.2.5 Potassium in Milk and Other Fluids

Compared to other dairy species, camel milk had less potassium (Wang et al. 2011) but more than in human milk (Shamsia 2009). At reverse, Soliman (2005) found similar concentrations in camel milk (1563.2 ± 28.5 mg/l) than in other species, except in human milk. On average, Mostafidi et al. (2016) found 1357 mg/l of potassium in camel milk from different regions of Iran with a variability between 1076 and 1623 mg/l. Hamed et al. (2016) found 2295 mg/l on average.

An important difference was observed between early and late lactation regarding potassium in camel milk (Mal et al. 2007): respectively, 1983 ± 19.5 and 2809.6 ± 55.9 mg/l. According to Bengoumi et al. (1998), the mean potassium content in camel milk is 2110 ± 294 mg/l, and the range of potassium concentration in camel milk in literature is between 520 and 1800 mg/l (Yagil and Etzion 1980; Sawaya et al. 1984; Abu-Lehia 1987; Elamin and Wilcox 1992; Shamsia 2009; Alwan and Zwaik 2014) although some values above 2000 mg/l were reported (Farag and Kebary 1992; Bengoumi et al. 1998; Mal et al. 2007; Elhassan et al. 2016). Lower concentrations were reported by Al-Wabel (2008), but the value in camel milk (133.7 ± 5.6) was comparable to that of sheep milk (127.4 ± 1.10) and significantly higher than in cow (113.7 ± 5.8) and goat milk (123.8 ± 9.9 mg/l). In the Bactrian camel, the reported values were 1824.7 ± 66.3 in wild Bactrian and between 1640 ± 17.2 and 1958.4 ± 29.9 mg/l in domestic Bactrian camel (Jirimutu et al. 2010).

As for sodium, potassium concentration in peritoneal fluid is like in the blood serum: 5.21 ± 0.66 mmol/l (Naeini and Nazifi 2001). It was the same in cerebrospinal fluid (4.8 ± 0.1) compared to the serum (5.2 ± 0.1 mmol/l) (Nazifi and Maleki 1998). Ahmed et al. (2009) found, respectively, 3.53 ± 0.07 in CSF and 3.74 ± 0.16 mmol/l of potassium in the blood serum.

The sweat output of potassium was much higher than that of sodium, with a ration K/Na ranging from 2 to 4. Moreover the potassium concentration in sweat was quite more important than in blood serum, 68–128 times higher (El-Zeiny 2011).

The camel urine contained a high quantity of potassium, 124.25 ± 10.78 mmol/l, i.e., 4 times more than in cattle urine (Kamalu et al. 2003). However, these values do not correspond to what Bengoumi (1992) observed: 4.7 ± 0.6 mmol/l only.

In the digestive tract, potassium concentration varied between 37.3 ± 15.5 in the rumen and 42.7 ± 13.0 mmol/l in the last part of the colon (Maloiy and Clemens 1980). While sodium concentrations increased in the small intestinal areas and diminished from cecum to rectal regions, potassium concentrations showed an inverse relationship to sodium. For other authors, potassium in ruminal fluid was at the concentration of 10.73 ± 0.83 mmol/l (Kamalu et al. 2003) or 27.5 ± 2.3 mmol/l (Kamal 2008). According to the last author, potassium level in rumen fluid increased significantly in the camel affected by pasteurellosis (39.8 ± 2.3) and acidosis (38.7 ± 3.9 mmol/l).

6.2.2.6 Potassium in Meat and Wool

With a concentration of 1008 mg/100 g, camel meat appeared rich in potassium (Mahmud et al. 2011). This was confirmed formerly by Kadim et al. (2006) with a range of 471.4–1053 mg/100 g. Even with lower values (411 ± 29.89), Siham and Daoud found higher potassium than in beef (323.2 ± 12.44) and goat meat (310.2 ± 8.76 mg/100 g). Raiymbek et al. (2013) reported that the potassium concentration in *longissimus dorsi* of camel meat was on average 369 mg/100 g. In different age groups and for different muscles, potassium in meat varied nonsignificantly between 751 and 859 mg/100 g (Ibrahim et al. 2017).

The concentration of potassium in the testis and epididymis appeared very high: 2620–2840 and 2180–2290 μ/g, respectively, according to season (Zia-Ur-Rahman et al. 2007).

Dromedary camel hair contained high level of potassium: 2247 μg/g in raw hair, 878 μg/g in fine fibers, and 1571 μg/g in coarse fibers (Helal 2015).

6.2.3 *Chloride*

Chloride is a monovalent anion (Cl^-), mainly present in extracellular fluids: about 65% of the total anions of blood plasma and other extracellular fluids. However, chloride exists in all tissues, and its metabolism is passive with rapid and practically total absorption in the gastrointestinal tract and excretion in the urine, feces, and skin. Chloride maintenance requirements are estimated at 12 g/100 kg of body weight. Chlorides are not subject to special regulation: their metabolism follows that of sodium. It contributes to the electric balance of sodium, potassium, or hydrogen proton exchange and has an important role in acid-base balance. It also plays a specific role in the transport of oxygen and carbon dioxide in the blood and the maintenance of digestive juice pH.

Plasma chloride concentration does not provide significant information on the hydro-electrolytic balance. Variations of plasma chloride concentration occur in the same trends as the plasma sodium concentration and the opposite of that of bicarbonates.

The main source of choride is the diet, and in many places, camels benefit from salt cure. The camel is characterized by a high salt tolerance, and its requirements are high compared to other herbivorous. The units are expressed in mmol/l.

6.2.3.1 Usual Values

According to Bogin (2000), the reference value for chloride in the blood serum was 115 ± 7 with a range of 106–123 mmol/l (Table 6.3). Such concentrations are slightly higher than in other domestic species: 97–109 (horse), 97–111 (cattle),

Table 6.3 Chloridemia in camel blood serum according to different authors (all values are expressed in mmol/l)

References	Mean (mg/100 ml)	SD	nb	Country
Barakat and Fatah (1970)	101	± 1	200	Egypt
Ghosal et al. (1973a)	120	± 2	43	India
Abdelgadir et al. (1979)	107–115	–	96	Sudan
Orliac (1980)	111	± 4	102	Algeria
Snow et al. (1988)	116	± 4	9	UAE
Elkasmi (1989)	117	± 5	60	Morocco
Evans et al. (1992)	115	± 1.3	5	UAE
Sarwar and Majeed (1997)	628.01	± 8.08	56	Pakistan
Ismail et al. (2003)	384.2	± 4.97	9	USA
Kataria and Kataria (2004)	382.3	± 14.2	83	India
Sarwar et al. (2004)	175.80	± 2.02	56	Pakistan
Osman and Al-Busadah (2003)	130.2	± 1.9	5	Saudi Arabia
Gutierrez et al. (2005)	115	± 4.1	16	Spain
Mohammed et al. (2007)	368.9	± 7.26	11	Nigeria
Bengoumi (2008)	105	± 3	10	Morocco
Mohri et al. (2008)	116.6	± 6.9	11	Iran
Al-Sobayil and Mousa (2009)	419.1	± 8.5	5	Saudi Arabia
Wernery et al. (2009)	314.9	± 18.7	764	UAE
Singh et al. (2015)	118.38	± 0.37	28	India
Al-Jassim et al. (2016)	119.5	± 0.42	12	Australia
Shawaf (2017)	131.3	± 2.5	7	Saudi Arabia
Zaher et al. (2017)	115.97	± 0.59	51	Egypt

95–103 (sheep), 99–110 (goat), and 102–109 mmol/l (llama) (Djegham and Belhadj 1986; Kaneko 1989; Bengoumi 1992).

An increase on plasma chloride concentration is observed in case of dehydration, salt poisoning, and reduction in glomerular filtration, and it induces acidosis. Decrease in plasma chloride concentration occurs during chloride depletion due to gastrointestinal disorders (diarrheas), and it induces alkalosis.

6.2.3.2 Physiological Variations

Usually, there was no strong effect of age, gestation, and lactation, but a slight sexual difference was reported (Barakat and Fatah 1971; Gupta et al. 1979; Salib and Soahir 1984).

Chloride concentration increased in a pregnant camel (113.6 ± 1.12) compared to nonpregnant (106.26 ± 0.99 mmol/l) (Khadjeh 1998).

In rumen fluid, chloride concentration varied from 11.22 to 12.87 g/l according to the type of diet (Alhendi and Amer 1998), while it was 114.5 ± 1.49 mmol/l in cerebrospinal fluid (Shawaf 2017).

Chloride in camel blood serum did not change with the breeding season (Zia-Ur-Rahman et al. 2007): 201 ± 7.3 and 190.6 ± 6.3 mmol/l in rutting and non-rutting months, respectively (Zia-Ur-Rahman et al. 2007). There was no effect of seasonal thermal stress on chloride level in blood plasma (Nazifi et al. 1999), but during the dry season, the dehydrated camel could present higher chloridemia (Barakat and Fatah 1971; Ghosal et al. 1973a).

Contrary to calcium and phosphorus, Nagpal et al. (2011) found a difference in blood chloride according to the level of crude protein in diet, the lower value being observed in the low protein diet (75.66 ± 1.81) and the highest in the medium protein diet (87.45 ± 3.09 mmol/l), but with values quite lower than in other references of the literature. Similar comment could be done on the results of Baraka et al. (2000) who found very low values of chloride in the blood serum: 31.47 ± 0.32 mmol/l with no effect of any diseases. At reverse, Kamal (2008) found lower chloride in the blood serum of a camel affected by trypanosomosis (41.3 ± 8.75), pasteurellosis (47.3 ± 7.25), skin necrosis (41.24 ± 8.5), frothy bloat (42.5 ± 8.2), or indigestion (35.3 ± 7.2) than in healthy animals (65.7 ± 6.2 mmol/l).

6.2.3.3 Effect of Dehydration

Chloride increased slightly during dehydration (Bengoumi 1992) passing from 112 ± 2 before water restriction to a peak of 146 ± 4 mmol/l after 14 days and returned to baseline 24 h after rehydration (Fig. 6.1). In the same time, urine excretion of chloride increased from 121 ± 51 to 368 ± 163 mmol/l after 4 days water restriction but decreased rapidly up to 48 ± 12 mmol/l after 14 days dehydration and remained stable after rehydration (Bengoumi 1992).

6.2.3.4 Chloride in Other Substrates

Shamsia (2009) reported higher chloride concentration in camel milk (1320 ± 75) than in human milk (630 ± 65 mg/l).

The testis is very rich in chloride whatever the breeding season: 4680 ± 125 and 4590 ± 130 μg/g in rutting and non-rutting season, respectively. At reverse, a seasonal effect was observed in the chloride concentration of the epididymis: 5125 ± 115 in rutting season and 4752 ± 80 in non-rutting one (Zia-Ur-Rahman et al. 2007).

Chloride concentration in cerebrospinal fluid is comparable to that of blood serum: 101 ± 3.6 and 98.5 ± 2.6 mmol/l, respectively (Nazifi and Maleki 1998). Synovial fluid at reverse contained less chloride (117.7 ± 13.28) than the corresponding blood serum (154.7 ± 33.72 mmol/l) without significant difference between the type of articulation (Al-Rukibat and Ismail 2014). Rumen fluid contained 35.25 ± 2.51 mmol/l chloride. This concentration increased significantly above 40 mmol/l in a camel affected by indigestion, pasteurellosis, skin disease, and trypanosomosis (Kamal 2008). In the different parts of the digestive tract, chloride varied from 12 ± 2.1 in the rumen to 15.3 ± 3.7 mmol/l with a maximum of 71.7 ± 14.4 mmol/l in the first part of the small intestine (Maloiy and Clemens 1980).

6.2.4 *Bicarbonates*

Bicarbonates, also known as alkaline reserve, originate from the dissociation of carbonic acid (H_2CO_3) itself resulting from the condensation of a molecule of water (H_2O) and a molecule of carbon dioxide (CO_2). Under physiological conditions, bicarbonates filtered by the kidney are reabsorbed in totality and mainly in the proximal tube of the glomerulus. To maintain the acid-base balance, the kidney ensures the reabsorption of bicarbonates and excretes acids resulting from the metabolism. Reabsorption begins with the active transfer of H^+ protons from the cell to the tubular lumen in exchange with a sodium ion (Na^+). In the tubular lumen, H^+ ion reacts with HCO_3^- to elaborate H_2CO_3 which will then be dehydrated to CO_2 that diffuses into cells. This dehydration is catalyzed by carbonic anhydrase. In cells, CO_2 is rapidly rehydrated to H_2CO_3, which dissociates into H^+ and HCO_3^-. Bicarbonates pass into the peritubular plasma, and H^+ protons are eliminated in the urine. By its concentration, the HCO_3^- /H_2CO_3 couple is the most important in the body. According to the Henderson-Hasselbach equation, the blood pH depends on the ratio of bicarbonates and carbonic acid concentrations (Kaneko et al. 2008).

Plasma concentration of bicarbonates in the dromedary camel (22–30 mmol/l) is similar to that in other domestic ruminants (17–29 mmol/l) (Bengoumi 1992; Kaneko et al. 2008).

Age, sex, castration, milking, and pregnancy have no significant effect on plasma sodium concentration in camels (Bengoumi 1992). During acute dehydration (30% of body weight loss), plasma concentration of bicarbonates increases slightly but remains with normal value interval. In addition, an important increase of urine excretion of bicarbonates occurs during dehydration confirming the tendency to alkalosis (Yagil et al. 1975; Bengoumi 1992). Physical exercise induces a transitory significant decrease of bicarbonates concentration in camels indicating postexercise acidosis as reported in horses (Bengoumi et al. 2006).

6.2.5 *Calcium*

Calcium is the main mineral of the animals in terms of required quantities. It plays an essential role as electrolyte in muscular (muscles' contractions), nervous (nerve conduction), circulatory (blood clotting), and digestive systems and overall in the skeleton building (which contains 99%, of the calcium in the body). It may be complexed by proteins in many metabolic reactions and is present in the body under different salts. The only source of calcium is the diet, and its absorption (mainly in duodenum) by the animal depends on the solubility of calcium salt available in the diet, on the ratio Ca/P in the diet, on intestinal pH, and on vitamin D level. The regulation of calcium is hormone-dependent (calcitonin and parathyroid hormone) through feedback mechanisms and vitamin D-dependent (El Khasmi and Faye 2011). Calcium excess or deficiency could provoke different types of non-specific or specific disorders, especially affecting the bones. So, the dosage of calcium is recommended in case of bone, neuromuscular, cardiovascular, and kidney disorders.

6.2.5.1 Usual Values

The reference values of calcemia in camel varied between 8.4 and 12.4 mg/100 ml, i.e., 2.1–3.1 mmol/l (Bengoumi 1992). Such values are comparable to those reported in the cattle (9.6–12.4 mg/100 ml) and llama (8.8–10.4 mg/100 ml), slightly lower than the sheep (11.6–12.8 mg/100 ml) and horse (11.4–13.6 mg/100 ml) (Kaneko 1989). Calcemia in the literature is expressed in mg/100 ml or in mmol/l. In order to compare all results, data expressed in mmol/l are converted into mg/100 ml (Table 6.4).

6.2.5.2 Physiological Variations

Age and Parity Effect

Usually, no effect of age is recorded (Faye and Mulato 1991; Rezakhani et al. 1997; Saeed et al. 2004), but a higher calcemia was reported by Al-Busadah (2003) in a camel calf less than 3 months old (11.2 $\pm$ 0.2) compared to older calves (9.6 $\pm$ 0.4 to 9.8 $\pm$ 0.4 mg/100 ml up to 12 months old). Similar figure was recorded later by the same author (Al-Busadah 2010) and by Barri et al. (2005). During the first year of their life, camel caves showed calcemia attaining its highest concentration (12.88 $\pm$ 0.8 mg/100 ml) at the age of 3 months; thereafter, its value stabilized with an overall mean of 10.4 $\pm$ 0.32 mg/100 ml (Hussein et al. 1992b). An age effect was however revealed by Elias and Yagil (1984) with higher calcemia in young camels compared to adult in relationship with the calcium transfer to the milk. However, in older animals, the lack of age effect was confirmed (Biagi and Salutini 1983).

No clear effect of parities, the values varying from 8.6 $\pm$ 1.8 at the third gestation to 12.0 $\pm$ 2.3 mg/100 ml at the second and fourth gestation (Ahmed et al. 2003). The same authors found a higher calcemia in low yielding camel (16.9 $\pm$ 1.2) than in high yielding (12.6 $\pm$ 0.5 mg/100 ml) in relationship with the calcium exportation into milk.

Table 6.4 Calcemia in camel according to different authors (all values are expressed in mg/100 ml)

References	Mean (mg/100 ml)	SD	nb	Country
Bath and Kohli (1961)	10.8	± 0.1	50	India
Höller and Hassan (1966)	9.2	± 0.8	50	Sudan
Soliman and Shaker (1967)	10.8	± 1.2	80	Egypt
Barakat and Fatah (1970)	12.4	± 0.09	200	Egypt
Yagil et al. (1975)	10.84	± 0.52	5	Israël
Whabi et al. (1979)	9.2	± 0.99	96	Sudan
Orliac (1980)	10.0	± 1.1	102	Algeria
Musa and Mukhtar (1982)	19.5	± 8.1	15	Sudan
Abu Damir et al. (1983)	9.1	± 1.0	17	Sudan
Biagi (1983)	9.62	± 0.74	200	Somalia
Al-Amrousi et al. (1984)	11.68	± 2.5	25	Saudi Arabia
Abbas and Musa (1986)	8.8	± 2.8	36	Sudan
Abdalla et al. (1988)	9.5	± 0.6	20	UAE
Bizzeti et al. (1988)	8.4	± 0.8	44	Somalia
Snow et al. (1988)	9.6–10.4	–	9	UAE
Faye (1989)	9.5	± 0.58	52	Djibouti
Abu-Damir et al. (1990)	9.7	± 0.6	7	Sudan
Faye et al. (1992)[a]	8.5	± 0.53	352	Djibouti
Liu et al. (1994)	8.68	± 0.76	57	China
Faye et al. (1995)	10.2	± 6.5	82	France
Saeed et al. (1995)	10.4	± 0.4	6	Pakistan
Rezakhani et al. (1997)	10.36	± 1.34	31	Iran
Sarwar and Majeed (1997)	11.28	± 0.2	56	Pakistan
Ayoub and Saleh (1998)	11.3	± 1.4	3	UAE
Chaudhary and Iqbal (2000)	10.03	± 1.27	16	UAE
Naeini and Nazifi (2001)	8.64	± 1.04	50	Iran
Ahmed et al. (2003) (young)	12.6	± 4.1	4	Sudan
Ismail et al. (2003)	9.02	± 0.38	9	USA
Al-Sultan (2003) (female)	7.45	± 0.71	62	Saudi Arabia
Osman and Al-Busadah (2003)	9.0	± 0.1	5	Saudi Arabia
Kataria and Kataria (2004)	10.34	± 0.99	83	India
Saeed et al. (2004)	11	± 0.4	82	UAE
Sarwar et al. (2004)	11.28	± 0.02	56	Pakistan
Barri et al. (2005)	10.48	± 0.3	30	Saudi Arabia
Gutierrez et al. (2005)	9.2	± 0.8	16	Spain
Al-Busadah (2007)	10.48	± 2.1	60	Saudi Arabia
Mohammed et al. (2007)	9.56	± 0.2	11	Nigeria
Al-Sultan (2008)	10.4	± 1.2	50	Saudi Arabia
Mohri et al. (2008)	11.2	± 0.1	11	Iran
Al-Shami (2009)	10.2	± 0.49	105	Saudi Arabia
Al-Sobayil and Mousa (2009)	8.6	± 0.1	5	Saudi Arabia
Nazifi et al. (2009)	9.0	± 0.16	20	Iran

(continued)

Table 6.4 (continued)

References	Mean (mg/100 ml)	SD	nb	Country
Saini et al. (2009)	8.63	± 0.46	3	India
Shukla et al. (2009) (male)	6.47	± 1.25	16	India
Wernery et al. (2009)	10.68	± 1.92	62,390	UAE
Aichouni et al. (2010)	8.96	± 2.4	48	Algeria
Ali et al. (2010)	8.6	± 0.7	15	Saudi Arabia
Eltahir et al. (2010)	11.2	± 2.08	30	Oman
Patodkar et al. (2010)	9.67	± 0.34	16	India
Shen and Li (2010)	10.4	± 1.6	15	China[b]
Al-Mujalli et al. (2011)	10.7	± 0.18	20	Saudi Arabia
Hekmatimoghaddam et al. (2011)	10.8	± 1.2	92	Iran
Sazmand et al. (2011)	10.4	± 0.1	93	Iran
Vyas et al. (2011)	9.64	± 0.22	27	India
Hussein et al. (2012)	9.4	± 1.0	209	Saudi Arabia
Babeker and Suleem (2013)	11.40	± 0.52	101	Sudan
El Khasmi et al. (2013)	10.8	± 0.8	17	Morocco
Tajik et al. (2015)	10.2	± 0.08	180	Iran
Youssef et al. (2015)	8.58	± 0.19	9	Egypt
Ali et al. (2016)	8.11	± 0.15	6	Saudi Arabia
Badakhshan and Mirmahmoudi (2016)	11.16	± 0.25	18	Iran

[a]Young camels
[b]Bactrian camel

Sex and Pregnancy Effect

No sex effect was recorded both in dromedary (Höller and Hassan 1966; Barakat and Fatah 1971; Biagi and Salutini 1983; Chiericato et al. 1983; Faye and Mulato 1991; Al-Sultan 2003; Saeed et al. 2004; Al-Busadah 2010; Patodkar et al. 2010; Babeker and Suleem 2013; Tajik et al. 2015) and in Bactrian camel (Omidi et al. 2014). However, a slight but significant difference was reported for pregnancy status, calcemia decreasing in pregnant camel (9.30 ± 0.06 mg/100 ml) compared to nonpregnant (9.77 ± 0.11 mg/100 ml) probably linked to the calcium requirements of the fetus (Khadjeh 1998). Similar result was reported by Muhammad et al. (2011) in Nigeria: 8.68 vs 10.52 mg/100 ml in pregnant and nonpregnant, respectively. In UAE, Saeed et al. (2009) reported also higher calcemia in a nonpregnant camel compared to pregnant: 10.5 ± 1.01 and 9.31 ± 0.77 mg/100 ml, respectively. However, Vyas et al. (2011) did not observe significant difference linked to the pregnancy status.

Lactation Effect

The change along the lactation is significant but very slight with calcemia 10.24 ± 0.04 at the early lactation and 10.04 ± 0.05 mg/100 ml at late lactation (Singh et al. 2015). With more blood sampling every 3 months throughout the lactation, Hussein et al. (1992a) found lower calcemia at the first sampling (3.48 ± 1.2) compared to the second sampling (4.45 ± 0.4), the third (4.57 ± 0.4), and the fourth one at the end of lactation (5.29 ± 0.4 mg/100 ml), but the overall mean (4.24 ± 0.4 mg/100 ml), quite lower than the observed values in other references (Table 6.1), was not significantly different than in non-lactating camel (4.81 ± 0.4 mg/100 ml). For Elias and Yagil (1984), the sudden transfer of calcium to the udder at the beginning of lactation leads to a slight hypocalcemia restored only 2 weeks later.

Seasonal Effect

No difference was observed in male dromedary during rutting (10.37 ± 0.42) and non-rutting (9.18 ± 0.24) season by Zeidan and Abbas (2003) contrary to Abdel-Salaam et al. (2011) who found lower calcemia in a male camel during non-breeding (9.09 ± 0.26 to 9.12 ± 0.51 mg/100 ml according to the months) than in breeding season (10.61 ± 0.38 mg/100 ml). Similar figure was reported by Zia-Ur-Rahman et al. (2007): 11.6 ± 1.2 in rutting season vs 8.5.0 ± 1.4 mg/100 ml in non-rutting one. Zeidan et al. (2008) in Egypt reported hypercalcemia during the breeding season of a female camel (21.56 ± 1.24) that is almost double than in non-breeding season whatever the season: 14.31 ± 1.01 and 16.11 ± 1.02 mg/100 ml in humid and dry season, respectively.

In some cases, a seasonal variation was observed (Abdel-Salaam et al. 2008) with higher value in winter (10.4 ± 0.04) than in summer (9.17 ± 0.03 mg/100 ml). Similar observation was done by Aichouni et al. (2013), Tajik et al. (2015), and Bargaâ et al. (2016). For El-Harairy et al. (2010), calcemia is significantly higher in summer (11.3 ± 0.52) and lower in autumn (6.86 ± 0.87 mg/100 ml) (Fig. 6.3), but

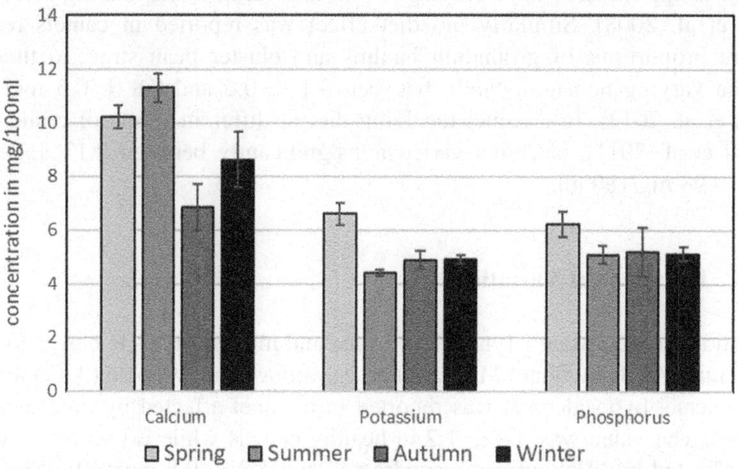

Fig. 6.3 Seasonal change of calcium, potassium and phosphorus in camel blood serum (calculated from after El-Harairy et al. 2010)

Nazifi et al. (1999) did not observe a significant effect of seasonal thermal stress, the calcemia being comparable in winter (10–10.4) and in summer (9.6 mg/100 ml on average). From their side, Babeker et al. (2011) found higher calcemia in dry hot summer, but the seasonal reported values (from 0.8 to 1.8 mg/100 ml) were out of the references.

At reverse, no significant difference was observed between wet and dry season in Ethiopia: 11.82 ± 0.62 and 9.58 ± 0.62 mg/100 ml, respectively (Desalegn et al. 2012). In Jordan, a slight difference was observed between blood serum collected in spring (6.85 ± 0.16) and summer (6.17 ± 0.11 mg/100 ml; Abdelrahman and Madanat 2014). In Kenya, the seasonal difference appeared quite important with calcemia at 12.8 ± 4.1 in dry season vs 3.8 ± 0.8 mg/100 ml only in wet season (Kuria et al. 2013). Amin et al. (2007) found higher calcemia at green season (8.8 ± 0.08) compared to dry one (8.12 ± 0.08 mg/100 ml) confirming the observations of Barakat and Fatah (1971). However, Bengoumi (1992) did not observe a significant change in calcemia during dehydration, the values varying between 9.2 and 10.4 mg/100 ml. Similar observations were done by Berlyne et al. (1978) and Mahmud et al. (1984).

Other Physiological Variations

No significant genetic variation (breed effect) was observed (Aichouni et al. 2010; Hussein et al. 2012).

A lower calcemia in camel after a long walking trip: 9.14 ± 0.06 vs 12 ± 0.19 mg/100 ml in camel (Abdel-Salaam et al. 2008). The camels submitted to heavy work presented also a lower calcemia (10.4 ± 0.04 mg/100 ml) compared to resting camels.

No effect of type of diet was observed on calcemia in growing camel calves: between 7.4 ± 0.2 and 7.5 ± 0.1 mg/100 ml according to the feeding composition (Omer et al. 2008). Similarly, no diet effect was reported in camels receiving different proportions of groundnut haulms and cluster bean straw in their feed, calcemia varying nonsignificantly between 8.1 ± 0.6 and 8.8 ± 0.5 mg/100 ml (Gupta et al. 2012). In a camel receiving diet at different levels of crude protein (Nagpal et al. 2011), calcemia varied nonsignificantly between 8.13 ± 0.84 and 9.63 ± 0.98 mg/100 ml.

6.2.5.3 Pathological Variations

No relationships between wryneck syndrome and mineral status is observed including calcium (Al-Sobayil and Mousa 2009) as well as with dystocia (Ali et al. 2016).

At reverse, hypocalcemia was reported in a camel affected by musculoskeletal disorders. The value was 9.8 ± 1.2 in healthy camels while it was 6.5 ± 0.9 in a camel affected by "Haboub-neck syndrome" and 5.8 ± 0.8 mg/100 ml in case of Bent-neck, two wryneck-like disorders reported in Sudan without determined specific causes (Mohamed 2004). Calcemia was recovered (around 10 mg/100 ml) after treatment including α-tocopherol and calcium borogluconate.

Calcemia was lower in a camel affected by reproductive disorders (Zaher et al. 2017): 10.65 ± 0.18 and 8.94 ± 0.06 mg/100 ml, respectively, in healthy and affected camels. It was reported by these authors that reduced serum calcium level might have an effect on the uterine motility and involution. No difference also was observed in camel submitting pre-slaughtering stress compared to unstressed animals (El Khasmi et al. 2011).

Compared to control camels (6.32 ± 0.6), calcemia increased highly in animals affected by acidosis (16.7 ± 1.7) and decreased drastically in case of lymphadenitis (3.72 ± 0.6 mg/100 ml) (Baraka et al. 2000).

The effect of trypanosomosis on camel calcemia is not clear (Karram et al. 1991). Many references are contradictory, probably because the status of the disease (clinical, subclinical, acute, chronic) could have a different impact, not only on calcemia but on all the electrolytes. Some authors considered that trypanosomosis did not change the calcium status of the camel (Chaudhary and Iqbal 2000; Sazmand et al. 2011) as well as theileriosis (Youssef et al. 2015). However, Baraka et al. (2000) found higher calcemia in a camel affected by trypanosomosis compared to a control camel, 12.5 ± 2.3 vs 6.32 ± 0.6 mg/100 ml. At reverse, Kamal (2008) found a quite lower calcemia (5.4 ± 0.54) in affected camels compared to healthy animals (11.6 ± 0.4 mg/100 ml). Hypocalcemia was also described by the same author in a camel affected by pasteurellosis, contagious skin disease, frothy bloat, and indigestion.

No variation due to goiter was observed (Abu-Damir et al. 1990).

A camel affected by jaw fracture, both males and females, had lower calcemia: 11.4 ± 1.16 compared to 17.4 ± 1.28 mg/100 ml in healthy animals (Al-Mujalli 2012). Similar observation was done by Kataria et al. (2013): 11.0 ± 0.2 in a healthy camel compared to 6.88 ± 0.12 mg/100 ml in a camel affected by fracture of the mandible.

In a camel exposed to monensin toxicosis, calcemia decreased significantly from 8.69 ± 0.08 to 7.25 ± 0.4 mg/100 ml (Al-Jassim et al. 2016).

6.2.5.4 Calcium in Milk

The concentrations reported in the literature are highly variable between authors, and the comparison is difficult to be done. This concentration is effectively dependent on the dehydration status of the animal (Yagil and Etzion 1980). On average in the literature, calcium concentrations in camel milk are between 30 and 257 mg/100 ml (Knoess 1976; Abu-Lehia 1987; Sawaya et al. 1984; Farah and Ruegg 1989; Elamin and Wilcox 1992; Hamed et al. 2016). According to Bengoumi et al. (1998), camel milk in a Moroccan camel contained 146.2 ± 24.8 mg/100 ml calcium, and Shamsia (2009) considered camel milk richer in calcium than human milk: 109 ± 7.50 vs 34 ± 3.50 mg/100 ml. The values are between 109 ± 4.5 and 120 ± 5.1 mg/100 ml in different camel breeds from Saudi Arabia (Mehaia et al. 1995). In Iran, the mean value in 25 camels was 94.4 mg/100 ml with an important regional variability, from 78.2 to 151.0 mg/100 ml (Mostafidi et al. 2016).

Significant lower values were reported in milk from the camel in an intensive farm (85.69 ± 1.30) compared to desert camel (114 ± 5.35 mg/100 ml) in Libya (Alwan and Zwaik 2014), while reverse was found in Sudan: 151.17 ± 3.3 vs 131.05 ± 3.9 mg/100 ml in intensive and desert system, respectively (Elhassan et al. 2016). In comparison to cow, goat, and sheep milk, Al-Wabel (2008) found a similar concentration of calcium: 69.9 ± 9.6 vs 66.1 ± 4.2 (cow), 75.1 ± 7.27 (goat), and 82.2 ± 11.3 mg/100 ml (sheep). From his side, Soliman (2005) reported a similar concentration in cow milk calcium (119.9 ± 0.69) than in cow milk (111.36 ± 4.36) but lower than in goat (130.28 ± 2.26) and buffalo (163.19 ± 4.56 mg/100 ml).

The Bactrian camel milk appeared richer in calcium than other dairy species (Wang et al. 2011), but with values quite lower than the other authors (60.75 ± 9.67 μg/l, i.e., 0.00607 mg/100 ml!); such results are doubtful. In Kazakhstan, on average, calcium in Bactrian and dromedary milk was 123.2 ± 29.2 mg/100 ml (Konuspayeva et al. 2008) with a significant difference between Bactrian (130.3 ± 28.7), dromedary (116.3 ± 27.3), and hybrid (125.7 ± 26.8 mg/100 ml; Faye et al. 2008). Cow, goat, and human milk had concentrations below 45 μg/l according to the same authors. In another reference (Jirimutu et al. 2010), including wild and domestic Bactrian camel from China, concentrations of 143.2 ± 4.7 mg/100 ml was recorded in wild Bactrian and from 122.9 ± 0.8 to 183.4 ± 1.87 mg/100 ml in different domestic Bactrian farms.

According to Mal et al. (2007), a higher milk calcium was observed at the late lactation (12th–13th month) than at early one (2nd–3rd month), respectively, 97.3 ± 0.51 and 94.06 ± 0.75 mg/100 ml. At reverse, El Khasmi et al. (2001) reported a higher calcium concentration in milk during the first week of milking: the concentration was 175 ± 16.4 mg/100 ml at the parturition and decreased to120.6 ± 16.6 mg/100 ml on the tenth day postpartum and did not change up to the end of the observation (day 30). Calcium excretion in milk is under vitamin induction: the injection of 1α-(OH)D_3 induced effectively a significant increase of milk calcium passing from around 110 to more than 140 mg/l (Riad et al. 1994).

6.2.5.5 Calcium in Other Substrates

In the urine of a normally hydrated camel, calciuria varied between 12.16 and 12.6 mg/l, while it increased up to 48.76 mg/l after 13 days rehydration (Bengoumi 1992). A value of 135 ± 46.9 mg/l was reported by Kamalu et al. (2003). In ruminal fluid, the same authors found a mean concentration of calcium in camel, ten times higher than in cattle rumen fluid, respectively, 45.7 ± 0.96 and 4.4 ± 9.0 mg/l. This concentration increased significantly from 71.3 ± 0.6 to 127 ± 9.2 mg/l in case of trypanosomosis and to 168.7 ± 6.4 mg/l in case of acidosis (Kamal 2008).

The calcium concentration in peritoneal fluid was lower than in serum: 3.68 ± 2.2 mg/100 ml (Naeini and Nazifi 2001). In cerebrospinal fluid (Nazifi and

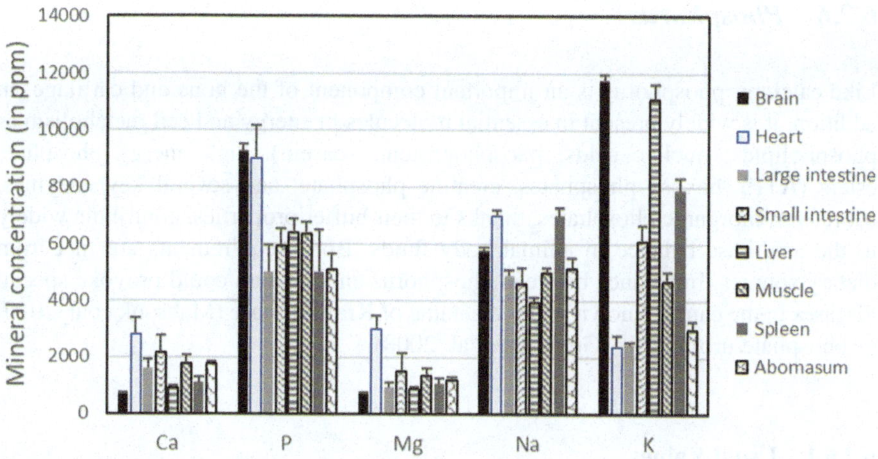

Fig. 6.4 Mineral concentrations in the main organs of the camel (calculated in ppm from Awad and Berschneider 1977)

Maleki 1998), calcium was in less proportion (2.8 ± 0.4) than in serum (10.4 ± 0.4 mg/100 ml). Similar observation was done by Ahmed et al. (2009): 5.65 ± 0.02 in CSF vs 8.12 ± 0.18 mg/100 ml in serum. In synovial fluid, calcium concentration was in lower concentration than in serum: respectively, 4.5 ± 7.72 and 8.4 ± 1.0 mg/100 ml (Al-Rukibat and Ismail 2014).

The determination of main minerals in camel organs was achieved by Awad and Berschneider (1977) in an Egyptian camel (Fig. 6.4).

Generally, the concentration of calcium in vital organs was higher in male than in female (Mustafa et al. 2012) and in a lower proportion than phosphorus.

In Bactrian camel, the calcium concentration in wool appeared high with value of 1989 ± 432 µg/g (Liu et al. 1994). The calcium concentration in wool was significantly variable according to the type of fiber, between 2961 and 4533 µg/g (Helal 2015).

The calcium concentration in camel meat varied from 12.56 ± 1.78 (Siham and Daoud 2015) to 27 mg/100 g (Mahmud et al. 2011), i.e., comparable value than in beef and goat meat. Similar results were reported by Kadim et al. (2006) with a range of 9.2–46.6 mg/100 g. With a mean of 5 mg/100 mg, the Bactrian camel meat should have less calcium than dromedary (Raiymbek et al. 2013). The calcium concentration in the muscle increased with the age, passing, for example, in *longissimus thoracis* from 13.3 to 24.3 mg/100 g, respectively, in young and adult camels (Ibrahim et al. 2017).

In the testis, the concentration of calcium was significantly higher in rutting season (980.0 ± 70.0 µg/g) than in non-rutting (770.0 ± 39.5 µg/g) contrary to the epididymis which contains similar values in both seasons, i.e., 895.0 ± 60.0 and 878.6 ± 30.6 µg/g, respectively (Zia-Ur-Rahman et al. 2007).

6.2.6 *Phosphorus*

Like calcium, phosphorus is an important component of the bone and cartilage. In addition, it is widely present in essential molecules in energy and cell metabolism as phospholipids, nucleic acids, phosphoproteins (casein), high energy phosphate esters (ATP), hexose phosphates, creatine phosphate, and several key enzymes. Moreover, inorganic phosphates, thanks to their buffer properties, contribute widely to the acid-base balance in animal body fluids. Body requirements are met from dietary sources. Imbalance between phosphorus and calcium could provoke special diseases in the camel, known under the name of Krafft disease (Mabrouk et al. 2010) or phosphate urolithiasis (Gutierrez et al. 2008).

6.2.6.1 Usual Values

In the review of Bogin (2000), phosphoremia was reported to be 5.2 ± 1.0 mg/100 ml with a range of 3.8–6.8 mg/100 ml. According to the data of the literature, phosphoremia is varying from 4.8 to 8.4 mg/100 ml (Table 6.5), i.e., 1.6–2.7 mmol/l (Bengoumi 1992). Similar values are reported in cattle (5.6–6.5 mg/100 ml), rather higher in sheep (4.8–10.8 mg/100 ml) and lower in horse (3.0–5.6 mg/100 ml). In llama, the range appeared wider: 3.0–10.8 mg/100 ml (Kaneko 1989). Phosphorus (as well as calcium) concentration in the blood serum is not affected by storage duration (Saeed et al. 1995).

6.2.6.2 Physiological Variations

Age and Parity Effect

The phosphoremia decreased slightly with the age passing from 7.16 ± 1.52 in camels before 3 years old to 6.85 ± 1.38 in 3–6-year-old camels and then 5.97 ± 1.09 mg/100 ml in adult camels (Rezakhani et al. 1997). Similar trend was recorded by Al-Busadah (2010), but the difference between camel calves (7.1 ± 0.56) and the adult females (6.21 ± 0.50) or male (6.14 ± 0.45 mg/100 ml) was not significant. In another reference (Omer et al. 2010), phosphoremia appeared slightly higher in suckling calves of less than 1 year old (4.0 ± 0.19) than in weaned calves (3.73 ± 0.42 mg/100 ml). At reverse, Saeed et al. (2004) and Tajik et al. (2015) did not reveal significant age or sex difference as well as Ahmed et al. (2003) who did not find parity effect or milk yield effect. Barri et al. (2005) as well as Al-Busadah (2003) did not reveal also significant differences all along the first year of life even if a trend to regular decrease occurred: from 6.7 ± 0.23 in camel calves 0–3 months old, 6.46 ± 0.6 in 3–6 months old, 6.26 ± 0.5 in 6–9 months old, and 5.58 ± 0.4 mg/100 ml in 9–12s months old. Elias and Yagil (1984) had already

Table 6.5 Phosphoremia in camel according to different authors (all values are expressed in mg/100 ml)

References	Mean (mg/100 ml)	SD	nb	Country
Bath and Kohli (1961)	6.9	± 0.9	30	India
Soliman and Shaker (1967)	6.3	± 0.3	80	Egypt
Barakat and Fatah (1971)	6.3	± 0.6	200	Egypt
Abdelgadir et al. (1979)	5.1	± 0.9	96	Sudan
Orliac (1980)	6.6	± 1.2	102	Algeria
Musa and Mukhtar (1982)	5.4	± 3.6	174	Sudan
Abu Damir et al. (1983)	5.1	± 0.7	17	Sudan
Biagi and Salutini (1983)	5.4	± 1.5	200	Somalia
Abbas and Musa (1986)	6.6	± 2.1	36	Sudan
Abdalla et al. (1988)	6.6	± 2.1	20	UAE
Al-Ali et al. (1988)	8.4	± 1.7	20	Saudi Arabia
Abu-Damir et al. (1990)	6.17	± 0.53	4	Sudan
Faye et al. (1992) (young)	7.9	± 1.04	352	Djibouti
Liu et al. (1994)	5.63	± 1.5	57	China[a]
Saeed et al. (1995)	6.2	± 0.3	6	Pakistan
Rezakhani et al. (1997)	5.97	± 1.09	31	Iran
Khadjeh (1998)	6.67	± 0.33	109	Iran
Nazifi and Maleki (1998)	5.88	± 0.3	21	Iran
Naeini and Nazifi (2001)	7.77	± 1.7	50	Iran
Ahmed et al. (2003) (young)	6.0	± 1.6	4	Sudan
Al-Sultan (2008) (young)	4.33	± 0.6	50	Saudi Arabia
Ismail et al. (2003)	5.51	± 1.13	9	USA
Osman and Al-Busadah (2003)	3.8	± 0.5	5	Saudi Arabia
Kataria and Kataria (2004)	4.98	± 0.57	83	India
Saeed et al. (2004)	6.6	± 1.0	82	UAE
Barri et al. (2005)	6.12	± 0.47	30	Saudi Arabia
Gutierrez et al. (2005)	5.02	± 0.7	16	Spain
Mohammed et al. (2007)	3.31	± 0.12	11	Nigeria
Mohri et al. (2008)	1.4	± 0.5	11	Iran
Al-Shami (2009)	3.7	± 0.2	105	Saudi Arabia
Al-Sobayil and Mousa (2009)	5.26	± 2.5	5	Saudi Arabia
Nazifi et al. (2009)	7.86	± 0.21	20	Iran
Saini et al. (2009)	4.38	± 0.22	3	India
Shukla et al. (2009) (male)	4.35	± 0.13	16	India
Wernery et al. (2009)	6.34	± 1.05	58,979	UAE
Ali et al. (2010)	6.7	± 0.3	15	Saudi Arabia
Chaudhary and Iqbal (2000)	4.39	± 0.61	16	UAE
Eltahir et al. (2010)	8.7	± 1.7	30	Oman
Patodkar et al. (2010)	5.64	± 0.57	16	India
Al-Mujalli et al. (2011)	7.66	± 0.6	20	Saudi Arabia
Hekmatimoghaddam et al. (2011)	6.1	± 1.5	92	Iran
Sazmand et al. (2011)	6.07	± 1.5	93	Iran

(continued)

Table 6.5 (continued)

References	Mean (mg/100 ml)	SD	nb	Country
Vyas et al. (2011)	3.13	± 0.24	27	India
El Khasmi et al. (2013)	4.64	± 0.92	17	Morocco
Abderahman and Madanat (2014)	3.33	± 0.05	25	Jordan
Ali et al. (2015)	3.96	± 0.7	10	Saudi Arabia
Singh et al. (2015)	4.75	± 0.43	10	India
Tajik et al. (2015)	9.1	± 2.4	180	Iran
Youssef et al. (2015)	5.11	± 0.22	9	Egypt
Ali et al. (2016)	7.11	± 0.59	6	Saudi Arabia
Badakhshan and Mirmahmoudi (2016)	9.81	± 0.27	18	Iran

[a]Bactrian camel

mentioned a higher phosphatemia in newborn camel calves in relation with a better intestinal absorption thanks to vitamin D_3 in higher quantity after parturition.

Sex and Pregnancy Effect

No significant effect of sex was reported. Phosphoremia would be higher in a nonpregnant camel (5.68 ± 1.24) compared to pregnant (4.61 ± 1.29 mg/100 ml) as for calcium (Saeed et al. 2009). However, Omidi et al. (2014) in Iran did not observe significant difference between pregnant and nonpregnant camel despite a quite lower value in the first one (4.58 ± 0.61) than in the second (7.08 ± 0.66 mg/100 ml).[2] Similar observation was reported in India (Vyas et al. 2011). During pregnancy, a hypophosphoremia was described at the end of gestation in relation with the development of fetal skeleton (Eltohamy et al. 1986).

Seasonal Effect

After breeding season, infertile she-camels had higher inorganic phosphorus (4.66 ± 0.22) than fertile camels (3.84 ± 0.22 mg/100 ml) (Saini et al. 2009). A higher phosphoremia (Zeidan and Abbas 2003) was observed in male during rutting season (5.26 ± 0.25) than in non-rutting time (3.15 ± 0.32 mg/100 ml). The higher phosphoremia in rutting season was also reported by Abdel-Salaam et al. (2011): 8.6 ± 05 vs 7.39 ± 0.54 to 7.52 ± 0.55 mg/100 ml according to the months in non-rutting season. In female, significant higher phosphoremia was also reported at

[2]The results are given by the authors in mmol/l, but the values as well as for calcium are corresponding to mg/100 ml.

breeding season (Zeidan et al. 2008), 7.48 ± 0.55 mg/100 ml, while it was 4.35 ± 0.65 and 5.59 ± 0.74 in non-breeding time, respectively, in humid and dry season.

Contrary to Desalegn et al. (2012), a seasonal difference is related by Tajik et al. (2015): 10 ± 0.3 vs 8.34 ± 0.8 mg/100 ml in summer and winter, respectively. A reverse observation was done by Aichouni et al. (2013) in Algeria with higher phosphoremia in winter (6.9 ± 0.4) than in summer (6.0 ± 0.7 mg/100 ml). For El-Harairy et al. (2010), the maximum phosphoremia was observed in spring (6.23 ± 0.48), and no difference was recorded in the other seasons, between 5.10 and 5.21 mg/100 ml on average (Fig. 6.1). At reverse, with almost constant value at around 5.5 mg/100 ml, no seasonal thermal stress was reported by Nazifi et al. 1999. Bargaâ et al. in 2016 in Morocco found a very slight difference between winter (7.43 ± 0.7) and summer (7.12 ± 0.7 mg/100 ml).

The seasonal variation observed in Kenya by Kuria et al. (2013) was quite important with phosphoremia 20 times higher in wet season (5.96 ± 2.7) than in dry season (0.27 ± 0.08 mg/100 ml), while Amin et al. (2007) in Sudan found 7.27 ± 0.09 and 6.0 ± 0.09 in green and dry season, respectively. Barakat and Fatah (1971) observed at reverse an increase of phosphoremia during dry season, contrary to calcemia. Indeed, Bengoumi (1992) observed a decrease of urinary flow during the dry season provoking a low excretion of the phosphorus.

Other Physiological Effect

A lower phosphoremia was observed in camel after a long trip of walking (5.3 ± 0.03) compared to resting camel (5.8 ± 0.08 mg/100 ml) (Abdel-Salaam et al. 2008). The type of diet seems to have no effect of phosphoremia of growing camel calves, the concentrations being between 3.7 ± 0.2 and 3.8 ± 0.2 mg/100 ml according to four different feeding components (Omer et al. 2008). Gupta et al. (2012), in their comparative trial assessing the effect of different proportions of groundnut haulms and cluster bean straw in the camel diet, did not find also significant effect, phosphoremia varying between 8.1 ± 1.4 and 9.9 ± 1.1 mg/100 ml. Similar observation was done by Nagpal et al. (2011). However, in a growing camel, a significant effect was observed after supplementation with concentrates on phosphoremia: 7.5–7.6 mg/100 ml in supplemented camels compared to 6.5–6.6 mg/100 ml in non-supplemented (Faye et al. 1992).

No breed effect occurred (Aichouni et al. 2010) although camel with black color coat (Majaheem breed) from Saudi Arabia presented a slight but significant phosphoremia (3.3 ± 0.2) than brown coat (Homor breed, 3.21 ± 0.1) and white coat (Waddha breed, 3.0 ± 0.1 mg/100 ml) (Hussein et al. 2012).

The concentration of phosphorus changed significantly after dehydration, increasing from 5.7 to 12.6 mg/100 ml after 2 weeks of water deprivation. This concentration returned to normal level 4 days after rehydration (Bengoumi 1992).

6.2.6.3 Pathological Variations

In dehydrated camels, phosphoremia decreased dramatically, passing from 4.9 ± 0.6 to 1.73 ± 0.32 mg/100 ml whatever the sex of the camel (Kataria and Kataria 2004). In sodium thiocyanate-induced hypothyroidism in camels, serum phosphorus decreased from 6.14 ± 1.03 after 1 month of treatment to 4.60 ± 0.77 mg/100 ml after 3 months (Barsham et al. 2005).

As for calcemia, a significant lower phosphoremia occurred in a camel affected by low reproductive performance (4.15 ± 0.36) compared to 5.41 ± 0.17 mg/100 ml in control animals (Zaher et al. 2017). Such disturbances in phosphorous level could be associated with reduced appetite, decreased body gain, altered estrus, decreased ovarian activity, and impaired reproduction. However, in male affected by impotentia generandi, no variability in phosphorus status was observed (Ali et al. 2015).

Serum phosphorus concentration of the camel seems to be affected by skin necrosis (hypophosphoremia: 3.4 ± 0.18) and lymphadenitis (hyperphosphoremia: 6.72 ± 0.4) compared to a healthy camel: 3.96 ± 0.37 mg/100 ml (Baraka et al. 2000).

In a camel intoxicated with monensin toxicosis (Al-Jassim et al. 2016), phosphoremia was significantly higher (8.26 ± 0.6) compared to control animals (5.73 ± 0.09 mg/100 ml). Phosphoremia decreased in case of intramuscular or intravenous injection of dexamethasone (Wasfi et al. 1989). There was apparently no impact of trypanosomosis on plasma phosphorus in camels (Chaudhary and Iqbal 2000; Sazmand et al. 2011) as well as theileriosis (Youssef et al. 2015). However, hyperphosphoremia was described in case of trypanosomosis by Baraka et al. (2000): 6.25 ± 0.6 mg/100 ml. At reverse, hypophosphoremia was described by Kamal (2008): 4.61 ± 0.77 in comparison to 7.75 ± 0.8 mg/100 ml in a healthy camel. Low phosphorus in the blood serum was also reported by the same author in case of skin necrosis and digestive disorders and especially in case of pasteurellosis (3.28 ± 0.65 mg/100 ml).

Slightly lower phosphoremia was observed in a camel affected by jaw fracture: 2.84 ± 0.52 vs 3.4 ± 0.52 in a normal camel (Al-Mujalli 2012). This difference was more marked in another study: 3.99 ± 0.09 and 6.96 ± 0.12 mg/100 ml in healthy and affected camel by mandible fracture, respectively (Kataria et al. 2013).

The presence of goiter did not change significantly the phosphoremia (Abu-Damir et al. 1990). The stress induced by long-distance transportation has no effect on plasma phosphorus concentration in camels (El Khasmi et al. 2011).

6.2.6.4 Phosphorus in Milk and Other Substrates

On average, camel milk contained 78.4 ± 10.3 mg/100 ml (Bengoumi et al. 1998), the range recorded in the literature being 34–100 mg/100 ml (Ahmed et al. 1977; Yagil and Etzion 1980; Abu-Lehia 1987; Farah and Ruegg 1989; Elamin and Wilcox 1992). In Kazakhstan, Konuspayeva et al. (2008) found a high mean level of phosphorus in camel milk including Bactrian and dromedary: 100.3 ± 21.7 mg/100 ml. However, Bactrian milk was richer (107.5 ± 17.7) than dromedary (91.5 ± 19.0), the hybrid milk being intermediary (106.7 ± 27.3 mg/100 ml; Faye et al. 2008).

Camel milk contained almost five times more phosphorus than human milk: 76 ± 2.55 vs 16 ± 0.85 mg/100 ml (Shamsia 2009). In a comparative study (Soliman 2005), camel milk appeared less rich in phosphorus (81.17 ± 3.08) than buffalo (111.36 ± 2.61), cow (95.03 ± 0.72), and goat (110.16 ± 1.61 mg/100 ml).

Reverse to calcium, Alwan and Zwaik (2014) found higher phosphorus in milk from intensive farm camel (89.02 ± 6.68) than in desert camel (65.15 ± 2.87 mg/100 ml) in accordance with Elhassan et al. (2016), respectively, 95.3 ± 6.5 and 79.86 ± 7.3 mg/100 ml, although these values are depending on the region.

As for calcium, phosphorus in milk was lower in early lactation (41.68 ± 0.55) than in late lactation (47.1 ± 0.5 mg/100 ml) (Mal et al. 2007). However, the phosphorus concentration in colostrum at the day of parturition was higher (110.6 ± 13.6 mg/100 ml), then decreased up to the tenth day of postpartum (82.1 ± 10.4 mg/100 ml), and then was stable at least for the first month of lactation (El Khasmi et al. 2001). As for calcium also, phosphorus increased significantly in milk after injection of vitamin D, passing approximately from 80 to 90 mg/100 ml (Riad et al. 1994).

Phosphorus in the urine was not strongly influenced by the hydration status of the camel, and the values were lowest, between 0.6 and 0.9 mg/l (Bengoumi 1992). At reverse, high phosphorus concentration in camel urine was reported by Kamalu et al. (2003): 350 ± 23 mg/l. In ruminal fluid, the same authors found 249.9 ± 12.0 mg/l, while Kamal (2008) reported lower concentration (41.8 ± 3.7 mg/l) even in case of clinical trypanosomosis (85.1 ± 6.5 mg/l).

In other substrates, reported values were 5.97 ± 1.9 mg/100 ml (peritoneal fluid), 3.71 ± 0.3 mg/100 ml (cerebrospinal fluid), and 5 ± 5.1 mg/100 l (synovial fluid) (Naeini and Nazifi 2001; Nazifi and Maleki 1998; Al-Rukibat and Ismail 2014, respectively). In Bactrian camel hair, phosphorus was assessed to 127 ± 24.6 μg/g (Liu et al. 1994). Camel meat contains 249.9–584 mg/100 g phosphorus (Kadim et al. 2006), 549 mg/100 g (Mahmud et al. 2011), 229.0 ± 67.0 mg/100 g (Raiymbek et al. 2013), and 176.0 ± 4.30 mg/100 g (Siham and Daoud 2015). The concentration increased with the age, for example, 352–412 mg/100 g, in *longissimus thoracis* with similar pattern in the other muscles (Ibrahim et al. 2017).

Phosphorus concentrations in vital organs (liver, kidney, and spleen) are much higher than calcium and magnesium (range from 30 to 80 mg/g), and males had higher values than females (Mustafa et al. 2012).

6.2.7 Magnesium

Magnesium is also a component of the bone and cartilage contributing to the structure of skeleton. It is an activator of several key enzyme systems, including kinases, mutases (transphosphorylation reactions), muscle ATPases, and the enzymes cholinesterase, alkaline phosphatase, enolase, isocitric dehydrogenase, deoxyribonuclease, and glutaminase.

Through its role in enzyme activation, magnesium (like calcium) stimulates muscle and nerve irritability (contraction) and is involved in the regulation of intracellular acid-base balance. Globally, its role in the carbohydrate, protein, and lipid metabolism is essential.

As with calcium and phosphorus, a proportion of the magnesium contained in plant foodstuffs may be present in the form of phytin (Ca or Mg salt of phytic acid). The main source of magnesium for camel is the diet. The differences in availability of magnesium in the natural grasses could explain the regional variability in camel magnesemia (Faye and Mulato 1991).

6.2.7.1 Usual Values

According to Bogin (2000), the normal value for magnesemia in camel is 2.4 ± 0.3 mg/100 ml with a range of 1.8–2.8 mg/100 ml (Table 6.6).

6.2.7.2 Physiological Variations

Serum magnesium concentration was not significantly different according to the sex of the animal: 1.72 ± 0.45 in female vs 1.3 ± 0.05 mg/100 ml in male camel (Al-Sultan 2003). Similar observation was done formerly by Faye and Mulato (1991), Al-Busadah (2010), and Tajik et al. (2015). Magnesemia seems to be higher in camels less than 3 months (4.5 ± 0.6) than in oldest ones (2.6–3.04 mg/100 ml) (Al-Busadah 2003). However, Omer et al. (2010) found lower magnesemia in suckling camel calves (1.76 ± 0.19) than weaned ones, more than 1 year (2.25 ± 0.33 mg/100 ml), and Al-Busadah (2010) reported no significant difference in magnesemia between camel calves (2.1 ± 0.14) and their mother (1.6 ± 0.12 mg/100 ml). For Hussein et al. (1992a), magnesemia was fluctuating slightly with age at the first year of life, giving an overall mean value of 2.86 ± 0.7 mg/100 ml.

An important seasonal variation was reported by Desalegn et al. (2012) with double value in wet season (2.77 ± 0.13) than in dry one (1.37 ± 0.13 mg/100 ml). This could be in relationship with the higher availability of magnesium in green fodder compared to dry one. There was no effect of rutting season on magnesemia in male (3.0 ± 0.6 and 2.1 ± 0.5 mg/100 ml in breeding and non-breeding season, respectively; Zia-Ur-Rahman et al. 2007).

Magnesemia was lower on early lactation than in late (Mal et al. 2007), 1.82 ± 0.02 vs 1.35 ± 0.03 mg/100 ml, respectively, but no difference was observed according to parity or milk yield (Ahmed et al. 2003). Hussein et al. (1992a) did not observe a significant change in serum magnesium throughout the lactation (overall mean: 2.55 ± 0.2) and no difference with non-lactating animals (2.43 ± 0.2 mg/100 ml). In contrary to Vyas et al. (2011), Al-Busadah (2010) found higher magnesemia in a pregnant camel (3.6 ± 0.40) than in lactating one

Table 6.6 Magnesemia in camel according to different authors (all values are expressed in mg/100 ml)

References	Mean (mg/100 ml)	SD or range	nb	Country
Soliman and Shaker (1967)	1.9	± 0.32	80	Egypt
Ghosal et al. (1973a)	2.0	1.4–2.6	20	India
Yagil et al. (1975)	2.53	± 0.27	5	Israël
Whabi et al. (1979)	2.5	± 0.28	96	Sudan
Abu Damir et al. (1983)	2.69	± 0.49	17	Sudan
Biagi (1983)	2.6	± 0.47	200	Somalia
Abdalla et al. (1988)	2.5	± 0.35	20	UAE
Faye (1989)	2.27	± 0.62	52	Djibouti
Faye et al. (1992)	2.1	± 0.3	352	Djibouti
Faye et al. (1995)	2.6	± 0.3	82	France
Khadjeh (1998)	–	1.8–2.3	109	Iran
Chaudhary and Iqbal (2000)	1.89	± 0.29	16	UAE
Ahmed et al. (2003) (young)	3.4	± 0.7	4	Sudan
Al-Sultan (2003) (female)	1.72	± 0.45	51	Saudi Arabia
Ismail et al. (2003)	3.12	± 0.26	9	USA
Osman and Al-Busadah (2003)	5.24	± 0.28	5	Saudi Arabia
Barri et al. (2005)	3.54	± 0.59	30	Saudi Arabia
Mohri et al. (2008)	2.18	± 0.2	11	Iran
Saini et al. (2009)	3.91	± 0.2	3	India
Wernery et al. (2009)	2.52	± 1.25	613	UAE
Ali et al. (2010)	2.5	± 0.04	15	Saudi Arabia
Eltahir et al. (2010)	3.32	± 0.6	30	Oman
Shen and Li (2010)	2.35	± 0.6	15	China[a]
Hekmatimoghaddam et al. (2011)	2.6	± 0.04	92	Iran
Sazmand et al. (2011)	2.62	± 0.05	93	Iran
Vyas et al. (2011)	3.45	± 0.05	27	India
Aichouni et al. (2013)	2.57	± 0.24	40	Algeria
Abderahman and Madanat (2014)	1.79	± 0.03	25	Jordan
Ali et al. (2015)	2.18	± 0.24	10	Saudi Arabia
Singh et al. (2015)	2.09	± 0.21	10	India
Tajik et al. (2015)	4.98	± 0.4	180	Iran
Ali et al. (2016)	2.33	± 0.2	6	Saudi Arabia

[a]Bactrian camel

(1.6 ± 0.12 mg/100 ml). Magnesemia was reported to be higher in free-grazing camels compared to indoor camels (Al-Shami 2009).

An important significant seasonal effect was described by Kuria et al. (2013) in India with a value of 15.2 ± 0.5 in dry season vs 1.2 ± 0.5 mg/100 ml which is surprising although an important difference was also reported in Morocco: 3.15 ± 0.48 in summer vs 1.94 ± 0.24 in winter (Bargaâ et al. 2016). From their side, Nazifi et al. (1999) reported a slight lower value in winter (2.18 ± 0.07 on average) than in summer (2.43 ± 0.2 mg/100 ml). Apparently, magnesium was the most sensitive mineral to seasonal variation.

Heavy work could increase magnesium concentration (4.1 ± 0.06 vs 3.43 ± 0.01 mg/100 ml in resting animals), but not after a long trip of walking: 3.23 ± 0.02 mg/100 ml (Abdel-Salaam et al. 2008). However, reversely, for Omer et al. (2008), magnesemia in growing camel calves was not influenced by the type of diet. No breed effect was reported in Saudi Arabia (Hussein et al. 2012), but at reverse, a breed difference was mentioned in the same country with higher magnesemia in Maghateer camel (white-coat camel) compared to Majaheem breed (black-coat camel), respectively, 3.77 ± 0.2 and 2.67 ± 0.17 mg/100 ml (Abdelrahman et al. 2013).

6.2.7.3 Clinical Variations

Under stress transportation, the dromedary camels showed a significant decrease of magnesium (1.21 ± 0.24) by comparison with values measured in the same animals before transportation (2.18 ± 0.23) (El Khasmi et al. 2013).

Monensin toxicosis was associated to a slight increase of serum magnesium from 2.09 ± 0.02 to 2.43 ± 0.07 mg/100 ml (Al-Jassim et al. 2016). There was no significant change in magnesemia in case of trypanosomosis (Chaudhary and Iqbal 2000; Sazmand et al. 2011).

6.2.7.4 Magnesium in Milk

The magnesium concentration in milk appeared slightly higher in colostrum (23.6 ± 3.1 mg/100 ml) than in milk, the concentration being stable after 10 days postpartum at 11.2 ± 2.2 mg/100 ml (El Khasmi et al. 2001). This concentration is comparable to that in the other dairy species but higher than in human milk (Wang et al. 2011), while for other authors, magnesium concentration in camel milk (6.7 ± 1.4 mg/100 ml) was lower than other species, except in human milk (Soliman 2005): for example, 29.56 ± 0.79 in buffalo milk, 13.42 ± 0.24 in cow milk, and 13.87 ± 0.11 mg/100 ml in goat milk. Shamsia (2009) found 14 ± 0.7 mg/100 ml in camel milk and 3.0 ± 0.35 mg/100 ml only in human milk. In Morocco, Bengoumi et al. (1998) found a mean value of magnesium at 10.8 ± 1.3 mg/100 ml and Mehaia et al. (1995) between 13.0 ± 1.1 and 11.6 ± 1.6 according to camel breed in Saudi Arabia. Similar values were recorded by Khan and Appanna (1964), Yagil and Etzion (1980), Sawaya et al. (1984), Abdel-Rahim (1987), Abu-Lehia (1987), Hassan et al. (1987), Farah and Ruegg (1989), Alwan and Zwaik (2014), and Hamed et al. (2016) in a range of 8–18 mg/100 ml, although lower values were recorded by Elamin and Wilcox (1992): 4.5 ± 0.9 mg/100 ml. However, higher concentration more than 20 mg/100 ml was sometimes recorded (Ahmed et al. 1977). Wild Bactrian camel contained more magnesium in its milk (11.13 ± 0.26) than in domestic ones (from 6.3 ± 0.1 to 9.25 ± 0.2 mg/100 ml; Jirimutu et al. 2010) probably in relation with their specific diet in the desert.

6.2.7.5 Magnesium in Other Substrates

Few data are regarding magnesium in tissues: 220 ± 60 μg/g in the liver and 200 ± 70 μg/g in the kidney (Abu Damir et al. 1983). From their part, Abdelrahman et al. (2013) reported a significant higher concentration in the liver of Majaheem camel (807.8 ± 47.4) than in Maghateer one (702.6 ± 47.4 μg/g), i.e., reverse to the serum where magnesemia was higher in Maghateer. The same authors did not find difference for the kidney, the magnesium concentrations being 478.6 ± 36.7 in Majaheem and 502.4 ± 36.7 ppm in Maghateer. The sex had an effect on magnesium concentration in the liver, kidney, and spleen with higher values in male compared to female (Mustafa et al. 2012).

In meat, the concentration of magnesium was 56 mg/100 g for camel (Mahmud et al. 2011) with a significant higher value in Majaheem breed (62.46) than in Maghateer (45.63 mg/100 g) (Abdelrahman et al. 2013). A range of 24.7–57.3 mg/100 g was reported by Kadim et al. (2006). Higher values were recorded in the Bactrian camel (251 mg/100 g on average) by Raiymbek et al. (2013). With a concentration of 90.16 ± 5.03, Siham and Daoud (2015) found more magnesium than in beef (37.6 ± 11.01) or goat meat (27.31 ± 4.57 mg/100 g). The magnesium did not vary significantly (from 35.6 to 44.4 mg/100 g) with the muscle and the age (Ibrahim et al. 2017).

The concentration in the testis is influenced by the rutting status but reverse to calcium, 70.0 ± 6.0 in rutting season vs 79.6 ± 5.8 μg/g in non-rutting, while no significant difference occurs in the epididymis, 98.0 ± 8.0 and 86.0 ± 6.0 μg/g, respectively (Zia-Ur-Rahman et al. 2007).

In cerebrospinal fluid, Nazifi and Maleki (1998) reported a lower concentration in magnesium (1.94 ± 0.1) than in the corresponding blood serum (3.64 ± 0.2 mg/100 ml). At reverse, Ahmed et al. (2009) found significantly more magnesium in CSF (2.28 ± 0.02) than in the corresponding blood serum (1.56 ± 0.14 mg/100 ml). Synovial fluid contained less magnesium than the blood serum, respectively, 1.4 ± 0.47 and 2.7 ± 0.63 mg/100 ml (Al-Rukibat and Ismail 2014). The camel urine contained more magnesium than cattle urine, respectively, 93.3 ± 9.0 and 40.6 ± 5.6 mg/l (Kamalu et al. 2003). In rumen fluid, a similar difference was observed: 10.9 ± 2.3 in camel vs 5.3 ± 8.9 mg/l in cattle (Kamalu et al. 2003).

6.3 Conclusion

The mineral parameters are good indicators, in general, both for nutritional and clinical point of view. Their regulation (through hormones notably) is complex, and their concentrations in biological fluid are dependent partly to the intake. Globally, the mineral metabolism in a camel is comparable to other species. However, the changes in electrolytic balance during dehydration process and the role of the kidney in this matter are under physiological mechanisms briefly described in this chapter which contribute to understand the adaptative advantage of the camel in desert areas.

References

Abbas B, Musa BE (1986) Analyses of blood sera and common browse plants from the Butana. ILCA, Group document SRC 12: 442–445

Abdalla OM, Wasfi IA, Gadir FA (1988) The Arabian race camel normal parameters. I. Haemogram, enzymes and minerals. Comp Biochem Physiol 90A:237–239

Abdelgadir SE, Wahbi AA, Idriss OF (1979) Biochemical studies of some blood and plasma constituents in camel (*Camelus dromedarius*). In: Cockrill WR (ed) Camel and camelids. Scandinavian Institute of African Studies, Uppsala, Sweden, pp 438–443

Abdel-Rahim AG (1987) The chemical composition and nutritional value of camel (*Camelus dromedarius*) and goat (*Capra hircus*) milk. World Rev Anim Prod 90A(2):237–239

Abdelrahman MM, Aljumaah RS, Ayadi M (2013) Variation of copper, zinc, manganese and magnesium in blood serum and tissues of two breeds of dromedary camels in Saudi Arabia. Asian J Anim Vet Adv 8(1):91–99

Abdel-Salaam AM, Mottelib AA, Ismail MN, Mohammed AA (2008) Effect of some physical factors on serum levels of calcium, phosphorus and magnesium in camels in Upper Egypt. J Camelid Sci 1:14–17

Abdel-Salaam AM, Zeidan AEB, Abdalla EB, Allam MWA (2011) Thermoregulation and blood biochemical changes in male dromedary camels during hot-humid and hot-dry environments under Egyptian conditions. J Camel Pract Res 18(2):297–304

Abderahman MM, Madanat A (2014) Seasonal variation in blood calcium, phosphorus, magnesium and some metabolites in camels (*Camelus dromedarius*). Res Opin Anim Vet Sci 4(11): 582–586

Abu Damir H, Tartour G, Adam SEI (1983) Mineral contents in livestock in Eastern Sudan. Trop Anim Health Prod 15:15–16

Abu-Damir H, Barri MES, Tageldin MH, Idris OF (1990) Clinical and subclinical colloid goiter in adult camels (*Camelus dromedarius*) at Kordofan region of Sudan. Br Vet J 146:219–227

Abu-Lehia IH (1987) Composition of camel milk. Milchwissenschaft 42:368–371

Ahmed AA, Awad YL, Fahmy F (1977) Studies on some minor constituents of camel's milk. Vet Med J 25:51–56

Ahmed MM, Abead SM, Barri MES (2003) Effect of physiological status on some macrominerals profile in the serum of female camels. J Camel Pract Res 10(2):107–110

Ahmed SH, Omer SAA, Gameel AA (2009) Some normal constituents in serum and cerebrospinal fluid in Sudanese camels (*Camelus dromedarius*). Assiut Vet Med J 55(123):163–170

Aichouni A, Jeblawi R, Dellal A, Hammou H, Aggad H (2010) Breed variation in blood constituents of the one-humped camel (*Camelus dromedarius*) in Algeria. J Camelid Sci 3:19–25

Aichouni A, Belhadia M, Aggad H (2013) Mineral indices in Algerian camels (*Camelus dromedarius*): effect of season. Camel Int J Vet Sci 1(1):29–36

Al-Ali AK, Husayni HA, Power DM (1988) A comprehensive biochemical analysis of the blood of the camel (*Camelus dromedarius*). Comp Biochem Physiol 89B(1):35–37

Al-Amrousi A, Hafiz M, Wasfi IA (1984) Some biochemical parameters of mature camels in Eastern provinces of Saudi Arabia. Assiut Vet Med J 13:121–124

Al-Busadah KA (2003) Some aspects of calcium and magnesium metabolism in camel calves. J Camel Pract Res 10(2):111–114

Al-Busadah KA (2007) Some biochemical and haematological indices in different breeds of camel in Saudi Arabia. Sci J King Faisal Univ 8(1):131–142

Al-Busadah KA (2010) Serum concentration of aluminum, calcium, magnesium and phosphorous in camels. Sci J King Faisal Univ (Basic Appl Sci) 11(1):161–167

Al-Haj Ali M, Adem A, Chandranath IS, Benedict S, Pathan JY, Nagelkerke N, Nyberg F, Lewis LK, Yandle TG, Nicholls GM, Frampton CM, Kazzam E (2012) Responses to dehydration in the one-humped camel and effects of blocking the renin-angiotensin system. PLoS One 7(5):e37299

Alhendi AB, Amer HA (1998) In vitro physico-chemical examination of ruminal fluids of Saudi camel. J Camel Pract Res 5(2):291–295

Ali A, Tharwat M, Al-Sobayil FA (2010) Hormonal, biochemical and hematological profiles in female camels (*Camelus dromedarius*) affected with reproductive disorders. Anim Reprod Sci 118:372–376

Ali A, Refaat D, Tharwat M, Al-Sobayil F (2015) Impotentia generandi in male dromedary camels: breeding soundness, haematology, biochemistry and testosterone level. Comp Clin Pathol 24: 727–731

Ali A, Derar D, Tharwat M, Zeitounb MM, Alsobyil FA (2016) Dystocia in dromedary camels: prevalence, forms, risks and hematobiochemical changes. Anim Reprod Sci 170:149–156

Al-Jassim RAM, Shahsavari A, Owen H, Khamas W (2016) Gross pathology, biochemistry and histology of selected organs of camels suffering from suspected monensin toxicosis in Australia. J Vet Sci Technol 7(3):1–5

Al-Mujalli AM (2012) Relationship between mineral status and jaw fractures in dromedary camels (*Camelus dromedarius*). Bulgarian J Vet Med 15(2):110–114

Al-Mujalli AM, Al-Naeem AA, Al-Ghamdi G, Al-Swailem A, Al-Yamani E, Shehata AM, Al-Dubaib MA, Hashad M, El-Lithy DA, Mahmoud OM, Alfayez M (2011) Cellular and biochemical blood profile in camels suffering from dubduba syndrome. Sci J King Faisal Univ (Basic Appl Sci) 12(2):165–172

Al-Qarawi AA (1997) Regulation of water and electrolyte metabolism during dehydration and rehydration in camels. Retrospective theses and dissertations. Iowa State University, USA. http://lib.dr.iastate.edu/rtd/11766

Al-Rukibat R, Ismail ZB (2014) Biochemical analysis of serum and synovial fluid in clinically normal young camels (*Camelus dromedarius*). Vet Sci Dev 4(5371):40–42

Al-Shami SA (2009) Comparative determination of sero-biochemical constituents in-door and free-grazing camels. J Anim Vet Adv 8(5):896–898

Al-Sobayil FA, Mousa HM (2009) Clinical and biochemical studies on wry-neck syndrome in camels in Saudi Arabia. J Camel Pract Res 16(2):179–182

Al-Sultan SI (2003) Studies of some normal biochemical parameters of Majaheem breed of camel (*Camelus dromedarius*) in Saudi Arabia. J Camel Pract Res 10(1):79–80

Al-Sultan SI (2008) Studies on hematological and certain serum biochemical values in young Magaheim dromedary camel at Al-Ahsa province. Vet Res 2(3-4):34–37

Al-Wabel NA (2008) Mineral contents of milk of cattle, camels, goats and sheep in the central region of Saudi Arabia. Asian J Biochem 3(6):373–375

Alwan OA, Zwaik HB (2014) Milk composition of Libyan Maghrebi camels (*Camelus dromedaries*) reared under farm and desert conditions. In: Proceedings of international conference on chemical, environment & biological sciences (CEBS-2014) Sept 17–18, 2014, Kuala Lumpur (Malaysia), pp 92–94

Amin ASA, Abdoun KA, Abdelatif AM (2007) Seasonal variation in blood constituents of one-humped camel (*Camelus dromedarius*). Pak J Biol Sci 10(8):1250–1256

Awad YL, Berschneider F (1977) Values for certain minerals and trace-elements in some tissues of the camel (*Camelus dromedarius*). Egypt J Vet Sci 14:31–35

Ayoub MA, Saleh AA (1998) A comparative physiological study between camels and goats during water deprivation. In: Alhadrami G, Ayoub MA (eds) Proceedings of the third annual meeting for animal production under arid conditions, vol 1. Al-Ain (UAE), 2–3/05/98, pp 71–87

Babeker EA, Suleem AE (2013) Observation of certain hematological and biochemical parameters in nomadic camels (*Camelus dromedarius*) in the Sudan. Univ Bakht Alruda Sci J 6:167–174
Babeker EA, Elmansoury YHA, Suleem AE (2011) The influence of seasons on blood constituents of dromedary camel (*Camelus dromedarius*). Online J Anim Feed Res 3(1):1–8
Badakhshan Y, Mirmahmoudi R (2016) Blood metabolites of one-humped camel (*Camelus dromedarius*) versus sheep during summer heat stress. Iranian J Vet Med 10(1):65–71
Banerjee S, Bhattacharjee RC (1963) Distribution of water body in the camel (*Camelus dromedarius*). Am J Phys 204:1045–1047
Baraka TA, El-Sherif MT, Kubesy AA, Illek J (2000) Clinical studies of selected ruminal and blood constituents in dromedary camels affected by various diseases. Acta Vet Brno 69:61–68
Barakat MZ, Fatah MA (1970) Biochemical analysis of normal camel blood. Zentbl Vet Med 17: 550–557
Barakat MZ, Fatah MA (1971) Seasonal and sexual variations of certain constituents of normal camel blood. Zentbl Vet Med 18:174–178
Bargaâ R, Lektib I, Farh M, El-Abbadir N, Belhgouari A, El-Khasmi M (2016) Seasonal variations of biochemical profile and its correlations with thyroid and adrenal gland hormones in Moroccan camel (*Camelus dromedarius*). Int J Agric Environ Res 2(6):1858–1872
Barri MES, Al-Busadah KA, Homeida AM (2005) Comparative calcium and magnesium status in adult and young camel (*Camelus dromedarius*). Sci J King Faisal Univ 6:129
Barsham MA, Elbagir NM, Barri MES (2005) Serum biochemical changes associated with experimentally-induced hypothyroidism in one-humped camels (*Camelus dromedarius*). J Camel Pract Res 12(2):117–121
Bath PL, Kohli RN (1961) Study of the normal calcium level of the camel. Indian Vet J 38:246–248
Bengoumi M (1992) Biochimie clinique du dromadaire et mécanismes de son adaptation à la déshydratation. Thèse de doctorat ès Sciences Agronomiques, I.A.V. Hassan II, rabat, Maroc, 184 p
Bengoumi M (2008) Tolerance study of Tiludronate in the dromedary camel. J Camelids Sci 1: 24–31
Bengoumi M, Faye B (2002) Adaptation du dromadaire à la déshydratation. Revue Sécheresse 13: 121–129
Bengoumi M, Riad F, Giry J, De La Farge F, Davicco M-J, Safwate A, Barlet JP (1993) Hormonal control of water and sodium in plasma and urine of camels during dehydration and rehydration. Gen Comp Endocrinol 89:378–386
Bengoumi M, Faye B, Tressol JC (1998) Composition minérale du lait de chamelle du sud marocain. Actes de l'atelier "Chameaux et dromadaires, animaux laitiers", Nouakchott, Mauritanie, 24–26 Oct 1994, Ed. CIRAD (coll. Colloques), pp 145–149
Bengoumi M, Moutaouakil F, De La Farge F, Faye B (2003) Seasonal variations of the plasma Thyroid hormone concentrations and the body temperature in the dromedary camel. J Camel Pract Res 10:115–119
Bengoumi M, Riad F, Moutawakil F, Davicco MJ, Coxam V (2006) Effects of the physical exercise on water and electrolyte metabolism in the dromedary camel. In: Proceedings of the first conference of the international society of Camelids research and development (ISOCARD), 15–17 April 2006, Al Ain, UAE
Berlyne GM, Shainkin-Kestenbaum R, Shany S, Finberg J, Yagil R (1978) Renal physiology of the dromedary camel. Nippon Jinzo Gakkai Ski 9:1015–1021
Biagi G (1983) Ulteriori indagini sul quadro sieroprotidemico del camelius dromedarius somalo. Ann Fac Med Vet Pisa 35:193–200
Biagi G, Salutini E (1983) Sul comportimento della calcemia, fosforemia e magnesemia nel *Camelus dromedarius* somalo di diversa eta e sesso. Ann Fac Med Vet Pisa 35:200–216
Bizzeti M, Farah ME, Gugluocci B, Demi S (1988) Rilievi enzymatici e minerali nel siero di *Camelus dromedarius* allevato in Somalia. Ann Fac Vet Univ Pisa 41:377–380
Bogin E (2000) Clinical pathology of Camelids: present and future. Rev Méd Vét 151(7):563–568

Charnot Y (1958) Répercussion de la déshydratation sur la biochimie et l'endocrinologie du dromadaire. Thèse es Sciences Naturelles, Université Paris VI, France

Charnot Y (1961) Equilibre minéral tissulaire dans la déshydratation du dromadaire. J Physiol 53:793–798

Chaudhary ZI, Iqbal J (2000) Incidence, biochemical and haematological alterations induced by natural trypanosomosis in racing dromedary camels. Acta Trop 77:209–213

Chiericato GM, Schiapelli MP, Warfa AA (1983) Caratteristiche del profile enzimatico e minerale del dromedarius (*Camelus dromedarius*). Clin Vet (Milan) 109:155–158

Desalegn T, Mohammed YK, Shimelis B (2012) Critical macro and micro minerals concentration in the blood serum of camel (*Camelus dromedarius*) in Jijiga district, Eastern Ethiopia. Livest Res Rural Devel 24(4). http://www.lrrd.org/lrrd24/4/desa24060.htm

Djegham M, Belhadj O (1986) Comportement de thermorégulation et résistance à la privation d'eau chez le dromadaire: variations saisonnières des profils biochimiques et hématologiques chez le dromadaire. Maghreb Vét 10:11–16

El Khasmi M, Faye B (2011) Parathyroid hormone-related peptide and vitamin D in phosphocalcic metabolism for dromedary camel. Iranian J Appl Anim Sci 1(4):205–213

El Khasmi M, Riad F, Safwate A, Hidane K, Faye B, Davicco MJ, Coxam V, Barlet JP (2001) Postpartum evolution of mammary secretion of minerals and 25-hydroxyvitamin D in lactating camels (*Camelus dromedarius*). J Camel Pract Res 8(2):131–135

El Khasmi M, Riad F, Safwate A, Tahri EH, Farh M, El Abbadi N, Coxam V, Faye B (2011) Effects of preslaughter stress on meat quality and phosphocalcic metabolism in camels (*Camelus dromedarius*). J Camelid Sci 3:33–38

El Khasmi M, Chakir Y, Riad F, Safwate A, Tahri EH, Farh M, El Abbadi N, Abouhaf R, Faye B (2013) Effects of transportation stress during the hot-dry season on some haematological and physiological parameters in Moroccan dromedary camels (*Camelus dromedarius*). J Life Sci 7(1):13–25

Elamin FM, Wilcox CJ (1992) Milk composition of Majaheim Camels. J Dairy Sci 75:3155–3157

El-Harairy MA, Zeidan AEB, Afify AA, Amer HA, Amer AM (2010) Ovarian activity, biochemical changes and histological status of the dromedary she-camel as affected by different seasons of the year. Nat Sci 8(5):54–65

Elhassan SM, Dowelmadina IM, El Zubeir I (2016) Variations in some macro minerals of camel milk as affected by management system, parity orders and stages of lactation. J Camelid Sci 9:54–62

Elias E, Yagil R (1984) Haematological and serum biochemical values in lactating camels (*Camelus dromedarius*) and their newborn. Refuah Vet 41(1):7–13

Elkasmi K (1989) Contribution à l'étude des protéines sériques et de quelques minéraux chez le dromadaire: influence de l'âge et du sexe. Mémoire de du 1er cycle, IAV HassanII, Rabat, Maroc

Eltahir YH, Ali HM, Mansour MH, Maghoub O (2010) Serum mineral contents of the Omani racing Arabian camel (*Camelus dromedarius*). J Anim Vet Adv 9(4):764–770

Eltohamy MM, Salama A, Youssef AEA (1986) Blood constituents in relation to the reproductive state in she-camel (*Camelus dromedarius*). Beitr Trop Land Vet Med 24:425–430

El-Zeiny WT (2011) Environmental stress and the loss of urea, sodium and potassium in the sweat of camels (*Camelus dromedarius*), under semiarid coastal conditions. J Camel Pract Res 18(2):287–296

Etzion Z, Meyerstein N, Yagil R (1984) Tritiated water metabolism during dehydration and rehydration in camel. J Appl Physiol 56:217–220

Evans DI, Rose RJ, Knight PK, Cluer D, Mannefield G (1992) Physiological responses during an incremental treadmill exercise test in the camel. In: Proceedings of the 1st international camel conference, Dubai, pp 223–227

Farag SI, Kebary KM (1992) Chemical composition and physical properties of camel's milk and milk fat. In: Proceedings of the 5th Egyptian conference of dairy sciences and technology, Egypt, pp 57–77

Farah Z, Ruegg MW (1989) The size distribution of casein micelles in camel milk. Food Microstruct 8:211–216

Faye B (1989) Statut nutritionnel du bétail dans la république de Djibouti. Ministère de la Coopération/INRA, Paris, 112 p

Faye B, Mulato C (1991) Facteurs de variation des paramètres protéo-énergétiques, enzymatiques et minéraux dans le plasma chez le dromadaire de Djibouti. Rev Elev Med Vét Pays Trop 44: 325–334

Faye B, Saint-Martin G, Cherrier R, Ali Ruffa M (1992) The influence of high dietary protein, energy and mineral intake on deficient young camel: II. Changes in mineral status. Comp Biochem Physiol 102A:417–424

Faye B, Ratovonanahary M, Chacornac JP, Soubre P (1995) Metabolic profiles and risk of diseases in camels in temperate conditions. Comp Biochem Physiol 112A:67–73

Faye B, Konuspayeva G, Messad S, Loiseau G (2008) Discriminant milk components of Bactrian camel (*Camelus bactrianus*), dromedary (*Camelus dromedarius*) and hybrids. Dairy Sci Technol 88:607–617

Ghosal AK, Appana TC, Dwaraknath PK (1973a) Seasonal variations in the blood constituents of the Indian camel (*Camelus dromedarius*). Indian J Anim Sci 43:642–644

Ghosal AK, Appana TC, Dwaraknath PK (1973b) A note on studies on the seasonal variation in serum electrolytes in Indian camel (*Camelus dromedarius*). Indian J Anim Sci 43:558–559

Ghosal AK, Apanna TC, Dwaraknath PK (1974) Seasonal variation in water compartment of the Indian camel. Br Vet J 130:47–49

Ghosal AK, Appana TC, Dwaraknath PK (1975) A note on the effect of short-term water deprivation on certain blood characteristics in the camel (*Camelus dromedarius*). Indian J Anim Sci 45:105–108

Gupta HK, Dwaraknath PK, Pareek PK (1979) A note on camel blood serum electrolytes during rutting and non-rutting seasons and after the exogenous testosterone treatment. Indian J Anim Sci 49:680–681

Gupta L, Kumar RA, Ghanshyam T, Rajesh D, Garg R (2012) Effect of feeding different proportions of groundnut haulms (*Arachis hypogaea*) and cluster bean straw (*Cyamopsis tetragonoloba*) on nutrient utilisation and serum biochemical parameters in dromedary camels. Trop Anim Health Prod 44:1689–1695

Gutierrez C, Corbera JA, Juste MC, Doreste F, Morales I (2005) An outbreak of abortions and high neonatal mortality associated with *Trypanosoma evansi* in dromedary camels in the Canary Islands. Vet Parasitol 130:163–168

Gutierrez C, Corbera J, Faye B (2008) Obstructive phosphate urolithiasis in a dromedary camel: a case study. J Camel Pract Res 15(1):77–79

Hamed H, El Feki AF, Gargouri A (2016) Effect of total and differential somatic cell counts, lactation stage and lactation number on lipolysis and physicochemical composition in camel (*Camelus dromedaries*) and cow milk. Iranian J Appl Anim Sci 6(4):777–782

Hassan YM, Hoeller H, Hassan IM (1968) Observations on the blood constituents of camels in the Sudan. Sudan J Vet Sci Anim Husb 9:464–474

Hassan AA, Hagrass AE, Soryal KA, Le Shabrawy SA (1987) Physicochemical properties of camel milk during lactation period in Egypt. Egypt J Food Sci 15(1):1–14

Hekmatimoghaddam S, Rasooli A, Sazmand A, Hamidinejat H, Jafari H, Nouri M (2011) Serobiochemical alternations in dromedary camels naturally infected with *Theileria* spp. in Iran. In: Proceedings of the XVth international congress on animal hygiene, vol 1, 07/2011, pp 455–457

Helal A (2015) Relationships among physical, chemical and industrial characteristics of different dromedary camel's hair types. J Am Sci 11(2):67–75

Höller HK, Hassan YM (1966) Bestimmung liniger blutbestandteile bei Kamelen im Sudan. Dtsch Tierärz Wschr 73:553–566

Hussein MF, Salah MS, Mogawer HH, Gar ElNabi AR (1992a) Effect of lactation on the haemogram and certain blood constituents of the dromedary camel. J Appl Anim Res 1:43–50

Hussein MF, Basmaeil SM, Bakkar MN, Gar-El-Nabi AR (1992b) Serum levels of some electrolytes and trace elements in camel calves during first year of life. J Appl Anim Res 2:13–18
Hussein YA, Al-Eknah MM, AL-Shami SA, Mandour MA, Fouda TA (2012) Coat color breed variation in blood constituents among indigenous Saudi Arabia camel strains. Mansoura Vet Med J 14(1):191–204
Ibrahim GA, Nour IA, Al-Maqbali R, Kadim IT (2017) Effect of age on concentrations of nutrients in four muscles of the Sudanese dromedary (*Camelus dromedaries*) camel. J Appl Anim Res 45(1):577–584
Ismail ZB, Qureshi T, Levy M, Khamas W, Nour A (2003) Preliminary report on some physiological parameters in pregnant she camel (*Camelus dromedarius*) in North America. J Camel Pract Res 10(2):125–129
Jirimutu B, Li J, Alam MS, Li H, Guo M, Zhang H (2010) Fatty acid and protein profiles, and mineral content of milk from the wild Bactrian camel (*Camelus bactrianus ferus*) in Mongolia. Milchwissenschaft 65:21–25
Kadim IT, Mahgoub O, Al-Marzooqi W, Al-Zadjali S, Annamalai K, Mansoor MH (2006) Effects of age on composition and quality of muscle longissimus thoracis of the omani arabian camel (*Camelus dromedarius*). Meat Sci 73(4):619–625
Kamal AM (2008) Some biochemical, hematological and clinical studies of selected ruminal and blood constituents in camel affected by various diseases. Res J Vet Sci 1(1):16–27
Kamalu TN, Okpe GC, Williams A (2003) Mineral contents of extracellular fluids in camel and cattle in the north-east Sahel region of Nigeria. Nigerian Vet J 24(1):13–20
Kaneko JJ (1989) Clinical biochemistry of domestic animals, 4th edn. Academic Press, New York
Kaneko JJ, Harvey JW, Bruss ML (2008) Clinical biochemistry of domestic animals, 6th edn. Elsevier, New York
Karram MH, Ibrahim H, Abd E-A, Manaa A (1991) Clinical and hematological changes in camels infected with trypanosomiasis and microfilaria. Assiut Vet Med J 25(49):118–127
Kataria N, Kataria AK (2004) Use of blood analytes in assessment of stress due to drought in camel. J Camel Pract Res 11(2):129–133
Kataria N, Maan R, Joshi A, Mathur S, Kataria AK (2013) Electrolyte and antioxidant profiling of dromedary camel (*Camelus dromedarius*) with mandible fracture. Anim Biol Anim Husb Bioflux 5(1):44–48
Khadjeh GH (1998) Concentration of serum electrolytes in pregnant and non-pregnant Iranian one-humped camels (*Camelus dromedarius*). J Camel Pract Res 5(2):289–290
Khan KU, Appanna TC (1964) Electrolytes in camel milk. Indian J Physiol Allied Sci 18:129–133
Knoess KH (1976) Assignment report on animal production in the Middle Awash Valley. Project ETH/75/001 Class W/K 3651, FAO, Rome
Konuspayeva G, Faye B, Narmuratova M, Meldebekova A, Loiseau G (2008) Variation factors of some minerals in camel milk. In: Faye B, Sinyavskiy Y (eds) Proceedings of international workshop, "Impact of pollution on animal products". Almaty (Kazakhstan), 27–30 Sept 2007, pp 125–132
Kuria SG, Tura IA, Amboga S, Walaga HK (2013) Status of minerals (*Camelus dromedarius*) in north eastern Kenya as evaluated from the blood plasma. Livest Res Rural Devel 25(8). http://www.lrrd.org/lrrd25/8/kuri25149.htm
Little A, Mc Enzie A, Morris RJH, Roberts J, Evans JV (1970) Blood electrolytes in the Australian camel. Aust J Exp Biol Med Sci 48:17–24
Liu ZP (2003) Studies on the haematology and trace element status of adult Bactrian camels (*Camelus bactrianus*) in China. Vet Res Commun 27:397–405
Liu ZP, Ma Z, Zhang YJ (1994) Studies on the relationship between sway disease of Bactrian camels and copper status in Gansu Province. Vet Res Commun 18:251–260
Mabrouk S, Khorchani T, Benromdhane M (2010) Bases épidémiocliniques de la maladie du Krafft chez le dromadaire (*Camelus dromedarius*) dans le Sud tunisien. Rev Elev Méd Vét Pays Trop 63(1–2):29–33

Macfarlane WV, Morris RJH, Howard B (1962) Water metabolism of merino sheep and camels. Austr J Sci 25:112–121

Mahmud HM, Abdulhamid HM, Locatelli A (1984) Water deprivation effects on the hematological and hematochemical pictures of *Camelus dromedarius*. Rev Elev Méd Vét Pays trop 37: 313–317

Mahmud T, Rehman R, Anwar J, Ali S, Abbas A, Salman M (2011) Minerals and nutritional composition of camel (*Camelus dromedarius*) meat in Pakistan. J Chem Soc Pak 33(6):835–838

Mal G, Sena S, Sahani MS (2007) Changes in chemical and macro-minerals content of dromedary milk during lactation. J Camel Pract Res 14(2):195–197

Maloiy GMO, Clemens ET (1980) Gastrointestinal osmolality electrolyte and organic acid composition in five species of East African herbivorous mammals. J Anim Sci 51(4):917–924

Mehaia MA, Hablas MA, Abdel-Rahman KM, El-Mougy SA (1995) Milk composition of Majaheim, Wadah and Hamra camels in Saudi Arabia. Food Chem 52:115–122

Mohamed HE (2004) A note on plasma antioxidant status in Sudanes camels (*Camelus dromedarius*) affected bu musculoskeletal disorders. J Camel Pract Res 11(1):65–66

Mohamed HE, Hussein AN (1999) Studies on normal haematological and serum biochemical values of the 'Hijin' camels (*Camelus dromedarius*) in Kuwait. Vet Res Commun 23:241–248

Mohammed AK, Sackey AKB, Tekdek LB, Gefu JO (2007) Serum biochemical values of healthy adult one humped camel (*Camelus dromedarius*) introduced into a sub-humid climate in Shika-Zaria, Nigeria. J Camel Pract Res 14(2):191–194

Mohri M, Moosavian HR, Hadian MJ (2008) Plasma biochemistry of one-humped camel (*Camelus dromedarius*): effects of anticoagulants and comparison with serum. Res Vet Sci 85:554–558

Mostafidi M, Moslehishad M, Piravivanak Z, Pouretedal Z (2016) Evaluation of mineral content and heavy metals of dromedary camel milk in Iran. Food Sci Technol (Campinas) 36(4): 717–723

Muhammad BF, Aliyu D, Njidda AA, Madigawa IL (2011) Some haematological, biochemical and hormonal profile of pregnant and non-pregnant she-camels (*Camelus dromedarius*) raised in a Sudan savanna zone in Nigeria. J Camel Pract Res 18(1):73–77

Musa BE, Mukhtar AM (1982) Studies on the normal haemogram and some blood electrolytes in camels (*Camelus dromedarius*). Sudan J Vet Sci Anim Husb 23:38–43

Mustafa AB, Haroun E, Abdelaati K, Alsharif SHM (2012) Effect of sex factor on macrominerals profile in vital organs of dromedary camels in Western Darfur, Sudan. In: Johnson EH et al (eds), Proceedings of the 3rd ISOCARD Conference, 29th January–1st February, 2012. Mascate (Sultanate of Oman), pp 280–281

Naeini AT, Nazifi S (2001) Biochemical and cytological properties of blood and peritoneal fluid in clinically healthy adult camels (*Camelus dromedarius*). J Camel Pract Res 8(2):123–126

Nagpal AK, Roy AK, Chirania BL, Patil NV (2011) Growth, nutrient utilization and serum profile in camel calves as affected by dietary protein levels. Indian J Anim Nut 28(2):166–171

Nazifi S, Maleki K (1998) Biochemical analysis of serum and cerebrospinal fluid in clinically normal adult camels (*Camelus dromedarius*). Res Vet Sci 65:83–84

Nazifi S, Gheisari R, Poorabbas H (1999) The influence of thermal stress on serum biochemical parameters of dromedary camels and their correlation with thyroid activity. Comp Haematol Int 9:49–53

Nazifi S, Oryan A, Bahrami S, Razavi SM (2009) Evaluation of haematological and serum biochemical parameters in Iranian camels (*Camelus dromedarius*) infected with haemotrophic Mycoplasma (Eperythrozoon) spp. Comp Clin Pathol 18:329–332

Omer SA, Agab H, Samad HAG, Turki IY (2008) Effect of feed type on some blood constituents of Sudanese growing camel (*Camelus dromedarius*) calves. Sudan J Vet Med Anim Husb 47:1–8

Omer SA, Salawa ME, Khougali H, Gussey AA, Samad HA (2010) Studies on some biochemical and haematological indices of Sudanese camels (*Camelus dromedarius*). Sudan University of Science Technology Bulletin, pp. 1–7

Omidi A, Sajedi Z, Montazer-Torbati MB, Mostafai M (2014) Metabolic profile of pregnant, non-pregnant and male two-humped camels (*Camelus bactrianus*) of Iran. Iran J Vet Med 8(4):235–242

Orliac D (1980) Contribution à l'étude de la biochimie sanguine de dromadaires et de chèvres sahariens. Thèse doct. Vét., Toulouse, 71, 31p

Osman TEA, Al-Busadah KA (2003) Normal concentrations of twenty biochemical parameters of she-camels, cows and ewes in Saudi Arabia. Pak J Biol Sci 6(14):1253–1256

Patodkar VR, Somkuwar AP, Parekar S, Khade N (2010) Influence of sex on certain biochemical parameters in Nomadic Camels (*Camelus dromedarius*) nearby Pune, in Maharashtra. Vet World 3(3):115–117

Raiymbek G, Faye B, Serikbaeva A, Konuspayeva G, Kadim IT (2013) Chemical composition of Infraspinatus, *Triceps brachii*, *Longissimus thoraces*, *Biceps femoris*, *Semitendinosus*, and *Semimembranosus* of Bactrian (*Camelus bactrianus*) camel muscles. Emir J Food Agric 25(4):261–266

Rezakhani A, Nazifi Habibabadi S, Maghrebi Ghojogh M (1997) Studies on normal haematological and biochemical parameters of Turkmen camel in Iran. J Camel Pract Res 4(1):41–44

Riad F, Bengoumi M, Davicco MJ, Safwate A, Barlet JP (1994) Influence of 1-α-hydroxycholecalciferol on calcium and phosphorus concentration in camel milk. J Dairy Res 61:567–571

Sackey AKB, Tekdek LB, Mohammed AK, Gefu JO (2007) Serum biochemical values of healthy adult one humped camel (*Camelus dromedarius*) introduced into a sub-humid climate in Shika-Zaria, Nigeria. J Anim Vet Adv 6:597–600

Saeed A, Afzal M, Saeed A (1995) Effect of storage on some constituents of camel serum. Austr Vet J 72:212–215

Saeed A, Hussain MM, Khan IA, Chand G, El-Yousuf RA (2004) Effect of sex and age on blood biochemical profile in camel. J Camel Pract Res 11(1):73–77

Saeed A, Khan IA, Hussein MM (2009) Change in biochemical profile of pregnant camels (*Camelus dromedarius*) at term. Comp Clin Pathol 18(2):139–143

Saini N, Singh N, Kiradoo BD, Pathak KML (2009) Comparative biochemical and mineral profile of female Indian dromedaries during breeding season. J Camel Pract Res 16(2):189–193

Salib AA, Soahir YS (1984) Foetal and maternal serum electrolytes during different stages of foetal growth in the camel. Acta Vet (Beograd) 34:77–82

Sarwar A, Majeed MA (1997) Interrelationships between 30 parameters of blood in normal one-humped camel in summer. J Camel Pract Res 4(1):35–39

Sarwar A, Majeed MA, Hur G, Khan LR (2004) Two transferases and four electrolytes in normal one-humped camel serum. J Camel Sci 1:57–61

Sawaya WN, Khalil A, Al-Shalhat H, Mohammad AI (1984) Chemical composition and nutritional quality of camel milk. J Food Sci 49:744

Sazmand A, Rasooli A, Nouri M, Hamidinejat H, Hekmatimoghaddam S (2011) Serobiochemical alterations in subclinically affected dromedary camels with *Trypanosoma evansi* in Iran. Pak Vet J 31(3):223–226

Sellaouti K (1984) Contribution à l'établissement des paramètres biochimiques sanguins du dromadaire. Thèse Méd. Vét., Sidi Thabet, Tunisie

Shamsia SM (2009) Nutritional and therapeutic properties of camel and human milks. Int J Gen Mol Biol 1(2):52–58

Shawaf T (2017) Analysis of serum and cerebrospinal fluid in clinically healthy dromedary camels (*Camelus dromedarius*). J Vet Sci Technol (Suppl) 8(4):63

Shen X, Li X (2010) Studies of "emaciation ailment" in the Bactrian camel. Afr J Biotechnol 9(49): 8492–8497

Shukla MK, Khan MJZ, Pathan MM, Siddiquee GM, Latif A (2009) Biochemical profile of uncastrated male Kutchi camel during the breeding season. J Camel Pract Res 16(1):213–215

Siebert BD, Macfarlane WV (1971) Water turnover and renal function of dromedaries in the desert. Physiol Zool 44:225–240

Siham AA, Daoud AAA (2015) A study in minerals concentrations in red meat. Int J Sci Res 4(8): 707–712

Singh S, Dedar RK, Legha RA, Bala PA, Patil NV (2015) Minerals and electrolytes profile in lactating and pregnant Indian camels. J Camel Pract Res 22(1):121–124

Snow DH, Billah A, Ridah A (1988) Effects of maximal exercise on the blood composition of the racing camel. Vet Rec 123:311–312

Soliman GZA (2005) Comparison of chemical and mineral content of milk from human, cow, buffalo, camel and goat in Egypt. Egyptian J Hosp Med 21:116–130

Soliman MK, Shaker M (1967) Cytological and biochemical studies on the blood of adult she camel. Vet Rec 123:311–312

Tajik J, Sazmand A, Hekmatimoghaddam S, Rasooli A, Mohammadzadeh YA (2015) Serum concentrations of some ions in clinically healthy camels (*Camelus dromedarius*). Eurasian J Vet Sci 31(4):204–208

Vyas S, Saini N, Kiradoo BD, Lukha A, Kishore N, Mal G, Pathak KML (2011) Biochemical and trace mineral profile in post-parturient dromedary camel (*Camelus dromedarius*). Indian J Anim Sci 81(6):586–587

Wang SY, Liang JP, Shao WJ, Wen H (2011) Mineral, vitamin and fatty acid contents in the camel milk of dromedaries in the Anxi Gansu China. J Camel Pract Res 18(2):273–276

Wasfi A, Gadir FA, Abdulla OM (1989) The acute effects of dexamethasone on some haematological parameters and serum biochemistry in the camel (*Camelus dromedarius*). Rev Élev Méd Vét Pays Trop 42(4):479–483

Wernery U, Abraham AA, Jyothi T, Abubakar Ali Y, George RM (2009) Mineral and vitamin contents in the blood of racing dromedaries in the United Arab Emirates. J Camel Pract Res 16(1):39–40

Whabi AA, Abdel Gadir SE, Neimat AA, Idris OF (1979) Plasma electrolytes and minerals of normal camels in the Sudan. In: Cockrill WR (ed) Camels and camelids. Uppsala, Suède, pp 431–437

Wilson RT (1989) Ecophysiology of the camelidae and desert ruminants. Springer, Berlin

Yagil R (1985) The desert camel: comparative physiological adaptation. Comparative animal nutrition. Karger, Basel

Yagil R, Berlyne GM (1976) Sodium and potassium metabolism in the dehydrated and rehydrated Bedouin camel. J Appl Physiol 41:457–461

Yagil R, Etzion Z (1979) The role of antidiuretic hormone and aldosterone in the dehydrated and rehydrated Bedouin camel. Comp Biochem Physiol 63A:275–278

Yagil R, Etzion Z (1980) Milk production of the dehydrated camel. Refuah Vet 37(1–2):58

Yagil R, Etzion Z, Berlyne GM (1975) Acid-base parameters in the dehydrated camel. Tijdsch Diergeneeskunde 100(20):1105–1108

Yagil R, Etzion Z, Ganani J (1978) Camel thyroid metabolism: effect of season and dehydration. J Appl Physiol 45:540–544

Youssef SY, Yasien S, Ali Mousa WH, Nasr SM, El-Kelesh EAM, Mahran KM, Abd-El-Rahman AH (2015) Vector identification and clinical, hematological, biochemical, and parasitological characteristics of camel (*Camelus dromedarius*) theileriosis in Egypt. Trop Anim Health Prod 47:649–656

Zaher H, El-Zahar H, Al Sharifi S, Shety T (2017) Alterations in hematological and biochemical parameters affecting the reproductive performance in female camels (*Camelus dromedaries*). Int J Vet Health Sci Res 5(1):155–160

Zeidan AEB, Abbas HE (2003) Physio-biochemical changes in the male dromedary camels during breeding (rutting) and non-breeding season. J Camel Pract Res 10(2):183–190

Zeidan AEB, Abd El-Salam AM, El-Malky OM, Ahmadi EA, Sarhan DMA, Daader AH (2008) Biochemical and histological changes of the ovary of the dromedary camel during breeding and non-breeding seasons. Egypt J Basic Appl Physiol 7:287–308

Zia-Ur-Rahman D, Ahmad N, Bukhari SA, Akhtar N, Haq IU (2007) Serum hormonal, electrolytes and trace element profiles in the rutting and non-rutting one-humped male camel (*Camelus dromedarius*). Anim Reprod Sci 101(1–2, 172):–178

Chapter 7
Trace Elements

Clinical trace element deficiencies or toxicities in the dromedary or Bactrian camel are rarely attested. Thereby, references are few, and until recently, standards were rather not defined. Regarding those elements, the characteristics of camel are the increasing of absorption capacities in scarcity periods (for example for copper, manganese and iron), a higher storage ability to anticipate the deficient periods (for example selenium), a certain tolerance to mineral excess by increasing the excretion, the maintenance of enzymatic activities in deficient seasons (copper and selenium especially). However, a certain sensitivity to selenium excess is observed.

However, suspicion of mineral deficiency related to health disorders in camel is old. In its founding book devoted to the camel diseases, Curasson (1947) wrote "poor mineral iron, copper etc., have repercussions on the health of animals in certain regions, but we don't know if they have an action in the camel." For several years, researches are confined to the determination of physiological "standards," but it is only recently that trace element metabolism has been studied.

7.1 The Biological Role of Trace Elements

Classically, essential trace elements are distinguished from the nonessential elements. The first are so called because their failure act objectively into functional disorder, and any supplementation at physiological dose prevents or cures this disorder (Mertz 1981). Essential trace elements are part of organic molecules having important biological activity. Those organic molecules are enzymes, hormones, vitamins, or metalloproteins. Some of those organic molecules have been determined in camel:

- Enzymes:
  - Copper in cytochrome oxidase, lysil oxidase, tyrosinase, and ceruloplasmin (Essamadi et al. 1998, 2002)

© Springer International Publishing AG, part of Springer Nature 2018
B. Faye, M. Bengoumi, *Camel Clinical Biochemistry and Hematology*,
https://doi.org/10.1007/978-3-319-95562-9_7

  - Zinc in carbonic anhydrase, aldolase, peptidase, alkaline phosphatases, the DNA and RNA polymerases, and dehydrogenases
  - Manganese in pyruvate carboxylases and superoxide dismutase (Fouad 2015)
  - Selenium in glutathione peroxidase (Bengoumi et al. 1998c)
- Hormones:
  - Iodine in thyroid hormones
- Vitamins:
  - Cobalt in vitamin B12
- Metalloproteins:
  - Iron in hemoglobin and myoglobin; selenium in different types of selenoproteins

Generally, the essential elements are only classically dosed in camels. Other minerals falling within this category of essential elements can be useful for various metabolisms, but their mode of action is rarely elucidated: chrome, tin, fluorine, nickel, silicon, and vanadium. At our knowledge, most of them have never (or rarely) been determined in the tissues of camel. It is the same, a fortiori, for nonessential elements or for whom a particular biological role has not yet been identified. This group can contain almost all the elements of the periodic table, with the exception perhaps of the lanthanides and actinides. The biological role of trace elements in camel is obviously identical to those of other animal species in general and of the ruminants in particular. However, their metabolism may be different.

7.2 The Interest for Laboratory Diagnosis

The determination of trace element concentrations in different biological materials presents a certain interest in the diagnosis of deficiency or intoxication. Severe deficiencies are listed, in general, in localized geographical situations (Mc Dowell et al. 1983; Schillhorn Van Veen and Loeffler 1990; Faye et al. 1991). However, sub-deficiencies are more common, but because their symptoms are not evident, they are not well known, a fortiori in the camel for which investigations of sub-deficiencies are rare.

In order to refine the diagnosis (primary deficiency due to inadequate intakes of the concerned elements; secondary deficiency by modification of the digestibility of the elements), analysis of plasma trace elements should be completed by an assessment of their concentrations in feeds (Lamand 1979).

Plasma is not the only substrate to be taken into account for the diagnosis of nutritional status in trace elements of animals. In particular, the liver tissue collected by biopsy in vivo may be a useful indicator of micronutrient stocks. But its routine use seems even more delicate, more expensive, especially for a species such as the dromedary (Cherrier et al. 1991). Generally, the diagnosis undertaken from the blood results has primarily a collective interest: A situation of deficiency (or poisoning) could be accepted as soon as a majority percentage of individuals have, within the herd, plasma values below the limit of deficiency (or above the permitted maximum physiological standard). The individual diagnosis, indeed, is operational only for clinically expressed deficiency or acute intoxication. Moreover, it is rare that the percentage of animals with anomalous values reaches 100%.

7.3 Sampling and Analytical Techniques

Biological substrates on which determinations can be made are plasma, milk, hair, saliva, feces, and urine. In practice, in the camel, only analyses on plasma were commonly used. With some exceptions, the preferential substrate is plasma, which requires the use of anticoagulant. Liquemine (N.D.) will be preferred to heparin (generally, more often used as anticoagulant). Liquemine is probably more expensive but does not contaminate during the determination of copper or zinc.

The separation of plasma should performed within 1 h after blood collection; otherwise the risk of hemolysis or dialysis of zinc from the red cells, five times richer in plasma zinc, is increased (Lamand 1987).

Blood sampling is made at the jugular vein by using a dripping needle (preferably than with syringe) and achieved with adapted equipment, avoiding contamination (in particular, tubes with stopper in latex are to be avoided because it could provoke zinc contamination). After centrifugation, plasma samples should be extracted and kept cold (temperature of the ice melting, possible in isotherm cooler for several days) and then frozen (−20 °C) before analysis at the laboratory.

Liver tissue sampling methodology has been described previously (see supra). Sampling on other organs such as the kidney, spleen, or muscle was performed on camels after slaughtering (Tartour 1969, 1975; El-Faer et al. 1991), but their clinical relevance is low, and the taking of sample in vivo is awkward, because it requires to intervene surgically. To our knowledge, the determination of the trace elements from samples of hair or saliva has been very rarely considered in the camel. Elsewhere, there are some references on the trace elements of milk composition.

Obviously, the determination techniques of trace elements in different substrates of the camel species do not differ from the other species. Various methods may be proposed in the literature and have changed for the last decades with the progress in sensitivity and precision of analytical equipment. This continuous improvement in analytical procedures may lead to some differences in the published results and to difficulties in the interpretations. In fact, it is difficult sometimes to compare the results of the literature. However, an important effort has been made to standardize the methods (Van de Wiel 2004).

Atomic absorption spectrophotometry was the main technique used for most of the trace elements in different substrates (Bellanger and Lamand 1975; Hocquellet 1974) or more recently on inductively coupled argon plasma-atomic emission spectrometer (ICP-AES) (Vanhoe 1993; Ebdon and Fisher 2000). Iodine and molybdenum were generally determined by colorimetry for a long time (Bellanger 1971).

Chromatographic methods have been proposed recently (Blazewicz et al. 2014; Jooste and Strydom 2010). Selenium was quantitated also by fluorimetry (Koh and Bensen 1983), but different types of spectrometry are used nowadays (Li and Yu 2016). Regardless of the method used, it is essential to use standard curve points (e.g., Accu TraceTM Reference Standard solutions from AccuStandard®—USA) and to apply international quality control standard #1 and Laboratory Performance Check Standard in order to assume the optimization and reliability of the results.

7.4 Biological Signification of Observed Values

7.4.1 Copper

7.4.1.1 Plasma or Serum Copper

The plasma copper concentration is a good indicator of copper intake. In sheep, the threshold value is obtained after 2 or 3 months of deficient diet containing 1.2 mg copper/kg DM (5 mg/kg DM is the minimum requirement for sheep) (Suttle et al. 1970). So, decrease in plasma copper concentration is the result of a more or less prolonged deficiency because copper is stored in the liver.

In ruminants, normal values for the plasma copper concentration are between 70 and 120 μg/100 ml (or 12 and 19 μmol/l). Most of the values observed in the camel are indeed between these two figures (Table 7.1).

In an important monitoring of racing camels in the UAE including 14,237 animals, the mean copper value was reported as 71 ± 17 μg/100 ml (Wernery et al. 2009). Some comparisons between species within the same geographic entity tend to show that camels have, on average, higher cupremia (Table 7.2). These results could be put in relation to the feeding behavior of camels which includes tree foliage rich in copper (Faye and Tisserand 1989; Rutagwenda et al. 1989). In fact, the trees and bushes are generally richer in copper than grasses (Tartour 1966; Faye et al. 1990).

In contrast, in identical feeding conditions (same diet), Bengoumi et al. (1998a) observed lower values of the cupremia of the dromedary camel compared to cattle. These authors have set up an experiment with an adaptation phase, a mineral supplementation phase, and a phase without complementation. For the duration of the test, the animals (camels and cattle) received the same basal diet (wheat straw + rice bran + molasses). Mineral supplementation was identical also during the second phase. Plasma copper concentrations observed for each of the phases of the experiment were, respectively, 44, 63, and 57 μg/100 ml on average for the camels and 106, 111, and 113 μg/100 ml for cattle. In these circumstances, it therefore appears that the dromedary camel regulates its cupremia to lower values than cattle (Fig. 7.1).

Table 7.1 Values of plasma or serum copper concentration in camel according to different authors (in µg/100 ml)

References	Mean values (µg/100 ml)	Range or SD	Number	Country
Moty et al. (1968)	83	±6.7	19	Egypt
Tartour (1975)	95.3	59–137	19	Sudan
Whabi et al. (1979)	118.3	±28.8	96	Sudan
Idris et al. (1980)	88.7	75–97	12	Sudan
Abu Damir et al. (1983)	92.6	66–129	17	Sudan
Abdelrahim (1983)	69	60–150	7	Sudan
Marx and Abdi (1983)	126.5	±26.8	67	Somalia
Faye et al. (1986)	107	66–151	53	Ethiopia
Eltohamy et al. (1986)	76.5	±4.3	45	Egypt
Abdalla et al. (1988)	75	59–90	25	UAE
El Kasmi (1989)	65	±21	60	Morocco
Faye and Mulato (1991)	60.7	7–122	52	Djibouti
Ghosal and Shekhawat (1992)	94.3	±3.2	122	India
Liu et al. (1994)[a]	86	±24	15	China
Faye et al. (1995)	65.4	±20.2	65	France
Bengoumi et al. (1995a)	102	±25	30	Morocco
Ma (1995)[a]	80.4	±29.7	20	China
Liu and Ma (1995a)[a]	67.3	± 12	10	China
Deen et al. (2004)	140	±5	33	India
Faye et al. (2005)	60.1	±13	240	UAE
Abu Damir et al. (2008)	69	±16	175	UAE
Shukla et al. (2009)	112.9	±0.44	24	India
Nafizi et al. (2009)	87	±9	30	Iran
Parekar et al. (2009)	85.2	±10.3	25	India
Eltahir et al. (2010)	75.5	55.8–110	30	Oman
Shen and Li (2010)[a]	93	±16	15	China
Ali et al. (2010)	70	±10	60	Saudi Arabia
Althamma et al. (2012)	66.5	±18.4	26	Saudi Arabia
Mustafa et al. (2012)	60.7	37–131	20	Sudan
Abdelrahman et al. (2013a)	64.3	57.5–71.2	30	Saudi Arabia
Pourjafar et al. (2014)	70	63–79	18	Iran
Faye et al. (2014)	76.1	31–121	16	Saudi Arabia
Li and Hai (2014)[a]	93	±16	5	China

[a]Bactrian camel

Table 7.2 Concentrations in plasma or serum copper (and number of sampled animals) in various species of domestic ruminants (in μg/100 ml)

Reference	Camel (*n*)	Cattle (*n*)	Goat (*n*)	Sheep (*n*)	Country
Moty et al. (1968)	83 (19)	64 (29)	–	82 (32)	Egypt
Tartour (1975)	95.3 (19)	73.8 (71)	78.9 (24)	85 (1)	Sudan
Abu Damir et al. (1983)	92.6 (17)	86.2 (30)	–	94.5 (36)	Sudan
Shekhawat (1983)	94.3 (122)	86.8 (29)	–	88.3 (34)	India
Faye and Grillet (1984)	45 (8)	37.2 (9)	41.8 (8)	24.7 (20)	Ethiopia
Faye et al. (1986)	107 (53)	64.5 (432)	89.2 (425)	95.1 (173)	Ethiopia
Faye et al. (1991)	60.7 (52)	73.8 (59)	94.5 (118)	87.2 (80)	Djibouti
Al-Busadah (2003)	113.5 (5)	70.2 (5)	–	95.6 (5)	Saudi Arabia

Fig. 7.1 Change in the plasma concentration of copper in camels (open circles) and cows (solid diamonds) (calculated from Bengoumi et al. 1998a)

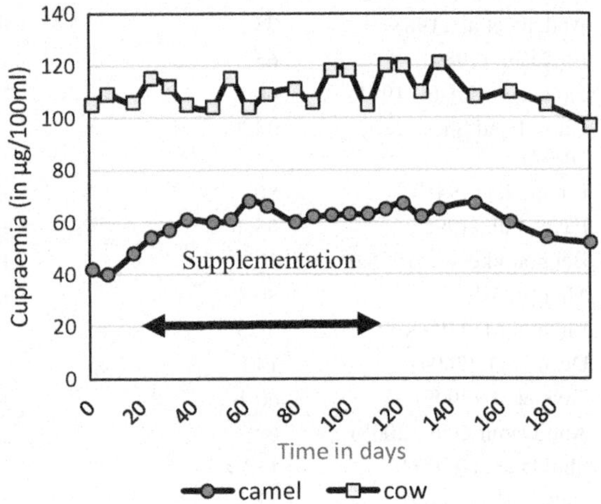

However, due to differences in feeding behavior referred above, such difference between species is not observed in field situations. Elsewhere, a seasonal variation was observed in Sudan (Mohamed 2004) with a significant higher value at the rainy season (67 ± 4.3 μg/100 ml) than in the dry season (55.6 ± 8.9 μg/100 ml) linked to the change in the type of diet.

Detectable deficiencies by below-normal cupremia were identified in the dromedary, only in the Horn of Africa, especially in the Rift Valley (Faye et al. 1991): 45 μg/100 ml (25–50) in Awash region (Faye and Grillet 1984) and 60.7 μg/100 ml (7–122) in different natural regions of Djibouti (Faye and Mulato 1991). These low values are linked to secondary deficiencies due to excess in the diet of antagonistic elements to copper such as sulfur and molybdenum. Indeed, these elements combine with copper to form the thiomolybdate of copper, totally inassimilable (Sanjabi et al. 2003). Sulfur-induced copper deficiency has been provoked in Bactrian camel leading to a sharp decrease of blood copper: 27 ± 3 μg/100 ml (Li and Hai 2014).

The lowest values ever reported in the camel were observed at Djibouti, particularly in herds feeding on leaves and twigs of mangrove from Obock coast, along the

Red Sea (Faye and Mulato 1991). This shrubby vegetation including woody species like *Avicennia marina* (consumed by camels) and *Ceriops tagal* (not consumed) is rich in major mineral elements such as salt (sodium chloride) but poor in trace elements, in particular copper, zinc, manganese, and selenium (Faye et al. 1992a; Faye 1993). The values observed in this context, therefore, reflect a primary deficiency which does not seem to provoke, however, spectacular clinical symptoms (Faye et al. 1993). The longevity of camels consuming those plants appears to be low (Godet et al. 1985), and the animals had rather poor general condition (Faye 1989), but no characteristic symptoms were observed.

In China, for Bactrian camel, Liu et al. (1994) described ataxic disease associated with secondary copper deficiency. Affected animals ($n = 20$) had an average plasma copper concentration of 28 μg/100 ml only, and a minimum value of 2 μg/100 ml was reported. Symptoms, histological lesions, and the results of blood tests were consistent with what is observed in enzootic ataxia in sheep for a long time (Chalmers 1974). In different regions of Sudan, copper deficiencies were reported with values between 28 and 31 μg/100 ml (Elrayah et al. 2010).

Significant changes in the plasma copper linked to sex are not reported in literature (Abdalla et al. 1988; Bengoumi et al. 1995a; Elrayah et al. 2010), except in the report of Faye et al. (2005) in the UAE where the plasma copper was significantly higher in female (61.9 μg/100 ml) compared to male (56.7). Variations during gestation were reported (Eltohamy et al. 1986; Liu et al. 1994), similar to those recorded in sheep: decrease of the cupremia at the end of gestation and return to normal after calving. This would be due to an active transfer of copper stored in the liver, from the mother to the fetus. Moreover, Tartour and Idris (1970) showed that liver copper content of fetus was higher at early embryonic life than newborn. Globally, cupremia would be lower during pregnancy than lactating period (Kuria et al. 2013), but it is not confirmed by other authors (Seboussi et al. 2004; Faye et al. 2005). Contradictory data on the effect of age are as follows: no significant difference reported (Faye and Mulato 1991; Ghosal and Shekhawat 1992; Bengoumi et al. 1995a; Hussein et al. 1982; Faye et al. 2005), higher cupremia in animals over 5 years (Marx and Abdi 1983) or over 10 years (Elrayah et al. 2010), and declining cupremia with parities (Kuria et al. 2013). For Pourjafar et al. (2014), copper concentration in serum is lower in young camels than in adults (4–6 years old) and then declines regularly in older animals. During the first year of life in newborn camel, serum copper increased from birth (at around 50 μg/100 ml) to 5 months old and then was stabilized at 63–76 μg/100 ml (Hussein et al. 1992). A breed difference was reported in Saudi Arabia with significant higher cupremia (71 μg/100 ml) in Majaheem breed (black-coat breed) compared to Maghateer (white-coat breed) (58 μg/100 ml), both breeds receiving similar diet (Abdelrahman et al. 2013a). Such breed difference was not observed by Faye et al. (2005) as well as by Deen et al. (2004).

The season effect was not reported, but the seasonal variation of copper intake according to its concentration in plant species selected by camel could have impact on the blood concentration (Shamat et al. 2009). Globally, despite its liver storage, plasma copper concentration could be considered as a good indicator of copper nutrition; the effect of copper supplementation has a significant effect on the copper status of the animal (Kinne et al. 2003; Abu Damir et al. 2008; Osman 2012; Kosanovic et al. 2014). In Sudan, with a diet enriched in concentrates, Abdel Rahim (1983) got plasma copper concentration (125 μg/100 ml) two times higher than in camels on free range (69 μg/100 ml). In Saudi Arabia, the use of rumen bolus improved the copper status of growing camel, the serum copper passing from 75 to 112 μg/100 ml (Ibrahim et al. 2016). Globally, the distribution of any mineral supplementation including copper may improve blood copper status of camel affected by rickets or osteomalacia (Liu and Ma 1995a; Liu 2005), but this improvement depends on the effect of antagonistic minerals as sulfur and molybdenum in the diet (Elhuda and Osman 2012).

Cupremia appears to increase with dehydration of the animal (Mohamed et al. 1984), but it could be a simple effect of hemoconcentration due to plasma volume losses. A decrease of the plasma copper was reported in case of testicular degenerative injury and of positive serology for brucellosis (Ahmed and Nada 1993). Moreover, serum copper is lower in case of infertility (Saini et al. 2009), and copper deficiency appeared to be a predisposing factor to septicemia in camel calves (Wernery et al. 2002).

Every inflammation accompanying or not an infectious disease can lead to a remarkable increase in plasma copper. A hypercupremia from inflammatory origin differs from the hypercupremia of toxic origin by the fact of its association, in the first case, to hypozincemia. However, if these changes were noted in sheep (Lamand and Levieux 1981) or goat (Faye et al. 1990), no data are currently available for the dromedary. No significant difference, for example, was revealed in camel affected by mastitis or not (Tuteja et al. 2004) or by internal parasite as echinococcosis (Heidarpour et al. 2012).

7.4.1.2 Liver Copper

Usually, the liver is the organ-storing copper. In fact, the diagnosis of copper deficiency becomes more relevant if added to blood tests is the assessment of levels in the liver tissue. The amounts of tissue obtained by biopsy are sufficient for determining copper concentration and, more generally, other mineral elements. However, there is little literature results of biopsy, as most samples are made on livers being removed at the slaughterhouse (Table 7.3).

Observed values vary considerably according to the data sources. The animal status (deficient or not) can explain a part of the observed variability. For example, in Chinese Bactrian camel, copper concentration in liver was ninefold more important in healthy camel (106 ± 11 ppm) compared to camel affected by "emaciation ailment" syndrome (14 ± 3 ppm) (Shen and Li 2010). However, methods of determination are not always specified, and it is likely that the observed differences between authors are partly related to differences in analytical procedures.

Two studies specifically were focused on the dynamics of the hepatic copper storage in deficient animals. In Djibouti, Faye et al. (1992a) studied the change in

Table 7.3 Liver copper concentrations (in ppm) according to different authors

Reference	Copper liver (ppm)	Sampling	Animal status (*n*)	Country
Tartour (1969)	163 ± 112	Slaughterhouse	Non-deficient (55)	Sudan
Khalifa et al. (1973)	154 ± 18	Slaughterhouse	Non-deficient (26)	Egypt
Tartour (1975)	275 ± 76	Slaughterhouse	Non-deficient (9)	Sudan
Hussein et al. (1982)	21.3 ± 10.6	Slaughterhouse	Non-deficient (36)	Egypt
Abu Damir et al. (1983)	50 ± 36	Slaughterhouse	Sub-deficient (36)	Sudan
Faye et al. (1992a)	41 ± 5	Biopsy	Deficient (5)	Djibouti
Faye et al. (1992a)	64 ± 16	Biopsy	Supplemented (3)	Djibouti
Faye et al. (1992a)	110 ± 57	Dead animals	Deficient (3)	Djibouti
Wensvoort (1992)	43 ± 38	Slaughterhouse	Non-deficient (5)	UAE
Wensvoort (1992)	42 ± 37	Dead animals	Non-deficient (24)	UAE
Ma Zhuo (1995)[a]	448 ± 213	Slaughterhouse	Ataxia (10)	China
Ma Zhuo (1995)[a]	538 ± 308	Slaughterhouse	Non-deficient (15)	China
Bengoumi et al. (1998a)	9.5 ± 14.5	Biopsy	Deficient (5)	Morocco
Liu (2003)[a]	230–538.2	Slaughterhouse	Non-deficient	China
Badiei et al. (2006)	51.5–72.6	Slaughterhouse	Non-deficient (26)	Pakistan
Abdelrahman et al. (2013a)	20 ± 1.3	Slaughterhouse	Non-deficient (30)	S. Arabia
Bakhiet et al. (2007)	103 ± 12.3	Slaughterhouse	Non-deficient (325)	Sudan
Ibrahim et al. (2013)	100.7 ± 5.2	Slaughterhouse	Non-deficient (100)	Sudan
Chafik et al. (2014b)	14.2 ± 6.1	Slaughterhouse	Non-deficient (30)	Morocco
Li and Hai (2014)[a]	13.6 ± 3.1	Biopsy	Deficient (15)	China
Li and Hai (2014)[a]	105.6 ± 11.2	Biopsy	Non-deficient (5)	China

[a]Bactrian camel

hepatic copper concentration on young deficient camels, according to the proposed type of trace element supplementation. Thus, mineral copper supplementation further increased the concentration of hepatic copper that animals also received a protein-energy supplementation.

In Morocco (Bengoumi et al. 1998a; Faye and Bengoumi 1997), the comparative study between cattle and camels receiving identical food intake is shown (Fig. 7.2):

1. Lower concentration in copper in the dromedary compared to cow (on average 2.6 times less).
2. Secondly, kinetics of copper storage and release significantly different (slower storage and faster release in the dromedary as in cows), primarily related to the most deficit status of camels in the specific case of quoted experimentation.

At reverse, in a large number of livers collected in slaughterhouses, Bakhiet et al. (2007) reported a significant higher value of copper concentration in camel liver (103 ± 12.3 ppm) compared to that of cattle (88 ± 9.8), sheep (65.5 ± 8.1), and goat (54.6 ± 4.1). No significant sexual difference was observed in the dromedary (Tartour 1969). Physical activity seems not to affect hepatic copper content, with no difference being observed between the racing camel and pack dromedary camel (41.0 ± 40.3 vs 43.7 ± 35.4 ppm according to Wensvoort 1992).

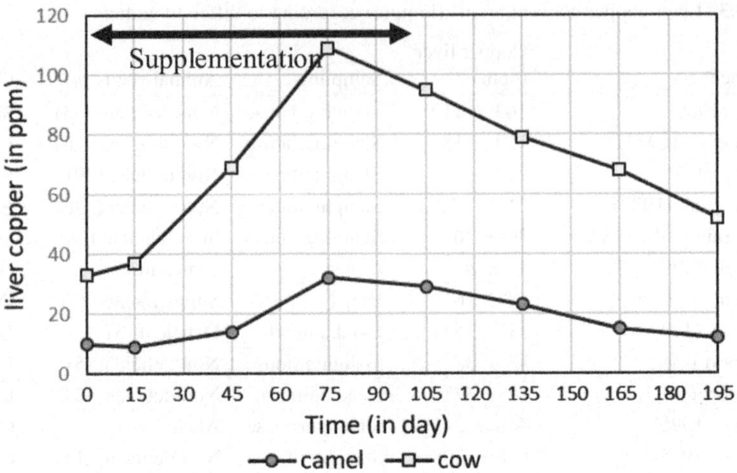

Fig. 7.2 Changes in liver copper levels in cows (solid diamonds) and camels (open circles) (calculated from Bengoumi et al. 1998a)

At reverse, it is noted differences related to the age of the animals. With an average concentration of 30 ± 25 ppm, hepatic copper in dead young camels levels appeared lower than those of dead adults: 42 ± 37 ppm (Wensvoort 1992). A similar difference was observed in animals from the slaughterhouse: 28 ± 22 in camels of less than a year vs 43 ± 39 ppm for mature females (Wensvoort 1992). On the other hand, in the fetus, the concentration of hepatic copper was significantly higher (93.4 ± 42.9 ppm according to the same author), which corroborates the hypothesis of an active transfer from the mother to the fetus especially during the first third of gestation (Tartour and Idris 1970).

In any case, the concentration of hepatic copper remains strongly influenced by nutritional status of the animal, and the hepatic storage capacity of the dromedary appears in light of the overall results of literature, rather lower than that of cattle and sheep. The use of bolus enriched in copper increased the copper concentration in liver from 19.7 to 48.7 ppm (Ibrahim et al. 2016). A breed effect was recently reported (Abdelrahman et al. 2013a). Elsewhere, Ma (1995) described lipomatous liver damage in Bactrian camels, associated with proven copper deficiency.

7.4.1.3 Copper in Milk

Data on milk copper content are scattered and rarely focused on mineral change according to the lactation stage or on the variation factors of the mineral composition. It is therefore difficult to have a precise idea of the biological standards. Indeed, very important gap is observed between the reported values, but globally copper content in camel milk is about 30 to 800 μg/100 ml. High variability is generally reported: In the study of Omer and Eltinay (2008) including 350 milk samples from different parts of the UAE, the range was 0.001–0.118 mg/100 g with an average of 0.04 mg/100 g. From after Abdelrahim (1987), the copper content for camel was similar to that of goat: 0.65 (0.10–0.72) for camel vs 0.63 mg/100 g (0.35–1.10 mg/100 g) for goat. In India, similar values were reported in camel milk (0.156 ± 0.21 mg/100 g) and cow, goat, or sheep milk (Saini et al. 2007). Sawaya et al. (1984) observed lower values (0.16 ± 0.02 mg/100 g), and Ahmed et al. (1977) reported that camel milk was 12 times richer in copper than cow's milk (0.49 mg vs 0.013 mg/100 g). Soliman in 2005 reported also higher value in camel milk (0.061 ± 0.0023 g/100 g) compared to cow (0.017), buffalo (0.04), or goat milk (0.04). In Saudi Arabia, Al-Wabel (2008) reported higher value in camel milk (0,161 mg/100 g) than in goat (0.057) or sheep (0.062) but similar to cow milk (0.180) as for Al-Awadi and Srikumar (2001). For Bengoumi et al. (1998b), the copper content of camel milk reached 1.13 ± 0.49 mg/100 g, content close to that reported by Gnan and Sheriba (1986). In Kazakhstan, the copper concentration in milk is assessed to 5–7 μg/100 ml (Diacono et al. 2008b; Konuspayeva et al. 2011) with significant regional differences. For Dell'Orto et al. (2000), trace element supplementation did not change significantly the copper content in milk: 0.037 mg/100 g in non-supplemented camel vs 0.040 in supplemented ones. In the Bactrian camel, values around 0.47 mg/100 mg DM was reported (Martynenko et al. 1977).

If contradictory data in the literature allow affirming that camel milk is relatively richer in copper than cow milk, there is no information in relation to nutritional (deficiency or toxicity) or inflammatory status (Mehaia et al. 1995). It also appears a wide variation between farms in mineral composition milk. This variability is resulting from dietary conditions which may strongly differ from one region to another (Rashed 1992). The composition of camel milk, in particular its richness in trace elements, led to propose a multi-year cure made from camel milk to treat human pulmonary tuberculosis (Urazakov and Bainazarov 1974).

7.4.1.4 Other Biological Substrates

Wool and Hair

Considered in the 1960s as a good indicator of the mineral status of domestic ruminants, and in particular of certain mineral poisoning, the analysis of the hair gradually was abandoned due to higher individual variability observed and especially to the risk of contamination during sampling (Burns et al. 1964), hair behaving as an ion-exchange resin. Indeed, hair captures easily environmental contaminations and does not reject them during washing. Therefore, analytical results are not sufficiently reliable (Lamand 1987). However, as a potential wool producer (Bakhat et al. 2003), the mineral composition of hair can occasionally provide additional useful information, particularly as it is recognized that copper deficiency has a direct effect on the wool quality (Ryder and Stephenson 1968), although the changes in mineral composition of hair are significant only for chronic deficiencies or intoxications during hair's growth. Unfortunately, the accessible literature remains low and is related exclusively to the Bactrian camel, best producer of wool as the dromedary.

Pilaris copper content would be a sensitive indicator of deficiency because of the correlation between liver copper and hair copper (Wang 1988). The limit of deficiency would be like cattle to 5.5 µg/g. In their study, Liu et al. (1994) observed hair copper level of 3.5 ± 1.0 µg/g in ataxic animals vs 6.4 ± 1.2 in non-deficient animals. On the other hand, Ma (1995) observed no difference between deficient animals (4.6 ± 1.5 µg/g) and non-deficient (4.7 ± 0.8 µg/g). Indeed, copper supplementation in diet has no significant effect on copper concentration in Bactrian wool (Liu and Ma 1995a). Copper in hair seems to be influenced by the pregnancy status, resulting in a significant decrease: 4.3 ± 1.3 µg/g in empty camel vs 3.0 ± 0.7 µg/g in pregnant females (Liu et al. 1994). There was no significant difference between male (4.95 ± 2.15 µg/g) and female (4.69 ± 1.87 µg/g) in dromedary hair (Badiei et al. 2006). However, high copper variability was observed according to the color and fiber diameter (Helal 2015): in coarse brown camel hair fibers, copper concentration was significantly higher (26.2 µg/g) than in the white coarse fibers (11.9 µg/g) and overall than in the finest fibers (8.2 µg/g).

Feces and Urine

Copper in feces or urine is only an indicator of levels of excretion in relation to levels of ingestion of the same element. So, the determination of fecal or urinary copper has of no interest, especially as the classical factorial method used to determine the true absorption coefficient is not applicable in the case of trace elements and net requirements being poorly known (Gueguen et al. 1988).

With a daily intake of 35 mg of copper per os, daily fecal excretion, calculated on the dry matter, varies from 2.5 to 6.2 ppm. With a daily intake of 275 mg (i.e., 7.8 times more), fecal excretion is multiplied by 13.8 (Faye et al. 1999) and varies between 27 and 50 ppm. Compared to cattle, the concentration of fecal copper in the dromedary is significantly higher, especially during mineral supplementation. Fecal excretion seems more important than cattle receiving the same basal diet, which leads to consider that the net copper requirement in the camel species are probably lower than those of other ruminants (Bengoumi et al. 1998a).

Urinary excretion of copper is insignificant as all positive ions. Urinary copper remains generally less than 0.0001 ppm (non-detectable by the usual measuring instruments). However, after intake of excessive copper, concentration up to 5 ppm has been observed in some animals (Faye and Bengoumi 1997).

Other Substrates

Other biological substrates have not routine clinical interest. However, some results of analyses are available from literature (Table 7.4).

Table 7.4 Copper level in different organs of camels according to different authors

Reference	Sampled tissue	Copper level (ppm)	Country
Tartour (1969)	Spleen	11.6 ± 6.0	Sudan
Tartour (1975)	Spleen	29.6 (6.4–43.0)	Sudan
Abu Damir et al. (1983)	Kidney	7.2 ± 3.5	Sudan
Ma (1995)[b]	Kidney	23.4 ± 9.9	China
Ma (1995)[b]	Kidney	11.3 ± 3.9[a]	China
Abdelrahman et al. (2013a)	Kidney	60.7 ± 21	Saudi Arabia
Chafik et al. (2014b)	Kidney	1.43 ± 0.14	Morocco
Abdelrahman et al. (2013a)	Muscle	14.4 ± 3.3	Saudi Arabia
El-Faer et al. (1991)	Muscle	0.8 (0.7–0.9)	Saudi Arabia
Ma (1995)	Muscle	8.7 ± 2.9	China
Ma (1995)[b]	Heart	11.3 ± 2.5	China
Chafik et al. (2014b)	Heart	2.06 ± 0.22	Morocco
Chafik et al. (2014b)	Lung	1.65 ± 0.49	Morocco
El-Faer et al. (1991)	Hump	0.3 ± 0.3	Saudi Arabia
Ma (1995)[b]	Brain	15.2 ± 2.0	China
Ma (1995)[b]	Bone	5.8 ± 1.9[a]	China
Ma (1995)[b]	Teeth	4.8 ± 0.3[a]	China

[a]Deficient camels
[b]Bactrian camel

Except in the kidney, no effect of the deficiency status of the animal is observed on the copper content in tissues (Ma 1995). Tartour (1969) did not reveal sexual difference on splenic copper content, and Abu Damir et al. (1983) noted no difference in renal copper between camel, sheep, and cow.

In synovial fluid, the copper concentrations determined in healthy and arthritic camels show no significant difference, 20.5 ± 0.5 vs 24.8 ± 0.1 μg/100 ml, respectively (Chalmeh et al. 2016).

7.4.1.5 Copper Toxicity

Copper intoxication was never described in the dromedary in the natural state, but a test of experimental poisoning was undertaken (Abu Damir et al. 1993). Daily intravenous injection of 200 mg of copper causes death within 8 days. An injection of 100 mg was long-term lethal (3 months) in one of the tested camels, the second one surviving until it was slaughtered 6 months after the start of the poisoning. The main symptoms observed were anorexia, frequent regurgitation of the gut contents, teeth gnashing, diarrhea, and lateral decubitus position before death. In case of chronic poisoning, losses of appetite, paleness of the mucous membranes, and progressive weakness were observed but no sign of jaundice as for sheep poisoning. The rate of copper in serum increases by a factor of 2–10 according to the animal.

In case of intoxication by anti-coccidian like nasarin, copper concentration in serum significantly increased (from 0.63 to 0.79 μg/100 ml), but the values stayed in normal range (Abu Damir et al. 2013).

7.4.2 *Zinc*

Tropical and subtropical forages are frequently deficient in zinc, and the potential deficiencies in the livestock only dependent on forage resources in the natural environment are common (Faye et al. 1986, 1990) in spite of unclear clinical expression of those deficiencies.

7.4.2.1 Plasma or Serum Zinc

Like copper, plasma zinc in all species of domestic ruminants oscillates between 70 and 120 μg/100 ml. In the dromedary, latest data argue in favor of the idea that a normal regulation of zincemia occurs at a lower level. The deficiency threshold could be around 40 μg/100 ml (Faye et al. 1992b; Bengoumi et al. 1995b, 1998a). Given that there is no zinc storage in the body, zincemia is an excellent early indicator of zinc deficiency, but the risks of contamination of the samples are frequent. Given these risks, we might wonder on some results of the literature indicating high values higher than 70 μg/100 ml (Table 7.5).

The values reported by Abdelrahim in 1983, for example, (284–309 μg/100 ml) are to be taken into consideration with caution. Some values reaching more than 1400 μg/100 ml in Bactrian camel by Ma (1995) and Liu and Ma (1995a) cannot be retained also because being probably results of analysis of whole blood. Indeed, the erythrocytes are very rich in zinc.

In comparative studies involving several species, in almost all cases, camel presented lower values which are validating the hypothesis of a different regulation of zinc metabolism in this species (Table 7.6).

Compared to cattle receiving the same basal diet, camel zincemia was twofold lower and did not appear influenced by mineral supplementation (Bengoumi et al. 1998a): with a daily zinc supply of 1000 mg, on average the serum zinc concentration changed from 35 to 36 μg/100 ml in camel vs 73 to 84 μg/100 ml in cow (Fig. 7.3). The low impact of zinc diet on plasma zinc concentration was confirmed by Fahmy et al. (2004).

The signification of a low zincemia is generally the sign of a deficient supply in zinc in the diet or an acute inflammatory reaction. This one being accompanied by hypercupremia, the association of high cupremia and low zincemia leads to a diagnosis of acute inflammatory process, linked to infectious diseases. For example, Ali et al. (2010) have observed a significant decline of zincemia in camels affected by reproductive disorders. Generally, plasma zinc concentration is lower in infertile camels, but the difference is not significant (Saini et al. 2009). Zincemia is also lower

Table 7.5 Values of plasma or serum zinc concentration in camel according to different authors (in µg/100 ml)

Reference	Mean values	Range or SD	Number	Country
Moty et al. (1968)	135	± 4.1	19	Egypt
Faye et al. (1986)	100.4	81–160	53	Ethiopia
Whabi et al. (1979)	118.3	± 28.8	96	Sudan
Abdalla et al. (1988)	41	37–46	25	UAE
Eltohamy et al. (1986)	93.4	± 4.2	45	Egypt
El Kasmi (1989)	107	± 3	60	Morocco
Faye and Mulato (1991)	46.2	± 17.2	52	Djibouti
Ghosal and Shekhawat (1992)	85.4	± 2.5	122	India
Singh et al. (1994)	101	± 4.1	8	India
Faye et al. (1995)	34,6	± 7.8	65	France
Bengoumi et al. (1995a)	34	± 7	30	Morocco
Liu (2003)[a]	14.9	± 3.1	30	China
Nafizi et al. (2009)	70	± 1.0	30	Iran
Shukla et al. (2009)	105.6	± 2.1	34	India
Parekar et al. (2009)	186.7	±15.1	25	India
Wernery et al. (2009)	50	± 10	6190	UAE
Eltahir et al. (2010)	106.8	±28.6	30	UAE
Ali et al. (2010)	107	± 10	15	Saudi Arabia
Mustafa et al. (2012)	24.5	± 15.8	20	Sudan
Heidarpour et al. (2012)	25.4	± 3.3	30	Iran

[a]Bactrian camel

Table 7.6 Concentrations in plasma or serum zinc (and number of sampled animals) in various species of domestic ruminants (in µg/100 ml)

Reference	Camel (*n*)	Cattle (*n*)	Goat (*n*)	Sheep (*n*)	Country
Moty et al. (1968)	135 (19)	144 (29)	–	160 (32)	Egypt
Shekhawat (1983)	85.4 (122)	86.8 (29)	–	94.8 (34)	India
Faye et al. (1986)	100.4 (53)	113.5 (432)	107.7 (173)	114.2 (425)	Ethiopia
Faye et al. (1991)	46.2 (52)	97.6 (59)	65.6 (118)	71.5 (80)	Djibouti
Bengoumi et al. (1998a)	38 (5)	83 (5)	–	–	Morocco
Al-Busadah (2003)	103.4 (5)	98.5 (5)	–	110.7 (5)	S. Arabia
Khamis et al. (2011)	104.4	96.7	–	–	Egypt

in parasitized animals (Heidarpour et al. 2012). However, Tuteja et al. (2004) reported an increase of plasma zinc concentration in camels affected by mastitis and a positive correlation between plasma zinc level and level of somatic cell count. The dosage of haptoglobin, specific protein of inflammation, could eliminate the hypozincemia of inflammatory origin, but there is no data on the literature regarding camel.

The variation of zincemia according to age and sex was not commonly studied (Faye et al. 2008b). Camel calves have generally higher zincemia (Elkasmi 1989;

Fig. 7.3 Change in the plasma concentration of zinc in camels (open circles) and cows (open square) (from Bengoumi et al. 1998a)

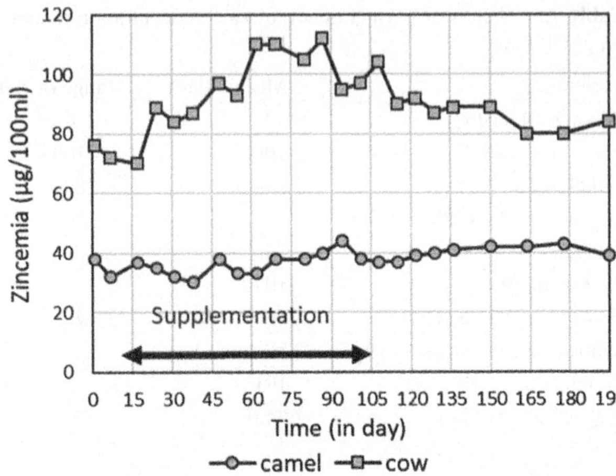

Bengoumi et al. 1995a), but zincemia is stable all along the first year of life (Hussein et al. 1992). For Faye and Mulato (1991) and Seboussi et al. (2004), the age is a discriminating parameter of plasma zinc. The higher zinc concentrations in the young camels could be due to the milk feeding, the milk providing wide quantity of zinc (Bengoumi et al. 1998b). At reverse, no significant difference was reported by Ghosal and Shekhawat (1992), Elrayah et al. (2010) as well as by Pourjafar et al. (2014).

No sexual difference was observed although a significant decrease of zincemia was reported at the end of gestation (Eltohamy et al. 1986; Sena et al. 2007), probably because of an active transfer to the fetus in the second phase of pregnancy (>6 months). For Seboussi et al. (2004), plasma zinc concentration is higher in male (24 μg/100 ml) compared to female (14.1 μg/100 ml). It is also higher in non-pregnant camels (26.8 μg/100 ml) compared to pregnant (10.5 μg/100 ml) and overall compared to lactating ones (1.8 μg/100 ml). Similar observation was reported in India (Vyas et al. 2011). Some breed differences were reported in Saudi Arabia (Abdelrahman et al. 2013a) with significant higher zinc serum concentration in Majaheem breed (black color camel) compared to Maghateer breed (white color camel): 82.3 vs 76.0 μg/100 ml. In India, significant difference was also observed between Jaisalmeri, Bikaneri, and Kacchi breeds (Deen et al. 2004), but with zinc values between 660 and 1025 μg/100 ml, the results are quite debatable. At reverse, in UAE, Seboussi et al. (2004) did not report significant difference between local breed and Sudanese breed. Zinc concentration in plasma decline as the number of parities increased (Kuria et al. 2013).

A proteo-energetic supplementation improves the zincemia of camel having normal values (Abdelrahim 1983) in accordance with the observations of Lamand (1985) in sheep. However, in deficient animals, Faye et al. (1992b) did not observe significant change of zincemia after proteo-energetic supplementation contrary to cupremia (Fig. 7.4).

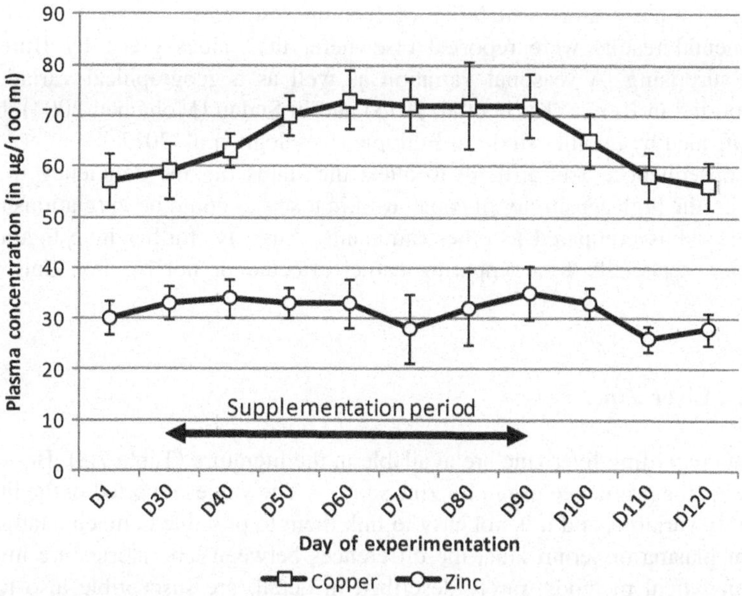

Fig. 7.4 Copper and zinc concentration in plasma of young camels receiving mineral and proteo-energetic supplementation between day 30 and day 90 (according to the experiment of Faye et al. 1992b)

Table 7.7 Mean values and standard deviation of cupremia and zincemia (in µg/100 ml) observed according to different types of supplementation in trace elements (from after Bengoumi et al. 1995b)

	Type of mineral supplementation			
Mineral	Cu + Zn by IM	Cu per os	Zn per os	Control
Copper	56 ± 8	55 ± 4	41 ± 3	44 ± 2
Zinc	53 ± 10	45 ± 2	52 ± 5	46 ± 2

Zinc supplementation per os contributes to an important decrease of cupremia. The reverse could be observed, but the effect is less marked (Bengoumi et al. 1995b). Indeed, a mutual negative interaction between copper and zinc (Table 7.7) during the absorption process in the intestinal mucous, phenomena was already observed in other species (Yu and Beynen 1994). In cattle, for example, a diet highly enriched in zinc decreases copper concentration in plasma until it reaches the deficient level (Towers et al. 1981). Such interaction is due to competition in transport of trace element at the absorption sites of the intestinal wall (Mc Dowell 1992).

The use of trace element bolus including zinc increases the plasma zinc concentration in growing camel (Alhidary et al. 2016). Burenbayar (1989) reported, in Bactrian camel, an important seasonal variability, zincemia being multiplied by 4 between December and May, passing from 41.2–43.7 to 143.1–160.5 µg/100 ml according to the type of supplementation. Probably, those seasonal changes are in relation to quality and availability of feeding resources. However, because the experimental results were reported elsewhere, the values given by Burenbayar appear surprising. A seasonal variation as well as a geographical variation was also reported in Kenya (Kuria et al. 2013) and in Sudan (Mohamed 2004), but it is not confirmed by another study in Ethiopia (Desalegn et al. 2012).

In consequence, it is difficult to attest the status of zinc deficiency in camel. However, the high sensitivity of camel to skin diseases could be in relationship with low zinc status compared to other ruminants. Anyway, further investigations are necessary, especially for comparing camel affected or not by any kind of skin diseases.

7.4.2.2 Liver Zinc

Few data regarding liver zinc are available in the literature (Table 7.8). Besides, the liver is not the privileged organ for zinc storage. The values reported in the literature are highly variable, and it is not easy to link them to possible deficient status. More than for plasma or serum zinc, the differences between laboratories are important since analytical methods, rarely described in detail, are susceptible also to differ among studies.

Contrary to liver copper, liver zinc in camel is not a good indicator of zinc deficiency. Besides, the mineral supplementation doesn't modify the concentration in liver zinc, contrary to copper (Fig. 7.4). Even, in camel, a slight decrease of liver zinc was observed in case of zinc supplementation per os (Bengoumi et al. 1998a, b, c). A proteo-energetic supplementation, associated or not to trace element supplementation, did not modify also the liver zinc concentration (Faye et al. 1992b). However, with supplementation by mineral bolus, a significant increase in liver zinc concentration was described, passing from 31.3 ppm (control group) to 42.8 ppm (growing camel receiving one long-acting bolus) and 52.1 ppm with two boluses (Alhidary et al. 2016).

Wensvoort (1992) reported a slight difference between hepatic zinc concentration in racing camel (163 ± 64 ppm) and pack camel (224 ± 86 ppm). According to the same author, a difference occurred between fetuses and dead young camel calves, respectively, 317 ± 117 and 389 ± 159 ppm. At reverse, there is no difference observed between young and adults in livers sampled at the slaughterhouse: 181 ± 58 and 182 ± 30, respectively (Wensvoort 1992). Contrary to copper the hypothesis of an active transfer of zinc in the fetus is not verified.

Comparatively to cattle receiving similar diet (Fig. 7.5), no difference is reported (Bengoumi et al. 1998a), for mean concentration of liver zinc and for the changes according to mineral supplementation containing 240 mg Cu and 1 g Zn/day as well.

For Ibrahim et al. (2013) and Bakhiet et al. (2007), liver zinc concentration in camel is quite lower than in sheep, goat, and cattle liver. However Al-Busadah (2003) and Khamis et al. (2011) did not confirm this. A significant difference was reported between Majaheem breed and Maghateer breed in Saudi Arabia (Abdelrahman et al. 2013a).

Table 7.8 Concentration in hepatic zinc (in ppm) according to different authors

Reference	Liver zinc (in ppm)	Type of sampling	Status of animals (*n*)	Country
Awad and Berschneider (1977)	143 ± 5	At slaughter	Non-deficient (9)	Egypt
Abu Damir et al. (1983)	40 ± 18	At slaughter	Non-deficient (36)	Sudan
Faye et al. (1992b)	116 ± 30	By biopsy	Deficient (5)	Djibouti
Faye et al. (1992b)	123 ± 31	By biopsy	Supplemented (3)	Djibouti
Faye et al. (1992b)	309 ± 78	Dead animals	Deficient (3)	Djibouti
Wensvoort (1992)	182 ± 30	At slaughter	Non-deficient (5)	UAE
Wensvoort (1992)	185 ± 75	Dead animals	Non-deficient (24)	UAE
Ma Zhuo (1995)[a]	136 ± 50	At slaughter	With ataxia (10)	China
Ma Zhuo (1995)[a]	139 ± 39	At slaughter	Non-deficient (15)	China
Bengoumi et al. (1998a)	43 ± 5	By biopsy	Deficient (5)	Morocco
Al-Busadah (2003)	149 ± 9.6	At slaughter	Non-deficient (5)	S. Arabia
Liu (2003)	150 ± 40	At slaughter	Non-deficient (50)	China
Bakhiet et al. (2007)	35 ± 1	At slaughter	Non-deficient (325)	Sudan
Ibrahim et al. (2013)	33 ± 2.3	At slaughter	Non-deficient (100)	Sudan
Chafik et al. (2014b)	11 ± 1.7	At slaughter	Non-deficient (30)	Morocco

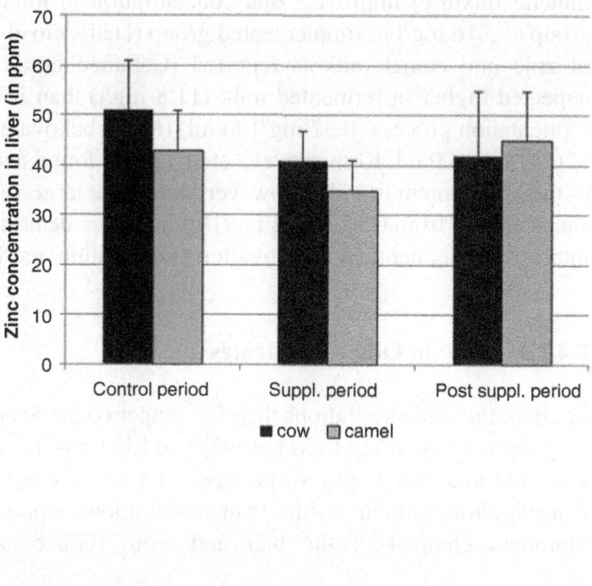

Fig. 7.5 Comparative zinc liver kinetic in camel and cow receiving similar diet and mineral supplementation (according to Bengoumi et al. 1998a)

Because of the lack of plasma and hepatic storage of zinc, the potential zinc storage form and excretion way have to be investigated.

7.4.2.3 Zinc in Milk

A high variability is observed between the authors. It is difficult to establish reference values. Contrary to copper, it has been reported no difference in zinc concentration in camel milk and in cow milk: 0.44 mg/100 g DM for camel vs 0.39 mg/100 g in cow (Sawaya et al. 1984). A similar observation was done by Wang et al. (2011) and Diacono et al. (2008a, b) in Bactrian camel from China and Kazakhstan, respectively. In wild Bactrian camel, zinc in milk is in similar quantity (0.64 mg/l) and comparable to domestic Bactrian (0.59 to 0.81 mg/l) but higher than cow (0.20 mg/l) living in the same area (Jirimutu et al. 2010). Abdelrahim (1987), comparing several dairy species, considered that the zinc concentration in camel milk (0.23 $\pm$ 0.06 mg/100 ml) was significantly lower than in goat milk (0.55 $\pm$ 0.02 mg/100 ml). Reporting about the whole milk, Bengoumi et al. (1998b) reported comparable values in camel (2.87 $\pm$ 0.8 mg/l) than in cow (2–5 mg/l according to Lamand (1974). However, other authors reported significant differences, for example, for Gorban and Izzeldin (1997), there is higher value in camel milk (4.9 mg/l) compared to cow milk, and for Soliman (2005) higher zinc value occurred in camel milk (0.51 mg/100 g) than other species. In Kazakhstan, if zinc concentration is higher in camel milk than cow milk, it is quite lower than horse milk (0.38 vs 0.23 and 3.6 mg/100 g, respectively) (Diacono et al. 2008a, b).

There are few data on the effect of feeding status or physiological stage on zinc concentration in camel milk. Contrary to copper, zinc supplementation (7000 ppm in mineral mixture) improved zinc concentration in milk from 2.52 mg/l in control group to 3.16 mg/l in supplemented group (Dell'Orto et al. 2000). An active transfer of zinc into camel milk is reported (Cattaneo et al. 2005). Zinc concentration appeared higher in fermented milk (11.8 mg/l) than in raw Bactrian milk used for fermentation process (0.37 mg/100 ml) (Meldebekova et al. 2008). With an average of 0.47 mg/100 ml, Konuspayeva et al. (2011) found a significant regional variation of the zinc content in milk. However, due to the large variation observed in different camel farms (from 0.15 to 7.4 mg/100 ml), zinc determination in milk is of limited interest for biochemical investigation (Elhardallou and El-Naggar 2016).

7.4.2.4 Zinc in Other Substrates

For hair, the same hesitations than for copper could be advanced. Moreover, data are very scarce. According to Ma (1995) and Liu (2003), Bactrian camel wool is rich in zinc, 147 and 146.9 ppm, respectively, i.e., on average 35 times more than copper concentration, similar results than observations reported on sheep, zinc being an important element of the hair and wool (Grace and Lee 1992; Sztych and

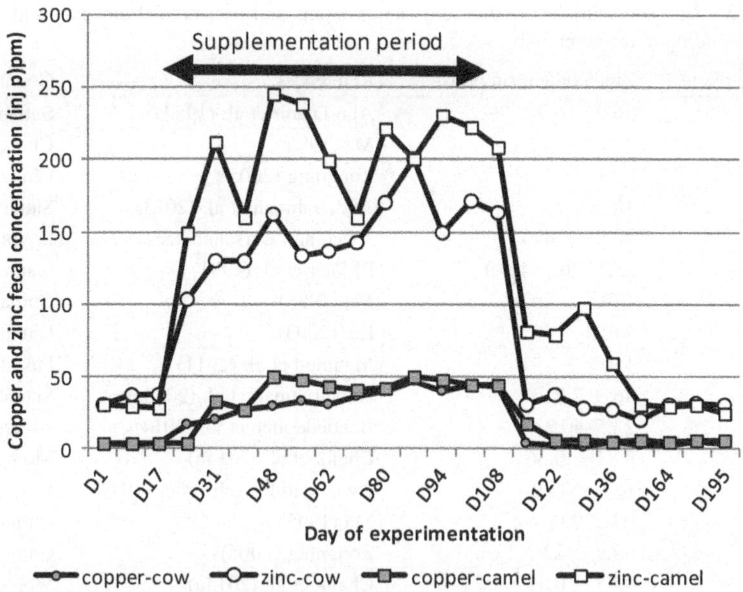

Fig. 7.6 Comparative changes in zinc and copper fecal excretion in camel and cow during mineral supplementation between day 22 and day 121 (from after Bengoumi et al. 1998a)

Soroczynska 1995). In camel, however, Singh et al. (1994) reported higher zinc values in hair: 280 ± 3 ppm. As for copper, high zinc variability was observed according to the color and fiber diameter (Helal 2015): In coarse brown camel hair fibers, zinc concentration was significantly higher (374.2 μg/g) than in the white coarse fibers (85 μg/g) and in the finest fibers (104.7 μg/g). The ratio Zn/Cu in camel hair varied from 5 to 12 according to the type of fiber.

Zinc concentrations in feces and urine reflected the level of ingestion in this element, and the clinical interest is limited for the same reasons than for copper.

With a daily supply of 290 mg zinc per os, the daily zinc excretion reported to the DM varied from 25 to 29 ppm. With a daily supply of 1290 mg (5 times more), the fecal excretion is multiplied by 7 (Faye and Bengoumi 1997). Compared to cattle receiving the same diet, the concentration in camel fecal zinc was significantly higher, particularly after mineral supplementation (Fig. 7.6).

After intensive zinc supplementation (1 g/day), the fecal concentration is high, contrary to cattle. Such results suggest that there is no tissue storage of zinc since the entire zinc intake after supplementation period is found in feces. As the whole, fecal excretion is finally higher in camel than in cow, in spite of a lower zincemia, this suggests lower net requirements of zinc than in other ruminants (Bengoumi et al. 1998a).

Urinary excretion, as for copper, is insignificant. Urinary concentration of zinc is generally less than 0.0001 ppm (non-detectable by usual apparatus) including in case of important zinc supply (Faye et Bengoumi 1997).

Table 7.9 Zinc concentration in different camel organs and tissues of dromedary and Bactrian camel according to different authors

Sample tissue	Zinc content (in ppm)	References	Country
Kidney	26.0 ± 6.9	Abu Damir et al. (1983)	Sudan
	83.6 ± 14.1	Ma (1995)[a]	China
	80.1	Zongping (2005)[a]	China
	18.3 ± 1.4	Abdelrahman et al. (2013a)	Saudi Arabia
Muscle	204.5 ± 47	Awad and Berschneider (1977)	Egypt
	38.1 (30.7–48.0)	El-Faer et al. (1991)	Saudi Arabia
	179.5 ± 40.5	Ma (1995)[a]	China
	179.3 ± 48.7	Liu (2003)[a]	China
	150	Mahmud et al. (2011)	Pakistan
	46 ± 2.2	Abdelrahman et al. (2013a)	Saudi Arabia
	23.5–40.2	Bendeddouch et al. (2014)	Algeria
	9.84 ± 0.36	Chafik et al. (2014a)	Morocco
Heart	187 ± 23.9	Awad and Berschneider (1977)	Egypt
	99 ± 23	Ma (1995)[a]	China
	98.9 ± 22.7	Zongping (2005)[a]	China
	4.85 ± 0.41	Chafik et al. (2014a)	Morocco
Hump	0.0	El-Faer et al. (1991)	Saudi Arabia
Lung	89.2 ± 30.6	Liu (2003)[a]	China
	96.7	Zongping (2005)[a]	China
	4.05 ± 0.15	Chafik et al. (2014a)	Morocco
Pancreas	88.0 ± 32.3	Liu (2003)[a]	China
Brain	52.4 ± 3.3	Awad and Berschneider (1977)	Egypt
	73.5 ± 8.9	Ma (1995)[a]	China
Spleen	112.0 ± 12.5	Awad and Berschneider (1977)	Egypt
	130.3 ± 30.1	Liu (2003)[a]	China
Bone	108.5 ± 42	Ma (1995)[a]	China
Teeth	100.3 ± 63	Ma (1995)[a]	China
Seminal liquid	126.6 ± 3.9 (μg/100 ml)	Singh et al. (1994)	India

[a]Bactrian camel

Other biological substrates are few and don't present routine clinical interest. Some results of the literature are reported in the Table 7.9.

Comparatively to the liver, kidney tissue may be poor in zinc (Abu Damir et al. 1983; Ma 1995), but there is no difference with other domestic ruminants. There are few variations between types of muscles: 1.10 to 1.61 mg/g (Rashed 2002). Last, if muscles and bones are globally rich in zinc, hump contains no trace (El-Faer et al. 1991).

In synovial fluid, the zinc concentrations appeared significantly higher in arthritic joints compared to healthy ones: 87.6 ± 1.3 vs 61.3 ± 9.9 μg/100 ml (Chalmeh et al. 2016). In a recent report (Kamili et al. 2018), zinc was determined in camel skin using Laser Induced Breakdown Spectroscopy. The zinc concentration was 115 ± 60 ppmin external skin face and 94 ± 82 in internal one.

7.4.3 *Iron*

Iron deficiency is not reported in ruminants grazing in natural conditions (Suttle 2010). So, the determination of plasma iron does not represent usual practical laboratory interest for the herbivorous, and its clinical relevance is low. In theory, the diagnosis of iron deficiency can be based also on the analyses of biochemical or hematological parameters such as hemoglobin (iron deficiency anemia screening), transferrin, or ferritin levels, but none of these parameters are specific.

7.4.3.1 Plasma Iron

In camel, available data on levels of iron in serum (S), plasma (P), or whole blood (WB) are presented in Table 7.10. However, information on the iron supply and requirements for the non-weaned camel calf are lacking although camel milk would only cover 26–34% of its iron needs (Sawaya et al. 1984). Globally, plasma or serum iron is comparable to that of other ruminants, i.e., between 70 and 120 µg/100 ml. Due to the higher concentration of iron in red blood cells, an insignificant hemolysis would highly increase plasma concentration. However, some references are above these values without clear clinical interpretation. In the whole blood, the values reported in the literature are between 40 and 60 mg/100 ml (except Liu and Ma 1995a who found more than 40 mg/100 ml which is not acceptable).

Rare interspecies comparisons let out a lower plasma iron in the dromedary compared to other domestic ruminants. Thus, Ghosal et al. 1976 report mean values for cattle, sheep, and camels, 213, 141, and 101 µg/100 ml, respectively. The same hierarchy is in Faye et al. (1986) with 162 µg/100 ml in cattle, 141 in sheep, 147 in goats, and 127 in camels. Similar observation is done by Al-Busadah (2003) in Saudi Arabia: 80.2 µg/100 ml in camel serum vs 168.4 in cow and 178.6 in sheep. Khamis et al. (2011) found also slightly higher values in cow (125.6 µg/100 ml) than in camel (119.0 µg/100 ml), similar to buffalo (115.9 µg/100 ml). Conversely, the rate of transferrin would be higher in the dromedary (Tartour and Idris 1970). Accordingly, camel transferrin has a saturation rate of iron around 30%, significantly lower than in other species of ruminants (Shekhawat et al. 1987).

Serum iron concentration would be influenced by physiological stage; it decreases slightly during pregnancy (Eltohamy et al. 1986) from 68 in early pregnancy to 64 µg/100 ml at the end of gestation and then increases to 91.4 in non-pregnant females. Similar changes, albeit insignificant, were observed on iron concentration in whole blood in the Bactrian camel (Liu et al. 1994). In fact, the effect of physiological stage in female camel is unclear: no difference between pregnant and non-pregnant camel (Vyas et al. 2011) or according to the breeding season (Saini et al. 2009) and significant higher serum iron concentration in non-pregnant camel (Faye et al. 2005).

Table 7.10 Serum (S), plasma (P), or whole blood (WB) iron concentration in camel according to different authors

Reference	Type	Mean values (*n*)	SD	Country
Bhattacharjee and Banerjee (1962)	S	121 μg/100 ml (10)	± 9.5	India
Moty et al. (1968)	S	186 μg/100 ml (19)	± 39	Egypt
Tartour (1969)	S	446.4 ppm[a] (45)	± 57	Sudan
Barakat and Abdel-Fattah (1970)	WB	44.4 mg/100 ml (200)	± 4.1	Egypt
Ghosal et al. (1976)	S	101 μg/100 ml (20)	± 4.6	India
Whabi et al. (1979)	P	98.5 μg/100 ml (96)	± 19	Sudan
Orliac (1980)	S	124 μg/100 ml (50)	± 30	Algeria
Marx and Abdi (1983)	S	92.9 μg/100 ml (102)	± 42	Somalia
Shekhawat et al. (1987)	S	110 μg/100 ml (20)	± 7	India
El Kasmi (1989)	P	85 μg/100 ml (–)	–	Morocco
Faye et al. (1986)	P	127 μg/100 ml	71–321	Ethiopia
Abdalla et al. (1988)	S	113 μg/100 ml (32)	± 46	UAE
Bengoumi et al. (1998b)	P	117 μg/100 ml (30)	± 33	Morocco
Ghosal and Shekhawat (1992)	S	107 μg/100 ml (122)	± 3	India
Liu et al. (1994)[b]	WB	41.1 mg/100 ml (72)	± 1.2	China
Ma (1995)[b]	WB	55.9 mg/100 ml (30)	± 10.3	China
Liu (2003)[b]	WB	56 mg/100 ml (30)	± 10.4	China
Deen et al. (2004)	S	321 g/100 ml (33)	± 35	India
Badiei et al. (2006)	S	249.2 μg/100 ml (26)	±74.3	Iran
Nafizi et al. (2009)	S	159,2 mg/100 ml (30)	± 88.2	Iran
Parekar et al. (2009)	S	132.5 μg/100 ml (25)	±12.8	India
Shukla et al. (2009)	S	117.6 μg/100 ml (30)	± 1.72	India
Wernery et al. (2009)	S	117.1 μg/100 ml (>10^5)	± 31.4	UAE
Eltahir et al. (2010)	S	119.8 μg/100 ml (–)	± 0.67	Oman
Mustafa et al. (2012)	S	169.3 μg/100 ml (20)	± 209.9	Sudan

[a]Reported to DM
[b]Bactrian camel

Like copper and zinc, there is an active transfer of iron from the mother to the fetus. However, according to several authors (Marx and Abdi 1983; Shekhawat et al. 1987; Ghosal and Shekhawat 1992), adults have a stronger iron level in serum than young camels (116 vs 98 μg/100 ml), but it was not confirmed by Saeed et al. (2004). In young camels, plasma iron level increases regularly in the young age and goes from simple to double between 1 month and 7 months but remains stable from the fifth month (Saeed et al. 1995).

Other results regarding physiological variation factors need further clarifications. El Kasmi (1989) observed no differences according to age or sex, which is not the case of Barakat and Abdel-Fattah (1971) for whom the iron content of whole blood is higher in the female in the rainy season and lowest in the dry season, compared to males. According to Ghosal and Shekhawat (1992), females have on average a significantly higher serum iron level (112.7 μg/100 ml) than males (100.1 μg/100 ml). At reverse, serum iron was significantly higher in male (231.1 μg/100 ml) than in female (177.8 μg/100 ml) in the UAE (Faye et al. 2005). For Tartour and Idris (1970), Saeed et al. (2004), and Badiei et al. (2006), the low observable difference in serum iron between males and females is insignificant. These results have been confirmed by Hussein et al. (1997) for bound iron, in contrast to the free iron which would tend increased in male adults.

On the other hand, race animals generally have higher rates (101.1 μg/100 ml) than pack animals (94.6 μg/100 ml) (Eltahir et al. 2010). The authors attribute this difference to genetic characteristics although no breed effect was reported elsewhere (Deen et al. 2004; Faye et al. 2005). Seasonal variation was reported, the serum iron being significantly lower in the dry season (85.9 μg/100 ml) than in the rainy season (97.2 μg/100 ml) (Marx and Abdi 1983).

Regarding health effect, serum iron is declining in the male affected by testicular degeneracy but also in the case of positive serology for brucellosis (Ahmed and Nada 1993). Significant decrease in the serum iron concentration was also observed in the case of parasitic infestation, averaging from 116 to 94 μg/100 ml (Ibrahim et al. 1982), or in case of theileriosis, from 126.1 to 70.5 μg/100 ml (Ismael et al. 2013) that it was not confirmed by Youssef et al. (2015). Iron concentration in serum increased in camel affected by mastitis or with high somatic cell count (Tuteja et al. 2004).

7.4.3.2 Liver and Spleen Iron

In most mammals, the target organ for iron storage is not the liver but the spleen that plays a key role in the erythrocytes' life cycle. However, several authors have provided some analyses giving an idea of the iron concentrations in the liver parenchyma (Table 7.11). Different studies agree that camel liver iron concentration is around 400–500 ppm.

The competition on the storage sites at the hepatic level between mineral elements helps to explain the negative correlation between iron concentration and observed copper by Tartour (1969).

Age is an important factor of variation for the liver iron concentration. Storage of iron in the fetus is considerable during the first half of pregnancy (Tartour and Idris 1970). The values observed in the fetus during autopsy of pregnant females are, on average, twice the rate recorded in the adult: 682 ± 74 ppm (Wensvoort 1992). In the calf, it was observed that mean values are higher at autopsy (1331 ± 824 ppm) than in the slaughterhouse (228 ± 85 ppm), but this large difference is due to exceptional rates (up to 2678 ppm) observed in a few animals (Wensvoort 1992).

No variation related to sex was reported despite a slight (nonsignificant) difference noted by Tartour (1969): 540 ± 256 ppm for males vs 604 ± 295 for females. The purpose of the animal seems not to affect the liver iron concentration: 388 ± 164 ppm for pack camels vs 300 ± 205 for racing camel (Wensvoort 1992).

Table 7.11 Liver iron concentration according to the literature (in ppm)

Reference	Mean values	SD	*n*	Country
Awad and Berschneider (1977)	460	± 85	9	Egypt
Tartour (1969)	558	± 266	55	Sudan
Wensvoort (1992)	388[a]	± 164	9	UAE
Wensvoort (1992)	302[b]	± 192	5	UAE
Ma (1995)	532[c]	± 221	25	China
Al-Busadah (2003)	295	± 21.6	5	Saudi Arabia
Badiei et al. (2006)	678	± 366	13	Iran
Ibrahim et al. (2013)	545	± 27.9	100	Sudan

[a]Samples at necropsy
[b]Samples at the slaughterhouse
[c]Bactrian camel

The liver iron concentration seems positively correlated with the serum iron, but it doesn't seem to be any significant relationship between liver iron and splenic iron which concentration range between 205 and 1641 ppm (mean = 629 ± 299 ppm) from after Tartour (1969). These values, not influenced by the sex of the animal, are lower than those observed in other species of domestic ruminants, and Tartour (1969) suggested that the liver plays a more important role than the spleen in the cycle of hemoglobin and iron metabolism. However, such interspecies difference was not confirmed later (Ibrahim et al. 2013), with iron concentration in sheep and goat liver (222.7 and 202.8 ppm, respectively, on average) being twofold lower than in camel liver (545 ppm). From his part, Al-Busadah (2003) did not find significant difference between camel, sheep, and cow liver.

7.4.3.3 Milk Iron

The iron requirements for young mammal are important, and milk remains virtually the single contribution to animals before weaning. Some species, and in some farming conditions, the low concentration of iron in the milk may be a considerable limiting factor. In camel, huge differences between the authors are listed, so that it is difficult to get an accurate opinion on iron intake by young animals. Some authors agree on iron concentration close to 0.30 mg/100 g, i.e., six times higher than in cow milk in the same ecological conditions (Sawaya et al. 1984) and values two times lower than that for goat (Abdelrahim 1987) or cow (Diacono et al. 2008b). In Bactrian camel, Wang et al. (2011) found five times more iron in camel milk than in cow one and eight times more than in goat one, but camel values (4.3 μg/100 ml) appeared very low compared to other data reported in the literature.

However, different concentrations are reported by Elamin and Wilcox (1992) and Bengoumi et al. (1998b), respectively, 28 and 341 mg/100 ml. In Bactrian camel from Kazakhstan, the reported concentrations for iron were 14.8 ± 5.3 mg/100 ml (Meldebekova et al. 2008) and 20.2 ± 12.4 mg/100 ml (Konuspayeva et al. 2008) without the effect of season or region of sampling. Similar values were reported in China: 21.6 mg/100 ml (Fantuz et al. 2016). In dromedary camel from Kenya, no effect of mineral supplementation was observed on iron concentration in milk, 10.0 vs 11.0 mg/100 ml in control and supplemented camel, respectively (Dell'Orto et al. 2000). It was reported 12–13 mg/100 ml in different regions of Egypt (Rashed 1998). Iron content is a discriminating component of the camel milk between Bactrian and dromedary camel, Bactrian milk being richer than dromedary one (Faye et al. 2008a).

The lack of standardization of analytical methods and especially the lack of precision in the mode of expression of results do not allow determining the standard values for mineral composition of milk. However, these results would lead to consider the camel milk as good source of iron for human consumers. Except for Elhardallou and El-Naggar (2016) for whom iron is in quite lower quantity in camel milk (0.49 mg/100 ml) compared to cow and goat milk, it appears that camel milk contains more iron (Gorban and Izzeldin 1997; Soliman 2005).

7.4.3.4 Iron in Other Biological Substrates

The iron does not appear as part of the importance of zinc or copper in the composition of wool. However, the hair iron content seems higher in the dromedary than in sheep: 327 ± 170 μg/g (Ma 1995) or 453 ± 66.6 (Badiei et al. 2006) vs 139 to 155 μg/g (Lewis et al. 1994). This content is subject to change according to the stage of gestation of the animal, from 546 ± 359 μg/g in the non-pregnant female to 241 ± 95 μg/g during gestation and then to 391 ± 192 μg/g during postpartum (Liu et al. 1994). Active transfer of iron, from the mother to the fetus, would be the origin of these changes. The iron content in wool is depending also on the type of fiber: the fine hair contained significantly more iron (1238 μg/g) than coarse fiber (554 μg/g) (Helal 2015). There is no difference due to the sex of the camel (Badiei et al. 2006)

In meat, the iron concentration oscillates between 1.16 and 1.35 mg/100 g according to the muscles, these values being comparable to those of beef (El-Faer et al. 1991). Similar results were reported by Dawood and Alkanhal (1995) (2.86–3.39 mg/100 g) and Elgasim and Alkanhal (1992) (1.94) or Badiei et al. (2006) (2.53 mg/100 g). For Bactrian camel, Ma (1995) reported different values, 20 times higher: 28 ± 7 mg/100 g in the muscles and 29 ± 9 mg/100 g in the heart. According to other results, dromedary camel meat contains approximately 25–51 mg/100 g iron fresh weight (Rashed 2002; Bekhit and Farouk 2013) or 16 mg/100 g dry meat (Mahmud et al. 2011). These differences can be attributed to the analytical techniques and to the expression of the results (reported or not to dry matter).

In Bactrian camel, concentrations of iron in the kidneys (460 ± 80 ppm), the brain tissue (170 ± 80 ppm), and bone (188 ± 120 ppm) seem comparable to other species (Ma 1995). In dromedary, reported values were 83 ppm in the kidney (Badiei et al. 2006), 81 ppm in the brain, and 485 ppm in the heart (Awad and Berschneider 1977).

The richer organ in iron is the spleen: 200 to 750 ppm (Badiei et al. 2006; Awad and Berschneider 1977). The synovial fluid in health joint contains 202.3 ± 18.0 μg/100 ml iron, while arthritic joint contains 226.0 ± 4.4 μg/100 ml which is not significantly different (Chalmeh et al. 2016).

Finally, the adipose tissue of the hump appears to be poor in iron: 0.31 ± 0.01 mg/100 g (El-Faer et al. 1991).

7.4.3.5 Iron Excretion

Fecal excretion is highly linked to the amount of iron present in the diet. In animals receiving a basal diet without mineral supplementation, Faye et al. (1999) found mean concentration of 80 mg/kg DM feces. This concentration increased significantly in animals receiving trace element supplementation (copper + zinc + manganese) averaged to more than 200 mg/kg DM feces. This change is similar to that seen in cattle fed with the same basal diet (Fig. 7.7).

This result suggests high interaction between iron and other trace elements. These, indeed, increase fecal excretion and thus decrease the apparent absorption. However, the depressing effect of other trace elements seems less important than in cow (Faye et al. 1999).

No data on urinary iron excretion is available in the camel, but it is probable that like other cations, the kidney is not the main way for excretion.

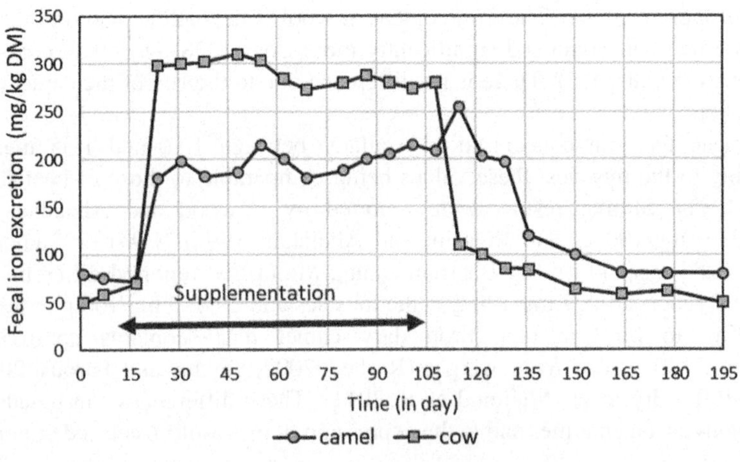

Fig. 7.7 Comparative change in fecal iron excretion according to the copper and zinc supplementation in the diet for camel and cow (according to Faye et al. 1999)

7.4.4 *Manganese*

Manganese can be a limiting factor for the diet of ruminants, and the risk of deficiency can locally be present due to the values observed in some tropical grasses (Faye et al. 1986, 1992a, b). However, in animal, dosage of manganese does not appear of clinical interest (Lamand 1987). Indeed, no tissue is a good indicator of manganese deficiency. Plasma concentration is easily contaminated. Moreover, the concentration in the tissues is not sensitive to dietary changes, including in camel (Alhidary et al. 2016) which makes very difficult deficiency evaluation, apart from clinical observation (McDowell 1992).

7.4.4.1 Plasma or Serum Manganese

Relatively few references regarding the concentration of plasma manganese in the dromedary are available. Moreover, those values are widely variables and have to be interpreted with caution. Usually, manganese is analyzed at the same time than copper and zinc. El Kasmi (1989) found an average of 174 μg/100 ml, with no effect of age or sex, and Saini et al. (2009) reported mean values of 75–77 μg/100 ml according to the fertility status of the female camel. In India, Shukla et al. (2009) reported also high value in male during rutting season (160 ± 0.75 μg/100 ml). These values are considerably higher than the concentrations in the order of 25–30 μg/100 ml advanced by Eltohamy et al. (1986), Al-Busadah (2003), Liu and Ma (1995b), Zongping (2005), Parekar et al. (2009), Khamis et al. (2011), and Desalegn et al. (2012). Values between 34 and 42 ± 10 μg/100 ml were also reported by Ma (1995) in China on the Bactrian camel.

Probably, these values should be reviewed as observed concentrations in other ruminants are generally less than 10 μg/100 ml (Lamand 1987). Other results seem more in agreement with the observations reported in other species, for example, 8.4 ± 1,2 μg/100 ml in 30 Moroccan camels (Bengoumi et al. 1995a), 0.16 μg/100 ml on 235 camels in the UAE (Faye et al. 2005), or 3.2 ± 0.8 μg/100 ml in 5 Indian camels (Sena et al. 2007). No effect of pregnancy, sex, or age has been reported (Faye et al. 2005; Vyas et al. 2011), but seasonal changes were recorded with a decrease at breeding season both for fertile and infertile females (Saini et al. 2009). Al-Busadah (2003) and Khamis et al. (2011) observed significant higher manganese level in camel compared to cow and sheep or buffalo. Significant breed difference was reported in Saudi Arabia (Abdelrahman et al. 2013a): 10 μg/100 ml in Majaheem vs 50 μg/100 ml in Maghateer.

7.4.4.2 The Liver Manganese

Hepatic concentration of manganese is generally lower in comparison with the other trace elements (Table 7.12) which generally does not distinguish the dromedary

Table 7.12 Liver concentration in manganese according to different authors (in ppm)

Reference	Mean values	SD	*n*	Country
Awad and Berschneider (1977)	7	± 2.1	9	Egypt
Abu Damir et al. (1983)	2.6	± 1.5	25	Sudan
Wensvoort (1992)	5.2[a]	± 3.0	9	UAE
Wensvoort (1992)	7.5[b]	± 1.4	5	UAE
Ma (1995)	5.6[c]	± 1.6	15	China
Al-Busadah (2003)	6.1	± 2.2	5	Saudi Arabia
Liu (2003)	5.03[c]	± 1.65	50	China
Khamis et al. (2011)	9.5	± 0.8	25	Egypt
Ibrahim et al. (2013)	6.9	± 0.86	100	Sudan
Abdelrahman et al. (2013a)	2.4	± 0.08	15	Saudi Arabia

[a]At necropsy
[b]At slaughterhouse
[c]Bactrian camel

from other species (Abu Damir et al. 1983; Al-Busadah 2003; Ibrahim et al. 2013). The liver parenchyma seems to be less rich in the fetus than in adults, but there is no difference associated with age. On the other hand, values recorded in racing camel appear significantly higher than in pack animals (10.3 vs 5.2 ppm according to Wensvoort 1992). In the Bactrian camel, Ma (1995) observed very low concentrations in the animals affected by ataxia: 0.8 ± 1.9 ppm.

A slight impact of the rumen bolus supplementation including manganese has been reported on the liver concentration, passing from 2.23 to 4.76 ppm (Alhidary et al. 2016).

7.4.4.3 Manganese in Milk

Manganese in milk does not reflect nutritional status in this element. Moreover, as for blood serum or plasma, results of the literature appear not homogeneous. With a concentration in the order of 20 µg/100 g DM (i.e., 0.32 mg/l), camel milk would be a little less rich than that of cow milk (Sawaya et al. 1984). This value is closed to that reported by Elhardallou and El-Naggar (2016): 0.57 and 0.40 mg/1 in two different camel farms from Saudi Arabia. In Egypt, Soliman (2005) reported a lower quantity of manganese in camel milk (0.13 mg/l) but comparable to that in the other species. However, this quantity of manganese in camel milk would provide 84% of the requirements of the young. Much higher values are advanced by Abulehia (1987) and Bengoumi et al. (1998b) with, respectively, 1930 and 1800 µg/l, while in cow milk, values range between 30 and 50 µg/l (Lamand 1974). Al-Wabel (2008), at reverse, did not find difference between manganese concentration on camel milk (1.3 mg/l) and other milks: 1.3 mg/l in cow, 1.12 in goat, and 1.14 in sheep. Rashed (1998) reported values between 3.2 and 7.8 mg/l according to different regions of Egypt. In Kazakhstan, Diacono et al. (2008a, b) found low quantity: 7 µg/100 ml, ten times less than in horse milk (66 μg/100 ml). Similar values are reported by Meldebekova et al. (2008) in Bactrian camel: 8.4 μg/100 ml.

7.4.4.4 Other Biological Substrates

With a mean manganese concentration around 4.5 ± 1.4 ppm (Ma 1995; Liu 2003), camel wool approximates sheep wool (Grace and Lee 1992). However, manganese is not described as an element playing a special role in wool synthesis (Ryder and Stephenson 1968) although it appears that fine fibers contained significantly more manganese (50 ppm) than coarse one (18 ppm) (Helal 2015).

No data on the manganese content in urine is available. Fecal manganese determination has been achieved in the study already cited as a reference for the other trace elements (Faye et al. 1999): with a basic ration ensuring maintenance requirements, fecal excretion was 67 ± 16 ppm; with a daily intake of 1 g of manganese, the fecal concentration in manganese fluctuated between 214 and 298 ppm, which is significantly higher than in cattle. However the absorption rate turned out to be higher in dromedary (67%) than in cattle (43%). The gap between the two species is even more important in the absence of supplementation (78% vs 32.5% for camel and cattle, respectively). Yet, it is recognized that manganese absorption is generally low, 95% of this element being excreted in cattle (Khalili et al. 1993). The dromedary camel therefore seems to present a greater metabolic efficiency for this element (Faye and Bengoumi 1997).

The other organs or tissues are rather poor in manganese, which does not differ the dromedary from other species. The kidneys contain 2.5 ppm (Ma 1995) to 2.9 ppm (Abu Damir et al. 1983), even less than 1 ppm according to Abdelrahman et al. (2013a) but with higher concentration in the cortex, 3.5 ppm (Liu 2003). The muscles and heart contain similar amount (1.2 to 2.5 ppm) according to Liu (2003) and Ma (1995). Those results are higher than those reported by El-Faer et al. (1991) and Abdelrahman et al. (2013a), 0.04–0.09 ppm and 0.23–0.27 ppm, respectively. The pancreas contains 2.8 ppm and the spleen, 1.12 ppm manganese (Liu 2003). The nervous tissue is at the same level as other soft tissue, i.e., 1.5–2.3 ppm (Liu 2003; Ma 1995). The bone and teeth are substantially richer, 4.6–5.2 ppm (Ma 1995). The richest organ in manganese is the ovary with 12 ppm on average (Liu 2003).

7.4.5 Selenium

Concerning selenium, there is some evidence to date of clinical deficiencies or toxicities, and up to recently, few available data on selenium requirements and metabolism were available in this species. However, recent findings on selenium metabolism in dromedary camel have been recently reported (Seboussi et al. 2008a, b, 2009a, b, 2010; Faye and Seboussi 2008, 2009).

7.4.5.1 Selenium in Blood or Serum

The mean concentration of blood/serum selenium reported in the literature for large animals was around 100 ng/ml, which is sufficient for the maintenance of suitable metabolic functions (Maas et al. 1990). In the dromedary from Morocco, Hamliri et al. (1990) observed in whole blood values varying according to age and sex, between 109 and 118 ng/ml, being thus similar to those reported in sheep in the same area. Similar figures were recorded by Liu et al. (1994) in China with concentrations varying from 97 to 112 ng/ml. In Sudan, Abdel Rahim (2005) reported values in whole blood varying between 25 and 53 ng/ml, while Elrayah et al. (2010) did not mention any sexual or age differences (between 118 and 128 ng/ml). Without specifying if it was whole blood or serum, Ma (1995) reported higher values on Bactrian camel: 274 to 288 ng/ml. The analytical method used could explain the observed differences, but the details of analysis procedures were not given in most of the publications.

Serum concentrations were 281 ng/ml on average in camels from the Sultanate of Oman (Faye, unpublished data), but sampled camels were suspected of selenium imbalance. In Morocco, in dromedaries receiving probably a low Se basal diet, the plasma selenium concentration was quite lower, about 21 ng/ml (Bengoumi et al. 1998c). In adult male camels in healthy conditions from Iran, reported selenium concentration in serum was 12.6 ng/ml only (Nafizi et al. 2009). In Saudi Arabia, serum Se values in young camels at the slaughterhouse varied between 5.3 and 131 ng/ml with 30% of samples higher than 100 ng/ml (Barri and Al-Sultan 2007). In the UAE, the mean value was 200 ± 90 ng/ml in animals with no Se supplementation (Seboussi et al. 2004). In recent experiments with different levels of Se supplementation, selenium content in serum for non-supplemented animals was on average 137 ± 18 ng/ml in non-pregnant, non-lactating camels (Seboussi et al. 2008a), 109.3 ± 33.1 ng/ml in pregnant females, and 103 ± 28.7 ng/ml at milking period (Seboussi et al. 2009a). The variability was higher, and values ranged between 12 and 200 ng/ml with an average of 100 ng/ml. However, in most of the reported values, the selenium status of the diet was unknown even if Se supplementation was not distributed to these animals. In some countries, the basal diet could be very deficient in selenium or, at reverse, with high quantity. In experiments achieved in the UAE on adult females mentioned above, pregnant (Seboussi et al. 2009a) and non-pregnant (Seboussi et al. 2008a) received 1.8 mg Se in the basal diet (6 kg of Rhodes grass—*Chloris gayana*—hay and 2 kg of concentrates), without any specific selenium supplementation. So, with approximatively 2 mg Se per day brought by the normal diet, the level of Se in camel blood and serum is comparable to that in the other species.

Comparing two Saudi breeds, Abdelrahman et al. (2013b) reported a significant difference, the concentration in Majaheem breed (black color) being double (147.1 ng/ml) than the Maghateer (white color, 73.3 ng/ml).

7.4.5.2 Selenium Deficiency

For a long time, selenium deficiency has been suspected to occur in camels kept in zoological parks affected by cardiomyopathy or myopathy (Finlayson et al. 1971; Wisner and Schotke 1975; Decker and McDermid 1977; Ozdemir et al. 2016), but no clinical descriptions and laboratory analysis have been made in these reports to confirm selenium. In China also, Liu et al. (1994) suspected selenium deficiency in cases of swayback in Bactrian camel. However, selenium deficiency with characteristic clinical signs has been recently reported in the UAE. Selenium deficiencies affect generally young animals and are responsible for white muscle disease, a degenerative muscle disease affecting muscle including the heart. Indeed, the most important lesions are degenerative myocarditis and discoloration of the skeletal muscle (Photo 7.1). In the UAE, soils and feedstuffs are generally considered deficient in selenium, and many cases of degenerative myocarditis (see below photo) are observed with histological lesions similar to those in cattle (El Khouly et al. 2001; Seboussi et al. 2004).

When skeletal muscles are affected, symptoms vary from mild stiffness to obvious pain upon walking to an inability to stand. Camel calves may tremble in pain when held in a standing position. When the problem occurs in newborns, they are born weak and unable to rise. Sudden exercise may trigger situation in older camel calves. When the disease affects the heart, the animal shows signs similar to pneumonia, including difficult breathing and fever with an elevated heart and respiratory rates. Anemia reported by El Khouly et al. (2001) was not in the three selenodeficient adult camels from Saudi Arabia (Al-Qarawi et al. 2001).

In sheep, selenium deficiency is diagnosed in blood below 50 ng/ml. Serum selenium level in deficient camels was also obviously below this limit: in camel calves, average level of Se serum in diseased cases was below 35 ng/ml (El Khouly et al. 2001) and between 0.8 and 3.7 ng/ml in 3-year animals (Al-Qarawi et al. 2001).

Photo 7.1 Degenerative myocarditis lesions in the heart of a 1-month-old camel calf

7.4.5.3 Selenium Toxicity (Selenosis)

At our knowledge, only experimental selenosis has been reported (Faye and Seboussi 2008; Seboussi et al. 2009b). First clinical signs appeared with a selenium supplementation of 8 mg/day. First physiological symptoms were an increase of the respiratory rate, pulse rate, and internal temperature up to 40 °C. Clinical signs occurred within 2 weeks with hair discoloration, followed by alopecia. With 16 mg/day, urinary excretion increased, and dark watery diarrhea was also observed. Loss of appetite and then of weight and weakness appeared. Tears with pale mucous were present as well as evidence of impaired vision. Dyspneic respiration and pain at auscultation appeared, and camels adopted the sternal decubitus position and tended to rest their neck extended. Salivation occurred, and finally camels showed no desire to eat and drink. Fissured pads were observed, leading to difficult walking (see below Photo 7.2).

After slaughtering, intoxicated camels showed paleness in all the abdominals, paleness of diaphragm and intercostal muscles, hydrothorax, and pulmonary emphysema. The heart, liver, and kidney were congested and necrotic (see below Photo 7.3). Brain edema was also observed.

The clinical symptoms observed in camel were in accordance with previous signs reported in chronic poisoning in other species (Casteel et al. 1985). After the liver, the kidney, particularly the cortex, retained the highest concentration. The heart as

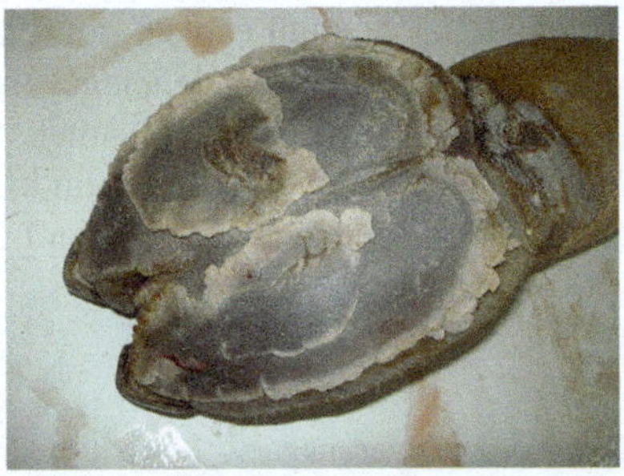

Photo 7.2 Fissured pads with necrosis on foot of camel receiving 16 mg Se/day

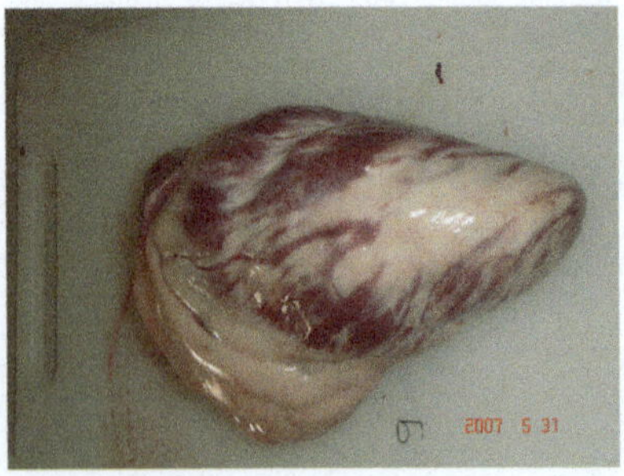

Photo 7.3 Heart discoloration and congestion in camel after daily supplementation of 8 mg selenium for 45 days

the target organ of selenium intoxication failed, leading to pulmonary edema and hydrothorax. The foot lesions with pad necrosis were comparable to those observed in alkali disease (chronic selenosis) in cattle (O'Toole and Raisbeck 1995) and horse (Raisbeck et al. 1993) in spite of the lack of hooves in camel.

Selenium deficiencies in animal, including camel, could also cause liver, heart, kidney, and skeletal muscle damages (El Khouly et al. 2001). Comparable necropsy lesions were reported in Se deficiency and intoxication. Lack or the excess of selenium seems to lead to similar cell damage.

The question of the poisoning threshold in camel has not been clearly determined. However, the supplementation threshold reported in other species is higher than the dietary levels in the studies performed on camel (Seboussi et al. 2008b, 2009b; Faye and Seboussi 2008), i.e., 0.051 to 0.095 mg/kg LW, which seems to show a high sensitivity of camel species to Se toxicosis. Selenium requirement and toxicity could be very close. For example, in intoxicated lambs with 4 mg/kg LW under sodium selenite form (four times higher than the camels receiving 16 mg Se daily in the trial of Seboussi et al. (2009a), the serum Se increased up to 274 ng/ml only in lambs (Tiwary et al. 2006), compared to 767 ng/ml observed in camel (Seboussi et al. 2009b).

7.4.5.4 Effect of Se Supplementation on Se Status in Camel

Few papers relate the impact of selenium supplementation on the mineral status of camel, and generally, doses applied for selenium deficiency control were those recommended for cattle. To our knowledge, the first trial achieved to assess the effect of selenium supplementation on the plasma selenium status was reported by Liu and Ma (1995a) in Bactrian camel. After 90 days supplementation with 30 g selenium pellet, serum selenium increased from 235 to 393 ng/l.

Later on, in the experiment of Bengoumi et al. (1998c), the selenium status of camels was compared with that of cattle with similar weight and receiving daily 2 mg Se per os under sodium selenite form for 2 months. Results showed a sharper increase of plasma selenium in camels (10 times the plasma level before supplementation) compared to cows (twice the starting level) (Fig. 7.8). As the magnitude of the decrease of plasma selenium concentration after stopping supplementation was similar to the previous increase, it was supposed that plasma (or serum) selenium concentration in camel was an extremely sensitive indicator of selenium intake. The fast selenium depletion at the end of supplementation period also indicated a better efficiency of selenium absorption and excretion in camel compared to cow.

In selenodeficient camels with muscular dystrophy, Al-Qarawi et al. (2001) gave an oral treatment with selenium—vitamin E (2.19 mg sodium selenite +50 mg vitamin E) by intramuscular injection at 0.5 mg/kg body weight for 2 consecutive days. Following treatment, selenium concentration rose from on average 2.3 ng/ml up to 23.7 ng/ml, i.e., with a similar trend to that observed by Bengoumi et al. (1998c). Indeed, selenium concentration was multiplied also by 10 after supplementation.

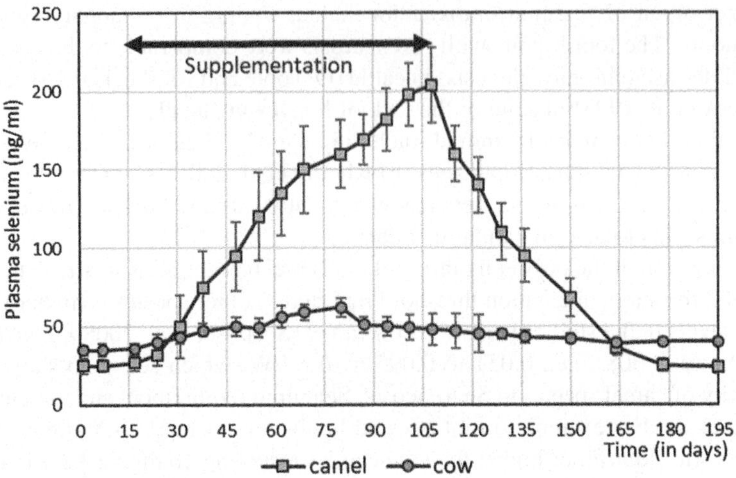

Fig. 7.8 Comparative change in plasma selenium concentration in cow (open square) and dromedary camel (open circle) receiving 2 mg/day selenium under sodium selenite form (Bengoumi et al. 1998c)

In several studies on the effect of oral selenium supplementation including adult females at different physiological stages or growing young camels (Seboussi et al. 2008a, 2009a, b, 2010; Faye and Seboussi 2008), different levels of supplementation were tested up to the toxic limit, from 2 up to 16 mg/day as sodium selenite form. All these experiments showed that camel is very sensitive to Se supplementation (these values in plasma could overpass 600 ng/ml in some cases). A slight decrease was observed at calving in orally supplemented dams due to the maternal transfer into milk (Faye et al. 2011), as well as after a single injection of selenium (Faye et al. 2014). Se serum concentrations in camel calves at parturition were 106.3 ± 26.5 and 273.2 ± 48.0 ng/ml in non-supplemented dams (0 mg/day) and supplemented ones (2 mg/day), respectively (Seboussi et al. 2009a). Supplementation with organic selenium was more efficient than with inorganic one: 82.1 ± 13.8 vs 39.0 ± 6.8 ng/ml (Faye et al. 2013).

Meta-analysis of the published data (Faye and Seboussi 2009) including oral supplementation from 0 to 16 mg/day showed clear linear relationships up to 4 mg and then a slight increase with a plateau after 12 mg/day (Fig. 7.9).

7.4.5.5 Selenium in Milk

In the experiment of Seboussi et al. (2009a), Se concentration in milk varied from 39.5 to 482.6 ng/ml with an average of 86.4 ± 39.1 ng/ml in the control group and 167.1 ± 97.3 ng/ml in the orally supplemented group. At birth, Se concentration in colostrum was threefold higher in the treated group: mean value 302 ± 95 vs 108.2 ± 43.9 ng/ml ($P < 0.001$). In both groups, Se milk concentration decreased,

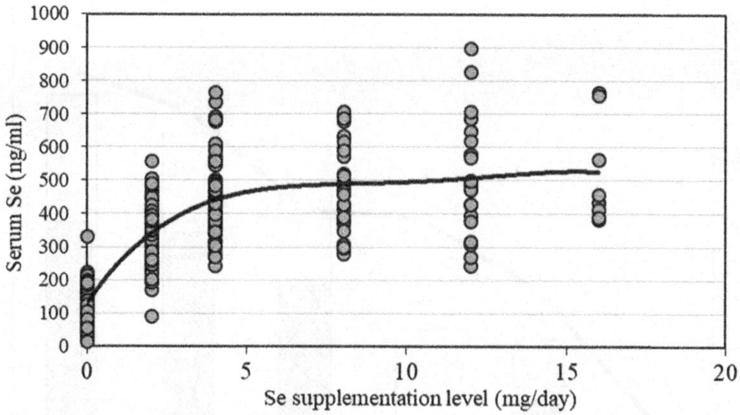

Fig. 7.9 Change in camel serum selenium according to the level of oral supplementation (from after Faye and Seboussi 2009)

and after the second milk sampling, no significant difference was observed. By considering Se concentration in colostrum and the status of the mothers and of their camel calves at parturition, positive correlations were observed with serum Se in mothers ($r = 0.659$; $P < 0.05$) and in calves ($r = 0.689$; $P < 0.05$). With female camels supplemented by Se injection before parturition, similar trend was observed: 59.1 ± 19.2 ng/ml in colostrum of control dams vs 93.2 ± 49.2 ng/ml for supplemented dams (Faye et al. 2014). Similar figure is reported in the experiment comparing selenium in colostrum after organic and inorganic supplementation (72.7 ± 28.9 vs 37.2 ± 7.1 ng/ml (Faye et al. 2013). Al-Awadi and Srikumar (2001) reported a much lower value (13.9 ± 2.4 ng/ml) than Seboussi et al. (2009a), but they did not mention the lactation stage. In a meta-analysis performed on cattle's data (Ceballos et al. 2009), it has been considered that selenium content increase in milk reached an average of 12.6 ng/ml only after oral Se supplementation at a dose of 3 mg/day under selenite form. In comparison, the apparent good efficiency of Se transfer in camel milk is confirmed.

7.4.5.6 Selenium Excretion

Very few data are available on fecal and urinary Se excretion in camel. According to the different trials achieved in the UAE (Seboussi et al. 2009a, b) with variable levels of Se supplementation in the diet, Se fecal excretion increased slowly up to 4 mg Se in the diet and then highly from 8 mg daily supplementation up to 16 mg (Fig. 7.10). Total fecal excretion varied from 637 ng/day in non-supplemented camels up to 4084 ng/day in camels receiving 16 mg Se/day in the diet. Total fecal excretion was comparable to urinary excretion when administering up to 4 mg supplementation, but the main part of Se excretion after 8 mg of supplementation was of fecal origin. Total urinary excretion varied from 518 ng/day (control groups) up to 179 ng/day

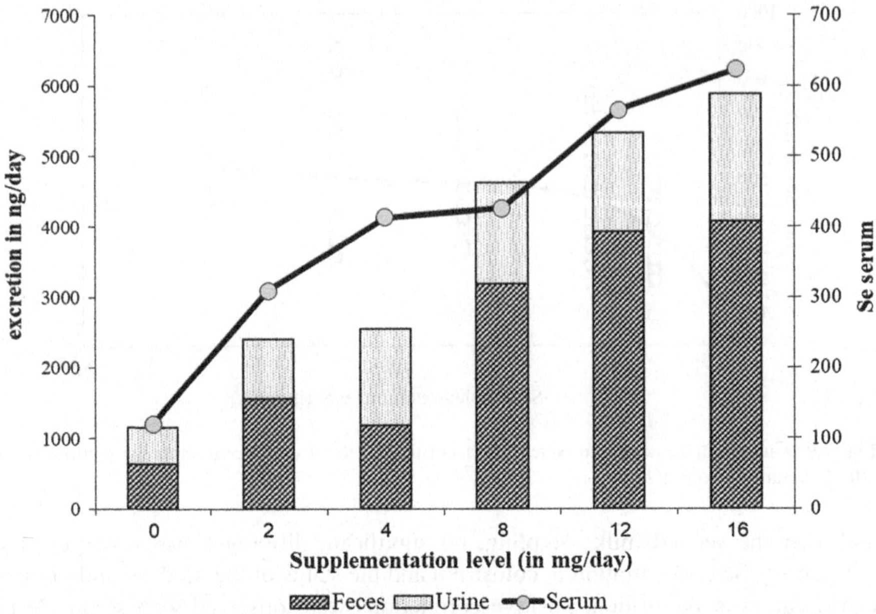

Fig. 7.10 Fecal and urine excretion of selenium according to the level of supplementation (data meta-analysis of Seboussi et al. 2008a, b, 2009a, b, 2010)

(16 mg Se supplemented group). Forty-five percent of excreted Se was in urine in non-supplemented animals vs 26–30% only in highly supplemented camels. Serum Se concentration was highly correlated with Se concentration in urine and feces content and total Se fecal excretion. Contrary to the observations reported in dairy cattle (Juniper et al. 2006), no linear effect of supplementation was observed in camel.

7.4.5.7 Se Storage in Organs

Selenium content determination in organs has rarely been reported because of its limited clinical interest. In wool of Bactrian camel from China, Liu et al. (1994) reported values between 140 and 190 µg/kg, depending on their physiological status. Similar results have been published by Ma (1995): 190–210 µg/kg.

In the experiment of Seboussi et al. (2010), the highest quantity of selenium was observed on average in the liver (2727 µg), kidney (807 µg), lung (443 µg), and heart (160 µg). Yet high quantity was also observed in the muscle (2513 µg). On average, whatever Se supplementation level, the kidney (1129 µg/kg), liver (921 µg/kg), hair (545 µg/kg), forelimb muscle (421 µg/kg), hind limb muscle (351 µg/kg), and the lung (308 µg/kg) had the highest Se concentrations. The highest quantity was reported in supplemented groups, but except in the hair, liver, kidney, and muscle, the quantity was not clearly linked to the Se supplementation level.

In Bactrian camel, only one reference is available for selenium concentration in organs (Ma et al. 1995). In this study, the kidney (3100–3900 μg/kg), liver, and heart (1100–1500 μg/kg), muscle, and brain (620–640 μg/kg) were organs with the highest Se concentrations. These values, except those of the liver, appeared much higher than those found by Seboussi et al. (2009b, 2010). Selenium supplementation increases significantly concentration in wool passing from 244 μg/kg to 496 μg/kg after 4 months of supplementation (Liu and Ma 1995a).

Considering the weight of the whole carcass and of different organs in camel, the total quantity of selenium in a camel of 200 kg carcass weight was around 100 mg with 90% in the muscle, 5.5% in blood, and 2.5% in the liver. Less than 1% was stored in the kidney.

As for serum, a breed difference was observed with higher values in Majaheem compared to Maghateer (Abdelrahman et al. 2013b). In arthritic joint, selenium concentration of synovial fluid is significantly lower than in healthy joint: 9.5 ± 0.14 vs 12.32 ± 0.26 ng/ml (Chalmeh et al. 2016).

7.4.6 *Cobalt*

Quantities of cobalt are very low in the plasma, and its dosage does not have clinical interest (Lamand 1987). However, in all herbivorous, cobalt is essential in the synthesis by bacteria in the rumen (or caecum) of cyanocobalamin or vitamin B12.

Clinical disorders attributed to a possible deficit of cobalt in the diet are not reported in the literature. Forage trees are not deficient in cobalt (Faye et al. 1986, 1990). Moreover, forage trees are dominant in the diet of camels in their traditional grazing areas during dry season (Faye and Tisserand 1989; Rutagwenda et al. 1989). However, low values of cobalt in grasses are reported in salty depressions in the Horn of Africa (Faye et al. 1990), which can theoretically cause cobalt deficit during the "salted cure."

7.4.6.1 Cobalt in Blood, Serum, or Plasma

Determination of serum or plasma cobalt concentration in camel is very rare. Some analyses were performed in China and in Mongolia on Bactrian camel. For Burenbayar (1989), blood cobalt concentrations are between 3.4 and 13.2 μg/100 ml according to the season and mineral content of the diet, but the method of analysis is not specified. By atomic absorption spectrophotometry, Liu et al. (1994) evaluated blood cobalt at 58 ± 39 μg/100 ml in non-pregnant females, 56 ± 53 μg/100 ml in pregnant camel, and 53 ± 39 μg/100 ml in the postpartum period, indicating the absence of physiological stage effect. Similar concentrations were reported by Liu (2003), 61 ± 12 μg/100 ml; Ma (1995), 59 ± 22 and 61 ± 12 μg/100 ml according to the health status of the animal; Liu and Ma (1995a), 31–44 μg/100 ml, and Shen and Li (2010), 56 ± 39 and 61 ± 12 μg/100 ml according to the presence or not of "emaciation ailment syndrome." Wide standard deviations suggest very high individual variability.

With a mean value of 11 μg/100 ml, Elrayah et al. (2010) did not observe regional, sexual, or age differences. With quite lower values, Badiei et al. (2006) also did not mention sexual difference (0.29 and 0.47 μg/100 ml in female and male, respectively). Also no significant breed difference was reported in India: 43 ± 4 μg/100 ml in whole blood of Jaisalmeri camel, 29 ± 5 in Bikaneri, and 40 ± 0.0 μg/100 ml in Kacchi camel (Deen et al. 2004). Cobalt concentration in whole blood was lower than in serum (93 ± 7 μg/100 ml). In their investigation, Faye et al. (2005) observed lower values, 0.08 μg/100 ml on average, with no significant variability due to age, sex, breed, or physiological stage. At reverse, Sena et al. (2007) reported a physiological stage effect with significant lower value in dry females (23 ± 3 μg/100 ml) compared to pregnant females (49 ± 3 μg/100 ml) or lactating ones (56 ± 2 μg/100 ml).

The status in plasma cobalt is influenced by the supplementation with bolus containing 8 ppm cobalt (Alhidary et al. 2016).

7.4.6.2 Cobalt in Other Substrates

The concentration of cobalt in the Bactrian wool is between 1 ± 0.28 μg/g (Ma 1995), 1.2 ± 0.7 μg/g (Liu et al. 1994), and 1.27–1.60 μg/g and does not vary according to the physiological stage. In Iranian dromedary, cobalt concentration is quite lower, 0.067–0.096 μg/g (Badiei et al. 2006), which is not in accordance with Helal (2015), 0.17 to 1.57 μg/g according to the type of fiber.

There is a lack of data on cobalt content in camel milk. In other tissues, Ma (1995) reported the following values (in ppm): liver, 0.81–0.89; kidney, 0.93–1; muscle, 1–1.2; heart, 0.79–0.81; and brain, 1.9–2. Lower values (in ppb) were reported by Badiei et al. (2006): 57–63 ppb in the liver according to sex, 53–67 ppb in the kidney, 40–49 in the spleen, 51–52 in the heart, and 43–52 in the muscle. In fine cobalt concentration is low in tissues and is between 0.6 and 2 ppm (Liu 2003).

In camel meat, cobalt concentration varies between 1.7 and 2.8 μg/g according to the type of muscle (Rashed 2002). The cobalt liver is slightly higher in camel (1.87 ± 0.35 ppm) compared to sheep (0.46 ± 0.22 ppm) or goat (0.30 ± 0.17 ppm) according to Ibrahim et al. (2013). The same value was observed in the liver by Bakhiet et al. (2007): 2.2 ppm (camel) vs 0.5 ppm (cattle), 0.7 ppm (sheep), and 0.4 ppm (goat). Liver concentration increases in case of supplementation passing from 2.46 in control animals to 3.93 ppm after long-acting trace mineral rumen (Alhidary et al. 2016). However, no effect of "ailment emaciation syndrome" was observed, 0.71 ± 0.36 vs 0.68 ± 0.21 ppm in affected and non-affected animals, respectively (Shen and Li 2010).

In camel wool, cobalt concentration is around 1.2 ppm (Shen and Li 2010), 0.07 ppm (Badiei et al. 2006), and 0.56 ppm (Helal 2015).

7.4.7 Iodine

Goiter, characteristic symptom of iodine deficiency in many species, has previously been reported in camel by several authors (Decker et al. 1979; Tageldin et al. 1985; Antoine-Moussiaux et al. 2005; Rejeb et al. 2012). However, iodine deficiency could occur without clinically observable increase in the size of the thyroid gland (see below Photo 7.4). Plasma iodide concentration is a better marker of iodine nutrition (Aumont et al. 1989) than plasma thyroid hormones that depend on too many factors of variation (Agarwal et al. 1989). Plasma iodine concentration varies between 50 and 114 ng/ml (mean, 82 ± 15) according to Bengoumi et al. (1998b), i.e., comparable values to those other domestic animals (Kaneko 1989).

The best diagnosis remains today the dosage of iodine in milk (Aumont and Tressol 1986). Milk iodine is well correlated to the plasma iodine concentration. Only one reference is available indicating high values in camel milk iodine, 98 ± 21 µg/l (Bengoumi et al. 1998b). Camel milk is richest in iodine than other ruminants (Haenlein 1980). However, the analyses published by Bengoumi et al. (1998b) included camels living in Laâyoune (South Morocco) near the Atlantic Ocean. Such a geographical situation contributes clearly to an important iodine intake. It would be useful to have data from areas far from the coasts. In environment rich in iodine, camel milk could contain excessive amount and become an important source of iodine for the human population, leading to chronic thyroid dysfunction as it was observed in Saharawi women from the refugee camp in Algeria (Aakre et al. 2015).

Indeed, according to Tageldin et al. (1985), the lack of iodine is common in pastures located in continental areas (case of Sudanese Darfur mentioned by Abu Damir et al. 1990) in contrast to the areas close to the coast where the rate of iodine in fodder is widely above the limit of deficiency (Faye et al. 1990; Bengoumi et al. 1998b). However, the exact role of goitrogen plants in camel is not well known as

Photo 7.4 Goiter observed in young camel in Niger around Agadez

Shouwia thebaïca, a Brassicaceae plant family known for containing antithyroid factors (Antoine-Moussiaux et al. 2005).

In similar environment, camel would be more sensitive to iodine deficiency than other domestic mammals, due to lower absorption (Abdel-Wahab and Osman 1971).

Blood iodine concentration increases (Etzion et al. 1987) from 112 to 124 ng/ml after 10 days of water deprivation. This would be due to a decrease in thyroid activity (hypothyroidism) and thus to a reduction in the use of iodine by the thyroid gland. Secretion of thyroid hormones and, in particular, of T3 is also decreased (Khanna et al. 1996). This would help to enhance adaptation of the camel to the lack of water supply and internal hypothyroidism resulting in a decrease of basal metabolism and the body temperature. Plasma iodine levels return to normal quickly after rehydration.

In Bactrian camel, Burenbayar (1989) reported seasonal and supplementation effect on plasma iodine.

7.4.8 Fluorine

Fluorine is characterized by its high chemical affinity to calcium and then to calcified tissues including bone and teeth. Fluorine is generally determined in biological substrates when chronic or acute intoxication is suspected. Fluorosis, the chronic intoxication, could affect plants, animals, and human in areas characterized by the presence of phosphate rocks in the soil (Kessabi et al. 1984). By erosion of the soil, fluorine compounds in excess contaminate plants and water (Laatar et al. 2003). Human and animals could be intoxicated by high fluorine level intake from the water and plants. Fluorine concentration in blood reflects fluorine content in alimentation (Cronin et al. 2000). However, in contaminated areas, camels could be less affected by osteo-dental fluorosis than cattle and human (Choubisa 2013).

Two kinds of fluorosis were described: hydrotelluric and industrial fluorosis. The first one is caused by high content of fluoride in the soil and water especially in phosphate rocks rich in fluoride. The second one is due to air emission of fluoride particle in the environment by phosphate plants and manufactories (Kessabi et al. 1984). Phosphate mines are common in desert area like Morocco, Tunisia, or Mauritania where camel are reared and their contamination could be important.

Because of the affinity of fluorine for teeth and bones, the main lesions due to intoxication are modification of color, structure, and orientation of teeth and also structure and texture of bones. Fluorosis has an important impact on animal health and welfare. The precocious grinding of teeth is responsible for the low production level of milk and meat and also precocious animal culling. The severity of lesions increases with fluorine quantity ingested and with the duration of exposure and toxicity of fluorine compounds. Intrinsic factors include species (herbivores are more sensible), breed, age, sex, individual, stress, and concomitant disease (parasitic and infectious diseases). Young animals (after weaning) are more sensitive to this intoxication compared to adults which oblige some farmers to keep these animals in non-contaminated areas during changing of teeth (Zouagui 1973). Fluorine toxicity on teeth being more dangerous during formation of adult teeth, toxic action is function of stage of teeth formation. That's why teeth lesions could be observed only on definitive teeth of adults exposed to high levels of fluorine during teeth formation. Bone lesions are observed when the maximal capacity of fluorine fixation by teeth is exceeded (Laatar et al. 2003).

It is generally admitted that the normal level of fluorine in plasma or serum is below 30 μg/100 ml. In two areas from Morocco, relatively closed to phosphate mines, camel plasma fluorine concentration was 6.4 ± 0.4 μg/100 ml and 4.5 ± 0.1 μg/100 ml (Boujdour and Lâayoune, respectively), whereas it was 4.2 ± 0.1 μg/100 ml in control areas (Diacono et al. 2008a, b). It was considered that water, soil, and fodders were not contaminated. In this survey, plasma fluorine was significantly higher in less than 1-year-old camels than older camels, but sex and physiological status do not have any significant effect (see below Photo 7.5).

Industrial cases of intoxication with fluoride in dromedary were described in Egypt (Karram et al. 1989), associated with sulfur poisoning, leading to a symptomatology well known with dental and bone lesions. Serum fluorine concentration was 30 μg/100 ml in animals located 25 km from the source of contamination and, respectively, 190 ± 30 in males and 125 ± 7 μg/100 ml in females.

In the Bactrian camel, in non-affected areas, Liu et al. (1994) recorded quite higher values ranging from 1290 to 1730 μg/100 ml, significantly lower than those reported by Ma (1995): 2210 to 2430 μg/100 ml.

In wool, fluorine content seems quite high, but significant discrepancies between authors are found, 19 to 26 ppm for Ma (1995) and 98 to 126 ppm for Liu et al. (1994). In other tissues, Ma (1995) argued the values below (in ppm), liver, 115 ± 31; kidney, 46 ± 34; muscle, 33 ± 12; heart, 133 ± 20; brain, 36 ± 6.5; bones, 56 ± 32 to 97 ± 11 according to the type of bone; and tooth, 91 ± 8.

Photo 7.5 Teeth discoloration in camel close to phosphate mine in Morocco

7.4.9 *Other Elements*

The other trace elements are rarely analyzed in clinical investigations. Due to the scarcity of data and obvious lack of consistency in the analytical methods (rarely described with precision) and in the results, substantial differences occurred in the published values.

7.4.9.1 Molybdenum

This element is generally associated to copper with which it is in competition. Excess of molybdenum associated with sulfur is known to depress digestibility of copper and lead to secondary copper deficiencies. Excess of molybdenum constitutes a potential problem for ruminants rather than deficiencies (McDowell et al. 1993). Molybdenosis cases have been suspected in camels grazing exclusively in bush composed of *Salvadora persica*, a plant particularly rich in molybdenum. Animals were affected with profuse and sickening diarrhea and showed signs of hepatic and kidney intoxication (Faye and Mulato 1991).

In the serum of the Bactrian camel, Liu et al. (1994) reported concentrations ranging from 19 to 23 μg/100 ml, Liu and Ma (1995b) at 43 μg/100 ml, and Shen and Li (2010) at 18 μg/100 ml. There is no effect of pregnancy on molybdenum (Liu et al. 1994) as well as of "ailment emaciation syndrome" (Shen and Li 2010). In dromedary, Badiei et al. (2006) reported serum concentrations in molybdenum of 5.3 μg/100 ml with no sexual difference as well as Faye et al. (2005) with an average of 2.9 μg/100 ml.

Camel milk would be quite rich in molybdenum with an average content of 0.66 mg/l (Martynenko et al. 1977), value confirmed by Saini et al. (2007): 0.70 mg/l. In wool, molybdenum content is very low, 0.41 ppm (Ma 1995). However, in animals suffering from ataxia, Liu et al. (1994) observed twice molybdenum in the wool of affected animals compared to non-affected (4.6 ± 1.7 vs 2.1 ± 0.9 ppm), while there is no significant difference from the blood assays. At reverse, Shen and Li (2010) did not observe difference between Bactrian camels affected or not by ailment emaciation syndrome (2.31 ± 1.72 vs 2.32 ± 0.81 ppm). The concentration of molybdenum in wool is not affected by the physiological status of animals (Liu et al. 1994). In dromedary, the concentration reported by Badiei et al. (2006) is quite lower: 0.08 ppm. For Helal (2015), concentration of molybdenum varied between 1.1 and 8.75 ppm according to the type of fiber.

In other tissues, Ma (1995) reported the following values (in ppm): liver, 6.8–7; kidney, 5–6.6; muscle, 4.5–5.9; heart, 1.4; brain, 5.4–5.7; bone, 1.1–5.3; and tooth, 2.9. Badiei et al. (2006) listed the following values (in ppm): 0.63 in the liver, 0.55 in the kidney, 0.60 in the spleen, 0.46 in the heart, and 0.42 in the muscle.

Finally, it is difficult to propose usual values for camel due to the scarcity and high variability of the available data in the literature.

7.4.9.2 Sulfur

Given the important requirements for the synthesis of proteins in ruminants, sulfur is considered to be an intermediate element between macronutrient and trace elements. However, sulfur requirements are poorly evaluated in cattle and a fortiori in camelids. Cases of poisoning by sulfur in the dromedary were reported in Egypt (Karram et al. 1989), which allows to have some references on the values of the plasma sulfur in case of excess in the environment. The farthest herds from the source of contamination (superphosphate factory) had serum sulfur concentration ranging from 449 ± 52 mg/100 ml for males to 503 ± 68 for females. Close to the polluting factory, these concentrations were 2085 ± 296 and 1882 ± 262 mg/100 ml for males and females, respectively. All symptoms of intoxication were present: emaciation, cachexia, respiratory distress, etc. In one camel from Saudi Arabia presenting chronic symptoms of polioencephalitis, high sulfur content of water from deep bore wells was incriminated as a possible cause, but there was no analysis of sulfur content in blood or tissues (Al-Swailem et al. 2009).

In Bactrian camel from China (Shen and Li 2010), sulfur in blood serum was significantly higher in animals affected by ailment (6.31 ± 1.7%) characterized by pica, emaciation, dyskinesia, deprived appetites, and anemia, compared to non-affected animals (4.12 ± 0.86%, i.e., 412 mg/100 ml).

Wild Bactrian camel milk contained 39.6 ± 0.8 mg/100 ml sulfur, i.e., similar to that in domestic Bactrian camel, 28.3–40.2 mg/100 ml (Jirimutu et al. 2010).

In muscles of non-affected animals, El-Faer et al. (1991) found significant levels of sulfur (from 55 to 65 mg/100 g according to the anatomical regions). Higher values (136 mg/100 g) were reported by Kadim et al. (2011).

Sulfur is naturally in high quantity in wool. It represents between 0.19% and 0.49% of the mineral content in dromedary wool according to the type of fiber, i.e., 1.83–4.84 mg/g (Helal 2015). In Bactrian camel, the content was 6.37 ± 2.3 mg/g in affected animals by "emaciation ailment" vs 4.67 ± 7.21 in non-affected (Shen and Li 2010).

In the liver, concentrations such as 1.32 ± 0.35 to 2.53 ± 0.36 mg/g were reported in Bactrian camels whether they be affected by "emaciation ailment" or not (Shen and Li 2010).

7.4.9.3 Bromide

Etzion et al. (1987) dosed bromide and observed values increasing during dehydration (55.3–58.9 µg/ml after 10 days of dehydration). This increase would have a sedative effect on the animal, which would contribute to the reduction of metabolism in the process of dehydration, sign of an adaptation to the deprivation of water.

7.4.9.4 Nickel

Recently, nickel is more or less considered as possible essential trace element. Indeed, microbial hydrogenases in the rumen of cattle contain this metal. In China, there is mention of disease ("roll disease") in the Bactrian camel, which would be due to poisoning with nickel (Tao et al. 1995). For Faye et al. (2005), mean nickel concentration in camel serum is 1.8 μg/100 ml with a significant effect of sex, values in male being higher (2 μg/100 ml) than in female (1.7 μg/100 ml).

In meat, nickel content varied from 0.5 to 3.8 μg/g according to the muscle (Rashed 2002). These values are higher than the average (0.25 μg/g) reported by Kadim et al. (2009).

In camel wool, nickel content is very high in fine fiber (81.5 ppm) compared to coarse brown (5.93) and coarse white fiber (6.73 ppm) (Helal 2015).

7.4.9.5 Lead and Cadmium

Lead and cadmium are generally determined together. They are heavy metals playing no biological role, and their presence in blood or milk is not acceptable when it is more than traces. Their dosage is interesting in case of excess in the environment, which may lead to proven cases of poisoning. Lead poisoning is known since ancient times in humans and most animals. However, lead poisoning was not identified, to our knowledge, in the camelids. Mainly lead and/or cadmium in milk was reported in camel. In blood serum, Faye et al. (2005) reported on average 1.5 μg/100 ml lead and 0.07 μg/100 ml cadmium in camel blood serum without any effect of age and sex.

Elamin and Wilcox (1992) measured the lead content in milk from Saudi Arabia to 18 mg/100 ml DM which seems considerable. Indeed, in polluted areas of Kazakhstan, Konuspayeva et al. (2009) revealed lower values: 2.5 ± 1.9 μg/ 100 ml. In another sampling, the reported values in milk were 3 μg/100 ml for lead and 0.2 μg/100 ml for cadmium (Konuspayeva et al. 2011). In Western Saudi Arabia, the cadmium values measured in milk was 0.7 and 8.9 μg/100 ml, (Elhardallou and El-Naggar 2016). In India, Saini et al. (2007) found 22 μg/ 100 ml in non-contaminated area.

In wool fiber, lead and cadmium could reach 2.14 and 0.02 ppm, respectively, with few differences between types of fiber (Helal 2015).

In Algeria, camel meat (in DM) contained between 2.01 and 3.21 μg/g lead and between 0.91 and 0.83 μg/g cadmium, i.e., values lower than in other species (Bendeddouch et al. 2014). In Morocco, these values were 0.71 and 0.12 μg/g, respectively, for lead and cadmium, while it was 1.33 and 0.25 μg/g in the liver, 0.86 and 0.023 μg/g in the lung, 1.04 and 0.069 μg/g in the heart, and 0.96 and 0.69 μg/g in the kidney (Chafik et al. 2014a). In Oman, lead content in camel meat was determined at 0.15 μg/g (Kadim et al. 2009).

Table 7.13 Some nonessential elements determined in serum, milk, and meat of camels according to different authors

Ag	Al	Au	As	B	Ba	Cr	Hg	Sr	Reference
Element in blood serum (µg/100 ml)									
	3.7		22.5	19.3	14.6	2		44	Faye et al. (2005), UAE
	220–1010								Al-Busadah (2010)
						7–9			Alhidary et al. (2018)
Element in camel milk (µg/100 ml)									
						19.3		15.1	Elhardallou and El-Naggar (2016)
Element in camel meat (ppb)									
							32		Bendeddouch et al. (2014)
300–900	3300 4200	90–180				3600 3800			Rashed (2002)
	1200 5800					100–300		200–300	El-Faer et al. (1991)
						80			Kadim et al. (2009)

7.4.9.6 Other Nonessential Elements

The other elements are rarely determined (Table 7.13) but could be interesting in case of poisoning like for the heavy metals.

In the muscle, strontium, chromium, and aluminum are detectable in trace amounts, meaning that they can be contaminants of meat. On the other hand, the hump does not contain practically rare minerals (El-Faer et al. 1991).

One reference is available regarding radionuclides (cesium137) involving Bactrian camel in Kazakhstan living close to wastage site of nuclear activities. With 90% confidence interval, two samples of raw milk and seven samples of powder milk had detectable radiological activity (>0.028 Bq), the maximum coming from South Kazakhstan being 0.294 ± 0.076 Bq/kg (Konuspayeva et al. 2011). The transfer coefficients to camel milk of Sr85, I131, Cs137, Pb210, Po210, and U238 varied between 1.3×10^{-4} (polonium) and 3.6×10^{-2} (iodine), but these values tend to be lower compared to cow, sheep, and goat milk (Al-Masri et al. 2014).

7.5 Conclusion

The regulation of trace element metabolism in camel seems turned into anticipating period with scarce resources (Faye et al. 2006). If the variability observed in the field or experimental conditions is relatively large and inconstant, camel presents different mechanisms marked by a certain resistance to mineral undernutrition which is one of the faces of its adaptation to desert life. These mechanisms include increasing of absorption capacity in scarcity periods, higher storage capacity, tolerance for minerals in excess, and maintenance of enzymatic activity in deficient period.

References

Aakre I, Bjoro T, Norheim I, Strand T, Barikmo I, Henhum S (2015) Excessive iodine intake and thyroid dysfunction among lactating Saharawi women. J Trace Elem Med Biol 31:279–284

Abdalla OM, Wasfi IA, Gadir FA (1988) The Arabian race camel normal parameters. I. Haemogram, enzymes and minerals. Comp Biochem Physiol 90A:237–239

Abdelrahim AG (1983) The relationships between the concentrations of serum Cu and Zn and the activities of the serum enzymes copper oxidase and alkaline phosphatase of the dromedary (*Camelus dromedarius*). J Arid Environ 6:265–268

Abdelrahim AG (1987) The chemical composition and nutritional value of camel (*Camelus dromedarius*) and goat (*Capra hircus*) milk. World Rev Anim Prod 18:9–12

Abdelrahim AG (2005) The relationship between whole blood selenium (Se) concentration and the activity of the seleno-enzyme, glutathione peroxidase (GSH-Px E.C.I.11.1.9) in camel (*Camelus dromedarius*). J Arid Environ 62:359–362

Abdelrahman MM, Aljumaah RS, Ayadi M (2013a) Variation of copper, zinc, manganese and magnesium in blood serum and tissues of two breeds of dromedary camels in Saudi Arabia. Asian J Anim Vet Adv 8(1):91–99

Abdelrahman MM, Aljumaah RS, Ayadi M (2013b) Selenium and iodine status of two camel breeds (*Camelus dromedaries*) raised under semi intensive system in Saudi Arabia. Italian J Anim Sci 12(14):83–87

Abdel-Wahab MF, Osman AM (1971) Iodine metabolism in domestic animals in the Sudan using I_{131}. Endokrinologie 58:198–204

Abu Damir H, Tartour G, Adam SEI (1983) Mineral contents in livestock in Eastern Sudan. Trop Anim Health Prod 15:15–16

Abu Damir H, Barri MES, Tageldin MH, Idris OF (1990) Clinical and subclinical colloid goitre in adult camels (*Camelus dromedarius*) at Kordofan region of Sudan. Br Vet J 146:219–227

Abu Damir H, Eldirdiri NI, Adam SEI, Howarth JA, Salih YM, Idris OF (1993) Experimental copper poisoning in the camel (*Camelus dromedarius*). J Comp Pathol 108:191–208

Abu Damir H, Abbas TA, Alhaj M (2008) Copper status in breeding and racing camels (*Camelus dromedarius*) and response to cupric oxide needle capsules. Trop Anim Health Prod 40 (8):643–648

Abu Damir H, Alhaj Ali M, Abbas TA, Omer EA, Al-Fihail AM (2013) Nasarin poisoning in the dromedary camel (*Camelus dromedarius*). Comp Clin Pathol 22:305–311

Abulehia IH (1987) Composition of camel milk, vol 42. Milschwissenschaft, pp 368–371

Agarwal SP, Khanna ND, Agarwal VK, Dwaraknath PK (1989) Circulating concentrations of thyroid hormones in pregnant camels (*Camelus dromedarius*). Theriogenology 31:1239–1247

Ahmed WM, Nada AR (1993) Some serum biochemical values of dromedary camels with impaired fertility. Pak Vet J 13:16–18

Ahmed AA, Awad YL, Fahmy F (1977) Studies on some minor constituents of camel's milk. Vet Med J Assiut Univ Egypt 25:51–54

Al-Awadi FM, Srikumar IS (2001) Trace elements and their distribution in protein fractions of camel milk in comparison to other commonly consumed milks. J Dairy Res 68:463–469

Al-Busadah KA (2003) Trace-elements status in camels, cattle and sheep in Saudi Arabia. Pak J Biol Sci 6(21):1856–1859

Al-Busadah KA (2010) Serum concentration of aluminum, calcium, magnesium and phosphorous in camels. Sci J King Faisal Univ (Basic & Appl Sci) 11(1):161–167

Alhidary IA, Abdelrahman MM, Harron RM (2016) Effects of a long-acting trace mineral rumen bolus supplement on growth performance, metabolic profiles, and trace mineral status of growing camels. Trop Anim Health Prod 48(4):763–768

Alhidary IA, Alsofi MA, Abdoun KA, Samara EM, Okab AB, Al-Haidary AA (2018) Influence of dietary chromium yeast supplementation on apparent trace elements metabolism in growing camel (*Camelus dromedarius*) reared under hot summer conditions. Trop Anim Health Prod 50 (3):519–524

Ali A, Thrawat M, Al-Sobayil FA (2010) Hormonal, biochemical and hematological profiles in female camels (*Camelus dromedarius*) affected with reproductive disorders. Anim Reprod Sci 118:372–376

Al-Masri MS, Al-Hamwi A, Amin Y, Safieh MB, Zarkawi M, Soukouti A, Dayyoub R, Voigt G, Fesenko S (2014) Radionuclide transfer from feed to camel milk. J Environ Radioact 132:8–14

Al-Qarawi AA, Abbas B, Haroun EM, Mahmoud OM, Al-Hawas A (2001) Clinicopathological investigation of selenium responsive myopathy in young adult camels. J Camel Pract Res 8:23–27

Al-Swailem A, Al-Dubaib MA, Al-Ghamdi G, Al-Yamani E, Al-Naeem A, Al-Mejali AM, Shehata M, Mahmoud OM (2009) High sulphur content of water from deep bore wells as a possible cause of polio-encephalitis in a camel. Bulgarian J Vet Med 1(4):265–270

Althamma OM, Bengoumi M, Faye B (2012) Selenium and copper status of camels in Al-Jouf area (Saudi Arabia). Trop Anim Health Prod 44:551–556

Al-Wabel NA (2008) Mineral contents of milk of cattle, camels, goats and sheep in the central region of Saudi Arabia. Asian J Biochem 3(6):373–375

Antoine-Moussiaux N, Faye B, Vias GF (2005) Tuareg ethnodiagnostic skill of camel diseases in Agadez area (Niger). J Camel Res Pract 12:85–93

Aumont G, Tressol JC (1986) Improved routine method for the determination of iodine in urine and milk. Analyst 111:841–843

Aumont G, Lamand M, Tressol JC (1989) Iodine nutrition in ewe: effect of low to high iodine intake on iodine content of biological fluids in pregnant and lactating ewes. Reprod Nutr Dev 29:113–125

Awad YL, Berschneider F (1977) Values for certain minerals and trace-elements in some tissues of the camel (*Camelus dromedarius*). Egypt J Vet Sci 14:31–35

Badiei KK, Mostaghni K, Pourjafar A (2006) Serum and tissue elements in Iranian camels (*Camelus dromedarius*). Comp Clin Pathol 15:103–106

Bakhat C, Metha MC, Sahani MS (2003) Annual hair yield attribute in indigenous camel breeds. Indian J Anim Sci 73(10):1189–1191

Bakhiet AO, Mohammed AA, Siham ESM, El-Badwi MAS (2007) Some trace-elements profile in the liver of camels, cattle, sheep and goats. Int J Trop Med 2(1):1–2

Barakat MZ, Abdel-Fattah M (1970) Biochemical analysis of normal camel blood. Z Vet Med (A) 17:550–557

Barakat MZ, Abdel-Fattah M (1971) Seasonal and sexual variations of certain constituents of normal camel blood. Zbl Vet Med A 18:174–178

Barri MES, Al-Sultan SI (2007) Studies on selenium and vitamin E status of young Megaheem dromedary camels at Al-Ahsa province. J Camel Pract Res 14:51–53

Bekhit AE, Farouk MM (2013) Nutritive and health value of camel meat. In: Kadim I, Maghoub O, Faye B, Farouk M (eds) Camel meat and meat products. CAB International, Oxfordshire, pp 205–223

Bellanger J (1971) Dosage des oligoéléments dans les fourrages. Ann Nutr Alim 25:59–96

Bellanger J, Lamand M (1975) Méthode de dosage du cuivre et du zinc plasmatique. Bull Techn CRZV Theix INRA 20:53–54

Bendeddouch B, Zellagi R, Bendeddouche E (2014) Levels of selected heavy metals in fresh meat from cattle, sheep, chicken and camel produced in Algeria. Ann Res Rev Biol 4(8):1260–1267

Bengoumi M, Faye B, El Kasmi K, Tressol JC (1995a) Facteurs de variation des indicateurs plasmatiques du statut nutritionnel en oligo-éléments chez le dromadaire au Maroc. I. Valeurs usuelles et variations physiologiques. Rev Elev Méd Vét Pays Trop 48:271–276

Bengoumi M, Faye B, Tressol JC, Bengoumi D (1995b) Facteurs de variation des indicateurs plasmatiques du statut nutritionnel en oligo-éléments chez le dromadaire au Maroc. 2. Effet d'une complémentation minérale. Rev Elev Med Vét Pays Trop 48:276–280

Bengoumi M, Essamadi K, Tressol JC, Faye B (1998a) Comparative study of copper and zinc metabolism in cattle and camel. Biol Trace Elem Res 63:81–94

Bengoumi M, Faye B, Tressol JC (1998b) Composition minérale du lait de chamelle du sud marocain. Actes de l'atelier "Chameaux et dromadaires, animaux laitiers", Nouakchott, Mauritanie, 24–26 Oct 1994, Ed. CIRAD (coll. Colloques), pp 145–149

Bengoumi M, Essamadi AK, Tressol JC, Chacornac JP, Faye B (1998c) Comparative effect of selenium concentration and erythrocyte gluthatione peroxidase activity in cattle and camels. Anim Sci 67:461–466

Bhattacharjee RC, Banerjee S (1962) Biochemical studies on Indian camel (*Camelus dromedarius*). II. Inorganic constituents of serum. J Sci Ind Res 21:106–107

Blazewicz A, Klatka M, Dolliver W, Kocjan R (2014) Determination of total iodine in serum and urine samples by ion chromatography with pulsed amperometric detection – Studies on analyte loss, optimization of sample preparation procedures and validation of analytical method. J Chromatogr B 962:141–146

Burenbayar R (1989) Supply of trace-elements for female camels in Mongolia. In: VIth International trace-element symposium, Leipzig, DDR, 2

Burns RH, Johnston A, Hamilton JW, Mc Colloch RJ, Duncan WE, Fisk HG (1964) Minerals in domestic wools. J Anim Sci 23:5–11

Casteel SW, Osweiler GD, Cook WO, Daniels G, Kadlec R (1985) Selenium toxicosis in swine. J Am Vet Med Assoc 186:1084–1085

Cattaneo D, Dell'Orto V, Fava M, Savoini G (2005) Influence of feeding on camel milk components. In: Faye B, Esenov P (eds) Proceedings of the international workshop, "Desertification combat and food safety: the added value of camel producers". Ashkabad (Turkménistan), 19–22 April 2004. "Vol. 362 NATO Sciences Series, Life and Behavioural Sciences". IOS Press, Amsterdam, pp 181–186

Ceballos A, Sanchez J, Stryhn H, Montgomery JB, Barkema HW, Wichtel JJ (2009) Meta-analysis of the effect of oral selenium supplementation on milk selenium concentration in cattle. J Dairy Sci 92:324–342

Chafik A, Eddoha R, Bagri A, Nasser B, Essamadi A (2014a) Distribution of trace elements and heavy metals in liver, lung, meat, heart and kidney of cattle, sheep, camel and equine slaughtered at Casablanca city-Morocco. Int J Sci Eng Res 5(2):294–303

Chafik A, Essamadi A, Eddoha R, Bagri A, Nasser B, Faye B (2014b) Trace elements and heavy metals in organs of camels (*Camelus dromedarius*) slaughtered in Casablanca city, Morocco. J Camel Pract Res 21(2):145–152

Chalmeh A, Badiei K, Pourjafar M, Nazifi S (2016) Synovial fluid antioxidant vitamins and trace elements in clinically healthy and arthritic joints of dromedary camels. Istanbul UnivVet Fak Dergisi 42(1):1–4

Chalmers GA (1974) Swayback (enzootic ataxia) in Alberta lambs. Can J Comp Med 38 (2):111–117

Cherrier R, Sumboro M, Faye B (1991) Note sur une méthode de biopsie chez le jeune dromadaire. Rev Elev Méd Vét Pays Trop 44:131–133

Choubisa SL (2013) Why desert camels are least afflicted with osteo-dental fluorosis ? Curr Sci 105 (12):1671–1672

Cronin SJ, Manoharan V, Hedley MJ, Loganathan P (2000) Fluoride: a review of its fate, bioavailability and risks of fluorosis in grazed-pasture systems in New Zealand. N Z J Agric Res 43:295–321

Curasson G (1947) Le chameau et ses maladies. Vigot Frères, Paris

Dawood A, Alkanhal MA (1995) Nutrient composition of Najdi- Camel Meat. Meat Sci 39:71–78
Decker RA, McDermid A (1977) Nutritional myopathy in young camel. J Zoo Anim Med 8:20–21
Decker RA, Hruska JC, McDerrid AM (1979) Colloid goiter in a newborn dromedary camel and an aborted fetus. J Am Vet Med Assoc 175:968–969
Deen A, Bathi A, Sahani MS (2004) Trace mineral profiles of camel blood and sera. J Camel Pract Res 11(2):135–136
Dell'Orto V, Cattaneo D, Beretta E, Baldi A, Savoini G (2000) Effect of trace element supplementation on milk yield and composition in camels. Int Dairy J 10:873–879
Desalegn T, Mohammed YK, Shimelis B (2012) Critical macro and microminerals concentration in the blood serum of camel (*Camelus dromedarius*) in Jijiga district, eastern Ethiopia. Livest Res Rural Dev 24(4). http://www.lrrd.org/lrrd24/4/desa24060.htm
Diacono E, Bengoumi M, Kessabi M, Abdendi E, Faye B (2008a) Hydrotelluric and industrial fluorosis survey in the dromedary camel the south of Morocco. In: Faye B, Sinyavskiy Y (eds) Proceedings of international workshop, "Impact of pollution on animal products" Almaty (Kazakhstan), 27–30 Sept 2007, pp 85–90
Diacono E, Meldebekova A, Konuspayeva G, Faye B (2008b) Plant, water and milk pollution in Kazakhstan. In: Faye B, Sinyavskiy Y (Eds), Proceedings of the international workshop, "Impact of pollution on animal products". Almaty (Kazakhstan), 27–30 Sept 2007, pp 107–116
Ebdon L, Fisher A (2000) Chapter 8: The use of ICP-AES as a detector for elemental speciation studies. Compr Anal Chem 33:227–248
El Kasmi K (1989) Contribution à l'étude des protéines sériques et de certains minéraux chez le dromadaire: influence de l'âge et du sexe. Mémoire de 1er cycle. Univ. Hassan II, Rabat, Maroc, 46 pp
El Khouly AA, Abbas TA, Moustafa T (2001) Myocardial dystrophy in camel calves in the United Arab Emirates (field cases). Emir J Agric Sci 13:11–17
Elamin FM, Wilcox CJ (1992) Milk composition of Marjheim camels. J Dairy Sci 75:3155–3157
El-Faer ZM, Rawdah TN, Attar KM, Dawson MV (1991) Mineral and proximate composition of the meat of the one-humped camel (*Camelus dromedarius*). Food Chem 42:139–143
Elgasim EA, Alkanhal MA (1992) Proximate composition, amino-acids and inorganic mineral content of Arabian camel meat: comparative study. Food Chem 45:1–4
Elhardallou SB, El-Naggar AY (2016) Determination of micro minerals in milk from farm and pasture-reared cow, goat and camel; using inductively coupled plasma-optical emission spectrometry. Orient J Chem 32(1):341–347
Elhuda N, Osman IED (2012) Effect of copper supplemented salt licks on total and TCA-soluble plasma copper concentrations in Omani camels. J Camel Pract Res 19(2):277–281
Elrayah HA, Barri MES, Abdelrahman SH (2010) Trace elements level in camels (*Camelus dromedarius*) western Sudan (Kordofan state). J Camel Pract Res 17(2):263–267
Eltahir YH, Ali HM, Mansour MH, Maghoub O (2010) Serum mineral contents of the Omani racing Arabian camel (*Camelus dromedarius*). J Anim Vet Adv 9(4):764–770
Eltohamy MM, Salama A, Youssef AEA (1986) Blood constituents in relation to the reproductive state in she-camel (*Camelus dromedarius*). Beitrage zür Trop Landwirtschaft und Vet Med 24:425–430
Essamadi K, Bengoumi M, Tressol JC, Chacornac JP, Faye B (1998) Comparative relationship of plasma copper concentration and ceruloplasmin activity of camel and cow. Trends Comp Biochem Physiol 5:211–220
Essamadi K, Bengoumi M, Zaoui D, Faye B, Bellenchi C, Musci G, Calabrese L (2002) Purification and partial characterization of camel (*Camelus dromedarius*) ceruloplasmin. Comp Biochem Physiol B 131:509–517
Etzion Z, Alfassi Z, Lavi N, Yagil R (1987) Halide concentration in camel plasma in various states of hydratation. Biol Trace Elem Res 12:411–418
Fahmy LS, Berbish EA, Teleb HM, Hegazy AA (2004) Effect of zinc supplementation on wound healing in camels. J Camel Sci 1:76–80

Fantuz F, Salimei E, Papademas P (2016) Macro- and micronutrients in non-cow milk and products and their impact on human health. In: Tsakalidou E, Papadimitriou K (eds) Non-bovine milk and milk products. Elsevier and AP Publications, London, pp 209–261

Faye B (1989) Statut nutritionnel du bétail dans la république de Djibouti. Ministère de la Coopération/INRA, Paris, 112 p

Faye B (1993) Mangrove, sécheresse et dromadaires. Revue Sécheresse 4:47–55

Faye B, Bengoumi M (1997) Données nouvelles sur le métabolisme des principaux éléments-traces chez le dromadaire. Rev Elev Méd Vét Pays Trop 50:47–53

Faye B, Grillet C (1984) La carence en cuivre chez les ruminants domestiques de la région d'Awash (Ethiopie). Rev Elev Méd Vét Pays Trop 37:42–60

Faye B, Mulato C (1991) Facteurs de variation des paramètres protéo-énergétiques, enzymatiques et minéraux dans le plasma chez le dromadaire de Djibouti. Rev Elev Méd Vét Pays Trop 44:325–334

Faye B, Seboussi R (2008) Experimental selenium intoxication in camel. Veterinaria 3:18–29

Faye B, Seboussi R (2009) Selenium in camel – A review. Nutrients 1:30–49

Faye B, Tisserand JL (1989) Problèmes de la détermination de la valeur alimentaire des fourrages prélevés par les dromadaires. Options méditerranéennes. Séries séminaire n°2, pp 61–65

Faye B, Grillet C, Tessema A (1986) Teneur en oligo-éléments dans les fourrages et le plasma des ruminants domestiques en Ethiopie. Rev Elev Méd Vét Pays Trop 39:227–237

Faye B, Kamil M, Labonne M (1990) Teneur en oligo-éléments dans les fourrages et le plasma des ruminants domestiques en République de Djibouti. Rev Elev Méd Vét Pays Trop 43:365–373

Faye B, Grillet C, Tessema A, Kamil M (1991) Copper deficiency in East African Rift Valley. Trop Anim Health Prod 23:172–180

Faye B, Saint-Martin G, Cherrier R, Ruffa A (1992a) The influence of high dietary protein, energy and mineral intake on deficient young camel: I. Changes in metabolic profiles and growth performance. Comp Biochem Physiol 102A:409–416

Faye B, Saint-Martin G, Cherrier R, Ali Ruffa M (1992b) The influence of high dietary protein, energy and mineral intake on deficient young camel: II. Changes in mineral status. Comp Biochem Physiol 102A:417–424

Faye B, Ratovonanahary M, Cherrier R (1993) Effet d'un facteur alimentaire sur la pathologie néonatale: résultats d'une enquête rétrospective sur la distribution de mangrove aux chamelons en République de Djibouti. Rev Elev Med Vét Pays Trop 46:471–478

Faye B, Jouany JP, Chacornac JP, Ratovonanahary M (1995) L'élevage des grands camélidés. Analyse des initiatives réalisées en France. INRA Prod Anim 8:3–17

Faye B, Bengoumi M, Tressol JC (1999) Comparative trace-element excretion in camel and cow. J Camel Res Pract 6:19–25

Faye B, Seboussi R, Askar M (2005) Trace elements and heavy metals in healthy camel blood of United Arab Emirates. J Camel Pract Res 12(1):1–6

Faye B, Bengoumi M, Seboussi R (2006) Metabolism of some minerals in camel: a face of the adaptation to harsh conditions. In: Proceedings of international scientific conference on camels (Part 4), Qassim University (Publications), 10–12 May 2006, Saudi Arabia, pp 1593–1615

Faye B, Konuspayeva G, Messad S, Loiseau G (2008a) Discriminant milk components of Bactrian camel (*Camelus bactrianus*), dromedary (*Camelus dromedarius*) and hybrids. Dairy Sci Technol 88:607–617

Faye B, Seboussi R, Askar M (2008b) Trace elements and heavy metals status in Arabian camel. In: Faye B, Sinyavskiy Y (eds) Proceedings of international workshop, "Impact of pollution on animal products". Almaty (Kazakhstan), 27–30 Sept 2007, pp 97–106

Faye B, Seboussi R, Alhadrami G (2011) Maternal transfer of selenium by blood and milk in camels. J Camelid Sci 4:30–39

Faye B, Saleh S, Konuspayeva G, Musaad A, Bengoumi M, Seboussi R (2013) Comparative effect of organic and inorganic selenium supplementation on selenium status in camel. J King Saud Univ Sci 26:149–158

Faye B, Althamma OM, Musaad A, Konuspayeva G, Bengoumi M (2014) Effect of selenium injection in pregnant camels on selenium status of their new-born and milk. Emir J Food Agric 26(4):342–348

Finlayson R, Keymer IF, Manton JA (1971) Calcific cardiomyopathy in young camels (*Camelus* spp.). J Comp Pathol 81:71–77

Fouad D (2015) Molecular modeling and phylogeny of the manganese superoxide dismutase from the camel, *Camelus dromedarius*. Pak J Zool 47(4):1015–1023

Ghosal AK, Shekhawat VS (1992) Observations on serum trace elements levels (zinc, copper and iron) in camel (*Camelus dromedarius*) in the arid tracts of Thar Desert in India. Rev Elev Méd Vét Pays Trop 45:43–48

Ghosal AK, Dwarakanath PK, Jatkar PR (1976) A note on serum iron levels in domestic animals of north-western Rajasthan. Indian J Anim Sci 46:449

Gnan SO, Sheriha MA (1986) Composition of lybian camel milk. Aust J Dairy Technol 41:33–35

Godet J, Guedda M, Said-Elmi M, Abbas D (1985) Note sur l'élevage camelin en république de Djibouti. Rapport section Sciences Humaines I.S.E.R.T, chapitre "pathologie et alimentation", Djibouti, pp 32–41

Gorban AMS, Izzeldin OM (1997) Mineral content of camel milk and colostrum. J Dairy Res 64:471–474

Grace ND, Lee J (1992) Influence of high zinc intakes, season, and staole site on the elemental composition of wool and fleece quality in grazing sheep. NZ J Agric Res 35:367–377

Gueguen L, Lamand M, Meschy F (1988) Nutrition minérale. In: Jarrige R (ed) Alimentation des bovins, ovins, caprins. INRA Publications, Paris, pp 95–111

Haenlein GF (1980) Mineral nutriton in goat. J Dairy Sci 63:1727–1748

Hamliri A, Olson WG, Johnson DW, Kessabi M (1990) Evaluation of biochemical evidence of congenital nutritional myopathy in the two-week prepartum fetuses from selenium-deficient ewes. J Am Vet Med Assoc 51:1112–1115

Heidarpour M, Mohri M, Borji H, Moghdass E (2012) Oxidative stress and trace elements in camel (*Camelus dromedarius*) with liver cystic echinococcosis. Vet Parasitol 187:459–463

Helal A (2015) Relationships among physical, chemical and industrial characteristics of different dromedary camel's hair types. J Am Sci 11(2):67–75

Hocquellet P (1974) Dosage du cobalt dans les aliments des animaux par spectrophotométrie d'absorption atomique avec flamme et sans flamme. Ann Falsif Expert Chim 67:495–511

Hussein MF, Al-Sobayel AA, Hassan HA (1982) Study of some aspects of blood chemistry in Saudi Arabian camels (*Camelus dromedarius*). Sudan J Vet Sci Anim Hus 23:68–72

Hussein MF, Basmaeil SM, Bakkar MN, Gar-El-Nabi AR (1992) Serum levels of some electrolytes and trace elements in camel calves during first year of life. J Appl Anim Res 2:13–18

Hussein MF, Al Mufanej SI, Mogarver HH, Le Nabi ARG, Sanad HH (1997) Serum iron and iron binding capacity in the Arabian camel (*Camelus dromedarius*). J Appl Anim Res 11:201–205

Ibrahim MS, Mohamed AR, El-Balkemy FA, Omran H, El Makawi MF (1982) Etude de l'activité de l'Imovec chez le chameau. Action antiparasitaire et incidence sur l'état général. Res Bull Univ Zagazig, Egypt, 14 p

Ibrahim IA, Shamat AM, Hussein MO, El-Hussein AM (2013) Profile of some trace elements in the liver of camels, sheep, and goats in the Sudan. J Vet Med 4 p. https://doi.org/10.1155/2013/736497

Ibrahim AA, Abdelrahman MM, Raafat MH (2016) Effect of long-acting trace mineral rumen bolus supplement on growth performance, metabolic profiles and trace element status of growing camels. Trop Anim Health Prod 48(4):763–768

Idris OF, Salih YM, AbdelGadir AW, Abdelgadir SE (1980) Toxicity of *Capparis tomentosa* administrated to camels. In: Workshop on camels, Khartoum (Soudan), dec. 79, IFS Publications, Stockholm, pp 373–386

Ismael AB, Swelum AA, Khalaf AF, Abouheif MA (2013) Clinical, haematological and biochemical alterations associated with an outbreak of Theileriosis in dromedaries (*Camelus dromedarius*) in Saudi Arabia. Pak Vet J 34(2):209–213

Jirimutu B, Li J, Alam MS, Li H, Guo M, Zhang H (2010) Fatty acid and protein profiles, and mineral content of milk from the wild Bactrian camel (*Camelus bactrianus ferus*) in Mongolia. Milchwissenschaft 65:21–25

Jooste PL, Strydom E (2010) Methods for determination of iodine in urine and salt. Best Pract Res Clin Endocrinol Metab 24(1):77–88

Juniper DT, Philipps RH, Jones AK, Bertin G (2006) Selenium supplementation of lactating dairy cows: effect on selenium concentration in blood milk, urine and faeces. J Dairy Sci 2006 (89):3544–3551

Kadim IT, Maghoub O, Al-Mazoorqi W, Khalaf SK, Mansour MH, Al-Sinani SSH, Al-Amri IS (2009) Effects of electrical stimulation on histochemical muscle fiber staining, quality and composition of camel and cattle Longissimus thoracis muscles. J Food Sci 74:S44–S52

Kadim IT, Al-Ani MR, Al-Maqbaly RS, Mansour MH, Maghoub O, Johnson EH (2011) Proximate, amino-acid, fatty acid and mineral composition of raw and cooked camel (*Camelus dromedarius*) meat. Br Food J 113:482–493

Kamili A, Faye B, Bengoumi M, Tligui NS, Besse Y (2018) Laser induced breakdown spectroscopy to dose zinc in camel skin (In press)

Kaneko JJ (1989) Clinical biochemistry of domestic animals, 4th edn. Academic Press, New York

Karram MH, Mottelib AA, Nafie THS, Sayed AS (1989) Clinical and biochemical studies in chronic fluorosis and sulphurosis in camels. Assiut Vet Med J 21(41):165–176

Kessabi M, Assimi B, Braun JP (1984) The effects of fluoride on animals and plants in the South Safi zone. Sci Total Environ 38:63–68

Khalifa H, Fouad MT, Awad YL, Georgy ME (1973) Application of the grey R.A. to the spectrophotometric determination of copper in the liver of the Egyptian camel. Microchem J 18:536–542

Khalili M, Lindgren E, Varvikko T (1993) A survey of mineral status of soil, feeds and cattle in the Selale Ethiopian highlands. II. Trace-elements. Trop Anim Health Prod 25:193–201

Khamis GF, El-Naser EMA, Aamer AA (2011) Assessment of some trace elements in healthy camel, cattle and buffalos. Assiut Vet Med J 57:159–169

Khanna D, Agarwal SP, Gupta ML (1996) Effect of water deprivation during summer and winter on thyroid hormones concentration in the Indian camel. Indian J Anim Sci 66(3):253–255

Kinne J, Nagy P, Wernery U (2003) Serum copper levels in dromedaries after long term exogenous copper supplementation. J Camel Pract Res 10(2):121–124

Koh TS, Bensen TH (1983) Critical re-appraisal of fluorometric determination of selenium in biological materials. J Assoc Off Anal Chem 66:918–926

Konuspayeva G, Faye B, Narmuratova M, Meldebekova A, Loiseau G (2008) Variation factors of some minerals in camel milk. In: Faye B, Sinyavskiy Y (eds) Proceedings of international workshop, "Impact of pollution on animal products". Almaty (Kazakhstan), 27–30 Sept 2007, pp 125–132

Konuspayeva G, Faye B, Loiseau G, Diacono E, Akhmetsadykova S (2009) Pollution of camel milk by heavy metals in Kazakhstan. Open Environ Poll Toxicol J 1:112–118

Konuspayeva G, Jurjanz S, Loiseau G, Barci V, Akhmetsadykova S, Meldebekova A, Faye B (2011) Contamination of Camel milk (heavy metals, organic pollutants and radionuclides) in Kazakhstan. J Environ Prot 2:90–96

Kosanovic M, Rao MV, Al-Shamisi NS, Tinson A (2014) Effect of supplementation with trace elements (copper, cobalt, zinc and selenium) on blood cell count in camels (*Camelus dromedarius*). J Trace Elem Anal 3(1):23–31

Kuria SG, Tura IA, Amboga S, Walaga HK (2013) Status of minerals (*Camelus dromedarius*) in north eastern Kenya asevaluated from the blood plasma. Livest Res Rural Dev 25(8). http://www.lrrd.org/lrrd25/8/kuri25149.htm

Laatar A, Mrabet D, Zakraoui L (2003) La fluorose en Afrique subsaharienne. Rev Rhum 70:178–182

Lamand M (1974) Les oligo-éléments dans le lait de vache et de brebis. Ann Rech Vét 5:281–289

Lamand M (1979) Le diagnostic des carences en oligo-éléments: l'analyse du sol ou de la plante? Bull Techn CRZV Theix INRA 35:27–36
Lamand M (1985) Influence of protein intake on per os zinc deficiency tratment in sheep. Ann Rech Vét 16:285–287
Lamand M (1987) Place du laboratoire dans le diagnostic des carences en oligo-éléments. Rec Med Vet 163:1071–1082
Lamand M, Levieux D (1981) Effects of infection on plasma levels of copper and zinc in ewes. Ann Rech Vét 12:133–136
Lewis DC, Sale PWG, Kemp DR, Michalk DL (1994) Management of nutients for pastures. Pasture management: technology for the 21st century. CSIRO Information Service, Melbourne, Australia, pp 38–50
Li WX, Hai WJ (2014) Studies of Sulphur-induced copper deficiency in the Bactrian camel. J Anim Vet Adv 13(14):886–890
Li X, Yu Z (2016) Determination of selenium in biological samples with an energy-dispersive X-ray fluorescence spectrometer. Appl Radiat Isot 111:45–49
Liu ZP (2003) Studies on the haematology and trace element status of adult Bactrian camels (*Camelus bactrianus*) in China. Vet Res Commun 27:397–405
Liu ZP (2005) Treatment and prevention of rickets and osteomalacia in Bactrian camel (*Camelus bactrianus*). J Camel Pract Res 12(2):99–104
Liu ZP, Ma Z (1995a) Studies on supplement of copper and selenium for Bactrian camels. J Jiangsu Agric College 19–22
Liu ZP, Ma Z (1995b) Studies on trace elements status in Bactrian camels. J Jiangsu Agric College:49–52
Liu ZP, Ma Z, Zhang YJ (1994) Studies on the relationship between sway disease of Bactrian camels and copper status in Gansu Province. Vet Res Commun 18:251–260
Ma Z (1995) Studies on sway disease of chinese bactrian camels. Epidemiological and aetiological aspects. Report of International Foundation for Science Project, Stockholm, Sweden, 17 p
Ma Z, Liu Z, Huaitao C (1995) The discovery of multiple lipomas in the liver of Bactrian camels and study of their relationship with trace elments. Acta Vet Zootech Sinica 26(1):87–92
Maas J, Parish SM, Hodgson DR (1990) Nutritional myodegeneration. In: Smith BP (ed) Large animal internal medicine. CV Mosby, St Louis, pp 1352–1357
Mahmud T, Rehman R, Anwar J, Ali S, Abbas A, Salman M (2011) Minerals and nutritional composition of camel (*Camelus dromedarius*) meat in Pakistan. J Chem Soc Pak 33(6):835–838
Martynenko NI, Yagodinskaya SG, Akhundov AA, Charyev KC, Khummedov O (1977) Contents of trace elements (copper, manganese, molybdenum) in cultured chal and camels' milk and their clinical significance. Zdravookhr Turkmenistana 3:20–22
Marx W, Abdi HN (1983) Serum levels of trace elements and minerals in dromedaries (*Camelus dromedarius*) in South Somalia. Anim Res Dev 17:83–90
Mc Dowell LR, Conrad JH, Ellis GL, Loosli JK (1983) Minerals for grazing ruminants in tropical regions. University of Florida, Gainsville, p 86
McDowell LR (1992) Minerals in animal and human nutrition. Academic Press, San Diego, p 517
McDowell LR, Conrad JH, Glen Hembry F (1993) Minerals for grazing ruminants in tropical regions, 2nd edn. University of Florida, Gainsville, p 77
Mehaia MA, Hablas MA, Abdel-Rahman KM, Le Mougy SA (1995) Milk composition of Majaheim, Wadah and Hamra camels in Saudi Arabia. Food Chem 52:115–122
Meldebekova A, Konuspayeva G, Diacono E, Faye B (2008) Heavy metals and trace elements content in camel milk and shubat from Kazakhstan. In: Faye B, Sinyavskiy Y (eds) Proceedings of international workshop, "Impact of pollution on animal products". Almaty (Kazakhstan), 27–30 Sept 2007, pp 117–123
Mertz W (1981) The essential trace-elements. Science 213:1332–1338
Mohamed HE (2004) The zinc and copper content of the plasma of Sudanese camels (*Camelus dromedarius*). Vet Res Commun 28:359–363

Mohamed MH, Mohamed AH, Locatelli A (1984) Water deprivation effects on the hematological and hematochemical pictures of *Camelus dromedarius*. Rev Elev Méd Vét Pays Trop 37:313–317
Moty IA, Mulla A, Zaafer SA (1968) Copper, iron and zinc in the serum of Egyptian farm animals. Sudan Agric J 3:146–151
Mustafa AB, Sayied AA, Atti KAA (2012) Trace minerals profile in wild pasture and in the blood serum of camel in Butana region. Res Opin Anim Vet Sci 2(5):329–333
Nafizi S, Mansourian M, Nikahval B, Razavi SM (2009) The relationship between serum level of thyroid hormones, trace elements and antioxidant enzymes in dromedary camel (*Camelus dromedarius*). Trop Anim Health Prod 41(1):129–134
O'Toole T, Raisbeck MF (1995) Pathology of experimentally induced chronic selenosis (alkali disease) in yearling cattle. J Vet Diagn Investig 7:364–373
Omer RH, Eltinay AH (2008) Chemical composition of camel's colostrum and milk in United Arab Emirates. Arab Univ J Agric Sci 16(1):127–134
Orliac D (1980) Contribution à l'étude de la biochimie sanguine de dromadaires et de chèvres sahariens. Thèse doct. Vét., Toulouse, 71, 31p
Osman NIED (2012) Effect of copper supplemented salt licks on total and TCA-soluble plasma copper concentrations in omani camels. J Camel Pract Res 29(2):1–5
Ozdemir O, Ciftçi MK, Hatipoglu F, Ortatatli M, Yavuz O, Kanat O (2016) Nutritional cardiomyopathy in a young camel (*C. dromedarius*). Eurasian J Vet Sci 32(1):52–54
Parekar SS, Sanjeev K, Mody SK, Kathirvelan C, Bhagwat SR (2009) Trace mineral bioprofile of male camels. Indian Vet J 86:1184
Pourjafar M, Badiei K, Nazifi S, Chalmeh A, Setayesh A, Naghib M (2014) Correlations between serum trace elements (sqelenium, copper and zinc) and antioxidant vitamins (vitamin A, E and C) in clinically healthy dromedary camels. J Fac Vet Med Istanbul Univ 40(1):7–13
Raisbeck MF, Dahl ER, Sanchez DA, Belden EL, O'Toole D (1993) Naturally occurring selenosis in Wyoming. J Vet Diagn Investig 5:84–87
Rashed MN (1992) Determination of trace elements in milk of some animals from Aswan (Egypt). Int J Environ Anal Chem 48:41–50
Rashed MN (1998) Trace elements in camel's milk from Aswan city and desert (Egypt). Actes de l'atelier "Chameaux et dromadaires, animaux laitiers", Nouakchott, Mauritanie, 24–26 Oct 1994, Ed. CIRAD (coll. Colloques), pp 205–210
Rashed MN (2002) Trace elements in camel tissues from a semi-arid region. The Environmentalists 22:111–118
Rejeb A, Amara A, Rezeigui H, Crespeau F, Delverdier M (2012) Etude anatomopathologique et hormonale du goitre chez le dromadaire (*Camelus dromedarius*) dans le sud tunisien. Rev Med Vet 163(5):242–249
Rutagwenda T, Lechner-Doll M, Kaske M, Engelhardt WV, Schultka W, Schwartz HJ (1989) Adaptation strategies of camels on a thornbush savannah pasture, comparison with other domestic animals. Options Méditerranèennes, Série Séminaires n°2, C. I. H. E. A. M., pp 69–73
Ryder ML, Stephenson SK (1968) Wool growth. Academic Press, London
Saeed A, Afzal M, Khan IA (1995) Haematological and serum constituents of camel calves during first 7 months of age. Indian J Anim Sci 65:297–301
Saeed A, Hussain MM, Khan IA, Chand G, El-Yousuf RA (2004) Effect of sex and age on blood biochemical profile in camel. J Camel Pract Res 11(1):73–77
Saini N, Bhati AK, Singh N, Tuteja FC (2007) Trace mineral and vitamin C content of camel milk: a comparative study. Vet Pract 8(1):20–21
Saini N, Singh N, Kiradoo BD, Pathak KML (2009) Comparative biochemical and mineral profile of female Indian dromedaries during breeding season. J Camel Pract Res 16(2):189–193
Sanjabi MR, Moeini MM, Telfer SB (2003) Thiomolybdate – The major factor on clinical copper deficiency. Acta Vet Scand 44(Suppl 1):258–259
Sawaya WN, Khalil JK, Al-Shalhat A, Al-Mohammed H (1984) Chemical composition and nutritional quality of camel milk. J Food Sci 49:744–747
Schillorn Van Veen TW, Loeffler IK (1990) Mineral deficiency in ruminants in Sub-saharian Africa: a review. Trop Anim Health Prod 22:197–205

Seboussi R, Faye B, Alhadrami G (2004) Facteurs de variation de quelques éléments trace (sélénium, cuivre, zinc) et d'enzymes témoins de la souffrance musculaire dans le sérum de dromadaire (*Camelus dromedarius*) aux Emirats Arabes Unis. Rev Elev Méd Vét Pays Trop 57 (1–2):87–94

Seboussi R, Faye B, Alhadrami G, Askar M, Ibrahim W, Hassan K, Mahjoub B (2008a) Effect of different selenium supplementation levels on selenium status in camel. Biol Trace Elem Res 123:124–138

Seboussi R, Alhadrami G, Askar M, Faye B (2008b) Effect of excess selenium on dromedary camel in the United Arab Emirates. In: Faye B, Sinyavskiy Y (eds) Proceedings of international workshop, "Impact of pollution on animal products". Almaty (Kazakhstan), 27–30 Sept 2007, pp 143–146

Seboussi R, Faye B, Askar M, Hassan K, Alhadrami G (2009a) Effect of selenium supplementation on blood status and milk, urine and fecal excretion in pregnant and lactating camel. Biol Trace Elem Res 128:45–57

Seboussi R, Faye B, Alhadrami G, Askar M, Bengoumi M, Elkhouly A (2009b) Chronic selenosis in camels. J Camel Pract Res 16(1):25–38

Seboussi R, Faye B, Alhadrami G, Askar M, Ibrahim W, Mahjoub B, Hassan K, Moustafa T, Elkhouly A (2010) Selenium distribution in camel blood and organs after different level of dietary selenium supplementation. Biol Trace Elem Res 133(1):34–50

Sena DS, Mal G, Sahani MS, Bhati A (2007) Comparative studies on micromineral profile in camels. Indian Vet J 84:698–700

Shamat AM, Babiker IA, Mukhtar AMS, Ahmed FA (2009) Seasonal and regional variations in mineral content of some important plant species selected by camels (*Camelus dromedarius*) in arid and semi-arid lands (ASAL) of Sudan. J Appl Sci Res 5(10):1676–1684

Shekhawat VS (1983) Some studies on serum trace mineral (zinc, copper and iron) levels of ruminants in arid tract of western Rajasthan. MVSc Thesis, Sukhadia University, Udaipur (Rajasthan), India

Shekhawat VS, Bathia JS, Ghosal AK (1987) Serum iron and total iron capacity in camel. Indian J Anim Sci 57:168–169

Shen X, Li X (2010) Studies of "emaciation ailment" in the Bactrian camel. Afr J Biotechnol 9 (49):8492–8497

Shukla MK, Siddiquee GM, Latif A, Parekar SS (2009) Plasma trace mineral concentration of Kutuchi camels. Indian J Vet Res 18(2):28–30

Singh AP, Sharma SN, Taneja M (1994) Status of zinc (*Camelus dromedarius*) with reference to blood serum, seminal plasma and hair. Indian J Anim Sci 64:750–751

Soliman GZA (2005) Comparison of chemical and mineral content of milk from human, cow, buffalo, camel and goat in Egypt. Egyptian J Hosp Med 21:116–130

Suttle NF (2010) Mineral nutrition of livestock, 4th edn. CABI, Cambridge, p 587

Suttle NF, Field AC, Barlow RM (1970) Experimental copper deficiency in sheep. J Comp Pathol 80:151–162

Sztych D, Soroczynska M (1995) Some chosen macro- and microelements in the wool of sheep in different stages of lactation. Ann Warsaw Agric Univ Anim Sci 31:47–53

Tageldin MH, Sid Ahmed El Sawi A, Ibrahim SG (1985) Observations on colloid goiter of dromedary camels in the Sudan. Rev Elev Méd Vét Pays Trop 38:394–397

Tao G, Geng C, Tao Y, Jiao H, Fan Y (1995) "Roll disease" of camels in Mongolia. Acta Vet Zootechn Sinica 26:541–544

Tartour G (1966) Study of the role of certain trace-elements in relation to the health of livestock in the Sudan. PhD Thesis, University of London

Tartour G (1969) Studies on the metabolism of copper and iron in the camel. Sudan J Vet Sci Anim Hus 10:14–20

Tartour G (1975) Copper status in livestock, pasture and soil in Western Sudan. Trop Anim Health Prod 7:87–94

Tartour G, Idris OF (1970) Studies of copper and iron metabolism in the camel fetus. Acta Vet Brno 39:397–403

Tiwary AK, Stegelmeier BM, Panter KE, James LF, Hall JO (2006) Comparative toxicosis of selenium selenite and sélénométhionine in lambs. J Vet Diagn Investig 18:61–70

Towers NR, Young PW, Wright DE (1981) Effect of zinc supplementation on bovine plasma copper. NZ Vet J 29:113–114

Tuteja FC, Dixit SK, Deen A, Bhati A, Sahani MS (2004) Mineral antioxidant status in serum and its relationship ,with somatic cell count in camel milk. J Camel Pract Res 11(1):59–62

Urazakov NU, Bainazarov S (1974) "Tushchibek", the first clinic in history for the treatment of pulmonary tuberculosis with cultured camels' milk. Probl Tuberk 2:89–90

Van de Wiel H (2004) Determination of elements by ICP-AES and ICP-MS. Horizontal-19, National Institute for Public Health and the environment (RIVM), Bilthoven, 46 pp

Vanhoe H (1993) A review of the capabilities of ICP-MS for trace element analysis in body fluids and tissues. J Trace Elem Electrolytes Health Dis 7(3):131–139

Vyas S, Saini N, Kiradoo BD, Lukha A, Kishore N, Mal G, Pathak KM (2011) Biochemical and trace mineral profile in post-parturient dromedary camel (*Camelus dromedarius*). Indian J Anim Sci 81(6):47–48

Wang JG (1988) Examination of clinical biochemistry. Hunan Science & Technology Press, Beijing, pp 227–279

Wang SY, Liang JP, Shao WJ, Wen H (2011) Mineral, vitamin anf fatty acid contents in the camel milk of dromedaries in the Anxi Gansu China. J Camel Pract Res 18(2):273–276

Wensvoort J (1992) Copper, iron, manganese and zinc concentrations in livers of race animals. Proceedings of 1st International Camel Symposium, Dubai, 2–7 Feb 91

Wernery U, Ali M, Kinne J, Abraham AA, Wernery R (2002) Copper deficiency: a predisposing factor to scepticaemia in dromedary calves. J Camel Pract Res 9(1):59–66

Wernery U, Abraham AA, Jyothi T, Abubakar Ali Y, George RM (2009) Mineral and vitamin contents in the blood of racing dromedaries in the United Arab Emirates. J Camel Pract Res 16 (1):39–40

Whabi AA, Abdel Gadir SE, Neimat AA, Idris OF (1979) Plasma electrolytes and minerals of normal camels in the Sudan. In: Cockrill WR (ed) Camels and camelids. Uppsala, Suède, pp 431–437

Wisner H, Schotke B (1975) White muscle disease at The Hellabrun Zoo in Munich. Vet Erkr Zoo Wild Tunis 1:717–720

Youssef SY, Yasien S, Al-Mousa WH, Nasr SM, El-Kelesh EAM, Mahran KM, Abd-El-Rahman AH (2015) Vector identification and clinical, haematological, biochemical and parasitological characteristics of camel (*Camelus dromedarius*) theileriosis in Egypt. Trop Anim Health Prod 47:649–656

Yu S, Beynen AC (1994) The combined effect of high iron and zinc intake on copper status in rats. Biol Trace Elem Res 42:71–79

Zongping L (2005) Studies on rickets and osteomalacia in Bactrian camels (*Camelus bactrianus*). Vet J 169(3):444–453

Zouagui H (1973) Contribution à l'étude de la fluorose chez les grands et les petits ruminants au Maroc. Thèse Méd Vét, Toulouse, France

Chapter 8
Vitamins

The 13 known vitamins (Table 8.1) are organic nutrients that all organisms require in small amount from their diet. Those nutrients are essential for the health, and their deficiency could be dramatic for animals and humans. Vitamins are classified according to their biological and chemical activities which are at the origin of their biochemical functions. They are also classified according to their liposolubility allowing the distinction between fat-soluble and water-soluble vitamins. They play different roles (cofactor of diverse enzymes, regulator of mineral metabolism and antioxidant activity, etc.) involved in the general metabolism of the organism (Bender 2003). Vitamins are now synthetized by industries and are included in feed supplement of livestock and poultry (mineral-vitamin powders and blocks), but in most of camel farming systems, only vitamins present in the natural diet are available for the animals.

Vitamin status of camel is not well known, and the number of references is quite low. However, the roles of vitamins in organism and, overall, the effect of deficiencies in specific vitamins should be quite similar than for other mammals.

8.1 Vitamin A (Retinol)

Vitamin A plays an important role in the vision and protection of mucosa and teguments. In consequence, hypovitaminosis A could affect the skin (hyperkeratosis) and the vision (crepuscular blindness). Vitamin A could play also a specific role in reproduction performances (Clagett-Dame and Knutson 2011), but this aspect has never been investigated in camel. The main source of vitamin A for camel is in the desert forages under the form of carotenoids (provitamin A).

© Springer International Publishing AG, part of Springer Nature 2018
B. Faye, M. Bengoumi, *Camel Clinical Biochemistry and Hematology*,
https://doi.org/10.1007/978-3-319-95562-9_8

Table 8.1 List of the vitamins and the main diseases linked to deficiency or overdose

Vitamin	Name	Solubility	Deficiency	Overdose
A	Retinol	Fat	Crepuscular blindness Hyperkeratosis	Hypervitaminosis A
B1	Thiamine	Water	Polioencephalomalacia, Beriberi	Drowsiness
B2	Riboflavin	Water	Glossitis and stomatitis	
B3	Niacin	Water	Pellagra	Liver damage
B5	Pantothenic acid	Water	Paresthesia	Diarrhea
B6	Pyridoxine	Water	Anemia and neuropathy	Nerve damage
B7	Biotin	Water	Dermatitis and enteritis	
B9	Folic acid	Water	Anemia	
B12	Cyanocobalamin	Water	Anemia and cachexia	
C	Ascorbic acid	Water	Scurvy	Hypervitaminosis C
D	Cholecalciferol	Fat	Osteomalacia	Hypervitaminosis D
E	Tocopherol	Fat	Sterility	Heart failure
K	Phylloquinone	Fat?	Bleeding diathesis	

8.1.1 *Vitamin A in Plasma*

The normal level of vitamin A in plasma is comprised between 30 and 70 μg/100 ml, but part of the observed variability could be due to the analytical method used. There are few references regarding β-carotene, but concentration in camel plasma is low, generally below 10 μg/100 ml.

The first reference on vitamin A levels in camel plasma was that of Ghosal et al. (1973) in India. The mean value of vitamin A in 25 healthy camels was 45.7 ± 4.9 μg/100 ml. Almost identical values were determined in UAE by Snow et al. (1992) (42.2 ± 0.02 μg/100 ml), Abbas and Ali (2001) (46.0 ± 4.9 μg/100 ml), and Stahl et al. (2006) (39.2 ± 6.6 μg/100 ml) and in Sudan by Mohamed (2006a) (47.9 ± 6.9 μg/100 ml). More recently, Ghadrdan-Mashhadi et al. (2013) using spectrophotometry method have reported mean values of 63.9 ± 4.7 μg/100 ml on 168 Iranian camels for vitamin A and 9 ± 1.1 μg/100 ml only for β-carotene, the most important carotenoid. Usually, camel plasma is limpid, almost like water, contrary to cow plasma which is yellowish. This limpidity of camel plasma is linked to its low level in carotenoid pigments, especially β-carotene (Snow et al. 1992). Comparing camel and cow plasma, Stahl et al. (2006) were not able to determine the quantity of β-carotene (below the detection limit of 0.32 μg/100 ml), while the concentration in cow plasma was 496 ± 88 μg/100 ml.

In 132 racing camels from UAE, Wernery et al. (2009) using HPLC reported lower values for vitamin A, 32.6 ± 12.6 μg/100 ml, which was close to the usual value proposed by Bogin (2000) for cattle, 30 ± 4 μg/100 ml. However, quite higher values were rarely reported. For example, Al-Senaidy (1998) observed

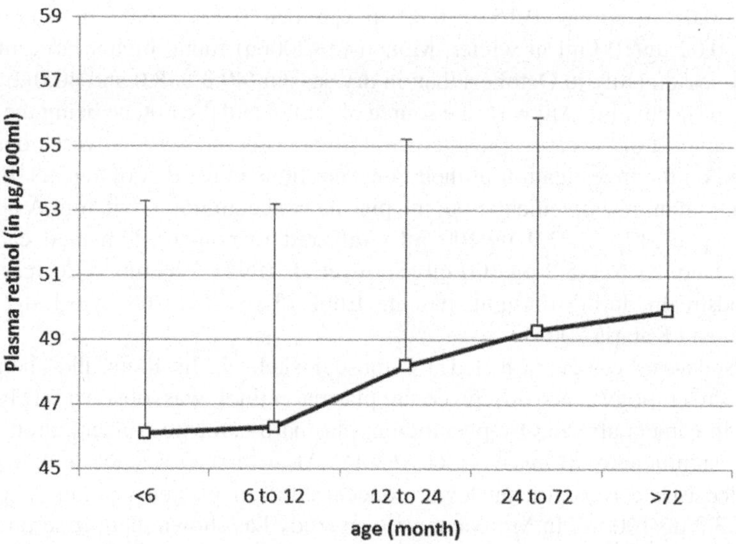

Fig. 8.1 Change in plasma retinol concentration with the age of camel (in month) [from after the data of Mohamed (2006a)]

173 ± 5.1 µg/100 ml for retinol and 21.5 ± 1.4 µg/100 ml for β-carotene in 14 Saudi Arabian camels.

There is no significant difference between adult and young camels, 44.7 ± 18.1 vs 37.8 ± 12.5 µg/100 ml (Homeida et al. 2010), although Ghadrdan-Mashhadi et al. (2013) and Mohamed (2006a) found a linear increase of retinol concentration with the age (Fig. 8.1). In Egypt, Baraka (2012) found higher values in adult (67.05 ± 4.04 µg/100 ml) than in young camel (49.8 ± 2.9 µg/100 ml). For β-carotene, the values were reverse: 3.39 ± 0.65 µg/100 ml in adult vs 10.1 ± 3.17 µg/100 ml in young camel. No sexual difference was also reported (Abbas and Ali 2001; Mohamed 2006a), although a slight difference was reported by Baraka (2012), 69.9 ± 3.7 vs 55.6 ± 4.1 µg/100 ml in male and female, respectively. However, for β-carotene, the concentration was lower in male than in female, 3.87 ± 1.18 vs 5.75 ± 1.02 µg/100 ml, respectively.

Night blindness (deficiency in vitamin A) in camel was described in the Horn of Africa (Faye and Mulato 1991; Agab et al. 1992, 1993). In Sudan, it represented 18% of the cases of diseases reported in summer, and the whole prevalence was 7.5% (Agab and Abbas 1999). In Eritrea, the prevalence of night blindness was 2.1% (Gebrehiwet 1999). In all the cases, night blindness appeared linked to the hot season with an almost complete disappearance of the symptoms during autumn, probably because the availability of green fodder as source of β-carotene is higher at this season. Yet, in the investigation of Ghadrdan-Mashhadi et al. (2013), concentrations of retinol and β-carotene were higher in summer (86.6 ± 5.6 and 12.9 ± 2.1 µg/100 ml, respectively) than in winter (42.7 ± 6.8 and 5.1 ± 0.8 µg/100 ml, respectively). Confirming lower levels, Baraka (2012) found β-carotene plasma values of

0.24 ± 0.02 in spring, 0.15 ± 0.02 in summer, 0.20 ± 0.02 in autumn, and 0.12 ± 0.02 µg/100 ml in winter. Mohamed (2006a) found higher concentrations in rainy season (July to October) than in dry season (49.3 ± 8.0 and 40.0 ± 6.0 µg/100 ml, respectively). Anyway, the source of retinol and β-carotene being the natural pasture in most of the cases, the interpretation of observed seasonal variations has to be based on the investigation of their concentrations in the diet of camels. In camel receiving vitamin supplementation, the plasma concentration in vitamin A appeared higher, up to 49.1 ± 7.9 µg/100 ml compared to non-supplemented camels in grazing land: 33.6 ± 5.2 µg/100 ml (Snow et al. 1992). Vitamin A decreased also in bloodstream during drought, passing from 75 ± 7.8 to 66 ± 6.1 µg/100 ml (Kataria and Kataria 2004).

In Sudanese camels affected by musculoskeletal disorders like bent neck (unknown etiology), no difference in plasma retinol was observed (Mohamed 2004). In camels affected by pneumonia, plasma β-carotene concentration did not change significantly (Elnisar et al. 2011). At reverse, a camel intoxicated by aflatoxicosis had a significant lower concentration of plasma vitamin A (retinol): 24.3 ± 3.2 µg/100 ml. In Morocco, a recent study has shown that the serum retinol concentration was significantly lower in camels affected by mange (35.3 ± 10.8) than in healthy camels (44.4 ± 5.9 µg 100 ml). The difference was more important in adults (males and females) over 8 years (Lyaktini et al. 2013). This could be linked to the protective function of vitamin A for teguments: a low vitamin A status could be a risk factor for mange. Vitamin A supplementation is generally proposed for mange treatment (Fassi-Fehri 1987; Palanivelrajan et al. 2015). There is no significant effect of digestive disorders as indigestion, acidosis, or diarrhea (Baraka 2012).

8.1.2 Vitamin A in Milk

The first reference of vitamin A in milk was Sawaya et al. (1984) in Saudi Arabia who have found 150 µg/l on average. In 20 dairy camels, Farah et al. (1992) have reported an average of 100 µg/l with a range of 50–140 µg/l. Vitamin A concentration is lower in camel milk compared to cow milk (Wernery 2003): 100–150 µg/l vs 170–380 µg/l in cow milk. Globally, vitamin A in camel milk is lower than in ruminants' milk as sheep, goat, cow, or buffalo (Claeys et al. 2014).

A significant difference was observed between camel colostrum and camel milk: 307 ± 132 µg/l vs 201 ± 100 µg/l, respectively (Stahl et al. 2006). In spite of high values reported in this last reference, vitamin A concentrations appeared lower than in cow milk (609 ± 256 µg/l). Vitamin A concentration seems higher in Bactrian milk than in dromedary, 970 µg/l, on average, with no significant change all along the lactation (Zhang et al. 2005).

The camel milk is also poor in β-carotene that explains the very white color of camel milk compared to cow milk. In camel colostrum, β-carotene concentration was 3.2 µg/l, but in milk, the quantity was below the detection limit (<0.32 µg/100 ml), while the concentration in cow milk was 996 µg/l on average.

Vitamin A was determined also in synovial fluid of arthritic camel, and no difference was found with healthy joint: 6.52 ± 0.1 and 6.55 ± 0.5 µg/100 ml, respectively (Chalmeh et al. 2016).

8.2 Vitamin B1 (Thiamine)

Thiamine or vitamin B1 is known in human since the seventeenth century because its deficiency provoked "beriberi," a neurological disease observed in Asia among people eating white rice only rather than complete rice. In ruminants, thiamine deficiency has been described as responsible of polioencephalomalacia (PEM), also called cerebro-cortical necrosis (CCN), provoking severe nervous troubles, sometimes fatal. Camels affected by PEM exhibit staggering gate and muscle tremors. In some cases, affected animals become blind and stay sternally recumbent (Abbas et al. 2008). PEM is caused by the overconsumption of simple carbohydrates and less fibers. That is why PEM was observed in racing camel from UAE because their diet is generally poor in fiber and rich in rapidly fermentable glucides provoking acidosis (Wernery and Kinne 2001).

8.2.1 Vitamin B1 in Blood

Usual values for thiamine concentrations in camel plasma vary between 30 and 90 µg/l. The first publication in camels was on the methodology of analysis of thiamine by HPLC: 57 ± 13 µg/l (Wernery et al. 2002).

Later on, based on more than 48,000 analyses, Wernery et al. (2009) stated that the usual value for plasma thiamine concentration was 48 ± 13 µg/l. In a study involving 1315 racing camels at different physiological stages and different ages, plasma thiamine concentration in newborn camel was 98 ± 18 µg/l (Abbas et al. 2008). This concentration decreased rapidly after parturition to reach a minimum value at weaning (36 ± 6 µg/l). In adult racing camels, the values varied between 53 ± 12 and 59 ± 15 µg/l according to age, the highest values being observed in 6-year-old camels (Fig. 8.2).

Concentrations were higher in non-racing camels (up to 74 µg/l). Elsewhere the thiamine concentration appeared higher in non-pregnant-non-lactating camels (72 ± 8 µg/l) than in pregnant females at the end of pregnancy (62 ± 8 µg/l) and in lactating camels (from 59 µg/l on average in postpartum to 48 µg at the end of lactation). Tinson et al. (1999) found similar values in non-pregnant, non-lactating, and non-racing camels with mean value of 74 µg/l.

Plasma thiamine concentration is probably decreasing in stressed camels (weaning, lactation, end of gestation) but overall in racing camels 2–4 years old which is the main population affected by PEM (Abbas et al. 2008). Affected camels

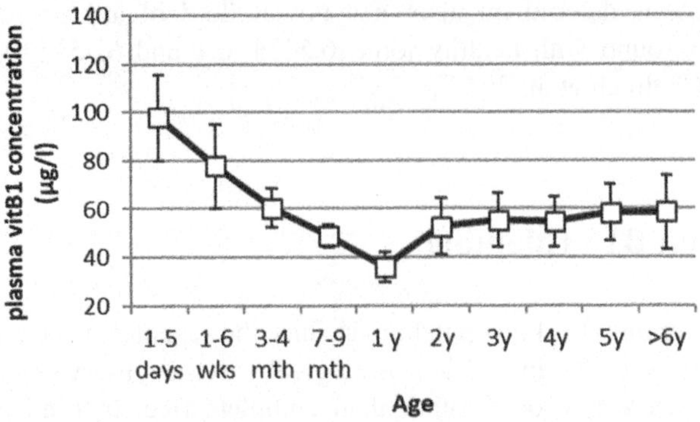

Fig. 8.2 Change in camel blood thiamine according to the age of racing camels (from after the data of Abbas et al. 2008)

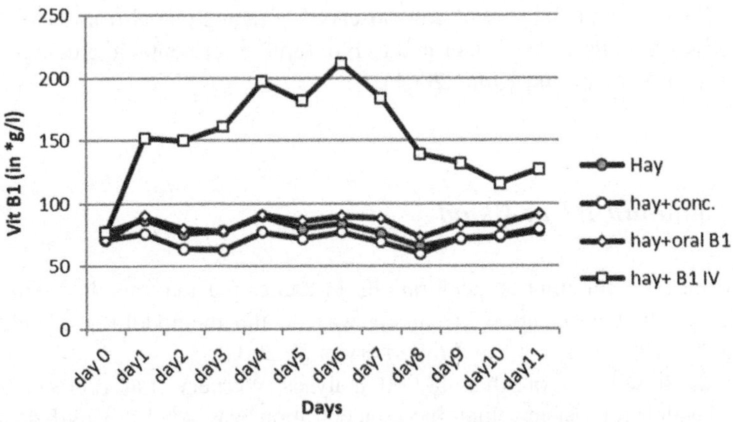

Fig. 8.3 Change in blood thiamine (in µg/l) in different diets (the group "hay+ B_1 IV" received one vitamin B1 injection at day 1) (from after the data of Tinson et al. 1999)

had a lower thiamine concentration (21 ± 10 µg/l) than non-affected camels. A rapid recovery is observed after thiamine injection (Milad and Ridha 2009).

According to Mohamed (2006b), plasma thiamine concentrations are higher in adults (89 ± 7 µg/l) than in yearlings (71 ± 9 µg/l) and in neonates (60 ± 4 µg/l). At reverse, no sex or breed difference was revealed. Values were significantly higher both in male and female during breeding season (84 and 80 µ/l, respectively) compared to non-breeding season (67 and 63 µg/l, respectively).

Plasma thiamine concentration is highly sensitive to vitamin supplementation by injection, values overpassing 200 µg/l after IV injection of 1 g thiamine (Tinson et al. 1999), while no significant effect was observed with oral supplementation (Fig. 8.3).

8.2.2 *Vitamin B1 in Milk*

The first publication on thiamine in camel milk was done by Sawaya et al. (1984) with a mean value of 33 μg/100 ml which is lower on average than in cow milk: 47 μg/100 ml (Lalic et al. 2014). According to Alhadrami (2003), the range of camel milk thiamine was 28–90 μg/100 ml. Zhang et al. (2005) found higher thiamine concentration in Bactrian milk during the first days of lactation: 51 μg/100 ml in colostrum vs 12 μg/100 ml in milk.

8.3 Vitamin B2 (Riboflavin)

Formerly known as vitamin G, riboflavin is participating in a large variety of flavoprotein enzyme reactions involved in a wide range of biological processes, notably in the degradation of free radicals. In animals, vitamin B2 deficiency, which can be fatal, results in growth failure and ataxia (Patterson and Bates 1989) and other symptoms like hair loss, corneal opacity, kidney degeneration, or inflammation of the intestinal tract. In camel, such symptoms were never strictly described in relation with riboflavin deficiency, and globally, the references are very scarce in this species. There is no data on plasma riboflavin concentration in camels.

In dromedary milk, only three references are available: 41.6 ± 1.6 μg/100 ml (Sawaya et al. 1984), 57 μg/100 ml with a range of 43–78 μg/100 ml (Farah et al. 1992), and 56 ± 11 μg/100 ml (Mehaia 1994). In Bactrian camel, concentration seems to be higher, 124 μg/100 ml (Zhang et al. 2005), without significant change all along lactation. Globally, riboflavin concentration in camel milk is lower than cow milk (156 μg/100 ml on average according to Farah et al. 1992).

In a camel affected by hematuria, the treatment based on vitamin B injection including riboflavin appeared efficient (Bhandare 2009).

8.4 Vitamin B3 (Niacin)

Niacin, also known as nicotinic acid or vitamin PP, is a precursor of nicotinamide adenine dinucleotide (NAD) and nicotinamide adenine dinucleotide phosphate (NADP) which play an essential role in metabolism of all nutrients (fat, carbohydrate, and protein) and nucleotic acids. Niacin deficiency provokes a severe disease called pellagra affecting the skin. To our knowledge, vitamin B3 deficiency was never described in camel although niacin deficiency was reported in young ruminants (Hopper and Johnson 1955).

Supply of niacin to ruminant animals comes from three main sources: (1) niacin present in the diet, (2) conversion of tryptophan to niacin, and (3) ruminal synthesis of niacin. Generally, niacin is widely distributed in feedstuffs, and its bioavailability is quite higher than in cereal grains and their by-products. Thus, niacin deficiency is more common in monogastric animals than in herbivorous (Luce et al. 1966). There is no data on plasma niacin concentration in camels.

In camel milk, only one publication of Sawaya et al. (1984) is available. They have found 461 ± 24 μg/100 ml which is higher than cow (107 μg/100 ml), goat (277 μg/100 ml), or buffalo (91 μg/100 ml) milk but similar to sheep milk (417 μg/100 ml) according to USDA Nutrients data.

8.5 Vitamin B5 (Pantothenic Acid)

Pantothenic acid is essential for animals to synthesize coenzyme A (CoA) and metabolize nutrients (proteins, carbohydrates, and fats). CoA plays a role in different metabolic cycles (as acetyl CoA) and in the biosynthesis of fatty acids, cholesterol, glycogen, and acetylcholine. Vitamin B5 deficiency is very rare and provokes similar symptoms than other vitamin B deficiencies (fatigue, apathy, impaired energy production). In non-ruminating animals, vitamin B5 deficiency could provoke nervous, gastrointestinal, and immune system disorders, loss of appetite, reduced growth rate, and skin lesions (Smith and Song 1996). However, in ruminating animals, synthesis of pantothenic acid by ruminal microorganisms is 20–30 times more than dietary contents. Consequently, vitamin B5 deficiency is never described in ruminant, and the pantothenic supplementation has no effect on growth performance of cattle. There is no reference on plasma pantothenic acid in camel. In cattle, the plasma values vary between 188 and 341 ng/l (Song et al. 1990).

In camel milk, Sawaya et al. (1984) reported 88 ± 22 μg/100 ml for vitamin B5 concentration which is quite lower than cow (362 μg/100 ml on average), goat (310 μg/100 ml), sheep (407 μg/100 ml), and buffalo milk (192 μg/100 ml).

8.6 Vitamin B6 (Pyridoxine)

Vitamin B6, and especially its active form, is involved in several metabolisms (amino acid biosynthesis and synthesis of histamine, hemoglobin, and neurotransmitters like serotonin, dopamine, epinephrine, norepinephrine, and γ-aminobutyric acid). It plays also important role as coenzyme in methionine and selenium metabolism as well as glucose and lipid metabolism.

Vitamin B6 deficiency has been described regularly in animals, provoking growth delay, dermatitis, convulsions, anemia, and alopecia. Due to the ruminal synthesis of vitamin B6, ruminant animals are not sensitive to this deficiency. However, young calves that do not have yet functional rumen might be affected when they are fed with milk substitute (Johnson et al. 1950).

As for the other vitamin B, no case of deficiency in vitamin B6 nor data on its plasma concentration is reported in camels.

For camel milk, concentration of vitamin B6 is around 50 µg/100 ml: 52.3 ± 11.5 for Sawaya et al. (1984) and 54 µg/100 ml on average for Zhang et al. (2005). These values are similar to those of other species: 36 (cow milk), 46 (goat milk), 60 (sheep milk), and 23 µg/100 ml (buffalo milk). The concentration in camel milk is constant all along the lactation (Zhang et al. 2005).

8.7 Vitamin B7 (Biotin)

Vitamin B7 is involved in cell growth and in metabolism of fats and amino acids. In human, it is used as supplement for strengthening hairs and nails. In grazing animals, biotin deficiency could occur in case of digestive disorders, especially in animals fed with high level of cereal grains. In ruminants, symptoms of deficiency are not pathognomonic: lethargy, dermatitis, anemia, and weak growth. Biotin deficiency in dairy cows is linked to lameness, and its supplementation can protect against hoof disorders (Badhauria et al. 2013) and improve dairy performance (Chen et al. 2011).

In camel plasma, biotin concentration is on average 32.8 ± 9.1 ng/100 ml (Snow et al. 1992) with an important effect of biotin content in the diet, values reaching 250 ng/100 ml a week after supplementation (Fig. 8.4). In cattle, plasma biotin varies between 30 and 80 ng/100 ml, but the concentration can increase 10 times in supplemented animals (Bhaudaria et al. 2015). There is no reference, at our knowledge, on biotin concentration in camel milk. In cow milk, biotin content is reported to be between 2 and 4 µg/100 ml, i.e., in higher quantity than in goat milk: 0.9–4.0 µg/100 ml (Woollard and Indyk 2013).

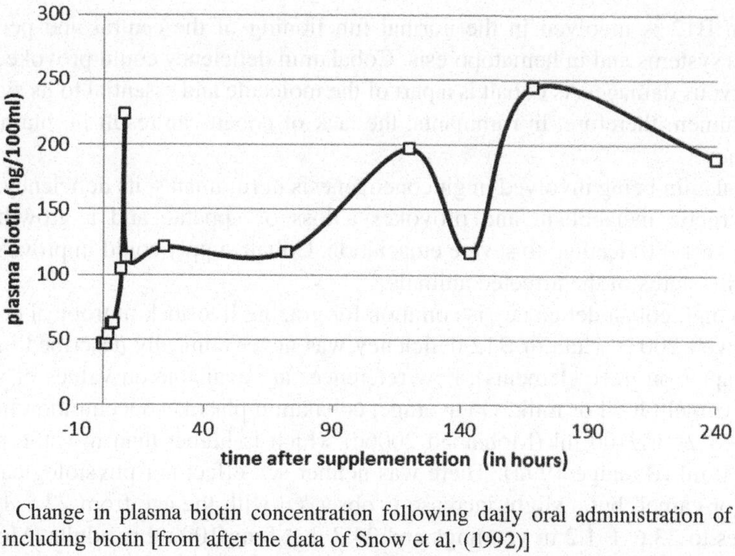

Fig. 8.4 Change in plasma biotin concentration following daily oral administration of vitamin mixture including biotin [from after the data of Snow et al. (1992)]

8.8 Vitamin B9 (Folic Acid)

Vitamin B9 is essential for production and maintenance of new cells and DNA and RNA synthesis. It is important during periods of active cell division, i.e., during growth and gestation. Deficiency in folic acid may result in anemia (by decreasing the number of red blood cells), diarrhea, and decrease in growth rate of young animals and fetuses during pregnancy. There is a high interaction between vitamins B9 and B12: an indirect lack of folates can be due to vitamin B12 deficiency. There is no case of vitamin B9 deficiency reported in camel, but probably folic acid supplementation would improve milk productivity as for dairy cattle (Girard et al. 2005).

The concentration of folic acid in camel plasma was reported to be 0.76 ± 0.33 µg/100 ml (Snow et al. 1992) that is quite lower than that in dairy cow without folate supplementation: 1.36–1.72 µg/100 ml (Ragaller et al. 2009). The variability of folic acid in camel plasma was investigated by Mohamed (2006c) who reported no significant sex effect, 0.37 ± 0.06 for female vs 0.44 ± 0.8 µg/100 ml for male, nor pregnancy or lactation status effect but a slight increase with the age, from 0.34 ± 0.04 (neonates) to 0.39 ± 0.8 (yearling) and 0.42 ± 0.09 µg/100 ml (adult).

In camel milk, concentration in folic acid was reported to be 0.41 ± 0.06 µg/100 ml (Sawaya et al. 1984). In comparison, cow milk concentration is on average ten times more important: 3.8–4.3 µg/100 ml (Ragaller et al. 2009).

8.9 Vitamin B12 (Cobalamin)

Vitamin B12 is involved in the normal functioning of the central and peripheral nervous systems and in hematopoiesis. Cobalamin deficiency could provoke important nervous damage. As cobalt is a part of the molecule and essential to its synthesis in the rumen, therefore, in ruminants, the lack of cobalt can result in vitamin B12 deficiency.

Cobalamin being involved in gluconeogenesis in ruminants, its deficiency affects the energetic metabolism and provokes a loss of appetite and a growth delay (Friesecke 1980) leading to severe emaciation. Cobalt supply could improve rapidly the health status of the affected animals.

Although cobalt deficiency is common for grazing livestock in tropical countries (McDowell 2003), vitamin B12 deficiency was never clinically described in camel (see chapter on trace elements). Few references are available on values of vitamin B12 in camel blood or milk. On average, cobalamin plasma concentration in camel is around 24 ng/100 ml (Mohamed 2006c) which is higher than in cattle plasma: 13.5/100 ml (Babidge 1992). There was neither sex effect nor physiological status effect for camel, but a slight increase is observed with the age from 22.5 ± 1.5 in neonates to 23.6 ± 1.2 in yearlings and 24.7 ± 1.8 ng/100 ml in adults (Mohamed 2006c).

In camel milk, cobalamin concentration is around 0.15 μg/100 ml (Sawaya et al. 1984) which is quite lower compared to cow milk (0.44 μg/100 ml) or sheep milk (0.71 μg/100 ml). According to some other authors, the concentration in cow milk is even higher (up to 1.05 μg/100 ml according to Matte et al. 2012), especially in colostrum where the concentration is 10 times higher (Marca et al. 1996).

8.10 Vitamin C (Ascorbic Acid)

Vitamin C or ascorbic acid is an important and essential water-soluble nutrient for animals as well as for humans. Ascorbic acid is a cofactor of many enzymes, especially in collagen synthesis, which explains severe symptoms of scurvy in case of vitamin C deficiency in human, formerly very common in long-sea sailors. In animals, including camel, enzymatic reactions are essential in wound-healing and overall in antioxidant activity. When animals are deficient in vitamin C, lipid peroxidation increases in the plasma and organs leading to oxidative damage of red blood cells, notably affecting their osmotic fragility (Minka and Ayo 2010), including in camel (Chakir et al. 2013). Globally, the vitamin C serves as radical scavenger and general antioxidant for cell metabolites including other vitamins (A and E) (Gershoff 1993). Ascorbic acid is also involved in mineral metabolism (Vannucchi 1991).

Vitamin C is probably the most studied vitamin in camel, and a comprehensive study was achieved by Mohamed (2002).

8.10.1 Vitamin C in Plasma

The first publication on plasma vitamin C concentration in camel was done by Soliman et al. (1975) and then by Snow et al. (1992) reporting a concentration of 0.45 ± 0.13 mg/100 ml, which is comparable to that in cattle (Hidiroglou et al. 1995). Similar results were reported by Mohamed (2002) (0.32–0.65 according to different physiological status of camels) and Stahl et al. (2006) (0.54 ± 0.11 mg/100 ml). In routine analysis by HPLC achieved in Emirates on hundred camels, an average of 0.40 ± 0.14 mg/100 ml was reported (Wernery et al. 2009). There is no significant effect of vitamin C supplementation on non-deficient racing camel (Snow et al. 1992) although there is an effect on the type of diet: camels receiving basal diet with alfalfa have higher ascorbic acid concentration in plasma than those fed with habitual diet from pasture, 0.54 ± 0.11 vs 0.39 ± 0.1 mg/100 ml, respectively (Mohamed et al. 2013). Supplementation using SC injection of 50 mg/kg seems more efficient than IV and oral administration, the maximum concentration obtained with subcutaneous route (2.8 mg/100 ml) (Elsheikh et al. 1998).

In Sudanese camels, sex effect was only observed in Bishari camel breed which also has lower values than other breeds (0.36 vs 0.65 in Arabi breed and

Table 8.2 Means and SD of the vitamin C concentrations in plasma and leukocytes of camel according to their reproductive status [from Mohamed et al. (2011)]

Reproductive phase	*n*	Plasma (mg/100 ml)	Leukocytes (mg/100 ml blood)
Non-estrus	280	0.313 ± 0.066^{a}	3.07 ± 0.134^{a}
Estrus	280	0.589 ± 0.188^{b}	4.03 ± 0.211^{b}
Pregnant	430	0.377 ± 0.081^{c}	3.27 ± 0.091^{c}
Lactating, non-pregnant	290	0.501 ± 0.068^{d}	5.2 ± 0.131^{d}
Non-pregnant, non-lactating	280	0.434 ± 0.070^{e}	4.56 ± 0.101^{e}

Different subscripts in the same column are significantly different ($P < 0.05$)

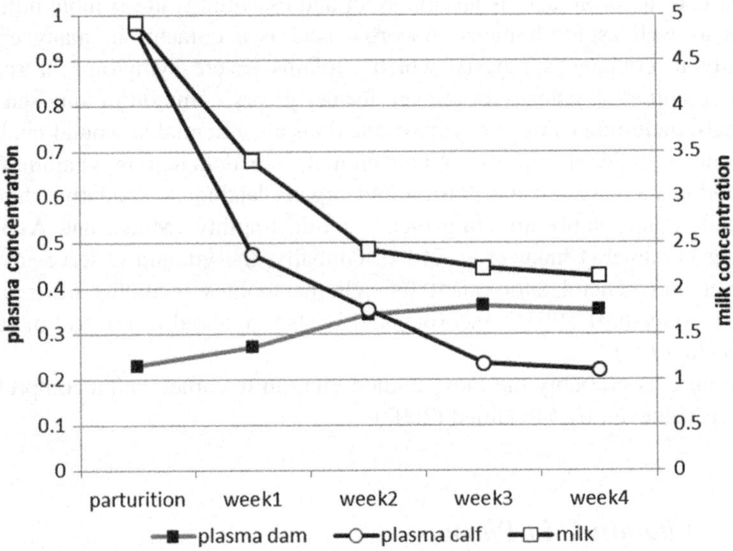

Fig. 8.5 Changes in vitamin C concentrations in plasma and milk (in mg/100 ml) during the first month postpartum [from after the data of Mohamed (2002)]

0.48 mg/100 ml in Anafi breed for males; 0.28 vs 0.64 and 0.41 mg/100 ml, respectively, for females), but the values decreased during breeding season: 0.51–0.39 for male and 0.42–0.20 mg/100 ml for female (Mohamed and Beynen 2002a). The reproductive status has a significant effect on the vitamin C concentration both in plasma (Table 8.2) with higher value during estrus (Mohamed et al. 2011).

Moreover, plasma vitamin C concentration increases significantly in dams after parturition while is decreased in their calves, in similar proportion than in milk (Mohamed 2002) (Fig. 8.5). Besides, significant lower values are reported in young camel compared to adult: 0.472 ± 0.007 vs 0.645 ± 0.04 mg/100 ml (Mousa et al. 2006).

As vitamin C is linked to stress, the health status affects obviously its concentration in plasma. For example, vitamin C concentration is significantly lower in camel positive to brucellosis (tested with Rose Bengal Test) than in negative: 0.30 ± 0.01

vs 0.42 ± 0.08 mg/100 ml, respectively (Mohamed et al. 2011). The presence of mastitis is also linked to an important decrease of ascorbic acid in plasma: from 0.44 ± 0.1 to 0.27 ± 0.09 mg/100 ml (Mohamed et al. 2005). But the most important depressive effect seems to be due to parasitism (helminthiasis, mange) and overall trypanosomosis (Mohamed and Beynen 2002b): 0.58 ± 0.12 mg/100 ml plasma in healthy camel vs 0.36 ± 0.09 in camel affected by helminthiasis, 0.29 ± 0.09 in camels suffering from mange, and 0.18 ± 0.04 in those affected by trypanosomosis.

8.10.2 *Vitamin C in Milk*

Camel milk is very well known as rich in vitamin C since the publication of Sawaya et al. (1984): 23.7 ± 2.65 mg/l. This concentration was confirmed later on by Farah et al. (1992): 37.4 mg/l with 26.2–61.1 range. On average, these values are 3–5 times higher than in cow milk (Stahl et al. 2006) or human milk (Wang et al. 2011).

In Chinese Bactrian camel, Zhang et al. (2005) found lower values in colostrum (10 mg/l) than in milk with an increase of the concentration significantly after a week to reach 29.6 mg/l after 90 days of lactation. In dromedary camel, lower concentration in colostrum was also reported: 35.6 ± 10.6 vs 52.5 ± 15.8 mg/1 in milk (Stahl et al. 2006). A slight but significant increase all along the lactation was also reported in Sudanese dromedary (Mohamed et al. 2005). Similar trend was observed in Bactrian camel from Kazakhstan, but with quite higher values varying from 48 to 256 mg/l (mean 184) with a tendency to increase all along the lactation with a maximum at week 31, i.e., during summer (Konuspayeva et al. 2010a). Such differences could be attributed to the species (mainly Bactrian camel) and sampling conditions based on analysis of fresh milk contrary to other references. It has been shown that freezer storage decreased the vitamin C concentration in camel milk (Wang et al. 2011) as well as human milk (Bank et al. 1985).

During the first week of lactation, Vitamin C concentration is stable for the first 6 days (between 19 and 32 mg/l) and starts to increase after a week up to 113 mg/l (Konuspayeva et al. 2010b). From their part, Mohamed et al. (2005) found similar values in colostrum than in milk (on average 44 mg/l) whatever the breed, but with significant differences between breeds.

Milk vitamin C concentration decreases in milk from camel affected by mastitis: 26.8 ± 4.4 vs 47.4 ± 5.2 mg/l in healthy camel (Mohamed et al. 2005).

8.10.3 *Vitamin C in Other Substrates*

Vitamin C was determined in camel organs showing very high concentrations in adrenal gland (151 ± 12.1 mg/100 g wet tissue) compared to the liver (60 ± 6.3), spleen (44 ± 4.6), kidney (17 ± 2.8), lung (13 ± 2.6), and heart (8 ± 1.3). There are neither seasonal nor sexual differences in these concentrations (Mohamed and

Beynene 2002a). In camel liver, similar data was published formerly in Egyptian camels: 58 mg/100 g with no sexual difference (Barakat and Abdallah 1965). As in plasma, liver ascorbic acid concentration decreases significantly in camel affected by parasitological disease: from 62.9 ± 2.0 mg/100 g wet tissue in healthy camels to 53.4 ± 1.9 for camel with helminthiasis, 50.2 ± 2.2 in mangy camel, and 33.2 ± 3.3 in camel with trypanosomosis (Mohamed and Beynen 2002b).

In camel urine, the quantity of vitamin C is higher in females (5.33 ± 0.25 mg/l) than in males (3.22 ± 0.97 mg/l) as it was observed in cattle (Mohamed and Beynen 2002a). The urine is the main way of vitamin C excretion.

The concentration of vitamin C in arthritic joint was significantly lower than in clinically healthy camel joint: 6.1 ± 0.24 vs 9.66 ± 0.49 μg/ml (Chalmeh et al. 2016).

8.11 Vitamin D (Cholecalciferol)

Vitamin D is a secosteroid contributing to the intestinal absorption of several main minerals (calcium, magnesium, phosphorus) and trace elements (iron, zinc). The most important form is cholecalciferol (vitamin D3) and ergocalciferol (vitamin D2). Generally, the diet is poor in vitamin D, and the main source is the biosynthesis in the skin using cholesterol as precursor. This synthesis is dependent on sun exposure and activation occurring in the liver and kidney. In the liver, vitamin D3 is converted to calcifediol (25-hydroxycholecalciferol abbreviated 25-OH-D_3) and vitamin D2 to 25-hydroxyergocalciferol. The more active form for stimulating intestinal absorption of calcium is the calcifediol after two hydroxylations are achieved in the liver: the calcitriol or 1,25-dihydrocholecalciferol abbreviated 1,25$(OH)_2$-D_3. This metabolite is analyzed in plasma or serum to assess the vitamin D status. It regulates calcium and phosphorus concentration in the blood, contributing to the bone remodeling (Holtrop et al. 1981). Due to its action on mineral metabolism, the lack of vitamin D leads to bone disorders especially during growth (osteomalacia, ricket).

In camel, few references are available on vitamin D, except on studies of phosphocalcic metabolism and hormonal regulation in camel (Riad 1995; El-Khasmi and Faye 2011).

8.11.1 *Vitamin D in Plasma*

Plasma 1,25$(OH)_2$-D_3 concentration in camel is 83.5 ± 4.5 ng/100 ml in non-lactating and non-pregnant camel (Riad et al. 1994). This value is on average ten times higher than the concentrations reported in other animals: 5–6 ng/100 ml in sheep (Ross et al. 1980) or 1–10 ng/100 ml in cattle (Horst et al. 1983; NRC 1987). Plasma 1,25$(OH)_2$-D_3 concentration did not change with pregnancy status of the camel (89.7 ± 6.4 ng/100 ml) but increased in lactating camel (123.8 ± 7.5 ng/100 ml) (Riad et al. 1994). These higher plasma levels enhance intestinal absorption of calcium which is also higher than in other ruminants (Riad et al. 1994) in order to ensure phosphocalcic requirements of the lactation. At birth, plasma concentrations of 25(OH)D_3 and 1,25$(OH)_2$-D_3 in newborn camels are low (5.82 ± 1.24 and 83.5 ± 6.2 ng/100 ml, respectively) compared to their dams (48 ± 6 and 151 ± 20.9, respectively). After a week, concentrations increase in the camel calves up to overpassing those of the dams for the two metabolites (El-Khasmi et al. 2000).

Seasonal variations were observed for plasma vitamin D concentration in camel. The highest levels of vitamin D were reported in February–July, while the lowest levels are observed in August–January (Mohamed 2008). Serum levels of 25(OH)D_3 in summer was almost twice than in winter (Shany et al. 1978). The changes in circulating vitamin D concentrations could be explained by the degree of coat coloration as it is the case for Arabi camels (white color) compared to Anafi with brown color (Mohamed 2008). The concentration of hydroxyl vitamin D is not influenced by transport stress, the values being on average 420 ng/ml before stress and 370 ng/ml after stress (El-Khasmi et al. 2011).

8.11.2 Vitamin D in Milk

In camel milk, 25-hydroxyvitamin D is higher after parturition (8.9 ± 0.6 ng/ml) but decreases to 1.2 ± 0.3 ng/ml after a week (El-Khasmi et al. 2001). This value confirms the higher vitamin D concentration in camel milk (Gnan and Sheriha 1986; Abu-Lehia 1987) than in other species. In cow milk, concentrations were 50.4 ± 4.1 pg/ml for vitamin D, 0.49 ± 0.05 ng/ml for 25(OH)D_3, and 9.7 ± 1.0 pg/ml for 1,25$(OH)_2D_3$ (Kunz et al. 1984).

8.12 Vitamin E (α-Tocopherol)

Vitamin E is a fat-soluble vitamin including four different molecules of tocopherols (the most bioactive being α-tocopherol). Vitamin E is recognized as a natural biological antioxidant. It is the first line of defense against peroxidation that is important for maintaining low tissue concentrations of peroxides. Vitamin E is also an essential component in the reproduction processes and performance of farm animals. Vitamin E acts in synergy with the selenium (as component of glutathione peroxidase). The main manifestation of vitamin E deficiency is white muscle disease (WMD). Young camel is susceptible to WMD, especially in the Gulf countries (Al-Qarawi et al. 2001; El-Khouly et al. 2001; Seboussi et al. 2004), but the references on the normal vitamin E level in camel serum and milk or on the variability of vitamin E status in this species were only recently explored (Seboussi et al. 2008, 2009a; Faye and Seboussi 2010). In Bactrian and dromedary camels in zoological garden, vitamin E deficiency was involved in extensive myocardial degeneration and necrosis with fine dystrophic calcification of the damaged cardiac muscle cells provoking sudden death (Finlayson 1971).

8.12.1 Vitamin E in Plasma

Based on 1064 samples, usual values for camel plasma is 194 ± 70 μg/100 ml (Wernery et al. 2009). Plasma concentration of vitamin E in camel varied between 0.2 and 486 μg/100 ml with a mean of 98 ± 67 μg/ml in the different experiments achieved by Seboussi et al. (2008, 2009a). Similar results were reported by Al-Senaidy (1996b) and Mousa et al. (2006). In cow, the usual value of vitamin E concentration is more than 200–250 μg/100 ml which is above deficiency limit (Maas et al. 1990; Leblanc et al. 2004). Values in camel are quite lower, both for α- and γ-tocopherol (Al-Senaidy 1996a). Comparing the serum of camel and cow, Stahl et al. (2006) have reported values ten times in cow (1168 ± 247 μg/100 ml) than in camel (161 ± 67 μg/100 ml).

Effect of age on plasma vitamin E concentration is contradictory. For Seboussi et al. (2009a), it is higher in adult (116 ± 81 μg/100 ml) than in young camels (82 ± 106 μg/100 ml. In 2-year-old camels, plasma vitamin E concentration (56 ± 22 μg/100 ml) was quite lower than in adult (Seboussi et al. 2010). Baraka (2012) found also higher values in adult (150 ± 120 μg/100 ml) than in young camel (110 ± 1 μg/100 ml). At reverse, Mousa et al. (2006) did not find difference: 270 ± 5 in young vs 252 ± 6 μg/100 ml in adult camel. Also Homeida et al. (2010) did not find significant difference between young (111 ± 19) and adult camels (126.2 ± 44.8 μg/100 ml). For Mohamed (2006a), plasma vitamin E concentration was higher in young camel calves less than 6 months (220 ± 90 μg/100 ml) than young animals 6–12 months (150 ± 10 μg/100 ml), while it was 230 ± 80 μg/100 ml in adults (Fig. 8.6).

Barri and Al-Sultan (2007) reported values between 30 and 165 μg/100 ml in young camels (3–4 months old) from Saudi Arabia. In apparently deficient camels affected by musculoskeletal disorders, vitamin E concentration in plasma was significantly lower (70–80 μg/100 ml according to the type of disorder) than in healthy camels: 250 ± 30 μg/100 ml (Mohamed 2004).

The effect of physiological status of the female adult was also investigated by Seboussi et al. (2008, 2009a). There was no difference between the pre-calving (112 ± 81 μg/100 ml) and post-calving period (120 ± 80 μg/100 ml). A slight nonsignificant decrease was observed at the peripartum time. A positive correlation was reported between plasma vitamin E and total food intake.

There is a very slight difference ($P < 0.05$) between sex: 201 ± 60 vs 199 ± 40 μg/100 ml in female and male, respectively (Mohamed 2006a). However, a slight difference was reported by Baraka (2012): 180 ± 2 μg/100 ml in male compared to 150 ± 2 μg/100 ml in female. A quite higher concentration is reported also at the rainy season compared to dry season: 230 ± 50 vs 140 ± 40 μg/100 ml, respectively (Mohamed 2006a).

In camels of 2 years old affected by chronic selenosis, the plasma vitamin E was on average 68 ± 36 μg/100 ml (Seboussi et al. 2009b), i.e., without significant changes compared to normal values. A tendency to the decrease of plasma vitamin E concentration in intoxicated camels with clinical signs was observed, but no

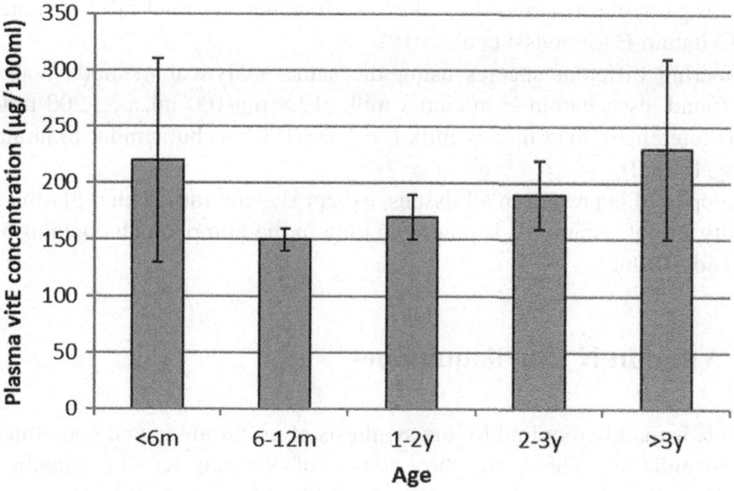

Fig. 8.6 Changes in plasma vitamin E in camels according to their age [from after the data of Mohamed (2006a)]

significant correlation was reported with the serum Se concentration. However, a high level of Se selenium seemed to depress the vitamin E level in plasma. Selenium (as component of GPx) and vitamin E are both antioxidants, both protecting cell membranes from oxidative damage. Due to this synergy, one can substitute for the other in a very small way. For instance, more Se is needed when animals are deficient in vitamin E.

Plasma vitamin E increased in case of diarrhea (260 ± 5 µg/100 ml) compared to camel with only simple indigestion (140 ± 3 µg/100 ml) or rumen acidosis (150 ± 2 µg/100 ml) (Baraka 2012).

There are some positive interactions with other blood parameters as PCV, hemoglobin, lymphocytes, and eosinophils (Faye and Seboussi 2010). Obviously, the vitamin E is included in blood cells where it strengthens the cell membranes, and it can ward off free radical attack from oxidation compounds (Bednarek et al. 1996).

8.12.2 Vitamin E in Other Substrates

The vitamin E concentration (determined by HPLC) in **camel milk** is similar to that of cow milk: 56 vs 60 µg/100 ml (Farah et al. 1992). Stahl et al. (2006) reported higher values in camel colostrum (137 ± 98) than in milk (33 ± 13), but in significant less quantity than in cow milk (171 ± 114 µg/100 ml). In camel milk fat is containing higher proportion of polyunsaturated fatty acids (Konuspayeva et al. 2008) and vitamin E absorption is decreasing with diet rich in polyunsaturated fatty acids. This particularity could stress the low efficiency in camel calf to maintain high level of vitamin E (Seboussi et al. 2010).

Comparing different species using the same analytical method, Wang et al. (2011) found less vitamin E in camel milk (12.8 mg/100 ml, i.e., 200 times than previous reference), than in cow milk (16.1 mg/100 ml) but similar to human milk (12.2 mg/100 ml).

γ-Tocopherol is present in all tissues, except skeletal muscle and plasma. Due to its affinity for fat, vitamin E is predominating in the hump which contains 50 mg/g (Al-Senaidy 1996b).

8.13 Vitamin K (Phylloquinone)

Vitamin K is mainly required for the synthesis of prothrombin and consequently for blood coagulation. There are three types of vitamin K: (1) vitamin K1 or phylloquinone synthetized by plants, fat soluble; (2) vitamin K2 or menaquinone, synthetized by the ruminal and intestinal bacteria; and (3) vitamin K3 or menadione, water-soluble precursor of vitamin K2.

Ruminants' requirements are met by (1) dietary intake and (2) microbial biosynthesis in the rumen and intestines. Ruminal microorganisms synthesize large amounts of vitamin K, which explains why ruminants do not need a dietary source of this vitamin except during intoxication by coumarin (main component of rodenticides).

The main symptom of vitamin K deficiency is impaired blood coagulation, but such deficiency is very rare in livestock, except in the case of antagonist such as dicoumarol which could be abundant in some plants like *Melilotus* sp. (clover). Affected animals can die from hemorrhage following a minor injury, or even from apparently spontaneous bleeding. However such deficiency was never reported in camel.

For camel, only one publication is available indicating plasma vitamin K1 concentration of 42 ± 11 ng/100 ml for adult camels and 18.4 ± 2.4 ng/100 ml for young camel less than 1 year (Homeida et al. 2010). The lower value in young camel is due to the low content of vitamin K in the milk, but no data is available for camel milk. Compared to other species, these values are higher: 31.5 ± 3.2 for cow, 19.5 ± 4.9 for sheep, and 20.1 ± 3.9 ng/100 ml (Homeida et al. 2010). These differences are explained by the type of plants grazed by camel, containing high amount of oil in g/kg dry matter.

8.14 Conclusion

Except for ascorbic acid, vitamins in camel are poorly investigated. The deficit in some vitamins could occur more often than suspected. Plasma values are comparable to those of other species except for vitamin D which is quite higher. In addition, vitamin C concentration in camel milk is also very high.

References

Abbas T, Ali B (2001) Retinol values in the plasma of the Arabian camel (*Camelus dromedarius*) and the influence of aflatoxicosis. Vet Res Commun 25(6):517–522

Abbas TA, Alhaj Ali M, Abu Damir H (2008) Thiamine (vitamin B1) status in the blood of young, pregnant, lactating and racing dromedary camels (*Camelus dromedarius*) in UAE. Revue Méd Vét 159:545–550

Abu-Lehia IH (1987) Composition of camel milk. Milchwissenschaft 42(6):368–370

Agab H, Abbas B (1999) Epidemiological studies on camel diseases in Eastern Sudan. World Anim Rev 92:42–51

Agab H, Abbas B, Le Horgne JM (1992) Vitamin A deficiency in camels. Sudan J Vet Sci Anim Hus 31:9–13

Agab H, Abbas B, Le Horgne JM, Saint-Martin G (1993) A note on vitamin A deficiency in camels in Sudan. Sudan J Vet Sci Anim Hus 32(1):9–14

Alhadrami GA (2003) Camel. In: Roginski H, Fuquay JW, Fox PF (eds) Encyclopedia of dairy science. Academic Press, London, pp 616–622

Al-Qarawi A, Abbas A, Haroun B, Mahmoud EM, Al-Hawas MA (2001) Clinicopathological investigation of selenium responsive myopathy in young adult camels. J Camel Pract Res 8:23–27

Al-Senaidy AM (1996a) Distribution of α- and γ-tocopherols within blood fractions of ruminants. Comp Biochem Physiol A 115(3):223–227

Al-Senaidy AM (1996b) Tocopherols in camel's plasma and tissues. Int J Vitam Nutr Res 66:210–216

Al-Senaidy AM (1998) Distribution of the fat soluble antioxidants (α-tocopherol, retinol and β-carotene) in the blood and other tissues of camel (*Camelus dromedarius*). Saudi J Biol Sci 5:64–77

Babidge WJ (1992) Plasma and liver vitamin b12 concentrations in cattle. Proc Aust Soc Anim Prod 19:388

Bank MR, Kirksey A, West K, Giacoia G (1985) Effect of storage time and temperature on folacin and vitamin C levels in term and preterm human milk. Am J Clin Nutr 41(2):235–242

Baraka TA (2012) Clinical evaluation of vitamin A, β-carotene, vitamin E and cortisol levels in health and selected diseases in camels (*Camelus dromedarius*). J Am Sci 8(1):106–111

Barakat MZ, Abdallah A (1965) The ascorbic acid content of edible liver. J Food Sci 30(2):185–187

Barri MES, Al-Sultan SI (2007) Studies on selenium and vitamin E status of young Megaheem dromedary camels at Al-Ahsa province. J Camel Pract Res 14:51–53

Bednarek D, Kondracki M, Cakała S (1996) Effect of selenium and vitamin E on white cells, serum concentration of several minerals and trace elements as well as immunologic parameters in calves. Deutsch Tierarzt Wochens 103:457–459

Bender DA (2003) Nutritional biochemistry of the vitamins. Cambridge University Press, Cambridge

Bhadauria P, Lathwal SS, Jadoun YS, Bhadauria SS (2013) Role of biotin in prevention of lameness in dairy cattle-review. Wayamba J Anim Sci 635–646

Bhadauria P, Lathwal SS, Jadoun YS, Gupta R, Devi I (2015) Variations in plasma biotin and mineral concentrations due to zinc-biotin supplementation in lame Karan Fries cows during peri-parturient period. Indian J Anim Sci 9(6):783–787

Bhandare SG (2009) Chronic hematuria in a camel: a case report. Vet World 2(4):148

Bogin E (2000) Clinical pathology of camelids. Present and future. Rev Med Vet 151:563–568

Chakir Y, El-Khasmi M, Farh M, Bargaâ R, Riad F, Abdallah S, El-Hassan T, Najia EA, Abouhafs R, Faye B (2013) Effects of vitamin E and vitamin C on hydrogen peroxide-induced hemolysis in Moroccan dromedary camels (*Camelus dromedarius*). Greener J Med Sci 3 (4):111–120

Chalmeh A, Badiei K, Pourjafar M, Nazifi S (2016) Synovial fluid antioxidant vitamins and trace elements in clinically healthy and arthritic joints of dromedary camels. Istanbul Univ Vet Fak Dergisi 42(1):1–4

Chen B, Wang C, Wang YM, Liu JX (2011) Effect of biotin on milk performance of dairy cattle: a meta-analysis. J Dairy Sci 94(7):3537–3546

Claeys WL, Verraes C, Cardoen S, De Block J, Huyghebaert A, Raes K, Dewettinck K, Herman L (2014) Consumption of raw or heated milk from different species: an evaluation of the nutritional and potential health benefits. Food Control 42:188–201

Clagett-Dame M, Knutson D (2011) Vitamin A in reproduction and development. Nutrients 3:385–428

El Khasmi M, Faye B (2011) Parathyroid hormone-related peptide and vitamin D in phosphocalcic metabolism for dromedary camel. Iranian J Appl Anim Sci 1(4):205–213

El Khasmi M, Riad F, Safwate A, Tahri EH, Farh M, El Abbadi N, Coxam V, Faye B (2011) Effects of preslaughter stress on meat quality and phosphocalcic metabolism in camels (*Camelus dromedarius*). J Camelid Sci 3:33–38

El Khouly AA, Abbas TA, Moustafa T (2001) Myocardial dystrophy in camel calves in the United Arab Emirates (field cases). Emir J Agric Sci 13:11–17

El-Khasmi M, Riad F, Safwate A, Bengoumi M, Hidane K, Davicco MJ, Coxam V, Faye B, Barlet JP (2000) Evolution comparée de quelques paramètres minéraux, de l'ostéocalcine, du 25(OH)2D chez la chamelle du sud marocain et son chamelon nouveau-né. Atelier international sur le chamelon, Ouarzazate, 24–26 Oct 1999, Maroc Rev Elev Méd Vét Pays Trop, 53, pp 115–119

El-Khasmi M, Riad F, Safwate A, Hidane K, Faye B, Davicco MJ, Coxam V, Barlet JP (2001) Postpartum evolution of mammary secretion of minerals and 25-hydroxyvitamin D in lactating camels (*Camelus dromedarius*). J Camel Pract Res 8(2):131–135

El-Nisar NA, Abd-Ellah MR, Khamis GF (2011) Evaluation of serum vitamin C, β-carotene and α-tocopherol status in pneumonia of camels. Comp Clin Pathol 21:1081–1085

Elsheikh HA, Mohamed HE, Mousa HM (1998) Bioavailability of ascorbic acid in camels (*Camelus dromedarius*) following administration by different routes. In: Proceedings of the third annual meeting for animal production under arid conditions, Al-Ain, vol 1, pp 180–186

Farah Z, Rettenmaier R, Atkins D (1992) Vitamin content of camel milk. Int J Vitam Nutr Res 62:30–33

Fassi-Fehri MM (1987) Diseases of camel. Rev Sci Tech Off Int Epiz 6(2):337–354

Faye B, Mulato C (1991) Facteurs de variation des paramètres protéo-énergétiques, enzymatiques et minéraux dans le plasma chez le dromadaire de Djibouti. Rev Elev Med Vét Pays Trop 44:325–334

Faye B, Seboussi R (2010) Variability of vitamin E concentration in camel plasma. J Camel Pract Res 16(2):157–161

Finlayson R (1971) Calcific cardiomyopathy in young camels (*Camelus* spp.). J Comp Pathol 81 (1):71–76

Friesecke H (1980) Vitamin B_{12} in animal nutrition. F. Hoffman-La Roche, Bale, p 45

Gebrehiwet T (1999) The camel, in Eritrea: an all-purpose animal. World Anim Rev 91:1–13

Gershoff S (1993) Vitamin C (ascorbic acid): new roles, new requirements. Nutr Rev 51:313–326

Ghadrdan Mashhadi AR, Sazmand AR, Karimiyan A, Hekmati Moghaddam SH (2013) Normal values and seasonal differences in the serum concentration of vitamin A and beta-carotene in the Iranian camel (*Camelus dromedarius*). Iran J Vet Med 7(2):91–94

Ghosal AK, Dwarkanath PK, Patney JML (1973) A note on plasma carotene and vitamin A levels in normal camel (*Camelus dromedarius*) of north-west Rajasthan. Indian J Anim Sci 43 (9):899–900

Girard CL, Lapierre H, Matte JJ, Lobley GE (2005) Effects of dietary supplements of folic acid and rumen-protected methionine on lactational performance and folate metabolism of dairy cows. J Dairy Sci 88:660–670

Gnan SO, Sheriha AM (1986) Composition of Libyan camel's milk. Aust J Dairy Technol 41:33–35

Hidiroglou M, Ivan M, Batra TR (1995) Concentrations of vitamin C in plasma and milk of dairy cattle. Ann Zootech 44(4):399–402

Holtrop ME, Cox KA, Clark MB, Holick MF, Anast CS (1981) 1,25-dihydroxycholecalciferol stimulates osteoclasts in rat bones in the absence of parathyroid hormone. Endocrinology 108:2293–2301

Homeida AM, Hussein MF, Al-Jumaah RS, Al-Haidary AI, Al-Shaikh MA, Albokhadaim I, Gar Elnabi A (2010) A comparative study on serum vitamin K_1, A and E levels in camels (*Camelus dromedarius*) and other animals. J Camel Pract Res 17(2):241–244

Hopper JH, Johnson BC (1955) The production and study of an acute nicotinic acid deficiency in the calf. J Nutr 56(2):303–310

Horst RL, Hove K, Littledike ET, Reinhardt TA, Uskovovic MR, Partridge JJ (1983) Plasma concentration of 1,25-dihydroxyvitamin D, 1,24R,25-trihydroxyvitamin D_3, and n1,25,26-trihydroxyvitamin D_3 after their administration to dairy cows. J Dairy Sci 66:1455–1460

Johnson BC, Pinkos JA, Burke KA (1950) Pyridoxine deficiency in the calf. J Nutr 40(2):309–322

Kataria N, Kataria AK (2004) Use of blood analytes in assessment of stress due to drought in camel. J Camel Pract Res 11(2):129–133

Konuspayeva G, Lemarie E, Faye B, Loiseau G, Montet D (2008) Fatty acid and cholesterol composition of camel's (*Camelus bactrianus*, *Camelus dromedarius* and hybrids) milk in Kazakhstan. Dairy Sci Technol 88:327–340

Konuspayeva G, Faye B, Loiseau G, Narmuratova M, Ivashchenko A, Meldebekova A, Davletov S (2010a) Physiological change in camel milk composition (*Camelus dromedarius*). 1. Effect of lactation stage. Trop Anim Health Prod 42:495–499

Konuspayeva G, Faye B, Loiseau G, Narmuratova M, Ivashchenko A, Akhmetsadykova S, Davletov S (2010b) Physiological change in camel milk composition (*Camelus dromedarius*). 2. Physico-chemical composition of colostrum. Trop Anim Health Prod 42:501–505

Kunz C, Niesem M, Lilienfeld Toal HV, Burmeister W (1984) Vitamin D, 25-hydroxyvitamin D and 1,25-dihydroxyvitamin D in cow milk, infant formulas and breast milk during different stages of lactation. Int J Vitam Nutr Res 54:141–148

Lalic J, Denic M, Sunaric S, Kocic G, Trutić N, Mitić S, Jovanović T (2014) Assessment of thiamine content in some dairy products and rice milk. CyTA J Food 12(3):203–209

LeBlanc SJ, Herdt TH, Seymour WM, Duffield TF, Leslie KE (2004) Peripartum serum vitamin e, retinol, and beta-carotene in dairy cattle and their associations with disease. J Dairy Sci 87:609–619

Luce WG, Peo ER, Hudman DB (1966) Availability of niacin in wheat for swine. J Nutr 88:39

Lyaktini MY, Ghalim N, Yacoubi B (2013) Comparaison de la concentration du rétinol sérique chez des dromadaires (*Camelus dromedarius*) atteints et indemnes de gale au Maroc. Rev Méd Vét 164(12):555–558

Maas J, Parish SM, Hodgson DR (1990) Nutritional myodegeneration. In: Smith BP (ed) Large animal internal medicine. CV Mosby, St Louis, pp 1352–1357

Marca MC, Ramos JJ, Saez T, Sanz MC, Verde MT, Fernandez A (1996) Vitamin B12 supplementation of lambs. Small Rumin Res 20:9–14

Matte JJ, Guay F, Girard CL (2012) Bioavailability of vitamin B12 in cows' milk. Br J Nutr 107:61–66

McDowell LR (2003) Minerals in animal and human nutrition. In: McDowell LR (ed.), 2nd ed. Elsevier, Gainesville, 660 p

Mehaia MM (1994) Vitamin C and riboflavin content in camels milk: effects of heat treatments. Food Chem 50(2):153–155

Milad KE, Ridha GS (2009) The occurrence of thiamine-responsive polioencephalomalacia in dromedary breeding camels in Libya: preliminary investigation of diagnosis. Iraqi J Vet Sci 23(Suppl I):119–122

Minka NS, Ayo JO (2010) Physiological responses of erythrocytes of goats to transportation and the modulatory role of ascorbic acid. J Vet Med Sci 72:875–881

Mohamed HE (2002) Vitamin C status in Sudanese camels. PhD dissertation, Utrecht University, Utrecht, 89 p

Mohamed HE (2004) A note on plasma antioxidant status in Sudanese camels (*Camelus dromedarius*) affected by musculoskeletal disorders. J Camel Pract Res 11(1):65–66
Mohamed HE (2006a) Factors affecting the plasma contents of retinol and alpha-tocopherol in camels (*Camelus dromedarius*). J Anim Vet Adv 5(3):301–303
Mohamed HE (2006b) Factors affecting plasma contents of thiamine and ascorbic acid in camels (*Camelus dromedarius*). J Anim Vet Adv 5(4):313–314
Mohamed HE (2006c) The plasma folate and vitamin B_{12} contents in camels (*Camelus dromedarius*). J Anim Vet Adv 5(1):1–2
Mohamed HE (2008) Vitamin D status in camels (*Camelus dromedarius*). Res J Biol Sci 3 (4):446–447
Mohamed HE, Beynen AC (2002a) Vitamin C concentrations in blood plasma, tissues and urine of camels (*Camelus dromedarius*) in Sudanese herds. J Anim Physiol Anim Nutr 86 (9–10):342–346
Mohamed HE, Beynen AC (2002b) Ascorbic acid content of blood plasma, erythrocytes, leukocytes and liver in camels (*Camelus dromedarius*) without or with parasite infections. Int J Vitam Nutr Res 72(6):369–371
Mohamed HE, Mousa HM, Beynen AC (2005) Ascorbic acid concentrations in milk from Sudanese camels. J Anim Physiol Anim Nutr 89(1–2):35–37
Mohamed HE, Alhaidary A, Beynen AC (2011) Ascorbic acid status of female camels during different phases of reproduction. Trop Anim Health Prod 43:279–281
Mohamed HE, Alhaidary A, Beynen AC (2013) Habitual diet and ascorbic acid status in Sudanese camels. Trop Anim Health Prod 45:887–888
Mousa HM, Omer OH, Ali BH, Al-Wabel N, Ahmed SM (2006) Antioxidant levels in tissues of young and adult camels (*Camelus dromedarius*). J Physiol Biochem 62:213–218
NRC (1987) Vitamin tolerance in animals. Chapter 2: Vitamin D. National Academies Press, Washington, DC, pp 11–22
Palanivelrajan M, Thangapandian M, Prathipa A (2015) Therapeutic management of sarcoptic mange in a camel (*Camelus dromedarius*). J Wildl Res 3(1):5–7
Patterson BE, Bates CJ (1989) Riboflavin deficiency, metabolic rate and brown adipose tissue function in sucking and weanling rats. Br J Nutr 61(3):475–483
Ragaller V, Hüther L, Lebzien P (2009) Folic acid in ruminant nutrition: a review. Br J Nutr 101:153–164
Riad F (1995) Régulation endocrinienne du métabolisme hydroélectrique et phosphocalcique chez le dromadaire. Thèse de Doctorat d'Etat es-sciences en physiologie animale. Université Hassan II-Mohammedia, Casablanca (Maroc), 225 p
Riad F, Bengoumi M, Davicco MJ, Safwate A, Barlet JP (1994) Influence of 1-α-hydroxycholecalciferol on calcium and phosphorus concentration in camel milk. J Dairy Res 61:567–571
Ross R, Care AD, Robinson JS, Pickard DW, Wheatherley AJ (1980) Perinatal 1,25-dihydrocholecalciferol in the sheep and its role in the maintenance of the transplacental calcium gradient. J Endrocrinol 87:17–18
Sawaya WN, Khalil JK, Al-Shalahat A, Al-Mohammed H (1984) Chemical composition and nutritional quality of camel milk. J Food Sci 49:744–747
Seboussi R, Faye B, Alhadrami G (2004) Facteurs de variation de quelques éléments trace (sélénium, cuivre, zinc) et d'enzymes témoins de la souffrance musculaire (CPK, ALT et AST) dans le sérum du dromadaire (*Camelus dromedarius*) aux Emirats Arabes Unis. Rev Elev Méd Vét Pays Trop 57:87–94
Seboussi R, Faye B, Alhadrami G, Askar M, Ibrahim W, Hassan K, Mahjoub B (2008) Effect of different selenium supplementation levels on selenium status in camel. Biol Trace Elem Res 123:124–138
Seboussi R, Faye B, Askar M, Hassan K, Alhadrami G (2009a) Effect of selenium supplementation on blood status and milk, urine and fecal excretion in pregnant and lactating camel. Biol Trace Elem Res 128:45–57

Seboussi R, Faye B, Alhadrami G, Askar M, Bengoumi M, Elkhouly A (2009b) Chronic selenosis in camels. J Camel Pract Res 16(1):25–38
Seboussi R, Faye B, Alhadrami G, Askar M, Ibrahim W, Mahjoub B, Hassan K, Moustafa T, Elkhouly A (2010) Selenium distribution in camel blood and organs after different level of dietary selenium supplementation. Biol Trace Elem Res 133(1):34–50
Shany S, Yagil R, Berlyne GM (1978) 25-hydroxycholecalciferol levels in camel, sheep and goat. Comp Biochem Physiol 59:139–140
Smith CM, Song WO (1996) Comparative nutrition of pantothenic acid. J Nutr Biochem 7 (6):312–321
Snow DH, Billah AM, Ridha A, Frigg M (1992) Plasma concentrations of some vitamins in camels. In: Allen WR, Higgins AJ, Mayhew IG, Snow DH, Wade JF (eds) Proceedings 1st International Camel Conference, Dubai, 2–6 Feb 1992, pp 335–338
Soliman MK, Toussef GW, Mansour SA (1975) Ascorbic acid content of erythrocytes, plasma, leukocytes and whole blood of healthy camels. Egypt J Vet Sci 12:107–110
Song WO, Smith CM, Wittwer C, Wyse B, Hansen G (1990) Determination of plasma pantothenic acid by indirect enzyme linked immunosorbent assay. Nutr Res 10(4):439–448
Stahl T, Sallmann HP, Duehlmeier R, Wernery U (2006) Selected vitamins and fatty acid patterns in dromedary milk and colostrum. J Camel Pract Res 13(1):53–57
Tinson AH, Gorde A, Kuhad KS, Singh K, Al-Masri J (1999) Study of plasma thiamine (B_1) levels in the racing dromedary camel in relation to source and route of administration of B_1 supplementation. J Camel Pract Res 6:287–294
Vannucchi H (1991) Interaction of vitamins and minerals. Arch Latinoam Nutr 41(1):9–18
Wang SY, Liang JP, Shao WJ, Wen H (2011) Mineral, vitamin and fatty acid contents in the camel milk of dromedaries in the Anxi Gansu China. J Camel Pract Res 18(2):273–276
Wernery U (2003) Novel observations on camel milk. In: Proceedings of the 9th Kenya camel forum, Giriftu, Giriftu Patoral Training Centre, Wajir District
Wernery U, Kinne J (2001) Endotoxicosis in racing dromedaries: a review. J Camel Pract Res 8 (1):65–72
Wernery U, Abraham A, Kinne J (2002) Evaluation of blood thiamine (Vit B_1) in racing dromedaries. J Camel Pract Res 9(1):27–29
Wernery U, Abraham AA, Jyothi T, Abubakar Ali Y, George RM (2009) Mineral and vitamin contents in the blood of racing dromedaries in the United Arab Emirates. J Camel Pract Res 16 (1):39–40
Woollard DC, Indyk HE (2013) Biotin analysis in dairy products. In: Preedy VR (ed) Food and nutritional components in focus No. 4., B vitamins and folate: chemistry, analysis, function and effects. The Royal Society of Chemistry, pp 377–395
Zhang H, Yao J, Zhao D, Liu H, Li J, Guo M (2005) Changes in chemical composition of Alxa Bactrian camel milk during lactation. J Dairy Sci 88:3402–3410

Chapter 9
Hormones

Hormones are a group of molecules produced by specific glandes (called endocrines), transported in the bloodstream to target distant organs where they regulate their physiology. By their action, they contribute to the regulation of metabolism and vital function including digestion, respiration, lactation, reproduction, sensory perception, nycthemeral rhythm, excretion, etc. The list of hormones with clinical interest in camel is similar to that in all mammals. In camel, the hormones contributed to the reproduction cycle. The seasonal physiology of reproduction in this species can be explained by the sexual hormones metabolism, notably in male where testosterone concentration in plasma is under complex seasonal influence. Thyroid hormones are also implied in the seasonal sexual activity. As easily stressed animal, camel is also sensitive to the hormones of the stress like cortisol. Other hormones are participating to the management of the fat storage (leptin) which is concentrated in the hump, a particularity of the species.

Contrary to many other biochemical parameters, there are no really reference values, but rather variability linked to the functional stage of the hormone release, especially as these molecules are involved in feedback mechanisms.

In the present chapter, two main groups of hormones are listed according to their functions:

- The hormones of reproduction (LH, FSH, testosterone, and estradiol) including prostaglandins, prolactin, and oxytocin
- Cortisol
- The hormones involved in the regulation of the metabolism (thyroid hormones) and of homeostasis (aldosterone and vasopressin)

© Springer International Publishing AG, part of Springer Nature 2018
B. Faye, M. Bengoumi, *Camel Clinical Biochemistry and Hematology*,
https://doi.org/10.1007/978-3-319-95562-9_9

9.1 Hormones of the Reproduction System

9.1.1 Testosterone

Testosterone is the main male sexual hormone responsible for the development of the testis, production of sperm, and typical sexual behavior. It is a steroid biosynthesized from cholesterol and secreted by testicles, in Leydig cells, of males and in an insignificant quantity by ovaries in females. In most of animals, testosterone contributes to protein synthesis including muscular growth that explains the higher weight and muscular development of males. Testosterone concentration in bloodstream is regulated by the hypothalamic-pituitary-testicular axis. Stimulation of testosterone synthesis depends on the release of gonadotropin-releasing hormone (GnRH) by the hypothalamus. GnRH acts by stimulating the pituitary gland leading to the increase of luteinizing hormone (LH) and follicle-stimulating hormone (FSH). In camel, the injection of 100 μg of GnRH increased significantly serum testosterone concentration from 2.8 ng/ml in control group to 11.9 ng/ml in treated group 2 h after the injection (Monaco et al. 2015). At reverse, immunization against GnRH reduced the serum testosterone concentration and libido (Ghoneim et al. 2012). Serum testosterone is usually analyzed by radioimmunoassay (RIA).

9.1.1.1 Seasonal Variation

Camel male is characterized by its seasonal reproductive performances. The seasonality is under complex physiological mechanisms depending on day length and external temperature (Bedrak et al. 1983). In all studies, serum testosterone concentration increases drastically from approximately 2 ng/ml in the non-rutting season to 24 ng/ml at the rutting season (Tibary and Anouassi 1997). The annual cycle has been described in camel males from Negev desert (Fig. 9.1) showing a clear relationship between testosterone level and season (Yagil and Etzion 1980).

The ratio "testosterone in rutting/testosterone in non-rutting season" is between 10 and 15 for most of the authors (Agarwal et al. 1987a; Dixit et al. 1987; Bono et al. 1993; Zia-Ur-Rahman et al. 2007; Deen 2008a), but variability is observed according to the countries and probably to the analytical methods. For example, in Egypt, Azouz et al. (1992) found lower values: 2.99 ± 0.19 ng/ml in rutting season vs 0.60 ± 0.09 ng/ml in non-rutting season, i.e., a ratio of 5:1 only. Similar ratio was reported in India: 1.46 ± 0.45 ng/ml vs 6.91 ± 1.63 in non-rutting and rutting season, respectively (Agarwal and Khanna 1993). Lower difference was reported in Egypt: 3.82 ± 0.38 ng/ml in rutting season vs 1.64 ± 0.15 ng/ml in non-breeding season (Zeidan and Abbas 2003). In Morocco, values in rutting and non-rutting season were 8.2 ± 2.1 and 2.1 ± 1.1 ng/ml, respectively (El Khasmi et al. 2011b). This seasonal variation is well illustrated in whole males compared to castrated ones (Fig. 9.2) (Shareha 1998).

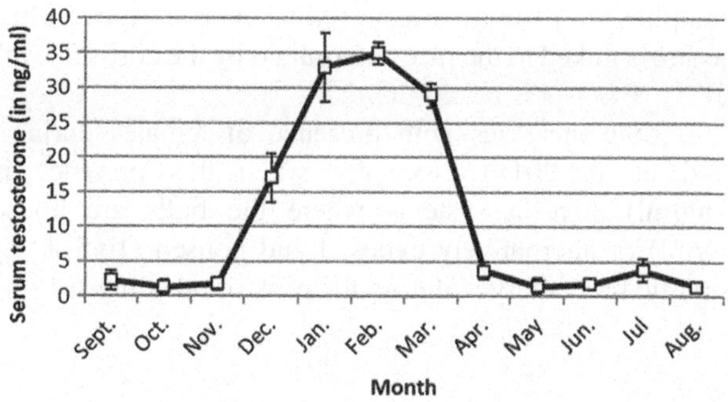

Fig. 9.1 Changes in serum testosterone concentrations in male camel all along the year (from Yagil and Etzion 1980)

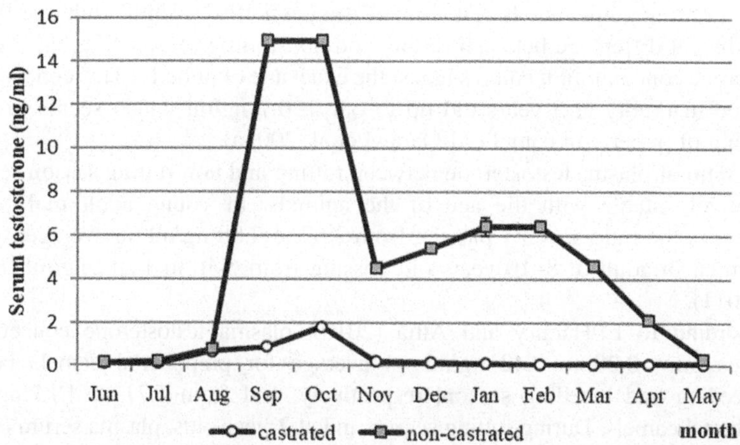

Fig. 9.2 Monthly changes in plasma testosterone in castrated and non-castrated camels (from after Shareha 1998)

At reverse, there are no relationships between erectile dysfunctions or phimosis with the testosterone concentration in serum (Ali et al. 2016; Derar et al. 2017). However, plasma testosterone concentration was lower in camel bulls affected by infertility (0.60 ± 0.11 in fertile vs 0.17 ± 0.24 ng/ml in infertile camel bulls) during the rutting season (Waheed et al. 2015). Serum testosterone concentration is also lower in case of trypanosomosis (4.8 ± 0.7 ng/ml vs 2.7 ± 1.5 ng/ml) in affected camel bulls (Al-Qarawi et al. 2004).

In Nigeria (Gombe and Oduor-Okelo (1977) as well as in Somalia (Bono et al. 1989), plasma testosterone concentration is negatively correlated with the temperature-humidity ratio. In post-racing mature male camels, lower plasma testosterone concentration were reported (0.118 ± 0.09 ng/ml) despite that the racing season is overlapping with rutting season (November to April). This low value is attributed to the stress linked to the race determined by the cortisol level (Abdel-Hadi and Wasfi 2001).

Plasma testosterone increases with presence of females during the breeding season (Fatnassi et al. 2016) especially when this presence is continuous (22.2 ± 1.9 ng/ml) than in systems where the bulls are housed separately (14.4 ± 2.0 ng/ml) or alternatively exposed and housed (16.5 ± 2.0 g/ml). This difference was particularly marked during the peak of breeding season (Bhakat et al. 2005).

9.1.1.2 Age Variation

Globally, it is admitted that immature male camels have similar testosterone level than non-rutting adult camels (Tibary and Anouassi 1997). Until adult age, there is no significant difference between rutting and non-rutting season (Fig. 9.3). Plasma testosterone concentration could suggest the exact age of puberty. This concentration increased markedly at 5 years old up to 6.8 ± 0.7 ng/ml which seems to be the beginning of puberty in camel (Al-Qarawi et al. 2001a).

The ratio of plasma testosterone between rutting and non-rutting season seems to be relatively stable with the age of the animals. In young adult male camels (4–6 years), the ratio was 3:1 passing from 2.62 to 7.88 ng/ml on average, while it was similar in adult of 8–10 years old passing from 6.46 to 14.52 ng/ml (El-Kon et al. 2011).

According to El-Harairy and Attia (2010), plasma testosterone concentration increases from 0.20 to 0.42 ng/ml on average for prepubertal camels between non-breeding and breeding season, respectively, and from 1.71 to 4.13 ng/ml in postpubertal camels. During rutting season and at 2-year-olds, plasma serum concentration was 0.12 ± 0.01 ng/ml, 0.73 ± 0.12 at 3-year-olds, 3.9 ± 0.06 at 4.5-year-olds (puberty), 5.0 ± 0.17 at 8-year-olds, and 6.3 ± 0.12 at 10-year-olds and then decreased at 13-year-olds down to 0.8 ± 0.12 ng/ml (El-Harairy and Attia 2010). In another

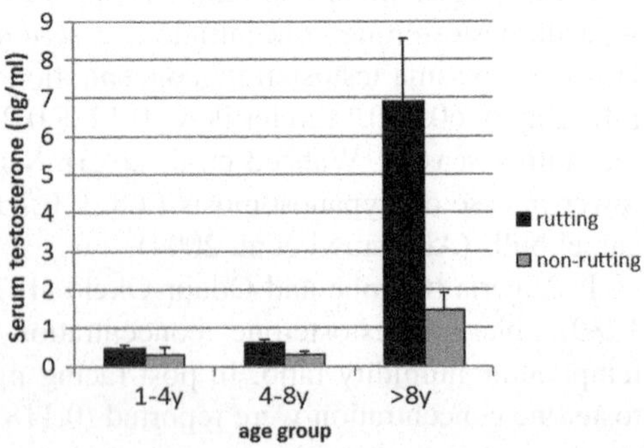

Fig. 9.3 Changes in testosterone level in camel according to age group and season (from after Agarwal and Khanna 1993)

study, plasma testosterone concentrations did not exceed 1.4 ng/ml before 3 years with a three- to fourfold increase in peripubertal camels 3–5 years old (3.2 ± 0.4 ng/ml) and mature (4.8 ± 0.6 ng/ml) camels (5–15 years old) followed by about 50% decrease (2.6 ± 0.3 ng/ml) in camels having more than 15 years (Al-Qarawi et al. 2000). Similar figure was reported later: 0.10 ± 0.01 ng/ml in immature camel, 0.50 ± 0.01 ng/ml in non-rutting adult, and 9.25 ± 0.22 ng/ml in rutting male (Al-Qarawi and El-Mougy 2008).

In peripubertal male camel, plasma testosterone concentrations (2.85–2.98 ng/ml) are not influenced by the type of diet in spite higher growth of the testicle size, but a seasonal effect is reported with lower values in winter (Al-Saiady et al. 2015). In another study, a slight effect of the diet was reported: 3.88 ± 0.08 ng/ml with enriched diet vs 3.65 ± 0.08 ng/ml with lower proteo-energetic diet (Al-Saiady et al. 2013).

9.1.1.3 Testosterone in Seminal Liquid or Semen

The concentration of testosterone is higher in testicular tissue than in plasma (El-Harairy and Attia 2010): in prepubertal camels, the testicular testosterone increased from 238.2 ng/g to 374.7 ng/g during non-breeding and breeding season, respectively, while these values were 296.3 and 391.7 ng/g, respectively, for postpubertal camel bulls. At 2 years old, testicular testosterone was 382.3 ng/g on average, 386.0 at 3 years old, 404.7 at 4.5 years old, and from 370.0 and 408.0 between 8 and 10 years old and then decreases to 366.0 at 13 years old (El-Harairy and Attia 2010). Globally, the level of intratesticular testosterone in camels should be 25–30 times greater than the level in peripheral blood (Al-Qarawi et al. 2001a): at 7 years old, the testosterone concentrations were 164.7 ± 16.8 ng/g and 6.8 ± 0.7 ng/ml in testicles and plasma, respectively.

In epididymal fluid testosterone concentration was 5.19 ± 1.69 ng/ml (Waheed et al. 2011). In seminal plasma, the concentration in testosterone was higher in fertile bulls (0.73 ± 0.06 ng/ml) compared to infertile camels (0.40 ± 0.08 ng/ml) (Waheed et al. 2015). Globally the concentration of testosterone in seminal fluid decreases in case of male reproductive disorders like azoospermia (Al-Qarawi and Belely 2004).

In camel urine, especially in racing camel where anabolic steroids could be used as doping products, endogenous testosterone was determined in 56 males with mean value of 12.3 ± 3.7 ng/ml and normal distribution (Abdel-Hadi et al. 1998).

9.1.1.4 Testosterone in Female Camel

There are few data on plasma testosterone concentration female camel (Tibary and Anouassi 1997). Plasma testosterone could change from 50 pg/ml up to 1000 pg/ml according to the size of follicle (Homeida et al. 1988). The parallel increase in follicle size and estrogen and testosterone concentration suggested that both hormones are secreted by the follicle as for other domestic species. A seasonal variation in female testosterone was reported (El-Harairy et al. 2010) with a significant higher value in winter (31.20 ± 1.48 pg/ml) than in spring (7.30 ± 0.63 pg/ml) and autumn (10.70 ± 0.32 pg/ml), the lower value being observed in summer (5.80 ± 0.43 pg/ml). The decrease of testosterone in summer season could be partly explained by the inhibitory effect of high prolactin levels at this time of the year corresponding to the peak of lactation (Musaad et al. 2013).

In **follicular fluid** collected in slaughterhouse, the mean testosterone concentration was 6.7 ± 1.9 ng/ml in non-estrogenic follicles vs 1.1 ± 0.4 ng/ml in estrogenic follicles (Basiouni 1999).

In **camel urine** of racing she-camels, the endogenous testosterone concentration was determined in low quantity (87.9 ± 140 pg/ml) with not-normal distribution (Abdel-Hadi et al. 1998).

9.1.2 Estradiol

Estradiol is a steroid hormone regulating estrus cycle and development of female genital tissues as well as the udder. It is also involved in fecundation process (ovulation, preparation of implantation of oocytes in the uterus), pregnancy, and parturition in combination with progesterone. Ovary follicles are the main producing tissue. Estradiol is biosynthesized from cholesterol; its plasma is changing according to the estrus cycle. The placenta is contributing to the increase of estradiol plasma concentration during pregnancy. In male, a few quantity of estradiol is produced by the Leydig cells. Determination of plasma estradiol is also based on radioimmunoassay (RIA).

The profile of estradiol and other estrogens (mainly estradiol-17β and estrone sulfate) during camel estrus cycle has been described by many authors (Elias et al. 1984a; Homeida et al. 1986, 1988; Agarwal et al. 1989a; Agarwal and Khanna 1993), and a complete review has been published by Tibary and Anouassi (1997).

9.1.2.1 Estrus Cycle Variation

On average, in camel, plasma estradiol concentration is higher during estrus (corresponding to the sexual receptivity). These higher values are in relationship with the presence of growing follicles. The maximum plasma estrogen concentration was observed at 2.9 ± 1.83 days of the estrus period (El-Wishy 1987). Three to 5 days after the estrus, the estradiol concentrations should decline. Values vary between 9 and 110 pg/ml (Tibary and Anouassi 1997). For Elias et al. (1984a), in early estrus, estradiol reaches a peak of 75 ± 7 pg/ml. The regression of follicles is followed by a decrease of estradiol concentration. Mean plasma estradiol concentration reached peak value (39.0 ± 1.8 pg/ml) when the dominant follicle measured 1.7 ± 0.1 cm and just after ovulation (Skidmore et al. 1996a). During ovulation, Ismail et al. (1998) found estradiol concentration of 34.20 ± 6.04 pg/ml.

Fig. 9.4 Changes in camel serum estradiol all along the estrus cycle (from after Homeida et al. 1988)

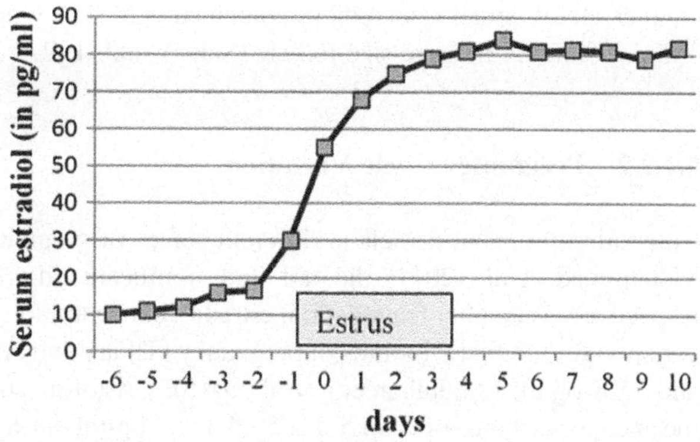

For Cristofori and Quaranta (1993), 17β-estradiol concentration in serum was 32.16 ± 9.9 pg/ml during anestrous phase, 44.4 ± 2.7 pg/ml during follicular development phase, and 94.4 ± 7.2 during estrus. For Homeida et al. (1988), plasma concentration increases from basal values 20 pg/ml before estrus to more than 80 pg/ml during15 days and then decreased steadily as the follicle regressed (Fig. 9.4). Lower values are reported by Ayoub et al. (2003): 29.7 ± 7.3 pg/ml during estrus and 5.72 ± 4.55 pg/ml a week after. During the growing follicular phase, serum estradiol concentration did not change significantly with different sized follicles: 228.6 ± 32 pg/ml with follicle 1.1–1.5 cm, 211.1 ± 31.9 pg/ml with follicle 1.6–2.1 cm, and 210.1 ± 30.0 with follicle 2.2–2.5 cm (El-Bahr et al. 2015). Similar results were reported by Ghoneim et al. (2013).

Plasma estradiol concentration remains higher along the breeding season in case of absence of mating. Mean values according the season (El-Harairy et al. 2010) were significantly higher in spring (56.15 ± 1.25 pg/ml) and winter (62.18 ± 1.16 pg/ml) than in summer (20.13 ± 1.02 pg/ml) and autumn (28.16 ± 1.31 pg/ml). This seasonal variation was already observed by Khaldoun (1993).

In prepubertal camel, a significant effect of the diet was reported with higher estradiol concentration in camel receiving diet richer in protein and energy: 38.16 ± 3.87 vs 28.73 ± 3.87 ng/ml. Moreover, estradiol concentration was higher in winter and spring in the same prepubertal she-camels (Al-Saiady et al. 2012).

Plasma estradiol concentration profile in Bactrian camel is similar to that of the dromedary but with lower concentrations: 26.8 ± 9.0 pg/ml during the follicular phase and 30.8 ± 5.1 pg/ml when the follicle was maximum size and then fell after ovulation to 19.0 ± 4.1 pg/ml (Xu et al. 1985).

Regarding the content of estradiol in **follicular fluid**, non-estrogenic follicles (detected by the ratio estradiol/testosterone) had significant lower concentration (4.5 ± 2.1 pg/ml) than in estrogenic follicles (31.9 ± 10.7 pg/ml) (Basiouni 1999). Estradiol concentrations in follicular fluid increase with the size of follicles (Afaleq et al. 2003). In small follicles (2–6 mm), 17β-estradiol concentration was 56.8 ± 1.2 pg/ml, while it was 138.0 ± 8.1 pg/ml in large follicles >7 mm (Zia-Ur-Rahman et al. 2008). In another study, small-size follicle contained 5.5 ± 4.7 ng/ml estradiol, while it was 145 ± 8.4 ng/ml in large-size follicle (Salem et al. 1997).

9.1.2.2 Pregnancy Cycle Variation

Comparing pregnant camels at different stages of pregnancy to non-pregnant camels, Muhammad et al. (2011) did not find significant difference: 27.03 vs 19.8 pg/ml, respectively. Yet, after fecundation, estradiol concentrations increase and show irregular variation for the first 6 months of pregnancy (Tibary and Anouassi 1997), between 34.5 and 45.8 pg/ml (Abdulkareem et al. 2015). Cristofori and Quaranta (1993) reported mean concentrations of estradiol at 59.9 ± 5.4 pg/ml during the first third of pregnancy, 71.2 ± 9.0 pg/ml at the second third, and 94.7 ± 3.3 pg/ml at the last third. According to Skidmore et al. (1996b), serum estrogen concentrations showed pronounced fluctuations during the first 100 days of gestation. Mean 17 β-estradiol concentrations increased at around day 50 to about 100 pg/ml and then remained relatively constant from day 90 to day 300. Agarwal et al. (1987b) reported comparative values: the mean estradiol levels increased progressively from a basal level of 20 pg/ml at 2 to 3 months of pregnancy to about 450 pg/ml at the final stages of gestation. A male fetus induced lower estradiol concentration in mother serum (76.5 ± 10.8 pg/ml) than female fetus (112.3 ± 19.6 pg/ml). The increase of estradiol at the end of pregnancy was also reported by Elias et al. (1984a): at the 10th month of pregnancy, estradiol concentration in serum rises to 338.3 ± 162.42 pg/ml and continues to rise until the 12th month, peaking at 606 ± 120.27 pg/ml (Elias et al. 1984a). A remarkable increase in plasma estradiol starts 2–4 days prior to parturition and peaking in excess of 200 pg/ml, 2 h before delivery (El-Belely 1994; Ayoub et al. 2003).

After parturition or abortion, the fall in estradiol could be linked to the feto-placental origin of estrogen (Agarwal et al. 1987b; Skidmore et al. 1996b).

In Bactrian camel, serum 17β-estradiol increased significantly from 11 months of pregnancy with peak mean concentrations of 617.47 ± 32.56 pg/ml at the 11.5 month (Zhao et al. 1998). In semen of Bactrian camel, 17β-estradiol seems higher in infertile camel (196 ng/ml) than in fertile (117 ng/ml), but the number of analyses was limited (Xu et al. 1993).

9.1.3 Progesterone

Progesterone is a steroid hormone biosynthesized from cholesterol by the ovaries, especially the corpus luteum and placenta. It plays different roles during the estrus cycle, embryogenesis and especially during the pregnancy (progesterone is sometimes called "the hormone of pregnancy"). It intervenes also in metabolism of corticosteroids. The physiological effect of progesterone is amplified in the presence of estrogens, notably in the development of udder. Progesterone is also implied in the sodium metabolism by decreasing natriuresis.

The main actions of progesterone are the preparation of the uterus in order to facilitate the implantation of fecundated ovule, decrease of maternal immune response to allow acceptance of future fetus, and inhibition of milk production during pregnancy. As for other sexual hormones, the determination of progesterone is based on RIA. The use of some commercial standardized kits for human would lead to wrong values in camels (Deen et al. 2001).

In camel, plasma progesterone concentrations were determined both during estrus cycle and pregnancy (Tibary and Anouassi 1997), but high individual variability was reported (El-Basheir et al. 2001).

9.1.3.1 Estrus Cycle Variation

In the absence of ovulation (provoked by mating in camel), plasma progesterone concentration remains below 1 ng/ml along the follicular cycle (Elias et al. 1984b; Marie and Anouassi 1986; Homeida et al. 1988; Cristofori and Quaranta 1993; Ayoub et al. 2003). This concentration increases 4–6 days after mating until 10–11 days reaching 2.4–6.1 ng/ml (Marie and Anouassi 1986) but with important individual variations (Marie and Anouassi 1987; Agarwal et al. 1991; Ayoub et al. 2003), and individual values up to 14 ng/ml can be observed (Agarwal et al. 1989a). Eleven to 14 days after mating or induction, progesterone decreases below 1 ng/ml in case of failure of ovulation or absence of conception. Plasma progesterone is a good indicator of ovulation and formation of corpus luteum (Tibary and Anouassi 1997).

In camel stimulation ovulation with GnRH increases plasma progesterone from basal value (0.32 ± 0.05) up to a peak (7.15 ± 0.97 ng/ml) by 7 days post-ovulation and then decreased to 0.91 ± 0.38 ng/ml by 11 days (Rawy et al. 2012).

In prepubertal she-camel, progesterone concentration varied between 0.07 and 0.226 ng/ml according to the season with higher value during autumn (Al-Saiady et al. 2012). In adult female camel, plasma progesterone concentration is quite higher in winter than in summer, the maximal activity being in March. Moreover, important variation within the day is observed (Khaldoun 1993).

9.1.3.2 Pregnancy Cycle Variation

Within the first week post-ovulation, there is no difference in progesterone concentration between pregnant and non-pregnant camel, but while it decreases in case of lack of pregnancy, maintaining corpus luteum in pregnant female, it is above 2 ng/ml along the gestation period (Marie and Anouassi 1987). The values could reach 3–9 ng/ml during the first month (Agarwal et al. 1987b; Elias et al. 1984b; Nagy et al. 2015) with a high individual variability (Vyas et al. 2010). On average, plasma progesterone concentration is ten times higher in pregnant female than in non-pregnant: 3.15 ± 5.38 ng/ml vs 0.35 ± 0.02 ng/ml (Cristofori and Quaranta 1993). In Nigeria, Muhammad et al. (2011) reported mean progesterone of 4.23 and 1.39 ng/ml in pregnant and non-pregnant camels, respectively.

Tail curling is the traditional way for camel farmers to diagnose early pregnancy. But false-positive detection is possible. In a monitoring achieved by Deen (2008b), among camels showing tail curling behavior (assumed pregnant), 68% only had plasma progesterone concentrations above 1 ng/ml confirming their pregnant status.

After the first month of pregnancy, the progesterone level is maintained up to the fifth month and then declines gradually until the parturition (Elias et al. 1984b). Different pattern was reported (Agarwal and Khanna 1993; Skidmore et al. 1996b), but in all the cases, fluctuations between 2 and 5.5 ng/ml occur along the pregnancy. These fluctuations could be linked to the secretory activity of corpus luteum or to the blood volume (Tibary and Anouassi 1997). Moreover, plasma progesterone level concentration could be higher in camel with male fetus compared to female fetus: 5.1 ± 0.6 vs 3.6 ± 0.3 ng/ml, respectively (Agarwal and Khanna 1993). The same authors have reported also age effect of the dam, with a maximum in female 5–10 years old.

In Bactrian camel, similar values were reported during gestation with mean concentrations varying from 3.0 ± 0.5 to 8.5 ± 4.8 ng/ml (Zhao et al. 1998).

In case of early embryonic mortality, the level of progesterone falls drastically after 1 month to less than 1 ng/ml (Vyas et al. 2010).

In the 48 h before parturition, significant decrease in plasma concentrations of progesterone occurs due to the regression of corpus luteum (El-Belely 1994). After calving, progesterone concentrations varied between 0.5 and 2.0 ng/ml on the day of calving and declined steadily thereafter to become undetectable by day 9. In the neonate, progesterone is undetectable (Agarwal et al. 1992). In another study, a peak of progesterone was observed at postpartum day 9 (3.24 ± 3.0 ng/ml) and then decreased gradually to a basal level (Derar et al. 2014).

9.1.3.3 Progesterone in Milk and Follicular Fluid

Determination of progesterone in milk could be useful for diagnosis of reproduction failure or pregnancy status as it is currently used in dairy cows (Ruiz et al. 1989). However, few references are available for camel milk. In pregnant camels, plasma progesterone concentration is more than 5 ng/ml, whereas in non-pregnant ones, it is below 1 ng/ml (Abdel Rahim and El-Nazier 1987). Milk progesterone was used also to monitor ovarian changes after parturition with an increase 3 weeks postpartum up to 25 ng/ml on average (Abdel Rahim 1989).

There was a slight significant difference in progesterone concentration in follicular fluid according to the size of follicle: 121 ± 4.2 in small follicles vs 141 ± 4.3 ng/ml in large ones (Zia-Ur-Rahman et al. 2008). Salem et al. (1997) reported 47.5 ± 2.3 ng/ml in small follicles vs 38.0 ± 2.4 ng/ml in large one and 53.2 ± 2.4 ng/ml in cystic follicles.

9.1.3.4 Progesterone in Semen

There was no significant difference in semen from fertile camel (0.82 ± 0.16 ng/ml) compared to infertile: 1.01 ± 0.79 ng/ml (Xu et al. 1993).

9.1.4 Gonadotropins: LH and FSH

Luteinizing hormone (LH) also called interstitial cell-stimulating hormone (ICSH) is a glycoprotein secreted by gonadotropic cells in the anterior pituitary gland. In male, it stimulates Leydig cell production of testosterone. In female, it is responsible for ovulation and development of the corpus luteum in synergy with FSH (follicle-stimulating hormone), another glycoprotein hormone secreted by the pituitary gland. FSH is involved in the maturation of germ cells both in male and female. As suggested by its name, it is responsible for the follicular growth. Its dosage in plasma is less informative than testosterone, estradiol, or progesterone.

In male, plasma LH concentrations fluctuate throughout the year between 0.2 and 1.10 ng/ml, while it is between 0.5 and 1.10 ng/ml in female with a nonsignificant seasonal pattern although a correlation between rainfall and LH levels in female was reported (Bono et al. 1993). At reverse, an effect of rutting season in male was reported: 0.39 vs 0.09 ng/ml in rutting and non-rutting season, respectively, for LH, and 0.49 vs 0.23 ng/ml, respectively, for FSH (Azouz et al. 1992). For Fat-Halla and Ismail (1980), only FSH has significant seasonal variation with higher values in winter. This is confirmed by Al-Qarawi and El-Mougy (2008) who found a significant effect of rutting on FSH (0.99 vs 0.69 ng/ml in non-rutting season), while mean LH concentrations were comparable in rutting (0.13 ng/ml) and non-rutting season (0.15 ng/ml).

In female camel, a peak of LH occurs 2–3 h after mating (2.9–19.1 ng/ml) and then decreases 6 h postmating (Marie and Anouassi 1987), but this timing is not observed in all the cases, and individual differences have been described (Bono et al. 1985; Cristofori et al. 1986). In Bactrian camel, a peak of LH (6.9 ± 1.0 ng/ml) is observed 4 h after insemination and decreases after 8 h, probably induced by semen (Xu et al. 1985) as it has been confirmed by the effect of injection of third fraction of seminal plasma which increases in vivo the LH concentration from 6.43 ± 0.14 before to 15.50 ± 2.64 ng/ml 6 h after injection (Zhao et al. 2001b).

On average, plasma LH concentration is higher in case of luteal cyst (0.097 ng/ml) or active ovaries (0.064 ng/ml) than in case of cystic ovaries (0.012 ng/ml), while FSH is quite higher with active ovaries (0.213 ng/ml) compared to cystic ovaries (0.039 ng/ml) (Hegazy et al. 2001).

LH concentrations change according to the energy level of the diet passing from 3.9–4.1 ng/ml in camel receiving normal diet (100% energy requirements) to 1.4–3.0 ng/ml with low-energy diet. The decreasing trend observed with FSH (3.5–4.1 ng/ml) is not significant (Khazali 2010).

In Bactrian camel, plasma LH concentrations varied significantly between 2.2 ± 0.9 and 20.3 ± 18.8 ng/ml during pregnancy, with the highest value on the day of artificial insemination. Then, LH concentration decreases, reaching a constant low level by the eighth month of pregnancy. Plasma FSH concentrations varied between 7.0 ± 0.2 and 28.9 ± 0.4 ng/ml during gestation, with the peak value being reached at 4.5 months, followed by a marked drop to 7.0 ± 0.2 ng/ml at 7.5 months (Zhao et al. 2001a).

The pituitary activity determined by the increase of FSH and LH occurs rapidly after parturition (Tibary and Anouassi 1997). LH was also analyzed in camel to determine the onset of puberty (Towhidi et al. 2001) with plasma concentration at 1.85 ± 0.24 ng/ml in 3-year-old and then 2.25 ± 0.30 ng/ml in 4-year-old camels.

In peripubertal camel, no seasonal effect was observed for LH (0.11–0.17 ng/ml) contrary to FSH, significantly lower in autumn (0.010 ng/ml) than in other seasons (0.16–0.21 ng/ml). Moreover, a diet more rich in protein and energy has a significant effect on FSH concentration, but not on LH (Al-Saiady et al. 2015).

In semen of fertile Bactrian camel, LH concentration is higher than in infertile: 2.3 ± 1.0 vs 1.0 ± 0.7 ng/ml, respectively (Xu et al. 1993).

9.1.5 Prostaglandins

Prostaglandins (Pg) are lipid compounds derivating from fatty acids having various hormone-like effects in all animals. Their effects are very including blood coagulation, inflammatory process, and reproduction.

During sexual cycle, no pulsatile releases of prostaglandin F 2α metabolite (PgFM) are detected in camel. In non-pregnant camels, the mean basal PgFM concentration on day 8 post induced ovulation was 30.0 ± 0.7 pg/ml and increased significantly to 58.8 ± 1.5 pg/ml on day 10 and then declined again significantly to 39.3 ± 1.0 pg/ml by day 12 (Skidmore et al. 1998). But the values could change according to the method of extraction (Skidmore et al. 1996c, 1998).

In pregnant camel, the secretion of prostaglandin F2α (PgF2α) is inhibited before parturition and then increases with a peak the day at parturition (Tibary and Anouassi 1997). Indeed, during pregnancy, values of PgFM (Skidmore et al. 1998) remained at low level (50–200 ng/ml) and then increased up to 1000 ng/ml from 50 days before parturition. An explosive increase up to 2000 ng/ml is observed the day of parturition (Skidmore et al. 1996b), but a wide individual variability is observed. El-Belely (1994) reported a tenfold increase to 5.4 ± 0.19 ng/ml in plasma concentrations of PgFM between days 10 and 1 before parturition, showing a sudden and large increase during 5 h prior to delivery. In an experiment, camels were intoxicated with endotoxin lipopolysaccharide (*Escherichia coli* serotype 055:B5) at different stages of pregnancy provoking abortion: the concentration of PGFM started to increase significantly 30 min post-injection rising from <0.1 ng/ml to >1.5 ng/ml and remaining high for 4 h (Al-Dughaym and Homeida 2010).

In semen of Bactrian camel, Xu et al. (1993) have determined PgE_1 and $PgF_2\alpha$, their concentration being 6.51 ± 0.62 ng/ml and 2.19 ± 0.39 ng/ml, respectively.

9.1.6 *Prolactin*

Prolactin (also named luteotropic hormone or luteotropin) is secreted by the pituitary gland. It plays a pivotal role in the stimulation of mammary glands during lactation and in milk excretion after parturition in addition to other metabolic functions.

Few publications are available in camel although the camel prolactin was purified (Martinat et al. 1990) and its structure described for long time (Martinat et al. 1991). In male, an increase of serum prolactin is observed during non-rutting season suggesting a role in inhibiting libido, probably in relationship with inhibition of FSH and LH secretion (Tibary and Anouassi 1997). Values in male serum were 3.03 ± 0.32 ng/ml at rutting season vs 8.2 ± 0.68 ng/ml at non-rutting (Azouz et al. 1992). In a more recent study, the reported values were, respectively, 0.9 ± 0.02 ng/ml during rutting season and 1.3 ± 0.03 ng/ml at non-rutting, while concentration in plasma of immature camels was 1.8 ± 0.06 ng/ml (Al-Qarawi and El-Mougy 2008).

Serum prolactin concentrations are significantly higher in healthy females than males: 20.4 ng/ml vs 13.9 ng/ml on average according to Kataria and Kataria (2010). The same authors observed quite higher values in stressed camels up to 100 ng/ml.

9.1.7 *Oxytocin*

Oxytocin is a peptide hormone produced by hypothalamus and released by the posterior pituitary gland in response to stretching of uterus at parturition and to stimulating udder at milking time. In spite of its use for stimulating milk ejection in reluctant she-camels, data on its concentration in blood are scarce in camel. Plasma concentrations varied around 10 µIU/ml (20 pg/ml) in spring and 5 µIU/ml (10 pg/ml) in summer for females. Values were slightly more elevated in males (Achaaban et al. 2002).

9.2 Hormones of the Adrenal System

9.2.1 *Adrenocorticotropic Hormone (ACTH)*

Adrenocorticotropic hormone (ACTH) or corticotropin is a polypeptide tropic hormone secreted by the anterior pituitary gland. Its main effect is the increasing production and release of cortisol, indicator of the stress by the cortex of the adrenal gland. It is also involved in circadian rhythm.

ACTH immunoreactive cells were identified and located on different parts of the anterior pituitary of camel (Quéré et al. 1985).

Usually, in camel experiment, ACTH is used in injection to assess the impact on cortisol release (Riad 1995; Sid-Ahmed et al. 2013). At reverse, few values are reported.

Plasma ACTH increases at the end of dehydration period (2 weeks) passing from around 11 to 16 pg/ml. At rehydration time, ACTH is maintained at higher level for 24 h (as for cortisol) and returns to basal level (Al-Qarawi 1997). Injection of ghrelin (the hormone of hunger) significantly increased ACTH secretion in prepubertal camels, passing from 0.9 pg/ml in control camels to 1.7 pg/ml after injection (Rashedi and Khazali 2010).

9.2.2 Cortisol

Stress in animals is accompanied by many mechanisms involving the increase of different physiological markers (respiratory rhythm, cardiac frequency, rectal temperature), hematological markers (blood formula, osmotic fragility), biochemical markers (mainly glucose), and hormonal markers (glucocorticoids and thyroid hormones). Cortisol is the main hormone secreted by adrenal gland involved in stress mechanism (Saleem 2006).

Cortisol is released in response to stress in all mammals. It contributes to increase glycemia through gluconeogenesis and to decrease the immune system. In camel, plasma cortisol was analyzed in relationship with rutting season in male, with water restriction or during transport before slaughtering.

9.2.2.1 Reproduction Cycle

In general, seasonal change of corticoid concentration was described, varying from 5.4 ± 3.0 to 62.0 ± 37 ng/ml in male camels and 7.5 ± 5.4 to 62.2 ± 21.3 ng/ml in female camels, the highest values being observed during rainy season (Bono et al. 1993). In Morocco, higher values were reported in winter (January) than in summer (May–June): 66.0 ± 13.2 vs 25.7 ± 6.7 ng/ml, respectively (Chakir 2016). At reverse, Baraka (2012), in Egypt, found higher cortisol value in summer than in winter, 38.6 ± 5.3 and 28.5 ± 4.8 ng/ml, respectively. In Israel, Elias and Weil (1989) found also higher concentration of cortisol in summer (8 ng/ml on average) than in winter (1.15 ng/ml on average).

However, the main effect is linked to the reproduction cycle, especially to the rutting season, as stressing period for the male: 3.2 ± 1.1 ng/ml in non-rutting adult vs 13.2 ± 4.04 ng/ml in rutting adult, while it is 2.7 ± 1.0 ng/ml only in prepubertal camels (Dixit et al. 1987). This difference was confirmed by Azouz et al. (1992) but with higher values: 9.6 ± 0.9 ng/ml in non-rutting season vs 36 ng/ml in rutting season.

High variability occurs during mating activity both in males and females (Elias and Weil 1989). Before mating, plasma cortisol concentration was 1.15 ng/ml, and 9.92 ng/ml, 10 min after coitus. These values were 1.08 ng/ml, 5.7 ng/ml, and 6.0 ng/ml, respectively, during the anestrus, 2, 15, and 30 min post coitus. A significant difference in the cortisol concentration is observed in pluriparous mated female camels during the breeding season before (8.0 ± 1.3 ng/ml) and immediately after mating. 45.0 ± 11.9 ng/ml. At *peripartum* period, maternal serum cortisol level rose suddenly from the basal value 11.4 ± 1.3 ng/ml 3 days prepartum to 45.3 ± 5.0 ng/ml at parturition. Immediately after birth, serum levels of cortisol in the newborn camels were 35.9–37.0 ng/ml and in the dams 57.0–60.0 ng/ml. Agarwal et al. (1992) found similar results: cortisol concentrations were high at parturition (25–30 ng/ml) in both the dams and their calves, then declined to 6–7 ng/ml in the dams, but became undetectable in the neonates by day 14 *postpartum*. According to Mohamed (2006), cortisol level was 121.6 ± 5.4 at day of parturition and then decreased to 30.1 ± 1.9 (day 3 *postpartum*) and 21.9 ± 1.0 ng/ml at day 5 *postpartum*. Cortisol serum level was 37.1 ± 1.4 ng/ml 1 day before weaning and then increased to 48.0 ± 1.5 at weaning and 69.5 ± 1.9 ng/ml the third day after weaning (Mohamed 2006).

A lower plasma value of cortisol was reported in high milk-yielding camel than in low yielding one: 7.8 ± 0.6 ng/ml vs 12.0 ± 0.9 ng/ml, respectively, probably in relationship with glucose metabolism (Zia-Ur-Rahman et al. 1998).

On average, cortisol concentrations in healthy camel are higher in male than in female (Kataria and Kataria 2010): 7.39 (27.03 nmol/l) vs 6.87 µg/ml (24.83 nmol/l) but these values are 1000 higher than those reported generally in camel. From his part, Baraka (2012) did not find sexual difference: 30.6 ± 3.7 ng/ml in male for 32.0 ± 5.7 ng/ml in female. This was already reported by Saeb et al. (2010) and Mohamed (2006).

Also no clear age effect was reported: 29.8 ± 4.2 vs 28.0 ± 6.0 ng/ml in young and adult camels, respectively (Baraka 2012).

9.2.2.2 Transport Stress

Preslaughtering stress, including transport, was widely addressed (Chakir 2016). Thus, after 2 h transportation during hot season, plasma cortisol increased from 137 ± 20 to 220 ± 30 ng/ml (El-Khasmi et al. 2013). This increase is correlated with the distance of transportation: 88.3 ± 19.4 ng/ml after 72–80 km, 152.4 ± 25.2 ng/ml after 160–170 km, and 231.7 ± 23.7 after 350–370 km (El-Khasmi et al. 2015). In the experiment of Saeb et al. (2010), cortisol concentration increased also from 13.8 ± 1.4 before transportation to 19.5 ± 1.7 after 1 h and to 24.1 ± 2.2 at the end of transportation (5 h) and then returned to initial values (11.7 ± 1.28 ng/ml) 24 h after arrival.

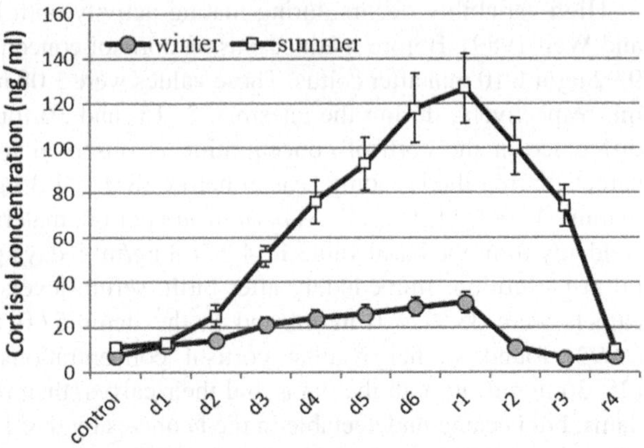

Fig. 9.5 Changes in cortisol concentrations in camel plasma during dehydration/rehydration cycle (from after Kataria et al. 2000)

9.2.2.3 Dehydration

In spite of the adaptation of camel to the thirst, dehydration is a factor of stress (Ziv et al. 1997). Plasma concentrations of cortisol increase regularly with the duration of dehydration with a stronger effect in summer season (Kataria et al. 2000). From a basal value of 7.0 ng/ml on average, a peak of 29 ng/ml is reached after 24 days of dehydration in winter. In summer, mean concentrations start from 11 ng/ml and increase up to 118 ng/ml after 13 days only of dehydration in summer (Fig. 9.5).

After rehydration, basic values around 10 ng/ml are reached in 48 h, but this back to initial values could be linked to dilution effect (Ziv et al. 1997).

Similar figure was reported more recently: plasma cortisol in normal hydrated camel (170 ± 10 ng/ml) increased threefold in dehydrated camel (730 ± 13 ng/ml) (Alhaj Ali et al. 2013).

9.2.2.4 Other Stresses

A slight difference is observed within the day, a higher cortisol concentration in plasma being observed in the morning than in the afternoon: 14.5 ± 1.24 vs 9.5 ± 0.9 ng/ml (Saeb et al. 2010).

In unhealthy camels, plasma cortisol increases sharply. In the study of Kataria and Kataria (2010), the basic level is multiplied by 5 on average. In Egypt, Baraka (2012) found values of 27.0 ± 6.1 ng/ml in case of simple indigestion, 36.8 ± 6.7 ng/ml in case of acidosis, 25.0 ± 3.9 ng/ml with rumen alkalosis, and 53.9 ± 15.1 ng/ml with diarrhea.

Plasma cortisol increases after injection of adrenocorticotrophic hormone, passing from 0.6–10.8 to 10.9–42.2 ng/ml (Sid-Ahmed et al. 2013).

9.2.2.5 Cortisol in Other Substrates

The blood sampling being stressing for the animal, the results reported after camel constraint could be biased. So, other types of sampling were suggested as fecal or hair sampling, as positive correlations were observed between the concentrations in plasma, hair, and feces in camel (Chakir 2016). Such, in hair cortisol concentrations were 0.93 ± 0.26 ng/g in January and 0.61 ± 0.08 ng/g in May–June, while in feces, it was 2.74 ± 0.14 and 1.42 ± 0.35 ng/g in January and May–June, respectively (Chakir 2016). However, cortisol in hair expressed more the accumulation of stress in the previous weeks rather than the immediate stress, for example, after transportation. Cortisol increases in feces after injection of adrenocorticotrophic hormone, passing from 286 to 2559 ng/g (Sid-Ahmed et al. 2013).

Cortisol has been also determined in camel urine before and after transport stress, the values passing from 6.56 ± 0.57 to 35.5 ± 1.12 ng/ml (El Khasmi et al. 2011a). Cortisol can be also determined in camel saliva. After injection of ACTH, cortisol concentration in saliva increased from 1.17 ± 0.09 ng/ml to 11.67–24.18 ng/ml, 2 h post-injection (Majchrzak et al. 2015).

9.3 Thyroid Hormones

The thyroid gland releases two main thyroid hormones, **triiodothyronine** (T3) and its prohormone, **tetraiodothyronine** or **thyroxine** (T4). Fractions T1 (monoiodotyrosine) and T2 (diiodotyrosine) stimulate the enzyme responsible of the conversion of T3 in T4. Usually, only T3 and T4 are determined in plasma. In camel, T1 and T2 represented 16.7% and 20.8%, respectively, of the iodine fraction during estrus and 15.2% and 20.7%, respectively, during pregnancy (Abdel-Wahab et al. 1974).

These hormones, which contain iodine, are mainly involved in the regulation of general metabolism. In consequence, iodine deficiency is linked to a decrease of thyroid hormone production and of thyroid swelling (goiter) which has been already described in camel (see chapter on trace elements). Moreover, T3 production being selenoenzyme dependent, dietary selenium is an important factor of thyroid hormone production.

Thyroid hormones contribute to increase basal metabolism, regulate bone metabolism in synergy with growth hormone, stimulate vitamins' metabolism, and increase sensitivity to adrenaline. They could be affected by health status of the organism, reproduction cycle, and dehydration status. In camel, the prevalence of thyroid disorders appears not negligible. For example, in Iran, it has been reported relatively high prevalence of hyperplasic goiter, degenerative changes, follicular cysts or atrophy, nodular hyperplasia, adenoma, carcinoma, or simple colloidal goiter (Yadegari et al. 2014; Yadegari 2015).

Secretion of the thyroid gland is stimulated by the **thyroid-stimulating hormone** (TSH), a glycoprotein hormone secreted by the pituitary gland. The half-life of TSH is very short (1 h). TSH is generally secreted in higher quantity in young animals during their growth. While TSH secretion by the pituitary gland is stimulated by **thyrotropin-releasing hormone** (TRH) secreted by the hypothalamus, **somatostatin** also secreted by the hypothalamus has an inhibiting effect on the release of TSH. T3 and T4 are measured using RIA method.

9.3.1 Thyroxine (T4) and Triiodothyronine (T3)

Plasma circulating concentrations of total T3 and T4 in pregnant camel ranged from 0.73 to 1.32 and 75.9 to 116.2 ng/ml, respectively (Agarwal et al. 1989b). Similar levels were reported in camels of different sex and status, and except for the youngest animals, ranging at 82–131 ng/ml for T4 and 0.68–1.07 ng/ml for T3 (Bengoumi et al. 1999). Other authors reported lower mean values, for example, 25 ng/ml for T4 and 5.5 ng/ml for T3 (Khazali 2009; Varshney et al. 1984; Agarwal et al. 1986). In Bactrian camel, no significant sex effect was reported: 94.1 ± 12.2 in male vs 115.9 ± 11.6 in non-pregnant female (Omidi et al. 2014a).

Mean values in adult male camel were reported at 122.7 ± 4.1 and 1.67 ± 0.71 ng/ml for total T4 and T3, respectively (Nazifi et al. 2009). The absence of sex effect was confirmed later on by Tajik (2013) and Tajik et al. (2013) for T4 (132.7 ± 5.9 vs 136.3 ± 5.9 ng/ml in male and female, respectively) and for T3 (2.3 ± 0.1 vs 2.4 ± 0.3 ng/ml). Elkhasmi et al. (1999) stated that the basal values of free plasma thyroid hormones in adult male and female camels were also similar: 19.2 ± 1.9 and 18.8 ± 2.3 pg/ml, respectively, for fT4, and 5.6 ± 0.9 and 5.5 ± 0.8 pg/ml for fT3. In Bactrian camel, no difference was found also for free T4 (1.67 ± 0.2 vs 1.65 ± 0.18 pg/ml in male and non-pregnant camel, respectively) and free T3: 0.68 ± 0.04 and 0.63 ± 0.04 pg/ml, respectively (Omidi et al. 2014b).

However, castration induced a slight significant decrease of T4 and T3 concentrations: from 131 ± 17 to 115 ± 13 ng/ml and from 1.07 ± 0.14 to 0.89 ± 0.13 ng/ml, respectively (Bengoumi et al. 1999).

9.3.1.1 Seasonal Effect

Several publications have been interested in the seasonal effect on thyroid activity, but the results are contradictory. Contrary to T3, T4 concentrations appeared significantly higher at the non-breeding season (summer) with a mean level of 100 ng/ml in summer and 80 ng/ml in winter. For T3, values varied between 1.1 and 1.25 ng/ml according to season without significant effect (Yagil et al. 1978). Similar seasonal difference was reported in Iran (Nazifi and Abassali 1998). Higher summer values of T4 were also reported by Nazifi et al. (1999): 74.8–81.8 ng/ml between December and February while it was 128.7–137.6 ng/ml between June and August. For T3, values were 1.24–1.30 ng/ml in winter and 1.69–1.72 in summer.

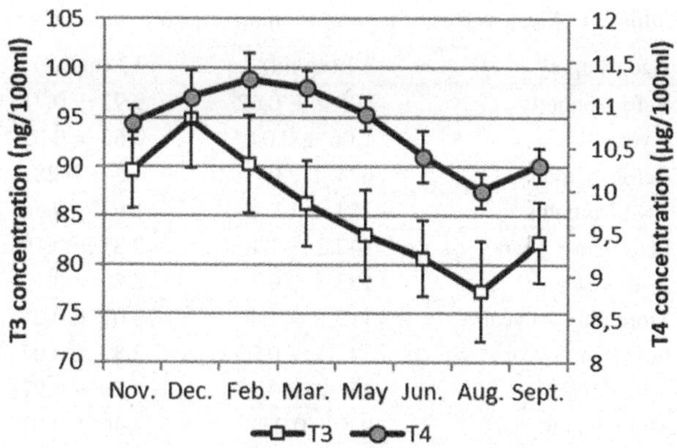

Fig. 9.6 Seasonal changes in plasma T3 and T4 concentrations in castrated camels all along the year (calculated from after Bengoumi et al. 2003)

This seasonal effect is confirmed in Morocco (Bargaâ et al. 2016) where plasma thyroxine and triiodothyronine were reported higher in winter than in summer: 210.8 ± 5.0 vs 120.7 ± 2.6 ng/ml and 2.63 ± 0.2 vs 1.0 ± 0.2 ng/ml, respectively. Higher values in winter were also reported by Tajik et al. (2013): 169.9 ± 7.0 in winter vs 89.8 ± 5.3 ng/ml in summer for T4 with similar pattern for T3, 2.61 ± 0.14 vs 1.95 ± 0.15 ng/ml, respectively. Usually, cold weather increases secretion of thyrotrophic hormones as it has been observed in other mammals. Bengoumi et al. (2003) have found a significant negative correlation between temperature gap (difference between morning and evening body temperature of camels) and T3 concentrations (Fig. 9.6). Moreover, histomorphometric studies of the thyroid gland seem to show significant seasonal differences (Abdel-Magied et al. 2000), the activity index and the thyroid size being higher in winter in relationship with a higher thyroid hormone secretion (probably free hormones): 13.4 ± 0.2 in winter vs 11.3 ± 0.2 pg/ml in summer for T4 and 0.12 ± 0.01 in winter vs 0.08 ± 0.01 pg/ml in summer for T3 (Rejeb et al. 2011).

At reverse, Zia-Ur-Rahman et al. (2007) found in male adult camel that concentrations of T4 and T3 are double during non-rutting season compared to rutting one, i.e., in winter: 116.7 ± 9.8 vs 84.1 ± 7.5 ng/ml for T4 and 2.06 ± 0.48 vs 0.45 ± 0.08 ng/ml for T3 (Zia-Ur-Rahman et al. 2007). This difference was attributed by the authors to the lower body metabolism of male during rutting season in relation with a lower food intake.

9.3.1.2 Effect of Age

The effect of age is complex and many references are contradictory. If most of the authors found higher concentrations in young camel (Table 9.1), reverse observations were reported in the literature (Afifi et al. 1979; Heshmat et al. 1984).

In baby camel at birth, free T4 (39.3 ± 3.2 pg/ml) is higher than in fetus at sixth month gestation (33.6 ± 3.2 pg/ml) as well as free T3 (respectively, 1.8 ± 0.1 and

Table 9.1 Mean values of thyroid hormones according to different age categories in the literature

Age category	T4 (ng/ml)	T3 (ng/ml)	Reference
Before puberty	20.12 ± 0.62	9.92 ± 0.71	Towhidi et al. (2001)
4 years	1.06 ± 0.04	0.62 ± 0.02	
Before 1 month	674 ± 23	4.5 ± 0.28	Bengoumi et al. (1999)
14–17 months	164 ± 12	1.63 ± 0.56	
Less than 2 years	142.2 ± 1.1	2.35 ± 0.19	Tajik et al. (2013)
2–10 years	133.6 ± 0.7	2.33 ± 0.12	
More than 10 years	112.7 ± 1.4	2.07 ± 0.2	
Less than 3 years	14.31 ± 0.23	2.82 ± 0.07	Rejeb et al. (2011)[a]
3–5 years	15.82 ± 0.23	3.70 ± 0.07	
More than 5 years	9.8 ± 0.23	2.46 ± 0.07	
Up to 5 years	92.47 ± 3.82	1.09 ± 0.05	Agarwal et al. (1989b)[b]
5–10 years	94.2 ± 2.4	1.09 ± 0.05	
More than 10 years	94.5 ± 3.0	1.09 ± 0.05	
At parturition	70	1.6	Agarwal et al. (1992)
3 weeks *postpartum*	110	2.2	
Baby at parturition	339 ± 48	–	El Khasmi et al. (2001)
Dam at parturition	54.6 ± 20	–	

[a]Free hormones
[b]Pregnant females

0.91 ± 0.3 pg/ml). After birth, fT4 concentrations decline regularly up to the 10th day at around 18.6 ng/ml, while fT3 increases significantly 24 h after parturition up to 11.8 pg/ml and then declines regularly at the 10th day postpartum to reach a plateau at around 6.5 pg/ml (El-Khasmi et al. 1999). In comparison, the values in the dam were lower at the sixth month of pregnancy compared to the fetus (respectively, 24.1 ± 2.2 and 10.9 ± 1.3 pg/ml for fT4 and fT3) and in the first days of postpartum: on average 8.5 and 1.7 pg/ml for fT3 and fT4. The thyroid hormones were in similar concentrations in the young camels and their dam from the 20th day postpartum, i.e., 17.8 and 5.6 pg/ml, respectively (El-Khasmi et al. 1999).

9.3.1.3 Changes During Pregnancy

Thyroid hormones play an important role in the modulation of metabolism during pregnancy, and any deficiencies could provoke different disorders affecting the fetus. On average, the difference in T4 and T3 concentrations between pregnant and lactating camels is not big: 97 ± 29 vs 82 ± 10 ng/ml for T4 and 0.86 ± 0.32 vs 0.68 ± 0.17 ng/ml for T3 (Bengoumi et al. 1999). In another recent reference, it is stated that pregnant camel has lower values for T4 (148 ± 24 vs 179 ± 19 ng/ml in non-pregnant) but not for T3 (Ahmed et al. 2016).

On average, values of T4 and T3 are high at the first month of pregnancy and then decline up to the 10th month before rising (not significantly) in the last period of

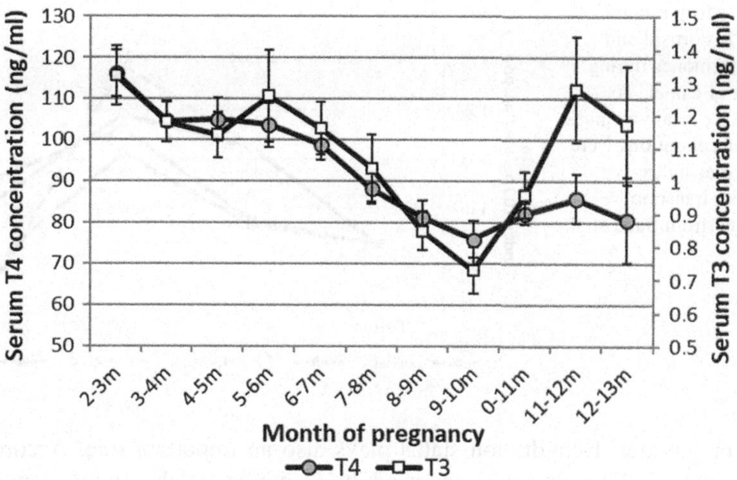

Fig. 9.7 Changes in total thyroid hormones T4 and T3 during camel pregnancy (calculated from Agarwal et al. 1989b)

pregnancy (Fig. 9.7). There was no difference between male and female fetus (Agarwal et al. 1989b).

Comparing female camels at the last trimester of pregnancy with non-pregnant camels, no significant differences were reported for any thyroid hormones except a slight difference for free T4 (Omidi et al. 2014a): the values were in pregnant and non-pregnant camels 110.7–128.7 ng/ml and 114–125.9 ng/ml, respectively, for T4; 0.02–0.03 and 0.03–0.04 ng/ml for T3; 8.98–11.98 and 11.47–13.97 pg/ml for fT4; and 0.037–0.051 and 0.032–0.045 pg/ml for fT3.

Similar results were reported on Bactrian camel where no significant difference was observed between pregnant and non-pregnant female, except for fT3 (Omidi et al. 2014b). Abdel-Wahab et al. (1974) did not find also a significant increase of T4 concentrations during pregnancy (99 ± 0.3 ng/ml) in comparison to the different phases of the estrus cycle (51.2–130.7 ng/ml).

9.3.1.4 Effect of Stress

Stress induces generally an increase of activity of hypothalamus-hypophysis-thyroid axis because of the increase of energy demands. Such, a high correlation between thyroid hormones and cortisol has been described in camel (Bargaâ et al. 2016). In camels transported before slaughtering, the increase of thyroid hormone concentrations is correlated with the duration of transport and with cortisol (Saeb et al. 2010) as shown in Fig. 9.8.

Heat stress is also responsible of the increase of thyroid hormones in the blood as suggested by Nazifi et al. (1999) and Yagil et al. (1978) for explaining the higher

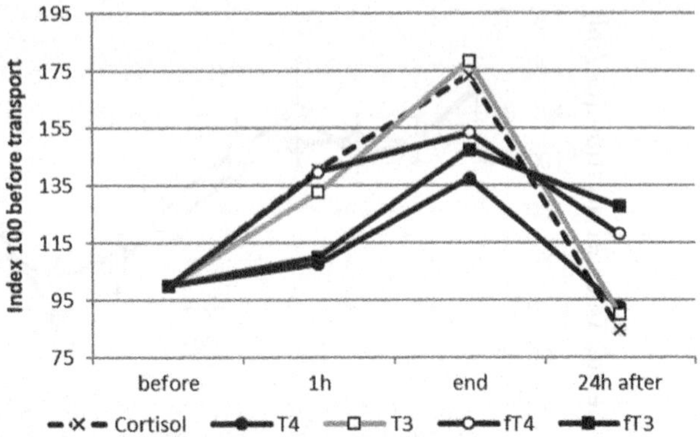

Fig. 9.8 Mean relative changes in cortisol and thyroid hormones during transport of camel. To compare the relative values, all the concentrations were calculated as index 100 before transport (calculated from Saeb et al. 2010)

values in summer. Dehydration status plays also an important role. According to Yagil et al. (1978), the thyroid gland is inhibited in case of dehydration, contributing to water preservation in the body by reducing basic metabolism. This mechanism is more efficient during summer time as suggested by Khanna et al. (1996). However, there is no apparent effect of low-energy diet on the thyroid hormone concentrations (Khazali 2009).

The maximal exercise has also a significant effect. Plasma T3 concentrations increased from 0.75 ± 0.07 ng/ml before exercise to 1.06 ± 0.11 ng/ml at the end of exercise (4 km race at 22.6 km/h) and continue to increase up to 1.21 ± 0.2 ng/ml 3 h after the end of exercise. For T4, the values were, respectively, 82 ± 3 and 87 ± 3.5 ng/ml before and at the end of exercise, and values continued to increase up to 130 ± 9 ng/ml 12 h after the exercise (Moutouakil and Bengoumi 2007).

9.3.1.5 Effect of Health Disorders

Obviously, plasma thyroid hormone concentration is a good marker of the thyroid disorders as goiter due to iodine deficiency. In iodine-deficient localities from Sudan, thyroid hormone concentrations in camel were effective slightly (Barsham et al. 2015). In induced hypothyroidism by sodium thiocyanate, T4 concentration decreased regularly passing from a mean of 56.5 after 1 month of experiment to 43.8 ng/ml after 3 months. For T3, mean values were, respectively, 0.28 and 0.11 ng/ml (Barsham et al. 2015). Free thyroid hormone concentrations also decrease in case of goiter: from 12.8 to 5.2 pg/ml for fT4 and 3.4 to 0.97 pg/ml for fT3 (Mohamed et al. 2006). In camels affected by clinical goiter, Abu-Damir et al. (1990) reported low T4 concentrations (88.9 ± 37 ng/ml) compared to camels affected by subclinical goiter (119.9 ± 18 ng/ml) and overall healthy camels (138.0 ± 23 ng/ml). For T3, the values were, respectively, 12.8 ± 3.9, 12.2 ± 2.5, and 20.1 ± 9.3 ng/ml.

In case of inactive ovary, T4 concentration increase considerably up to 237.6 ng/ml (Abdel-Wahab et al. 1974). In camels infected with larvae of *Cephalopina titillator*, serum T4 concentration is divided by two and serum T3 by four

(Abdelrahman 2010), but the reported values in healthy camels appeared quite different than the other references: 152 µg/ml for T4 and 126 ng/ml for T3. Similar pattern was published formerly by El-Bassiony et al. (2005) but also with quite higher levels than other references: T4 declined after infection by *Cephalopina titillator* from 116.2 to 88.8 µg/ml and T3 from 120 to 25.8 ng/ml.

Several studies have mentioned also an effect of infection with *Trypanosoma evansi* (Al-Qarawi et al. 2001b). A decrease is observed in infected camels. For example, in Sudan, normal levels were reported to be in the range of 38.8–46.5 ng/ml for T4 and 0.62–0.99 ng/ml for T3, while the values were 25.6–39.5 and 0.19–0.78 ng/ml, respectively, in infected camels, the treatment with Cymelarsan improving these concentrations after 1 month post-injection (Abdelsalam et al. 2003). The T4 concentration decreased from 123.3 ± 3.3 in healthy camels to 104.7 ± 7.2 ng/ml in infected ones and T3 from 3.49 ± 0.24 to 2.42 ± 0.14 ng/ml, respectively (Sazmand et al. 2011).

9.3.1.6 Thyroid Hormones in Other Substrates

T4 camel milk is low at parturition (2.6 ± 0.8 ng/ml) and increases up to day 7 and then remains stable at least up to 3 weeks between 5 and 6 ng/ml (El Khasmi et al. 2001).

In **camel semen**, in two consecutive ejaculates, T4 activity was, respectively, 27.7 ± 7.0 and 54.0 ± 20.0 ng/ml, while T3 activity was 1.06 ± 0.5 and 1.46 ± 0.04 ng/ml (Zia-Ur-Rahman et al. 2001). Moreover, these concentrations were higher in semen collected from young camels compared to old ones.

9.3.2 Thyroid-Stimulating Hormone

In camel, some references are available for TSH concentration of TSH in blood (Table 9.2), but the variability in the reported values cannot allow their convenience for biological investigations.

9.4 Other Hormones

9.4.1 Melatonin

Melatonin is produced by the pineal gland in animals and is responsible of the regulation of nycthemeral cycle and its relation with seasonal reproduction in many mammals.

The first publication reporting dosage of melatonin in camel was that of Vyas et al. (1997). Concentration was low during the period of light (5 pg/ml) and

Table 9.2 Reference values in camel TSH according to different authors

Reference	Camel status	Values	Country
Bengoumi et al. (1999)	All age and physiological status	Not detectable	Morocco
El-Bassiony et al. (2005)	Healthy	14 ± 3 μIU/l	Egypt
	Infested *C. titillator*	12 ± 2 μIU/l	
Mohamed et al. (2006)	Healthy	3 ± 1 μIU/l	UAE
	With goiter	10 μIU/l	
Abd El-Rahman (2010)	Healthy	18 ± 3 IU/l	Libya
	Infested *C. titillator*	12 ± 2 IU/l	
Rejeb et al. (2011)	<3 y	140 ± 0.01 μIU/l	Tunisia
	3–5 y	250 ± 10 μIU/l	
	>15 y	90 ± 10 μIU/l	
	Male	180 ± 10 μIU/l	
	Female	140 ± 10 μIU/l	
	Winter	190 ± 10 μIU/l	
	Summer	130 ± 10 μIU/l	
Omidi et al. (2014a)	Pregnant	10 ± 1 μIU/l	Iran[a]
	Non-pregnant	5 ± 1 μIU/l	
	Male	10 ± 1 μIU/l	
Omidi et al. (2014b)	Pregnant	300–500 μIU/l	Iran
	Non-pregnant	375–525 μIU/l	
Ahmed et al. (2016)	Pregnant	15 ± 1 μIU/l	Egypt
	Non-pregnant	15 ± 0.1 μIU/l	

[a]Bactrian camel

increased in period of darkness with peak varying from one camel to another from 50 to 233 pg/ml within 15 min of darkness. A peak amplitude between 20 and 200 pg/ml was also reported by El-Allali et al. (2005) during night, while no difference was observed between camels during the day, where melatonin values were under the limit of detection (10 pg/ml). This basal daylight value was not influenced by season contrary to the onset of melatonin secretion after darkness being significantly delayed of approximately 2 h during summer solstice as compared with the winter solstice. An effect of age was also observed, the peak of melatonin being higher in young animals than in adults. Comparing immature, non-rutting and rutting camels, Al-Qarawi and El-Mougy (2008) reported also higher peaks in young camel (231.4 ± 2.36 pg/ml) than adults with significant difference at rutting (36.44 ± 1.3 pg/ml) and non-rutting season (102.36 ± 4.6 pg/100 ml).

Moreover, it has been demonstrated in camel that ambient temperature may have an effect also on the circadian rhythm, but the range of melatonin concentrations in plasma still varies between less than 10 pg/ml at the basal level and 50–200 pg/ml at the peak of secretion (El-Allali et al. 2013).

9.4.2 *Insulin*

Insulin is a peptide hormone produced by beta cells of the pancreatic Langerhans islets (β-cells) playing a pivotal role in the regulation of glucose by stimulating its absorption from the blood into fat, liver, and skeletal muscle cells. Its secretion is stimulated by the increase of glucose in the bloodstream. At reverse, hypoglycemia inhibits secretion of insulin. Reversely, glucagon, another peptide hormone secreted by α-cells of Langerhans islets in the pancreas, has hyperglycemic function by stimulating glycogenesis in the liver. Insulin is the only hypoglycemic hormone.

In camel, few references are available on blood insulin, but due to the effect of camel milk on glycemia regulation in diabetic man, several investigations were achieved to determine insulin in milk. In an experiment to test the effect of **glucagon** injection on blood glucose, the basal value in blood was determined at 0.85 ± 0.11 ng/ml (Abdel-Fattah et al. 1999).

9.4.2.1 Concentrations in Plasma

The basal concentration of camel plasma insulin (5 ± 1 μU/ml, 173.5 ± 34.7 pg/ml) is lower than in sheep (Elmadhi et al. 1997), ponies, and pig (Kaske et al. 2001). Slightly higher references in 138 camels in Tunisia were 6.94 ± 0.45 μU/ml (Souilem et al. 1999). After glucose administration, the peak plasma insulin decreases considerably lower in camels (19 ± 4 μU/ml) than in sheep (48 ± 6 μU/ml) and ponies (43 ± 19 μU/ml). Basal level is reached after 4 h in camels (Fig. 9.9) and ponies vs 1 h and 40 min in sheep (Elmadhi et al. 1997).

Compared to other herbivores, insulin response of camel is less responsive which would explain the relative higher glycemia observed in this ruminant species.

In a monitoring over a period of 310 days after parturition, Wernery et al. (2006a) reported serum insulin values between 5 and 30 μU/ml with a mean of 12.77 ± 7.62 μU/ml, the higher values occurring after 4 months of lactation, i.e., after the peak of lactation.

Surprisingly, quite higher value of plasma insulin in immature camels, with an average of 3 ng/ml (i.e., 86 μU/ml), was reported (Rashedi and Khazali 2010). In several trials on insulin challenge and glucose tolerance test in suckling camel calves

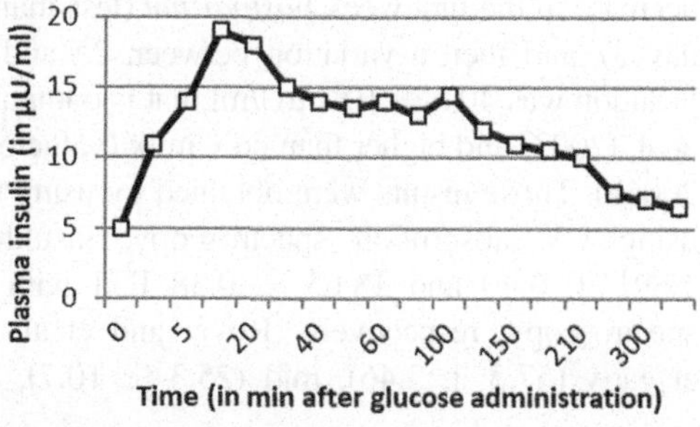

Fig. 9.9 Effects of an intravenous glucose load on plasma concentrations of insulin in camel (from after Elmadhi et al. 1997)

(Diaz-Medina 2017), the basal values of plasma insulin were 4.67 ± 2.5 μU/ml, and then a peak of 1720 ± 806 μU/ml after injection of 4.6 μg/kg BW of insulin was observed in calves of 15 days old. This peak reached 7377 ± 1710 μU/ml in calves at 132 days old. Values returned to baseline after 20 min. During glucose test (injection of 0.25 g/kg BW of glucose), insulin in plasma passed from 4.03 ± 0.7 to 35.7 ± 3.5 at the peak in youngest calves (15 days) and 11.3 ± 3.4 μU/ml in the oldest (132 days) showing a decreasing responsiveness to induced hyperglycemia with the age.

Depending on season, age, and gender, Al-Suhaimi et al. (2009) reported range of camel plasma insulin between 11.0 ± 0.01 and 75.6 ± 0.08 μU/ml. Concentrations were higher in summer than in winter for male, but it was the reverse for female. With an average of 5.8 ± 1.4 μU/ml, no clear impact of dehydration was observed (Siam et al. 1993).

Plasma insulin concentration can be modulated by the energy level of the diet: in camel fed with low-energy diet (25% of the normal diet) and receiving injection of galanin, an orexigenic peptide, plasma insulin concentrations decreased from 45 to 25 ng/ml with 1 μg/kg LW galanin injection and from 25 to 16 ng/ml with 2 μg/kg LW injection (Khazali 2009).

9.4.2.2 Concentration in Milk

The traditional belief in the "camel countries" is that regular consumption of camel milk helps in prevention and control of diabetes, and many papers were published to assess this effect on rat (Ali Khan et al. 2013), dog (Sboui et al. 2010), or human diabetic patients (Agrawal et al. 2002). It has been proved that camel milk could increase serum insulin concentration in patients (Ejtahed et al. 2015).

In consequence, several authors investigated the potential antidiabetic agents present in the camel milk. It has been supposed that insulin was in higher quantity in camel milk than other milks. Yet, the total quantity was around 50 μU/ml, comparable to sheep milk, but higher than cow milk (30 μU/ml). However, camel colostrum contained less insulin (around 130 μU/ml) than cow colostrum (400 μU/ml) or sheep colostrum (350 μU/ml) (Zagorski et al. 1998). For other authors, camel milk contains higher quantity of insulin: 150 IU/l (Agrawal et al. 2002). In a measurement over one lactation, Wernery et al. (2006a) observed a high value after parturition (400 μU/ml), a rapid decrease in the first week *postpartum* (less than 200 μU/ml at day 2 and 100 μU/ml at day 3), and then a variation between 25 and 75 μU/ml. The average all along the lactation was 40.5 ± 10.7 μU/ml that is comparable to the 52 μU/ml cited by Agrawal et al. (2003) and higher than cow milk (value of 23 μU/ml was cited by Wernery et al. 2006b). Those results were obtained by using radioimmunoassay (RIA). Recently, by using UV absorption spectroscopy, insulin concentration in camel milk was 17.91 ± 0.40 and 18.65 ± 0.38 IU/l with standard addition method and direct spectroscopy, respectively (Royatvand et al. 2013). Insulin in milk was measured at early (57.3 ± 3.46), mid (25.3 ± 10.7), and late lactation (47.26 ± 9.8 μU/l)

(Diaz-Medina 2017). Such lactational change is probably linked to dilution effect, the lowest concentration being observed at the peak of milk production.

Insulin concentration in milk decreases with pasteurization and storage at different cold temperatures (Wernery et al. 2006b): starting from 41.9 ± 7.38 μU/ml in raw milk, insulin concentration decreased by 7% after pasteurization and 13% after storage at 4 °C, while freeze-drying and boiling resulted in 19% and 26% reduction, respectively.

Later on, it was supposed that camel insulin had different structure leading to be resistant to stomach enzymes. Yet, camel insulin differs from human insulin by four mutations and from bovine and buffalo by just one mutation. None of the mutations affect specificity toward digestive enzymes. Thus, when camel insulin comes in contact with the proteases of digestive track, it should be digested like other mammalian insulin unless otherwise protected (Malik et al. 2012), for example, in casein micelles and such protected from digestion and proteolysis in the upper gastrointestinal tract: it is encapsulated in nanoparticles that facilitate its absorption (Mullaicharam 2014; Kula and Tegegne 2016).

9.4.3 Leptin

Leptin is known as hormone of satiety secreted by adipocytes. This peptide contributes to the regulation of appetite by inhibiting hunger. Leptin has an opposite effect of the hormone ghrelin, named the hunger hormone. Both hormones act on the hypothalamus to ensure energy homeostasis.

In camel, leptin was assessed in relationship with hump size and dynamics (Faye et al. 2001; Bengoumi et al. 2005). Camel leptin and its receptors have been characterized in the adipose tissue, mammary gland, and liver (Sayed-Ahmed et al. 2005). At our knowledge, there was no reference on ghrelin hormone in camel except one assessing the effect of its injection on insulin and ACTH concentration (Rashedi and Khazali 2010).

In experiment with sequences of underfeeding and overfeeding for several days, camel plasma leptin varied between 2.74 and 5.44 ng/mL on average with maximum concentration at 7.95 ng/ml (Delavaud et al. 2013). As previously observed in other ruminant species, plasma leptin concentration in camels is modulated by energy feeding level: when energy intake decreased from 134% to 17% of maintenance energy requirement, it induced a significant decrease in plasma leptin by 28%. In another experiment with dehydration/rehydration period sequence achieved by the same authors, values varied between 1.71 and 3.35 ng/mL with a maximum value at 8 ng/ml. Leptin concentration decreased by 17% in dehydrated camels, showing that dehydration specifically decreased plasma leptin.

In young camels slaughtered at different ages between 2–3 and 10–11 months, leptin concentrations in plasma increase regularly from 1.1 to 3.1 mg/ml (Fig. 9.10) and were positively correlated with back-fat thickness (Al-Azraqi 2007), but

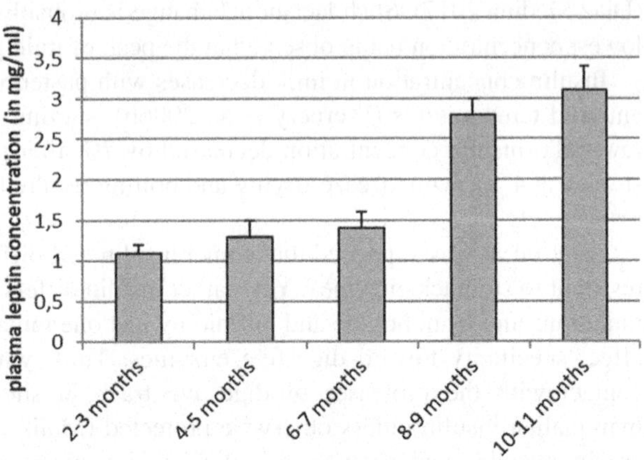

Fig. 9.10 Changes in plasma leptin concentrations according to age at slaughtering of young camels (from after Azraqi 2007)

surprisingly, concentration is expressed in mg/ml and not in ng/ml like for the previous reference.

A positive exponential correlation was also reported between leptin concentration in plasma and the mean volume of adipocytes from hump biopsies, the mean leptin concentration being 3.98 (2.0–8.3) ng/ml (Delavaud et al. 2004).

Different seasonal variations were reported according to the sex and age of camel. Leptin concentration in plasma of male camel was higher in summer than winter for young camel 1–3 years old (2.14 ± 0.44 vs 1.41 ± 0.48 ng/ml, respectively), while there was no difference for young camels less than 1 year (2.15 ± 0.56 vs 2.31 ± 0.94 ng/ml) and for adult (2.4 ± 0.62 vs 2.77 ± 0.86 ng/ml). In female, values were lower in summer than in winter for young camel (2.46 ± 0.72 vs 3.8 ± 0.17 ng/ml) and overall young camel 1–3 years old (1.22 ± 0.05 vs 4.73 ± 0.55 ng/ml), but not for adult (1.59 ± 0.56 vs 1.13 ± 0.19 ng/ml). However, fat condition of the animals was not reported (Al-Suhaimi et al. 2009).

9.4.4 Hormones for Regulation of Phosphocalcic Metabolism

This group of hormones included calcitonin and parathyroid hormone. Calcitonin (also named thyrocalcitonin) is a polypeptin hormone secreted by the thyroid gland. Calcitonin regulates calcium and phosphorus, contributing to reduce blood calcium and phosphorus at the opposite of parathyroid hormone or parathormone (PTH), also secreted by thyroid. Calcitonin acts by inhibiting osteoclast activity in bones and increasing urine excretion of calcium and phosphate. PTH, also polypeptide hormone, contributes to bone modeling and to maintain blood calcium by withdrawing it from bones. It is used as biomarker of the bone formation process and was suspected to play a role in bone disorders as Krafft disease (Mabrouk et al. 2010).

In camel, the injection of PTH or PTH-rp (PTH-related peptide) is accompanied by an increase of calcium and phosphorus excretion in milk (Riad 1995; Riad et al.

1995). In young camel, injection of PTH-rp stimulates intestinal absorption (El-Khasmi et al. 2000a).

9.4.4.1 Calcitonin

Few date on calcitonin in camel are available. Calcitonin being linked to calcium and phosphorus metabolism, it could be used to diagnose bone disorders as osteomalacia (Adeghate and Pallot 1996).

Calcitonin concentration in camel serum was 80.53 ± 10.8 pg/ml in healthy males, 129.33 ± 8.8 pg/ml in non-pregnant females, and 168.34 ± 5.8 pg/ml in pregnant females. The higher level in females is linked to the influence of steroid hormones as progesterone and estradiol. In case of skin wounds, concentration increased up to 203.4 ± 6.9 pg/ml (Kataria and Kataria 2004a). No relationship with bone fracture in racing camel was observed (Alshamsi et al. 2015).

9.4.4.2 Parathyroid Hormone

Average usual values of **PTH** in camel serum are 1.94 ± 0.03 ng/ml. It is slightly but significantly higher in pregnant (2.10 ± 0.02 ng/ml) and non-pregnant females (1.90 ± 0.05 ng/ml) than in males (1.81 ± 0.03 ng/ml). In case of drought, affected camels presented higher values up to 2.87 ± 0.04 ng/ml (Kataria and Kataria 2004b).

Plasma PTH concentration is not affected by stress linked to road transportation (3.5–4.1 ng/ml) (El Khasmi et al. 2011b).

9.4.4.3 Osteocalcin

In healthy camel, the values in plasma vary between 1.5 and 47 pg/ml, and no effect of water restriction was observed (Bengoumi et al. 1996). The values reported later by El-Khasmi et al. (2000b, 2005) were quite higher and expressed in ng/ml. The values were higher in newborn camel calf than their dams at birth (3.4 ± 0.3 vs 0.7 ± 0.3 ng/ml) and remained different until 30th day postpartum (5.2 ± 0.5 vs 0.7 ± 0.3 ng/ml). In the dams, the concentration increased up to 2.3 ± 0.5 on the fourth day postpartum and declined to the initial value at the seventh day postpartum. These values are lower than in cows (Davicco et al. 1992). There was no significant change between prepartum and parturition period (Tharwat and Al-Sobayil 2015) as well as in racing camel before and after race: 27.34 ± 12.5 ng/ml at rest and 30.31 ± 14.3 ng/ml after race (Al-Sobayil 2008).

A circadian rhythm of osteocalcin has been reported (Al-Sobayil 2010). The mean concentration of osteocalcin over the 24 h period was 29.16 ± 4.83 ng/ml with minimum concentrations at 13:00 h (22.5 ± 3.1 ng/ml) and maximum at 18:00 h (41.4 ± 3.5 ng/ml).

The plasma concentration of osteocalcin is lower in racing camel affected by bone fracture compared to healthy animals: 0.7 ± 0.5 ng/ml vs 1.8 ± 1.1 ng/ml (Alshamsi et al. 2015).

9.4.5 Hormones for Regulation of Water and Main Electrolytes

This group included aldosterone and vasopressin.

Aldosterone is a steroid hormone secreted by the adrenal cortex of the adrenal gland. It is the main mineralocorticoid hormone implicated in the regulation of sodium in plasma and potassium in extracellular compartment. It contributes to the reabsorption of sodium and excretion of potassium in the kidney and consequently influences indirectly water retention or loss, blood pressure, and volume. These functions are opposite to **atrial natriuretic factor** secreted by the heart. Regulation of aldosterone production is complex and involves stimulating (angiotensin II, ACTH, ions K) and inhibiting factors (ANF, dopamine). Aldosterone is part of the renin-angiotensin system which plays the main role in the maintenance of blood pressure and water/electrolytes homeostasis.

Vasopressin, also named antidiuretic hormone (ADH) or arginine vasopressin (AVP), is a peptide hormone stored in the posterior pituitary gland and released in the bloodstream (Adamsons et al. 1956; El-May et al. 1987). Its main role is to regulate the retention of body water and indirectly salts in blood and increase peripheral vascular resistance. Plasma ADH concentration increases water reabsorption in kidney's nephrons.

Plasma half-life of these two hormones is short (around 20 min). Due to the adaptation of camel to water restriction and rapid rehydration, the role of those hormones in water regulation in camel in desert condition has been regularly investigated for long time (Finberg et al. 1978; Yagil and Etzion 1979). These authors suggested that ADH, besides its water reabsorptive function in the kidney, initiates aldosterone secretion which then increases absorption water and salt in the colon. When camels are rehydrated, aldosterone and ADH secretion decreases, but to prevent hemodilution, aldosterone increases 24 h following drinking (Yagil and Etzion 1979).

9.4.5.1 Aldosterone

References regarding aldosterone concentration in camel plasma involved two groups of publications. A first group reported that basal values in normal hydrated camel vary between 2 and 6 ng/ml, and a second group, expressing the values in pg/ml, reported values between 10 and 60 pg/ml. Yet, the same method for determination (radioimmunoassay) was used.

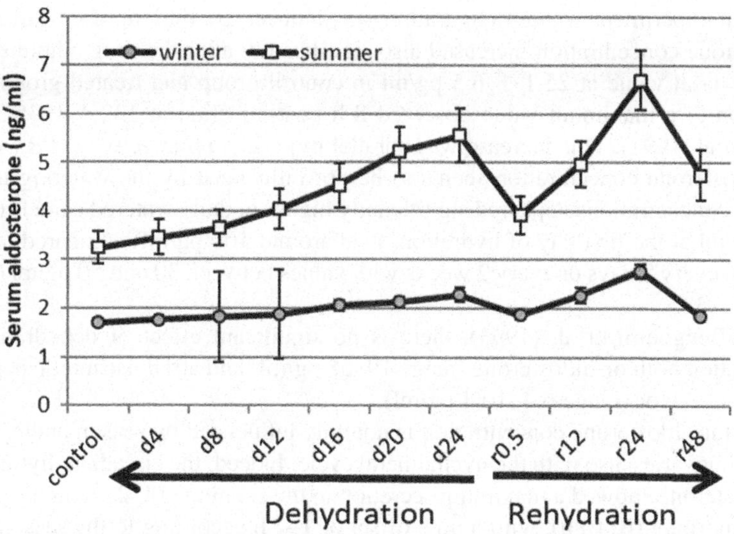

Fig. 9.11 Changes in aldosterone concentrations in camel plasma during dehydration/rehydration cycle (from after Kataria et al. 2000). The values are expressed in ng/ml

For Yagil and Etzion (1979), the normal values in hydrated camels were 2.7 ± 0.52 ng/ml in winter, 2.8 ± 0.52 in spring, and 3.9 ± 0.74 in summer. In dehydrated camel, these values were, respectively, 1.2 ± 0.22, 4.5 ± 1.27, and 6.3 ± 1.6 ng/ml. After 24-h rehydration, the values were 8.0 ± 2.48 in spring and 21.4 ± 2.04 ng/ml in summer. Similar figure was reported later by Kataria et al. (2000), aldosterone concentrations varying from 1.71 ± 0.08 ng/ml in winter and 3.25 ± 0.033 ng/ml in summer in normal watered camels.

After 24-day dehydration, values were 2.28 ± 0.14 ng/ml in winter, and after 13-day dehydration in summer, it reached 5.56 ± 0.55 ng/ml. These values were still high after 48-h rehydration: 1.85 ± 0.093 and 4.75 ± 0.52 pg/ml in winter and summer, respectively (Fig. 9.11).

However, in a more recent study, quite lower values (expressed in pg/ml and not in ng/ml as previously) were observed: serum aldosterone increased fivefold after 1-week dehydration passing from 42.4 ± 19.4 pg/ml to 191.8 ± 52.3 pg/ml and remained at high level after 48 h of rehydration, 151.3 ± 19.4 pg/ml (Abdoun et al. 2010).

Similar values were reported also by Al-Qarawi (1997): from basal value at 30.4 ± 1.6 to 137.4 ± 3 pg/ml after 15-day dehydration and then increased up to 220.0 ± 5.0 pg/ml within 30 min post-rehydration before returning to basal value after 4 h. After 20-day dehydration, aldosterone concentration passed only from around 16 to 26 pg/ml in the experiment of Alhaj Ali et al. (2012), and no significant difference was observed with camel treated with losartan (receptor blocker of aldosterone and vasopressin).

In an experiment where dehydration was induced by diuretic drug furosemide, aldosterone concentration increased also significantly after the drug administration: from a basal value at 25.1 ± 6.5 pg/ml in control group and treated group before injection to a maximum value observed 8 h post-injection at 132.4 ± 35.5 pg/ml (Riad et al. 1994). This increase was parallel to plasma renin activity (PRA).

Aldosterone concentration seems to be not influenced by the watering interval, but the concentrations appeared significantly higher in daily watered camels (around 540 pg/ml at the first day of hydration, then around 100 pg/ml) compared to camel watered every 8 days or every 2 weeks with values between 30 and 50 pg/ml (Bekele et al. 2013).

For Bengoumi et al. (1993), there is no significant effect of dehydration and rehydration both on aldosterone (range 19–58 pg/ml) and atrial natriuretic peptide—ANP—or factor (range 6.4–10.4 pg/ml).

Plasma aldosterone concentration is not only influenced by season and hydration status, but varies also with the nycthemeral cycle. Indeed, the circadian rhythm study of aldosterone showed a maximum secretion in the evening (18–22 h) and minimum in the morning (6–10 h), with a time offset of 1–2 h according to the season. Mean values range from 14 to 25 pg/ml (i.e., again expressed in pg and not in ng/ml) with amplitude of 5.6 ± 1.3 pg/ml in October, 7.82 ± 1.41 pg/ml in December, 8.83 ± 1.12 pg/ml in March, and 7.21 ± 1.25 pg/ml in June (Khaldoun et al. 2002).

An effect of physical effort was also described, aldosteronemia increasing from around 25 to a maximum of 66.4 ± 6.5 pg/ml after 4 km racing (Riad 1995). A significant higher value is observed also in pregnant camel compared to non-pregnant: 65.1 ± 8.2 vs 25.8 ± 3.4 pg/ml (Riad 1995).

9.4.5.2 Vasopressin (ADH)

In most of the cases, vasopressin is determined in parallel to aldosterone. Usual values in normal hydrated camel were around 0.1 to 1 pg/ml: 1.17 ± 0.45 (Yagil and Etzion 1979), 0.3 ± 0.2 (Bengoumi et al. 1993), 1.7 ± 0.2 (Benlamlih et al. 1993), 0.2 ± 0.1 (Riad et al. 1994), 0.18 ± 0.07 (Riad 1995), 0.7 ± 0.0 (Al-Qarawi 1997), 1.0 ± 0.2 (Alhaj Ali et al. 2012), and 1.0 ± 0.3 pg/ml (Bekele et al. 2013).

After dehydration, ADH increases significantly in plasma, 5.16 ± 1.82 pg/ml, and decreases rapidly after 1-h rehydration down to 0.21 ± 0.11 ng/ml (Yagil and Etzion 1979). This increase occurs from the fourth day of water privation up to 5.3 ± 2.2 pg/ml and remains elevated until the 13th day of privation (5.7 ± 2.2 pg/ml). It decreases rapidly 2 h after rehydration (2.2 ± 1.3 pg/ml) and reaches the basal values after 7 days rehydration at 0.6 ± 0.3 pg/ml (Bengoumi et al. 1993). Plasma ADH is correlated with plasma and urine osmolarity.

After 15 days dehydration, Al-Qarawi (1997) reported ADH concentration in camel plasma at 7.4 ± 0.2 pg/ml and returning to basal level 4 h after rehydration.

Injection of diuretic drug (furosemide) provokes a significant increase of ADH, 4 h after injection from 0.2 to 0.8 pg/ml (Riad et al. 1994).

As for aldosterone, ADH increased in case of physical effort passing from 0.18 ± 0.07 to 2.31 ± 0.6 pg/ml after 4 km racing and returning to basal level 3 h after racing (Riad 1995).

9.5 Conclusion

Sexual hormones are not really parameters to be investigated routinely for nutritional or clinical purpose. They are useful to understand the reproductive cycle of the camel, the most characteristic of this species being the presence of a seasonal activity. Corticoid hormones are useful to assess stress status of the camel face to a new situation (housing, transport, climate), while the other hormones of the regulation make it possible to understand the mechanisms of adaptation.

References

Abdelrahman S (2010) Prevalence and pathology of nasal myiasis in camels slaughtered in El-Zawia province-Western Libya: with a reference to thyroid alteration and renal lipidosis. Global Vet 4(2):190–197

Abdel Hadi AA, Wasfi IA, Osman MA, Boni NS (1998) Urinary testosterone equivalent levels in mature male and female Racing camels. In: Proceedings of the third annual meeting for animal production under arid conditions, United Arab Emirates University Publications, vol 1, pp 197–203

Abdel Rahim SEA (1989) The use of milk progesterone analysis to monitor reproductive activity in the camel (*Camelus dromedarius*). Br Vet J 145(1):23–27

Abdel Rahim SEA, El-Nazier AE (1987) Estimation of progesterone level in camels' (*Camelus dromedarius*) milk and its application in pregnancy diagnosis. Br Vet J 143(6):555–559

Abdel-Fattah M, Amer H, Ghoneim MA, Warda M, Megahed Y (1999) Response of one-humped camel (*Camelus dromedarius*) to intravenous glucagon injection and to infusion of glucose and volatile fatty acids, and the kinetic of glucagon disappearance from the blood. J Vet Med A 46:473–481

Abdel-Hadi AA, Wasfi AI (2001) 17β-Estradiol and testosterone level in post-racing plasma of mature female and male racing camels. Emir J Agric Sci 13:33–38

Abdel-Magied EM, Taha AA, Abdalla AB (2000) Light and electron microscopic study of the thyroid gland of the camel (*Camelus dromedarius*). Anat Histol Embryol 29:331–336

Abdel-Wahab MF, Abdo MS, Megahed YM, El-Mougy SA (1974) Thyroxine content in the thyroid gland of domestic animals. Part VII. Iodinated tyrosines and thyronines in the serum of she-camel at various reproductive stages. Endocrinology 63(1):116–121

Abdelsalam OA, Idris OF, Elbagir NM, Seri HI (2003) Effect of cymelarsan and *T. evansi* infection on thyroid hormones (T4 and T3) profile in female camels (*Camelus dromedarius*). In: Proceedings of the 7th science congress, Egyptian Society for cattle diseases, 7–9 Dec 2003. Assiut, Egypt, pp 330–333

Abdoun KA, Alameen A, Elmagbol W, Makkawi T, Al-Haidary A (2010) Effects of hydration status on osmolarity and minerals profiles of serum and forestomach liquor in dromedary camel (*Camelus dromedarius*). J Camel Pract Res 17(2):235–240

Abdulkareem TA, Al-Rawi HM, Abdul-Rahaman YT (2015) Plasma profile of progesterone, estradiol-17β and some blood biochemical attributes during different gestation periods in Iraqi female dromedary camels (*Camelus dromedarius*). Emir J Food Agric 27(8):643–649

Abu-Damir H, Barri MES, Tageldin MH, Idris OF (1990) Clinical and subclinical colloid goiter in adult camels (*Camelus dromedarius*) at Kordofan region of Sudan. Br Vet J 146:219–227

Achaaban MR, Forsling ML, Schroter RC, Oushine A (2002) The neurohypophyseal hormone profiles in camel (*Camelus dromedarius*). J Camel Pract Res 9(2):115–119

Adamsons K, Stanford JR, Engel L, Van Dyke HB, Schmidt-Nielsen B, Schmidt-Nielsen K (1956) The distribution of oxytocin and vasopressin (antidiuretic hormone) in the neurohypophysis of the camel. Endocrinology 58(2):272–278

Adeghate E, Pallot DJ (1996) Innervation of the pancreas of the one-humped camel (*Camelus dromedarius*) by neuropeptide-Y, galanin, calcitonin-gene-related peptide, atrial natriuretic peptide and cholecystokinin. Neuropeptides 30(5):420–424

Afaleq AA, Hozab AA, Basiouni GF, Homeida AM (2003) Hormonal content of the ovarian follicular fluids in the female camels (*Camelus dromedarius*) during the four seasons of the year. J Prod Dev 8:423–431

Afifi AA, Kraft W, Arif H (1979) Values of R-T3, T4, total T3 and cholestrol of some farm animals in Egypt. Indian Vet J 56:16–18

Agarwal SP, Khanna ND (1993) Endocrine profiles of the indian camel under different phases of reproduction. In: Saint-Martin G (ed) Proceedings of the workshop "is it possible to improve the reproductive performance of the camel?". IEMVT Publ., coll. Etudes et synthèses de l'IEMVT n°41, Paris (France), 10–12/9/90, pp 77–99

Agarwal SP, Khanna ND, Agarwal VK, Dwaraknath PK (1986) Thyroidal status of male camels during breeding and non-breeding season. Indian J Anim Sci 56:1036–1038

Agarwal SP, Agarwal VK, Khanna ND, Dwaraknath PK (1987a) Profiles of steroid hormones in male camels. Indian J Anim Sci 57:656–661

Agarwal SP, Khanna ND, Agarwal VK, Dwaraknath PK (1987b) Circulating levels of estrogen and progesterone in female camel during pregnancy. Theriogenology 28(6):849–859

Agarwal SP, Agarwal VK, Khanna ND, Dwaraknath PK (1989a) Serum estrogen and progesterone levels in camel (*Camelus dromedarius*) during estrus cycle. Indian Vet J 66:605–608

Agarwal SP, Khanna ND, Agarwal VK, Dwaraknath PK (1989b) Circulating concentrations of thyroid hormones in pregnant camels (*Camelus dromedarius*). Theriogenology 31(6):1239–1247

Agarwal SP, Rai AK, Khanna ND (1991) Serum progesterone levels in female camels during oestrus cycle. Indian J Anim Sci 61:37–39

Agarwal SP, Rai AK, Khanna ND (1992) Hormonal studies in postpartum female camels and their neonates. Theriogenology 38(4):735–747

Agrawal RP, Swami SC, Beniwal R, Kochar DK, Kothari RP (2002) Effect of camel milk on glycemic control, risk factors an diabetes quality of life in type-1 diabetes: a randomized prospective controlled study. Int J Diabetes 22:70–74

Agrawal RP, Swami SC, Beniwal R, Kochar DK, Sahani M, Tuteja FC, Ghouri SK (2003) Effect of camel milk on glycemic control, risk factors an diabetes quality of life in type-1 diabetes: a randomized prospective controlled study. J Camel Pract Res 10(1):45–50

Ahmed YF, Mahmoud KG, Kandiel MM, Nawito MF, Abdelraziq AMA (2016) Histomorphometry aspect of thyroid gland and biochemical profile in pregnant and non-pregnant dromedary camels. Afr J Biotechnol 15(10):370–375

Al-Dughaym AM, Homeida AM (2010) Effect of endotoxin administration in pregnant camels. Saudi J Biol Sci 17:101–103

Alhaj Ali M, Adem A, Chandranath IS, Benedict S, Pathan JY, Nagelkerke N, Nyberg F, Lewis LK, Yandle TG, Nicholls GM, Frampton CM, Kazzam E (2012) Responses to dehydration in the one-humped camel and effects of blocking the renin-angiotensin system. PLoS One 7(5):e37299

Alhaj Ali M, Kazzam E, Amir N, Nyberg F, Adem A (2013) Effect of dehydration and blockade of angiotensin II AT1 receptor on stress hormones and anti-oxidants in the one-humped camel. BMC Vet Res 9:232–240

Ali Khan A, Alzohairy MA, Mohieldein AH (2013) Antidiabetic effects of camel milk in streptozotocin-induced diabetic rats. Am J Biochem Mol Biol 3:151–158

Ali A, Derar D, Al-Sobyil FA, Zeitoun MM, Hassanein KM, Al-Howas A (2016) Phimosis in male dromedary camels: clinical findings and changes in the hemogram, nitric oxide metabolites, and testosterone concentrations. Theriogenology 85(9):1576–1581

Al-Qarawi AA (1997) Regulation of water and electrolyte metabolism during dehydration and rehydration in camel. Retrospective theses and dissertations, paper 11766, Iowa State University, Ames, IA, 99 p

Al-Qarawi AA, El-Belely MS (2004) Intratesticular morphometric, cellular and endocrine changes in dromedary bulls exhibiting azoospermia. Vet J 167(2):194–201

Al-Qarawi AA, El-Mougy S (2008) Seasonality and the melatonin signal in relation to age as correlated to the sexual cycle of the one-humped male camel (*Camelus dromedarius*). Biol Rhythm Res 39(2):131–142

Al-Qarawi AA, Abdel-Rahman HA, El-Belely MS, El-Mougy SA (2000) Age-related changes in plasma testosterone concentrations and genital organs content of bulk and trace elements in the male dromedary camel. Anim Reprod Sci 62(4):297–307

Al-Qarawi AA, Abdel-Rahman HA, El-Belely MS, El-Mougy SA (2001a) Intratesticular morphometric, cellular and endocrine changes around the pubertal period in dromedary camels. Vet J 162(3):241–249

Al-Qarawi AA, Abdel-Rahman HA, El-Mougy SA (2001b) Impairment in the pituitary-thyroid axis of the *Camelus dromedarius* infected with *Trypanosoma evansi*. Dtsch Tierarzlt Wschr 108:172–174

Al-Qarawi AA, Omar HM, Abdel-Rahman HA, El-Mougy SA, El-Belely MS (2004) Trypanosomiasis-induced infertility in dromedary (*Camelus dromedarius*) bulls: changes in plasma steroids concentration and semen characteristics. Anim Reprod Sci 84(1–2):73–82

Al-Saiady MY, Mogawer HH, Al-Mutairi SE, Bengoumi M, Faye B, Musaad A, Gar-Elnaby A (2012). Effect of different feeding regime on body weight, ovaries size, developments and blood progesterone level in pre-pubertal she-camel (*Camelus dromedarius*). J Anim Vet Adv 11(19): 3522–3526

Al-Saiady MY, Mogawer HH, Al-Mutairi SE, Bengoumi M, Musaad A, Gar-Elnaby A, Faye B (2013). Effect of different feeding regime on body weight, testicular size developments, and testosterone level in pre-pubertal male camel (*Camelus dromedarius*). Afr J Agric Res 8(22): 2631–2636

Al-Saiady MY, Mogawer HH, Al-Mutairi SE, Bengoumi M, Musaad A, Gar-Elnaby A, Faye B (2015) Factors affecting feed intake, body weight, testicular size and testosterone, follicle stimulating hormone (FSH) and luteinizing hormone (LH) serum concentrations in peri-pubertal male camels. Afr J Agric Res 10(14):1709–1713

Alshamsi NS, Ksiksi TS, Ashraf SS (2015) Altered serum enzymes and biochemical levels in Arabian racing camels with bone fractures. J Anim Plant Sci 25(4):1072–1080

Al-Sobayil FA (2008) Influence of exercise on bone formation and bone resorption biomarkers in sera of racing dromedary camel. J Camel Pract Res 15(1):43–48

Al-Sobayil FA (2010) Circadian rhythm of bone formation biomarkers in serum of dromedary camels. Res Vet Sci 89(3):455–459

Al-Suhaimi E, Abu-se'da K, Akbar A (2009) Relationship of leptin hormone with insulin and glucose in Arabian camel (*Camelus dromedarius*). J Camel Pract Res 16:201–207

Ayoub MA, El-Khouly AA, Mohamed TM (2003) Some hematological and biochemical parameters and steroids hormone levels in the one-humped camel during different physiological conditions. Emir J Agric Sci 15(1):44–55

Azouz A, Ateia MZ, Shawky H, Zakaria AD, Farahat AA (1992) Hormonal changes during rutting and the non-breeding season in male dromedary camels. In: Allen WR, Higgins AJ, Mayhew IG, Snow DH, Wade JF (eds) Proceedings of the 1st international camel conference, Dubai. R&W Publications, Newmarket, pp 169–171

Azraqi AA (2007) Relationship of serum leptin concentration to fat deposition in slaughtered young camels. J Anim Vet Adv 6:16–19
Baraka TA (2012) Clinical evaluation of vitamin A, β-carotene, vitamin E and cortisol levels in health and selected diseases in camels (*Camelus dromedarius*). J Am Sci 8(1):106–111
Bargaâ R, Lektib I, Farh M, El-Abbadir N, Belhgouari A, El-Khasmi M (2016) Seasonal variations of biochemical profile and its correlations with thyroid and adrenal gland hormones in Moroccan camel (*Camelus dromedarius*). Int J Agric Environ Res 2(6):1858–1872
Barsham MA, Elbagir NM, Barri MES (2015) Hormonal and biochemical alterations in naturally occurring and induced goiter in camels (*Camelus dromedarius*). Sud J Sci Technol 16 (Suppl):89–96
Basiouni GF (1999) Relationship between concentration of ovarian steroids and insulin-like growth factor-1 in the follicular fluids of the camel (*Camelus dromedarius*). J Camel Pract Res 6 (2):301–305
Bedrak E, Rosenstrawuch A, Kafka M, Friedlander M (1983) Testicular steroidogenesis in the camel (*Camelus dromedarius*) during the mating and non-mating seasons. Gen Comp Endocrinol 52:255–264
Bekele T, Olsson K, Olsson U, Dahlborn K (2013) Physiological and behavioral responses to different watering intervals in lactating camels (*Camelus dromedarius*). Am J Physiol Regul Integr Comp Physiol 305:R639–R646
Bengoumi M, Riad F, Giry J, De La Farge F, Safwate A, Davicco MJ, Barlet JP (1993) Hormonal control of water and sodium in plasma of camels during dehydration and rehydration. Gen Comp Endocrinol 89:378–386
Bengoumi M, Robins SP, De La Farge F, Coxam V, Davicco MJ, Barlet JP (1996) Water restriction and bone metabolism in camels. Reprod Nutr Dev 36:545–554
Bengoumi M, Moutaoukil F, De La Farge F, Faye B (1999) Thyroidal status of the dromedary camel: effect of some physiological factors. J Camel Pract Res 6(1):41–43
Bengoumi M, Moutaoukil F, De La Farge F, Faye B (2003) Seasonal variations of the plasma thyroid hormone concentrations and the body temperature in the dromedary camels. J Camel Pract Res 10:115–119
Bengoumi M, Tabarani A, Sghiri A, Faulconnier Y, Faye B, Chilliard Y (2005) Effects of overfeeding and underfeeding on body weight, lipid content and cellularity in the dromedary camel. Anim Res 54:383–393
Benlamlih S, Dahlborn K, Oukessou M (1993) Blood plasma kinetics of arginine-vasopressin in camels. Acta Physiol Scand 147:341–342
Bhakat C, Raghavendra S, Sahani MS (2005) Effect of different management conditions on rutting behavior of Indian dromedary camel. Emir J Food Agric 17:1–13
Bono G, Dahir AM, Ahmed JM (1985) Seasonal variations of LH response to GnRH treatment in camels (*Camelus dromedarius*). J Reprod Fertil 73:335–339
Bono G, Dahir AM, Comin A, Jumale MA (1989) Plasma LH, cortisol and sex steroid variations in camels (*Camelus dromedarius*) in relation to seasonal climatic changes. Anim Reprod Sci 21:101–113
Bono G, Dahir AM, Comin A, Jumale MA (1993) Seasonal effects on the endocrine reproductive activity of the dromedary camels. In: Saint-Martin G (ed), Proceedings of the workshop "is it possible to improve the reproductive performance of the camel?". IEMVT Publ., coll. Etudes et synthèses de l'IEMVT n°41, Paris (France), 10–12/9/90, pp 107–115
Chakir Y (2016) Réponses physiologiques au stress de pré-abattage chez le dromadaire (*Camelus dromedarius*). Thèse n° CED00/2016, Université Hassan de Casablanca, Maroc, 217 p
Cristofori F, Quaranta G (1993) Niveaux sériques des stéroïdes chez le dromadaire femelle au cours des différentes phases du cycle sexuel et de la gestation. In: Saint-Martin G (ed) Proceedings of the workshop "is it possible to improve the reproductive performance of the camel?", IEMVT Publ., coll. Etudes et synthèses de l'IEMVT n°41, Paris (France), 10–12/9/90, pp 65–75
Cristofori P, Aria G, Seren E, Bono G, Ahmed S, Nur HM (1986) Endocrinological aspects of reproduction in the female camel. Rev Mond Zootechn 57:22–25

Davicco MJ, Rémond B, Jabet S, Barlet JP (1992) Plasma osteocalcin concentrations in cows around parturition. The influence of a regular versus a very short dry period. Reprod Nutr Dev 32:313–319

Deen A (2008a) Testosterone profiles and their correlation with sexual libido in male camels. Res Vet Sci 85(2):220–226

Deen A (2008b) Evaluation of tail curling test for diagnosis of pregnancy in female camels through scientific means. J Camel Pract Res 15(2):219–222

Deen A, Mal G, Sahani MS (2001) Applicability of commercial progesterone analysis kits standardised on human serum/plasma for progesterone analysis in camel. J Camel Pract Res 8 (2):221–226

Delavaud C, Bengoumi M, Faye B, Sghiri A, Faulconnier Y, Tabarani A, Chilliard Y (2004) Plasma leptin measurement in the camel and its relationships to adiposity and feeding level. In: 5th International Congress on farm Animal endocrinology, Budapest, 3–7 July 2004, p 19

Delavaud C, Bengoumi M, Faye B, Levieux D, Chilliard Y (2013) Plasma leptin, glucose and non-esterified fatty acid variations in dromedary camels exposed to prolonged periods of underfeeding or dehydration. Comp Biochem Physiol A:177–185

Derar D, Ali A, Al-Sobayil F (2014) The postpartum period in dromedary camels: uterine involution, ovarian activity, hormonal changes, and response to GnRH treatment. Anim Reprod Sci 151:186–193

Derar D, Ali A, Tharwat M, Al-Sobayil F, Zeitoun MM (2017) Erectile dysfunction in male dromedary camels: clinical findings and changes in the nitric oxide metabolite, cardiac troponin I and testosterone concentrations. Theriogenology 89:201–205

Diaz-Medina E (2017) Studies on identification, lactation and rearing of dromedaries in the Island of Fuerteventura (Canary Islands). PhD thesis in Animal and Feeding Sciences. University Autonomous of Barcelona (Spain)

Dixit VP, Mehta SN, Georgia GC, Signal SP (1987) Serum testosterone and cortisol in rutting and non-rutting Indian male camel. Indian J Anim Sci 57:1191–1193

Ejtahed HS, Naslaji AN, Mirmiran P, Yeganeh MZ, Hedayati M, Azizi F, Movahedi AM (2015) Effect of camel milk on blood sugar and lipid profile of patients with type 2 diabetes: a pilot clinical trial. Int J Endocrinol Metab 13(1):e21160

El Khasmi M, Riad F, Safwate A, Bengoumi M, El Abbadi N, Faye B, Coxam V, Davicco MJ, El Alaoui K, Barlet JP (2001) Thyroxine and insulin like growth factor-I and thyroxine in milk and plasma of camels (*Camelus dromedarius*). J Camel Pract Res 9(1):53–58

El Khasmi M, Riad F, Safwate A, El Abbadi N, Farh M, Faye B, Coxam V (2005) La chamelle allaitante face au stress calcique: une fonction endocrine adaptée aux conditions désertiques. Revue Sécheresse 16(3):261–267

El Khasmi M, Riad F, Safwate A, Tahri EH, Farh M, El Abbadi N, Coxam V, Faye B (2011a) Effects of preslaughter stress on meat quality and phosphocalcic metabolism in camels (*Camelus dromedarius*). J Camelid Sci 3:33–38

El Khasmi M, Riad F, Safwate A, Farh M, Belhouar A, Hidane K, El Abbadi N, Coxam V, Faye B (2011b) Circulating levels of 25-hydroxyvitamin D and testosterone during the rutting and non-rutting periods in Maroccan dromedary camels (*Camelus dromedarius*). Emir J Food Agric 23(4):368–374

El Khasmi M, Chakir Y, Riad F, Safwate A, Tahri EH, Farh M, El Abbadi N, Abouhaf R, Faye B (2013) Effects of transportation stress during the hot-dry season on some haematological and physiological parameters in Moroccan dromedary camels (*Camelus dromedarius*). J Life Sci 7 (1):13–25

El Khasmi M, Chakir Y, Bergaâ R, Barka K, Lektib I, El-Abbadi N, Belhouari A, Faye B (2015) Impact of transport distance on stress biomarkers levels in dromedary camel (*Camelus dromedarius*). Emir J Food Agric 27(6):507–512

El-Allali K, Achaaban MR, Vivien-Roels B, Bothorel B, Tligui NS, Pevet P (2005) Seasonal variations in the nycthemeral rhythm of plasma melatonin in the camel (*Camelus dromedarius*). J Pineal Res 39:121–128

El-Allali K, Achaâban MR, Bothorel B, Piro M, Bouâouda H, El-Allouchi M, Ouassat M, Malan A, Pévet P (2013) Entrainment of the circadian clock by daily ambient temperature cycles in the camel (*Camelus dromedarius*). Am J Physiol Regul Integr Comp Physiol 304:R1044–R1052

El-Bahr SM, Ghoneim IM, Waheed MM (2015) Biochemical and hormonal analysis of follicular fluid and serum of female dromedary camels (*Camelus dromedarius*) with different sized ovarian follicles. Anim Reprod Sci 159:98–103

El-Basheir HM, Idris OF, Abdalla YH, Babikir AE (2001) Serum progesetrone profile to monitor the reproductive activity of camels. Adv Reprod 5(3):5–7

El-Bassiony GM, Al-Sagair OA, El-Daly ES, El-Nady AM (2005) Alterations in the pituitary-thyroid axis in camel *Camelus dromedarius* infected by larvae of nasal bot fly *Cephalopina titillator*. J Anim Vet Adv 4(3):345–348

El-Belely MS (1994) Endocrine changes, with emphasis on 13,14-dihydro-15-keto-prostaglandin F2a and corticosteroids, before and during parturition in dromedary camels (*Camelus dromedarius*). J Agric Sci 122(2):315–323

El-Harairy MA, Attia KA (2010) Effect of age, pubertal stage and season on testosterone concentration in male dromedary camel. Saudi J Biol Sci 17:227–230

El-Harairy MA, Zeidan AEB, Afify AA, Amer HA, Amer AM (2010) Ovarian activity, biochemical changes and histological status of the dromedary she-camel as affected by different seasons of the year. Nat Sci 8(5):54–65

Elias E, Weil S (1989) Serum cortisol levels in camels (*Camelus dromedarius*) during the reproductive cycle. Comp Biochem Physiol A Physiol 94(4):787–790

Elias E, Bedrak E, Yagil R (1984a) Estradiol concentration in the serum of the one humped camel (*Camelus dromedarius*) during the various reproductive stages. Gen Comp Endocrinol 56:258–264

Elias E, Bedrak E, Yagil R (1984b) Peripheral blood levels of progesterone in female camels during various reproductive stages. Gen Comp Endocrinol 53(2):235–240

El-Khasmi M, Derouiche AF, Riad F, Benouhoud M, Davicco MJ, Coxam V, Safwate A, Barlet JP (1999) Hormones thyroïdiennes iodées libres plasmatiques chez le dromadaire. Rev Elev Méd Vét Pays Trop 52(1):71–76

El-Khasmi M, Riad F, Safwate A, Bengoumi M, Hidane K, Davicco MJ, Coxam V, Faye B, Barlet JP (2000a) Effet du peptide apparenté à l'hormone parathyroïdienne (PTHrp) sur l'absorption intestinale du D-xylose chez le chamelon nouveau-né du Sud marocain. Rev Elev Méd Vét Pays Trop 53(2):120–121

El-Khasmi M, Riad F, Safwate A, Bengoumi M, Hidane K, Davicco MJ, Coxam V, Faye B, Barlet JP (2000b) Evolution compare de quelques paramètres minéraux, de l'ostéocalcine, du 25(OH) D et du 1,25(OH)^{2}D chez la chamelle du Sud marocain et son chameleon nouveau-né. Rev Elev Méd Vét Pays Trop 53(2):115–119

El-Kon II, Heleil BA, Mahmoud SA (2011) Effect of age and season on the testicular sperm reserve and testosterone profile in camel (*Camelus dromedarius*). Anim Reprod 8(3/4):68–72

Elmadhi B, Sallmann HP, Fuhrmann H, Von Engelhardt W, Kaske M (1997) Comparative aspects of glucose tolerance in camels, sheep, and ponies. Comp Biochem Physiol 118A(1):147–151

El-May M, El-May A, Bouzakoura C, Dubois PM (1987) Immunocytochemical evidence for vasopressin and oxytocin pathways in the hypothalamo-hypophyseal axis of the camel (*Camelus dromedarius*). Gen Comp Endocrinol 66(2):266–273

El-Wishy AB (1987) Reproduction in the female dromedary (*Camelus dromedarius*): a review. Anim Reprod Sci 15(3–4):273–297

Fat-Halla MM, Ismail AA (1980) Seasonal variations in gonadotropins of one-humped male camel (*Camelus dromedarius*). In: Proceedings of the 9th international congress on animal reproduction and artificial insemination, 16–20 June 1980. III. Symposia (free communications), Madrid, Spain

Fatnassi M, Padalino B, Monaco D, Hamed NM, Khorchani T, Lacalandra GM, Hammadi M (2016) Effect of continuous female exposition on testosterone level and sexual behavior in male dromedary camel. Rev Régions Arides 39(1):49–62

Faye B, Bengoumi M, Viateau E, Tourret M, Chilliard Y (2001) Adipocyte patterns of adipose tissue in camel hump and kidney. J Camel Res Pract 8:29–33

Finberg JP, Yagil R, Berlyne GM (1978) Response of the renin-aldosterone system in the camel to acute dehydration. J Appl Physiol Respir Environ Exerc Physiol 44(6):926–930

Ghoneim M, Waheed MM, Al-Eknah MM, El-Bahr SM (2012) Immunization against GnRH in the male camel (*Camelus dromedarius*): effects on sexual behavior, testicular volume, semen characteristics and serum testosterone concentrations. Theriogenology 78(5):1102–1109

Ghoneim IM, Waheed MM, El-Bahr SM, Al-Heider A, Al-Eknah MM (2013) Comparison of some biochemical and hormonal constituents of oversized follicles and preovulatory follicles in camels (*Camelus dromedarius*). Theriogenology 79:647–652

Gombe S, Oduor-Okelo D (1977) Effect of temperature and relative humidity on plasma and gonadal testosterone concentrations in camels (*Camelus dromedarius*). J Reprod Fertil 50:107–108

Hegazy AA, Ali A Al-Eknah M, Ismail S (2001) Studies on pituitary-ovarian axis in the female camel with special reference to cystic and inactive ovaries. Document n°2029, Deanship of Scientific Research, King Faisal University, Saudi Arabia, 10 p

Heshmat HA, Taha A, Ismail AA, Sami MBA (1984) Levels of thyroid hormones in the plasma of pregnant camels (*Camelus dromedarius*). Indian J Anim Sci 54:663–665

Homeida AM, Khalil MG, Taha AM (1988) Plasma concentrations of progesterone, oestrogens, testosterone and LH-like activity during oestrus cycle of the camel (*Camelus dromedarius*). J Reprod Fertil 85:593–598

Homeida AM, Musa BE, Mohammed FHA (1986) Induction of oestrus in camels by the use of oestradiol benzoate. Small Ruminant and Camel Group (ILCA) 12:21–24

Ismail AA, Siam AA, El-Nahla A, Abuzead SMM (1998) Synchronization of estrus in the she-camel. In: Proceedings of the third annual meeting for animal production under arid conditions, United Arab Emirates University Publications, 1, pp 96–107

Kaske M, Elmadhi B, Von Engelhardt W, Sallman HP (2001) Insulin responsiveness of sheep, ponies, miniature pigs and camels: results of hyperinsulinemic clamps using porcin insulin. J Comp Physiol B 171:549–556

Kataria N, Kataria AK (2004a) Serum calcitonin levels in dromedaries. J Camel Pract Res 11 (1):35–38

Kataria N, Kataria AK (2004b) Serum C-terminal parathyroid hormone levels in dromedaries. J Camel Pract Res 13(1):37–39

Kataria N, Kataria AK (2010) Can prolactin be a measurable marker of stress in dromedaries? Slov Vet Res 47(3):133–138

Kataria N, Kataria AK, Agarwal VK, Garg SL, Sahni MS, Singh R (2000) Effect of water restriction on serum aldosterone and cortisol in dromedary camel during winter and summer. J Camel Pract Res 7(1):1–7

Khaldoun M (1993) Potentialités reproductrices du dromadaire femelle: adaptation à l'environnement. In: Saint-Martin G (ed) Proceedings of the workshop "is it possible to improve the reproductive performance of the camel?". IEMVT Publ., coll. Etudes et synthèses de l'IEMVT n°41, Paris (France), 10–12/9/90, pp 37–50

Khaldoun M, Khaldoun T, Mellado M, Cambar J, Brudieux R (2002) Circadian rhythm in plasma aldosterone concentration and its seasonal modulation in the camel (*Camelus dromedarius*) living in the Algerian Sahara desert. Chronobiol Int 19(4):683–693

Khanna D, Agarwal SP, Gupta ML (1996) Effect of water deprivation during summer and winter on thyroid hormones concentration in the Indian camel. Indian J Anim Sci 66(3):253–255

Khazali H (2009) Effect of the Galanin on growth hormone, thyroid hormones and insulin in young castrated *Camelus dromedarius* fed different levels of their energy requirement. J Appl Sci 9 (15):2822–2828

Khazali H (2010) Intravenous orexin reduces LH secretion in castrated Camelus dromedaries fed a sub-maintenance diet. Asian Australas J Anim Sci 23(1):41–46

Kula JT, Tegegne D (2016) Chemical composition and medicinal values of camel milk. Int J Res Stud Biosci 4(4):13–25

Mabrouk S, Khorchani T, Benromdhane M (2010) Bases épidémiocliniques de la maladie du Krafft chez le dromadaire (*Camelus dromedarius*) dans le Sud tunisien. Rev Elev Méd Vét Pays Trop 63(1–2):29–33

Majchrzak YN, Mastromonaco GF, Korver W, Burness G (2015) Use of salivary cortisol to evaluate the influence of rides in dromedary camels. Gen Comp Endocrinol 211:123–130

Malik A, Al-Senaidy A, Skrzypczak-Jankun E, Jankun J (2012) A study of the anti-diabetic agents of camel milk. Int J Mol Biol 30:585–592

Marie M, Anouassi A (1986) Mating-induced luteinizing hormone surge and ovulation in the female camel (*Camelus dromedarius*). Biol Reprod 35:792–798

Marie M, Anouassi A (1987) Induction of luteal activity and progesterone secretion in the non-pregnant one-humped camel (*Camelus dromedarius*). J Reprod Fertil 80:183–192

Martinat N, Anouassi A, Huet JC, Nespoulous C, Pernollet JC, Combarnous Y (1990) Purification and characterization of glycosylated and non-glycosylated forms of prolactin from the dromedary (*Camelus dromedarius*). Comp Biochem Physiol B 97(4):667–674

Martinat N, Huet JC, Nespoulous C, Combarnous Y, Pernollet JC (1991) Determination of the primary and secondary structures of the dromedary (*Camelus dromedarius*) prolactin and comparison with prolactins from other species. Bioch Biophys Acta Protein Struct Mol Enzymol 1077(3):339–345

Mohamed HE (2006) Factors affecting cortisol status in camels (*Camelus dromedarius*). J Anim Vet Adv 5:307–309

Mohamed TM, Al-Khouli AA, Abbas TA, El-Badawy AS (2006) Clinical hypothyroid goiter in *Camelus dromedarius* (Case Report). 14 pp. Retrieved from https://www.academia.edu/3658185/Goiter_in_Camels_2006

Monaco D, Fatnassi M, Padalino B, Aubé L, Khorchani T, Hammadi M, Lacalandra GM (2015) Effects of a GnRH administration on testosterone profile, libido and semen parameters of dromedary camel bulls. Res Vet Sci 102:212–216

Moutouakil F, Bengoumi M (2007) Plasma thyroid hormones levels after maximal exercise in dromedary camel (*Camelus dromedarius*). J Camel Pract Res 14(1):75–78

Muhammad BF, Aliyu D, Njidda AA, Madigawa IL (2011) Some haematological, biochemical and hormonal profile of pregnant and non-pregnant she-camels (*Camelus dromedarius*) raised in a Sudan savanna zone in Nigeria. J Camel Pract Res 18(1):73–77

Mullaicharam AR (2014) A review on medicinal properties of Camel milk. World J Pharm Sci 2 (3):237–242

Musaad A, Faye B, Abu-Nikhela A (2013) Lactation curves of dairy camels in an intensive system. Trop Anim Health Prod 4:1039–1046

Nagy P, Faigl V, Reiczigel J, Juhasz J (2015) Effect of pregnancy and embryonic mortality on milk production in dromedary camels (*Camelus dromedarius*). J Dairy Sci 98:975–986

Nazifi S, Abassali PM (1998) Seasonal variations in hematological parameters in camel and their correlation with thyroid activity. J Fac Vet Med Univ Tehran 53:73–76

Nazifi S, Gheisari R, Poorabbas H (1999) The influence of thermal stress on serum biochemical parameters of dromedary camels and their correlation with thyroid activity. Comp Haematol Int 9:49–53

Nazifi S, Mansourian M, Nikhaval B, Razavi SM (2009) The relationship between serum level of thyroid hormones, trace elements and antioxidant enzymes in dromedary camel (*Camelus dromedarius*). Trop Anim Health Prod 41(1):129–134

Omidi A, Sajedi Z, Torbati MBM, Nik HA (2014a) Lipid profile and thyroid hormone status in the last trimester of pregnancy in single-humped camels (*Camelus dromedarius*). Trop Anim Health Prod 46(4):609–614

Omidi A, Sajedi Z, Montazer Torbati MB, Mostafai M (2014b) Metabolic profile of pregnant, non-pregnant and male two-humped camels (*Camelus dromedarius*) of Iran. Iran J Vet Med 8 (4):235–242

Quéré M, El May A, Bouzakoura C, Dubois MP, Dubois PM (1985) Evidence for somatostatin-like immunoreactivity in discrete cells of the anterior pituitary of camel (*Camelus dromedarius*). Gen Comp Endocrinol 60(2):187–195

Rashedi R, Khazali H (2010) The effect of intravascular injection of ghrelin on the mean plasma concentrations of insulin and ACTH in immature camels fed with diets containing different levels of their energy requirements. Physiol Pharmacol 13(4):386–396

Rawy MS, Derar DR, El-Sherry TM, Megahed GA (2012) Characterisation of follicular and luteal blood flow in female dromedary camel induced to ovulate using GnRH analogue. J Camel Pract Res 19(2):269–275

Rejeb A, Amara A, Rekik M, Rezigui H, Crespeau F (2011) Histomorphometry and hormone secretion of the thyroid gland of the dromedary (*Camelus dromedarius*). J Camelid Sci 4:10–22

Riad F (1995) Régulation endocrinienne du métabolisme hydroélectrolytique et phosphocalcique chez le dromadaire. Thesis for State doctorate in Animal Physiology, University of Casablanca, Morocco, 225 p

Riad F, Bengoumi M, Giry J, Davicco MJ, Safwate A, Barlet JP (1994) Renin-aldosterone axis and arginine-vasopressin responses to sodium depletion in camels. Gen Comp Endocrinol 95:240–247

Riad F, Bengoumi M, Tressol JC, Davicco MJ, Coxam V, Barlet JP (1995) Endocrine regulation of calcium and phosphorus concentration in camel's milk. Ann Zootech (Suppl) 44:306

Royatvand S, Fallah Hoseini H, Ezzatpanah H, Sekehchi M (2013) Determination of insulin concentration in camel milk using ultra violet–visible absorption spectroscopy. J Food Biosci Technol 3:53–60

Ruiz FJ, Oltenacu PA, Smith RD (1989) Evaluation of on-farm milk progesterone tests to determine non pregnant cows and to prevent insemination errors. J Dairy Sci 72:2718–2727

Saeb M, Baghshani H, Nazifi S, Saeb S (2010) Physiological response of dromedary camels to road transportation in relation to circulating levels of cortisol, thyroid hormones and some serum biochemical parameters. Trop Anim Health Prod 42:55–63

Saleem AKY (2006) Some factors affecting cortisol level and meat quality in camels. In: Proceedings of the international scientific conference on camels. Qassim University, Part 4, pp 2088–2096

Salem HAH, Serur BH, Amer HA (1997) Oestradiol, progesterone and thyroxine in follicular fluids of normal, cystic and atretic follicles of non-pregnant camels in Saudi Arabia. J Camel Pract Res 4(1):81–83

Sayed-Ahmed A, Rudas P, Bartha T (2005) Partial cloning and localisation of leptin and its receptor in the one-humped camel (*Camelus dromedarius*). Vet J 170:264–267

Sazmand A, Rasooli A, Nouri M, Hamidinejar H, Hekmatimoghaddam S (2011) Serobiochemical alterations in subclinically affected dromedary camels with *Trypanosoma evansi* in Iran. Pak Vet J 31(3):223–226

Sboui A, Khorchani T, Djegham M, Agrebi A, Elhatmi H, Belhadj O (2010) Anti-diabetic effect of camel milk in alloxan-induced diabetic dogs: a dose-response experiment. J Anim Physiol Anim Nutr 94(4):540–546

Shareha AM (1998) Effect of damaging brown gland (poll-gland) on plasma blood testosterone concentration in male camels. In: Proceedings of the third annual meeting for animal production under arid conditions, United Arab Emirates University Publications, 1, pp 122–125

Siam AA, Mona MA, Wafaa EM, Ismail A (1993) Plasma levels of glucose and insulin in camels during dehydration. J Vet Sci 9(3):93–96

Sid-Ahmed OE, Sanhouri A, Elwaseela BE, Fadallah I, Elazhari GE, Möstl E (2013) Assessment of adrenocortical activity by non-invasive measurement of faecal cortisol metabolites in dromedary camels (*Camelus dromedarius*). Trop Anim Health Prod 456:1453–1458

Skidmore JA, Billah M, Allen WR (1996a) The ovarian follicular wave pattern and induction of ovulation in the mated and non-mated one-humped camel (*Camelus dromedarius*). J Reprod Fertil 106:185–192

Skidmore JA, Billah M, Allen WR (1996b) Patterns of hormone secretion throughout pregnancy in the one-humped camel (*Camelus dromedarius*). Reprod Fertil Dev 8(5):863–969

Skidmore JA, Payne JH, Lamming GE, Allen WR (1996c) Control of luteolysis in the one-humped camel (*Camelus dromedarius*). In: Proceedings of the 13th international congress animal reproduction, Sydney, 30/06–4/07/96, AI 216 (Abstract)

Skidmore J, Starbuck GR, Lamming GE, Allen WR (1998) Control of luteolysis in the one-humped camel (*Camelus dromedarius*). J Reprod Fertil 114:201–209

Souilem O, Chine O, Alguemi C, Gogny M (1999) Etude de la glycémie chez le dromadaire (*Camelus dromedarius*) en Tunisie: résultats préliminaires. Rev Méd Vét 150(6):537–542

Tajik J (2013) Correlations of serum concentrations of thyroid hormones with some serum biochemical parameter in clinically healthy camels (*Camelus dromedarius*). Online J Vet Res 17(8):407–415

Tajik J, Sazmand A, Moghaddam SHH, Rasooli A (2013) Serum concentrations of thyroid hormones, cholesterol and triglyceride, and their correlations together in clinically healthy camels (*Camelus dromedarius*): effect of season, sex and age. Vet Res Forum 4(4):239–243

Tharwat M, Al-Sobayil F (2015) Serum concentrations of acute phase proteins and bone biomarkers in female dromedary camels during the periparturient period. J Camel Pract Res 22(2):271

Tibary A, Anouassi A (1997) Theriogenology in camelidae. Anatomy, physiology, pathology and artificial breeding. Actes Ed., IAV HassanII, Rabat, Morocco, 489 p

Towhidi A, Rostami F, Chamani M, Khazali H (2001) Endocrine profile of female camels to determine onset of puberty. Adv Reprod 5(3):1–2

Varshney VP, Khanna ND, Pande JK, Tandon SN (1984) Studies of the status of thyroid activity of the arid zone in India. J Vet Physiol Allied Sci 3:52–54

Vyas S, Ravault JP, Faye B, Chemineau P (1997) The circadium rythm of melatonin secretion in camel (*Camelus dromedarius*). Rev Elev Med Vét Pays Trop 50:261–263

Vyas S, Kishore N, Bissa UK, Mal G (2010) Serum progesetrone analysis by commercially available EIA kit to monitor ovulation and conception in dromedary camels. J Camel Pract Res 17(1):79–83

Waheed MM, Al-Eknah MM, El-Bahr SM (2011) Some biochemical characteristics and preservation of epididymal camel spermatozoa (*Camelus dromedarius*). Theriogenology 76(6):1126–1133

Waheed MM, Ghoneim IM, Alhaider AK (2015) Seminal plasma and serum fertility biomarkers in dromedary camels (*Camelus dromedarius*). Theriogenology 83(4):650–654

Wernery U, Johnson B, Tawfig Ishmail W (2006a) Insulin content in raw dromedary milk and serum measured over one lactation period. J Camel Pract Res 13(2):89–90

Wernery U, Nagy P, Bhai I, Schiele W, Johnson B (2006b) The effect of heat treatment, pasteurization and different storage temperatures on insulin concentrations in camel milk. Milchwissenschaft 61 (1):25–28

Xu YS, Wang HY, Zeng GQ, Jiang GT, Gao YH (1985) Hormone concentrations before and after semen-induced ovulation in the Bactrian camel (*Camelus bactrianus*). J Reprod Fertil 74:341–346

Xu YS, Gao YH, Zeng GQ (1993) Studies on the mechanism of ovulation in Bactrian camel. Hormonal profile in semen with special emphasis on their relation to ovulation. In: Saint-Martin G (ed) Proceedings of the workshop "is it possible to improve the reproductive performance of the camel?". IEMVT Publ., coll. Etudes et synthèses de l'IEMVT n°41, Paris (France), 10–12/9/90, 17–24

Yadegari M (2015) Evaluation of prevalence of the types of thyroid disorders using ultrasound and pathology of one humped camel in Iran (*Camelus dromedarius*). Int J Biol Biomol Agric Food Biotechnol Eng 9(2):154–157

Yadegari M, Azizi S, Khamesipour F (2014) Evaluation of prevalence of the types of thyroid disorders using ultrasound and pathology of one-humped camel (*Camelus dromedarius*). Kafkas Univ Vet Fak Derg 20(4):605–611

Yagil R, Etzion Z (1979) The role of antidiuretic hormone and aldosterone in the dehydrated and rehydrated camel. Comp Biochem Physiol A Physiol 63(2):275–278

Yagil R, Etzion Z (1980) Hormonal and behavioural patterns in the male camel (*Camelus dromedarius*). J Reprod Fertil 58:61–65

Yagil R, Etzion Z, Ganani J (1978) Camel thyroid metabolism: effect of season and dehydration. J Appl Physiol 45:540–544

Zagorski O, Maman A, Yafee A, Meisles A, Van Creveld C, Yagil R (1998) Insulin in milk – a comparative study. Int J Anim Sci 13:241–244

Zeidan AEB, Abbas HE (2003) Physico-biochemical changes in the male dromedary camels during breeding (rutting) and non-breeding season. J Camel Pract Res 10(2):183–190

Zhao XX, Zhang Y, Chen BX (1998) Serum progesterone and 17β-estradiol concentrations during pregnancy of Bactrian camel. Theriogenology 50(4):595–604

Zhao XX, Zhang Y, Chen BX (2001a) Preliminary observations on the concentrations of serum luteinizing hormone and follicle-stimulating hormone during gestation in Bactrian camels (*Camelus bactrianus*). Vet Res Commun 25(2):153–160

Zhao XX, Li XL, Chen BX (2001b) Isolation of ovulation-inducing factors in the seminal plasma of Bactrian camel (*Camelus bactrianus*) by DEAE-cellulose chromatography. Reprod Domest Anim 36(3–4):177–181

Zia-Ur-Rahman D, Straten MV, Haq IU (1998) Blood biochemical, hormonal profiles and milk composition of low and high yielding camel. Actes de l'atelier "Chameaux et dromadaires, Animaux laitiers" 24–26 Oct. 1994. Nouakchott (Mauritanie), Ed. CIRAD, 1998 (coll. Colloques), pp 163–165

Zia-Ur-Rahman D, Akhtar N, Ahmad N, Ali S, Haq IU (2001) Endocrine profile of female camels to determine onset of puberty. Adv Reprod 5(3):10a, 2 p

Zia-Ur-Rahman D, Ahmad N, Bukhari SA, Akhtar N, Haq IU (2007) Serum hormonal, electrolytes and trace element profiles in the rutting and non-rutting one-humped male camel (*Camelus dromedarius*). Anim Reprod Sci 101(1–2):172–178

Zia-Ur-Rahman D, Bukhari SA, Ahmad N, Akhtar N, Ijaz A, Yousaf MS, Haq IU (2008) Dynamics of follicular fluid in one-humped camel (*Camelus dromedarius*). Reprod Domest Anim 43 (6):664–671

Ziv A, Sod-Moriah UA, Creveld C, Yagil R (1997) Is water restriction a stress for camels (*Camelus dromedarius*)? J Camel Pract Res 4(2):217–221

Chapter 10
General Conclusion

The objective of the present book was not to give an exhaustive guide for the clinical pathology and nutrition investigations in camel as a particular species. Mechanisms of biochemical regulation, metabolism of the biochemical parameters, and their nutritional or clinical significations in camel are not fundamentally different than in other species. Available references on usual values of blood, serum, or plasma parameters and incidentally in other substrates (milk, urine, digestive fluid, etc.) for a species which represents barely 1% of the domestic herbivorous biomass are still insufficient to propose a clear interpretation of the laboratory results. The ambition of the present book was to give a large (if not exhaustive) panel of references with the default of such approach, some references being questionable, methodologies may not being homogeneous, environmental or experimental context not sufficiently described, statistical procedures not being implemented according to the right rules, etc.

However, our choice has been to gather all data on usual values and the factors of variations of the biochemical parameters used for camel vets and scientists in order to explore nutritional and clinical status. We can bring out here some highlights (among others) regarding the specificity of camel clinical and nutritional pathology:

1. Predominance of polynuclear neutrophils in its white cell formula
2. Maintenance of hematocrit in case of physical effort
3. Osmolality resistance
4. Relative hyperglycemia
5. Absence of ketone bodies
6. Low plasma cholesterol in plasma concentration
7. Susceptibility to hyperuremia
8. Thermoresistance of alkaline phosphatases
9. Maintenance of metalloenzyme activities in case of mineral deficiency
10. Maintenance of electrolyte balance in dehydrated animals
11. Low plasma zinc concentration in supplemented camel
12. Higher sensitivity to selenium toxicosis

© Springer International Publishing AG, part of Springer Nature 2018

B. Faye, M. Bengoumi, *Camel Clinical Biochemistry and Hematology*,
https://doi.org/10.1007/978-3-319-95562-9_10

13. Richness of camel milk in vitamin C
14. Richness of plasma in vitamin D
15. Seasonal pattern of testosterone
16. Decreasing insulin responsiveness with age of the young calves to induce hyperglycemia

Finally, we hope that this book will offer to the camel scientists and students the ability to fill the gaps, to investigate the less understood mechanisms, and to probe the most promising parameters. Overall, we expect that the present review will represent a step to undertake new investigations rather than to repeat what is already known.

Annex: Usual Values of Main Biochemical Parameters in Camel Plasma

Main parameters	Usual values	Main parameters	Usual values
Hematology		*Macro-minerals and electrolytes*	
Red blood cells	6–10 × 10^6/mm^3	Sodium	140–180 mmol/l
Hematocrit	25–30%	Potassium	3.5–6.3 mmol/l
Hemoglobin	9.3–15.5 g/dl	Chloride	106–123 mmol/l
MCV	30–45 fl	Bicarbonates	22–30 mmol/l
MCH	12–18 pg	Calcium	8.4–12.4 mg/100 ml
MCHC	40–50 g/dl	Phosphorus	3.8–8.4 mg/100 ml
White blood cells	10.5–15.5 × 10^3/mm^3	Magnesium	1.8–2.8 mg/100 ml
Lymphocytes	29–63%	*Trace elements*	
Neutrophils	0–1%	Copper	70–120 μg/100 ml
Eosinophils	1.5–13.8%	Zinc	50–100 μg/100 ml
Monocytes	1–11.6%	Iron	70–120 μg/100 ml
Basophils	<1%	Manganese	3–8 μg/100 ml
Platelets	230–360 × 10^3/mm^3	Selenium	50–150 ng/ml
Fibrinogen	200–400 mg/100 ml	Cobalt	30–60 μg/100 ml
Energetic parameters		Iodine	50–120 ng/ml
Glucose	60–140 mg/100 ml	Fluorine	4–6 μg/100 ml
Ketone bodies	0.001–0.01 mmol/l	*Vitamins*	
Cholesterol	18–150 mg/100 ml	Vitamin A	20–50 μg/100 ml
Triglycerides	10–80 mg/100 ml	Vitamin B1	35–60 μg/l
Phospholipids	12–50 mg/100 ml	Vitamin B3	400–500 μg/100 ml
Nitrogen and protein parameters		Vitamin B7	20–50 ng/100 ml
Blood urea nitrogen	8–30 mg/100 ml	Vitamin B9	0.5–1 μg/100 ml
Uric acid	0.2–2 mg/100 ml	Vitamin B12	20–30 ng/100 ml
Creatinine	0.8–2 mg/100 ml	Vitamin C	0.3–0.6 mg/100 ml
Total proteins	6.3–8.3 g/100 ml	Vitamin D	70–90 ng/100 ml
Albumin	25–45 g/l	Vitamin E	50–400 μg/100 ml
Globulin	20–50 g/l	Vitamin K	20–60 ng/100 ml

(continued)

© Springer International Publishing AG, part of Springer Nature 2018

B. Faye, M. Bengoumi, *Camel Clinical Biochemistry and Hematology*,
https://doi.org/10.1007/978-3-319-95562-9

Main parameters	Usual values	Main parameters	Usual values
Bilirubin	0.5–8.6 mg/l	*Hormones*	
Haptoglobin	0.1–0.6 g/l	Testosterone	2–35 ng/ml
Fibrinogen	2.2–3.6 g/l	Estradiol	9–110 pg/ml
Enzymes		Progesterone	1–9 ng/ml
ASAT	37–131 U/l	Prolactin	1–10 ng/ml
ALAT	6–25 U/l	Cortisol	3–30 ng/ml
ALP	32–110 U/l	Thyroxine (T4)	80–130 ng/ml
LDH	337–2620 U/l	Triiodothyronine (T3)	0.5–1 ng/ml
GGT	8–28 U/l	Melatonin	5–250 pg/ml
CK	40–120 U/l	Insulin	100–230 pg/ml
GLDH	0–97 U/l	Leptin	2.5–8 ng/ml
Ceruloplasmin	15–50 U/l	Calcitonin	70–180 pg/ml
GSH-Px	15–36 U/l	PTH	1.9–2.1 ng/ml
SOD	1400–1800 U/l	Osteocalcin	20–40 ng/ml
		Aldosterone	2–5 ng/ml
		Vasopressin	0.1–1 pg/ml

The manufacturer's authorised representative in the EU is Springer Nature Customer Service Centre GmbH, Europaplatz 3, 69115 Heidelberg, Germany. If you have any concerns regarding our products, please contact ProductSafety@springernature.com

Printed and bound by CPI Group (UK) Ltd, Croydon, CR0 4YY
25/09/2026
02222855-0016